Debdatta Ratna, Bikash Chandra Chakraborty
Polymer Matrix Composite Materials

Also of interest

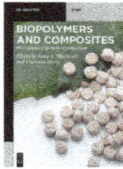

Biopolymers and Composites.
Processing and Characterization
Samy A. Madbouly, Chaoqun Zhang (Eds.), 2021
ISBN 978-1-5015-2193-5, e-ISBN 978-1-5015-2194-2

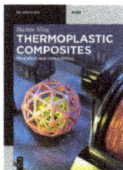

Thermoplastic Composites.
Principles and Applications
Haibin Ning, 2021
ISBN 978-1-5015-1903-1, e-ISBN 978-1-5015-1905-5

Plant and Animal Based Composites
Kaushik Kumar, J. Paulo Davim (Eds.), 2021
ISBN 978-3-11-069521-2, e-ISBN 978-3-11-069537-3

Polymers and Composites Manufacturing
Kaushik Kumar, J. Paulo Davim (Eds.), 2020
ISBN 978-3-11-065193-5, e-ISBN 978-3-11-065504-9

Debdatta Ratna, Bikash Chandra Chakraborty

Polymer Matrix Composite Materials

Structural and Functional Applications

DE GRUYTER

Authors
Dr. Debdatta Ratna
E-2/G 2 Mohan Puram
Kansai Section 421501
Maharashtra
India
ratnad29@hotmail.com

Dr. Bikash Chandra Chakraborty
Flat No. 46 B Wing
No. 36, Sector-5 Progressive Ambar CHS
Navi Mumbai 400709
Maharashtra
India
bikash051954@gmail.com

ISBN 978-3-11-078148-9
e-ISBN (PDF) 978-3-11-078157-1
e-ISBN (EPUB) 978-3-11-078170-0

Library of Congress Control Number: 2023940079

Bibliographic information published by the Deutsche Nationalbibliothek
The Deutsche Nationalbibliothek lists this publication in the Deutsche Nationalbibliografie;
detailed bibliographic data are available on the internet at http://dnb.dnb.de.

© 2024 Walter de Gruyter GmbH, Berlin/Boston
Cover image: XXLPhoto/iStock/Getty Images Plus
Typesetting: Integra Software Services Pvt. Ltd.
Printing and binding: CPI books GmbH, Leck

www.degruyter.com

Preface

This book is dedicated to polymer matrix composites especially fibre reinforced composites. In the last few decades, we have witnessed some remarkable developments in the field of composite materials. The composite materials are used to not only improve the performance of a component or system but also made user to rethink about the entire design. In a recent book on *Recent Advances and Applications of Thermoset Resins* written by one of the authors (D. Ratna), the area of thermoset composites is briefly covered. The objective of the present work is to provide the readers a self-sufficient book on polymer matrix composites covering different type of matrices like thermoplastic, rubbers or elastomer and thermoset resins. We have also covered the experimental characterization method used for composites and their life time assessment which is very important from the application point of view.

We have divided this book into seven chapters. It is started with general introduction to composite materials. The thermoplastics, rubber and thermoset resin-based composites are covered in Chapters 2, 3 and 4, respectively. Chapter 5 deals with elaboration on smart composites which have drawn considerable attention in recent years. The techniques/instrumentations (their principle) used to characterize a polymer matrix composite are covered in Chapter 6. Chapter 7 exclusively deals with the lifetime assessment of composites materials. With such a broad technical contents covering the basic concepts and recent advances on polymer matrix composites, especially fibre-reinforced composites, we are sanguine that this book will serve as a useful textbook-cum-handbook for the students, researchers, engineers and R&D scientists from academia, research laboratories and industries. It will be extremely useful for the scientists and researchers to make a knowledge-base in the subject as well as to plan their future works because we have not only presented the review of the recent advances in this book but also highlighted the future directions of research in the various areas of polymer matrix composites.

We would like to dedicate this book to our late parents. We are indebted to the members of our families for their patience and for always being the source of inspiration, without which this book would not have been a reality. Dr. Ratna would like to place on record his sincere thanks to his wife (Sujata) and sons (Saptarshi and Debarshi). Dr. Chakraborty thanks his wife (Mitali), son (Abhishek) and daughter (Anwesha) for their encouragement and support. We are thankful to our colleagues Sri Ramakant Khushwaha and Shri Prakash Vislavath for their help in preparing the book. We would also like to sincerely thank the publication team of De Gruyter for their support for publishing this book.

Debdatta Ratna
B. C. Chakraborty

https://doi.org/10.1515/9783110781571-202

Contents

preface —— V

About the authors —— XIII

Chapter 1
Introduction to composite materials —— 1
Description of abbreviations —— **1**
1.1   Introduction —— **1**
1.1.1   Classification of composites —— **4**
1.1.2   Constituents of polymer composites —— **5**
1.2   Polymer matrices —— **6**
1.2.1   Interpenetrating polymer network (IPN) —— **6**
1.2.2   Ionomeric elastomers —— **14**
1.3   Reinforcement —— **24**
1.3.1   Natural fibres —— **26**
1.4   Strength and toughness of composites —— **37**
1.4.1   Tensile cracking —— **37**
1.4.2   Interlaminar stresses —— **40**
1.4.3   Prediction of strength and toughness —— **42**
1.5   Fracture mechanics —— **49**
1.5.1   Fracture toughness —— **50**
1.5.2   Environmental stress cracking (ESC) —— **63**
             References —— **67**

Chapter 2
Thermoplastic matrix composites (TMC) —— 73
Abbreviations —— **73**
2.1   Thermoplastic polymer —— **73**
2.2   Particulate thermoplastic composite —— **75**
2.2.1   Effect of filler: mechanical property —— **77**
2.2.2   Effect of filler: thermal conductivity —— **80**
2.2.3   Example —— **80**
2.2.4   Electrical properties —— **81**
2.3   Short-fibre composite —— **86**
2.3.1   Mathematical models —— **87**
2.3.2   Example —— **90**
2.3.3   Natural short-fibre composites —— **96**
2.3.4   Bio-degradable composites —— **97**
2.4   Nanocomposite —— **98**
2.4.1   Thermoplastic elastomer nanocomposite —— **103**

2.5   Rheological properties —— **105**
2.5.1   Thermoplastic melts —— **105**
2.5.2   Rheology of thermoplastic composites —— **109**
2.5.3   Mathematical models —— **110**
2.5.4   Example —— **111**
2.6   Processing —— **116**
2.6.1   Extrusion —— **116**
2.6.2   Blow moulding —— **120**
2.6.3   Fibre spinning —— **122**
2.6.4   Additive manufacturing —— **124**
2.7   Conclusion —— **127**
           References —— **128**

Chapter 3
Thermosetting polymer-based composites —— 131
Description of abbreviations —— **131**

3.1   Thermoset matrix —— **132**
3.1.1   General-purpose resins —— **133**
3.1.2   High-temperature resin —— **137**
3.2   Kinetics of thermoset curing —— **148**
3.2.1   Isoconversion kinetics —— **148**
3.2.2   Integral form of model-free kinetics —— **151**
3.2.3   Advanced isoconversion models —— **153**
3.2.4   Prediction of T_g —— **160**
3.3   Classification of thermoset composites —— **164**
3.4   Processing of FRP composites —— **165**
3.4.1   Wet lay-up moulding —— **165**
4.3.2   Resin transfer moulding —— **166**
3.4.3   Pultrusion —— **168**
3.4.4   Filament winding —— **170**
3.4.5   Resin film infusion bolding —— **172**
3.4.6   Autoclave moulding —— **173**
3.5   Toughened thermoset composites —— **176**
3.6   Thermoset nanocomposites —— **178**
           References —— **183**

Chapter 4
Elastomer-based composites —— 189
Description of abbreviations —— **189**

4.1   Elastomer matrices —— **189**
4.2   Particulate rubber composite —— **192**
4.2.1   Carbon black —— **192**

4.2.2   Graphite — **193**
4.2.3   Mineral particulate fillers — **194**
4.3   Properties of particulate composites — **197**
4.3.1   Hardness — **198**
4.3.2   Elastic modulus — **199**
4.3.3   Filler effect on reinforcement — **202**
4.4   Short fibre rubber composites — **206**
4.4.1   Mathematical predictions — **209**
4.4.2   Examples — **211**
4.5   Steel chord reinforced rubber composite — **216**
4.5.1   Adhesion of steel cord to rubber — **217**
4.5.2   Mathematical models — **218**
4.5.3   Examples — **219**
4.6   Elastomer nanocomposite — **222**
4.6.1   Elastomer /nanoclay composites — **223**
4.6.2   Elastomer /carbon nanotube composites — **226**
4.6.3   Elastomer /graphene nanocomposites — **227**
4.7   Rheology of elastomer composites — **229**
4.7.1   Viscoelastic properties — **230**
4.7.2   Dynamic viscoelasticity — **234**
4.7.3   Effect of filler inclusion — **237**
4.8   Processing of elastomer composites — **243**
4.9   Application of elastomer composite — **246**
4.9.1   Vibration damping — **246**
4.9.2   EMI shielding and absorbing elastomer — **248**
4.9.3   Examples — **252**
4.10   Conclusion — **254**
           References — **255**

Chapter 5
Smart composites — 262
Description of abbreviations — **262**
5.1   Self-healing Composites — **263**
5.2   Smart structural composites — **266**
5.2.1   Passive constrained layer damping (PCLD) — **266**
5.2.2   Magnetic CLD — **270**
5.2.3   Structural health monitoring (SHM) — **271**
5.3   Electrically conducting composites — **272**
5.3.1   EMI shielding — **276**
5.3.2   Microwave-absorbing composites — **283**
5.4   Application of smart composites — **294**
5.4.1   Biomedical application — **296**

5.4.2   Industrial application —— **297**
5.4.3   Application in defence —— **299**
5.5   Conclusions —— **301**
            References —— **302**

Chapter 6
Experimental techniques —— 306
Description of abbreviations —— **306**
6.1   Introduction —— **307**
6.2   Mechanical testing —— **309**
6.2.1   Universal testing machine (UTM) —— **309**
6.2.2   Tensile test —— **311**
6.2.3   Flexure test —— **312**
6.2.4   Compression test —— **316**
6.2.5   Shear test —— **318**
6.2.6   Stress, strain and modulus in shear —— **321**
6.3   Impact testing —— **322**
6.3.1   Izod and Charpy tests —— **322**
6.3.2   Drop weight impact test —— **324**
6.3.3   Post-impact testing —— **326**
6.4   Dynamic mechanical analysis —— **329**
6.5   Thermal analysis —— **331**
6.5.1   Differential thermal analysis (DTA) —— **332**
6.5.2   Thermogravimetric analysis (TGA) —— **333**
6.5.3   Thermomechanical analysis (TMA) —— **336**
6.5.4   Differential scanning calorimetry (DSC) —— **338**
6.5.5   Dielectric analysis (DEA) —— **340**
6.6   Morphological study —— **342**
6.6.1   Light microscopy —— **342**
6.6.2   Scanning electron microscopy —— **344**
6.6.3   Transmission electron microscope —— **347**
6.6.4   X-ray diffraction analysis —— **349**
6.7   Conclusion —— **351**
            References —— **352**

Chapter 7
Lifetime estimation of polymer matrix composite —— 353
Description of abbreviations —— **353**
7.1   Introduction —— **353**
7.2   Physical ageing and life estimation —— **355**
7.2.1   Basic features —— **356**
7.2.2   Example of creep —— **357**

7.2.3   Relaxation spectra —— 359
7.2.4   Theory of physical ageing —— 360
7.2.5   Example —— 368
7.2.6   Conclusion —— 371
7.3   Chemical ageing —— 371
7.3.1   Thermal degradation kinetics —— 372
7.3.2   Isoconversion kinetics —— 374
7.3.3   Differential form of isoconversion kinetics —— 374
7.3.4   Peak rate method —— 376
7.3.5   Integral form of isoconversion model-free kinetics —— 378
7.3.6   Advanced isoconversion kinetics —— 381
7.3.7   Accuracy of kinetics analysis —— 385
7.3.8   Estimation of lifetime —— 387
7.3.9   Example —— 388
7.3.10   Conclusion on predictive Lifetime —— 394
7.4   Lifetime estimation by creep and relaxation —— 394
7.4.1   Stress relaxation and creep —— 395
7.4.2   Time-temperature superposition models —— 396
7.4.3   Time-temperature-stress superposition —— 397
7.4.4   Time-temperature superposition in stress relaxation —— 398
7.4.5   Shift factor —— 399
7.4.6   Master curve —— 399
7.4.7   Lifetime prediction —— 401
7.4.8   Critical considerations —— 401
7.4.9   Creep strain —— 402
7.4.10   Creep study: master curve —— 403
7.4.11   Life estimation from creep —— 404
7.4.12   Ageing study under stress —— 405
7.4.13   Bailey Criteria —— 406
7.4.14   Lifetime estimation —— 408
7.4.15   Example —— 408
7.4.16   Solution —— 409
7.4.17   Conclusion —— 411
           References —— 413

Index —— 417

About the authors

Dr. Debdatta Ratna has been working as a scientist at the Naval Materials Research Laboratory (Defence Research and Development Organization, DRDO), Ambernath, Maharashtra, India, for the last 28 years and presently heads the polymer science and technology department. He obtained his MSc in chemistry, MTech in materials science and engineering, and a PhD in polymer science from the Indian Institute of Technology, Kharagpur. He was a visiting scientist to Monash University, Australia, on BOYSCAST Fellowship in 2000, sponsored by the Department of Science and Technology, India. He was also a visiting scientist to Technical University, Kaiserslautern, Germany, on a prestigious Alexander von Humboldt Fellowship from 2006 to 2008. He has also visited a number of foreign universities and institutions in the United States, Australia, and Germany to deliver lectures. Dr. Ratna has carried out extensive work on toughened epoxy composites, nanocomposites, interpenetrating polymer networks, shape memory polymers, shape-healing polymers and composites, etc. He has published more than 100 technical papers in reputed international journals and authored four books. He has served as a reviewer for several international journals and research grant councils of Hong Kong, Czech Republic, and the United States. He has received laboratory level and national awards for product development and technology innovation. His name has figured in top 2% of Indian scientists in the global list compiled by Stanford University, USA.

Dr. B. C. Chakraborty graduated in Chemical Engineering from Jadavpur University (India) and earned a PhD in Chemical Engineering from IIT Bombay (India). He served on the Defence Research and Development Organisation in Mumbai and Ambernath (India) till 2016 and retired as an "outstanding scientist." He thereafter served as Director, Polymer Nano Technology Centre in BSA Crescent Institute of Science & Technology (Chennai, India) for one year. Dr Chakraborty has vast research experience in dynamic viscoelasticity, acoustic polymers, blends, IPNs, composites and nano composites. He has carried out extensive work on vibration damping materials and developed a number of products. He has received several prestigious awards from the Defence Research and Development Organization (DRDO) and is a member of various committees for several DRDO laboratories. He has published approximately 60 papers in varoius international journals and has 10 patents to his credit.

https://doi.org/10.1515/9783110781571-204

Chapter 1
Introduction to composite materials

Description of abbreviations

Abbreviation	Description	Unit
ABS	Acrylonitrile-Butadiene-Styrene	
AFP	Automated Fibre Placement	
A-GFRP	Aramid-Glass Fibre-Reinforced Plastic	
APTES	Aminopropyl triethoxysilane	
ATBN	Amine-terminated butadiene-nitrile	
CTBN	Carboxyl-terminated butadiene-nitrile	
CTPEGA	Carboxyl-terminated poly(ethylene glycol adipate)	
CTPEHA	Carboxyl-terminated poly(2-ethyl hexyl acrylate)	
DCB	Double Cantilever Beam	
DMA	Dynamic mechanical Analysis/Analyser	
DMC	Dough Moulding Compound	
ENF	End Notch Flexure	
ENR	Epoxidized natural rubber	
ESC	Environmental Stress Cracking	
ESCR	Environmental Stress Cracking Resistance	
ETBN	Ethoxy-terminated butadiene-nitrile	
FA	Fly ash	
FEM	Finite Element Method	. . . .
FRP	Fibre-Reinforced Plastic	
GPa	Giga Pascal	10^9 N/m^2
GPTES	Glycidoxy propyl triethoxysilane	
GRP	Glass-Reinforced Plastic	
HNT	Halloysite nanotube	
HTBN	Hydroxyl-terminated butadiene-nitrile	
ILSS	Interlaminar Shear Stress	N/m^2
LEFM	Linear Elastic Fracture Mechanics	
MEKP	Methyl ethyl ketone peroxide	
MLG	Multi layer graphene	
MMF	Micromechanics of Fracture	
MoPTMS	Methacryloxy propyltriethoxysilane	
MPa	Mega Pascal	10^6 N/m^2
MTBN	Mercaptan-terminated butadiene-nitrile	
MWCNT	Multiwall carbon nanotube	
NBR-PVC	Nitrile Rubber-Polyvinyl Chloride	
PPS	Poly(phenylene sulphide)	
PTBN	Phenol-terminated butadiene-nitrile	
SENB	Single end notch bend	
SEN-TPB	Single end notch-three-point bending	
SiC	Silicon carbide	
SIF	Stress Intensity Factor	MPa.m$^{0.5}$

https://doi.org/10.1515/9783110781571-001

(continued)

Abbreviation	Description	Unit
SSA	Sewage Sludge Ash	
WLF	Williams-Landel-Ferry	
XCT	X-Ray-Computed Tomography	
FRP	Fibre-reinforced plastic	
PMC	Polymer matrix composite	
MMC	Metal matrix composite	
CMC	Ceramic matrix composite	
RTM	Resin transfer moulding	
VARTM	Vacuum-assisted resin transfer moulding	
FRC	Fibre-reinforced composite	
SRIM	So-called structural RIM	
RFI	Resin film infusion	
ASTM	American Society for Testing and Materials	
SBS	Short beam shear	
DMA	Dynamic mechanical analysis	
TGAP	Trifunctional epoxy resin	
GFRP	Glass fibre-reinforced plastic	
UPE	Unsaturated polyester	
GFRT	Glass fibre-reinforced thermoset	
CFRT	Carbon fibre-reinforced thermoset	
BFRT	Besalt fibre-reinforced thermoset composite	
KFRT	Kevlar fibre-reinforced thermoset	
RIM	Reaction injection moulding	
CFRP	Carbon fibre-reinforced plastic	
TGDDM	Tetraglycidylether of 4,4′ diaminodiphenyl methane	
DGEBA	Diglycidyl ether of bis-phenol-A	
ILSS	Interlaminar shear stress	
TGAP	Triglycidyl p-amino phenol	
LCM	Liquid composites moulding	
CFRC	Carbon fibre-reinforced composite	
CNT	Carbon nanotube	
CNF	Carbon nanofibre	
CVD	Chemical vapour deposition	
DCB	Double cantilever beam	
CFRC	Carbon fibre-reinforced composite	
CVD	Chemical vapour deposition	
CSRP	Core-shell rubber particle	
SEM	Scanning electron microscope	

1.1 Introduction

The word "composite" means that two or more materials are combined together in a certain order on a macroscopic level to form a new material with different and useful properties. The properties of constituent materials of a composite remain intact within the composite, which makes composite materials distinct from materials that are simply a mixture of ingredients. Composite materials are characterized by distinguishable interfaces unlike polymer blend or alloys, which are homogeneous on macroscopic level. A composite material comprises two phases: the primary phase is a bulk material known as matrix and the secondary phase is a reinforcing filler of some types such as fibres, whiskers or particles, and so on. The secondary phase remains embedded into the primary phase (matrix). Wood is a classic example of a natural composite where cellulose fibres act as reinforcement in a lignin matrix. People have worked in the past in order to reshape the raw materials they discovered in nature to create the items they needed. The development of composite material is a continuation of similar effort in a sophisticated way, because composite materials are not just available off the shelf but engineered specifically to fit a particular application [1, 2].

Although composite materials have been used to resolve technological issues for a long time, it was only in the 1970s, when polymer-based composites were introduced, it could draw the attention of industries, leading to technological proliferation in this area. Since then, composite materials have been extensively used as common engineering materials and are designed and manufactured for a number of applications both in civil and defence sectors, e.g. consumer goods, automotive components, sporting goods, aerospace parts, naval components, etc. In the last few decades we have witnessed some remarkable developments in composite materials. As we know our civilization's history has been defined by the materials and moved from the Stone Age, to the Bronze Age, to the Iron Age. It is certainly realistic to say that in continuation we have entered into composite age, because composites are not only used to improve the performance but has also made users rethink about the entire design.

Polymer composites are known for their high specific stiffness and strength, dimensional stability, adequate electrical properties and excellent corrosion resistance. Due to higher specific strength of composite materials compared to metal, the weight component for a composite product will be much lower compared to metallic ones for a given design load. Thus, use of composite materials offers technological advantages like easy transportability, high payload for vehicle, low stress for rotating parts, high ranges for rockets and missiles. Such advantages have made composite materials very attractive for application in both civil and defence sector. Replacement of steel and aluminium components with composite can save 70 to 80% and 30 to 50% in component weight, respectively. At the same time, polymer composites are corrosion-free, and hence there is no issue in using them in marine or industrial corrosive environment. In addition, composites offer better dampening effect compared to conventional materials like metal and ceramics. The growth in composite usage is also attributed to increased awareness regarding product performance, the need for energy savings through reduced product weight, significant noise and vibration dampening leading to increased competition in the global market for lightweight composite-based components. Noise and vibration dampening is not only necessary to reduce human discomfort and component fatigue but also necessary from stealth point of view. In naval platforms, the vibration generated by the machineries radiate as sound in sea water, which is called underwater noise (UWN). Naval platforms can be detected through the detection of UWN by an enemy SONAR. Better damping properties of composites helps to reduce the radiated UWN in naval platforms. Therefore, a composite has the potential for reduced fuel consumption, reduced UWN for marine platforms and increased range with passenger aircraft and enhances the stealth performance for military aircrafts due to very high transmissibility of electromagnetic radiation [3, 4].

1.1.1 Classification of composites

Composite materials can be classified in various manners as shown in Fig. 1.1. On the basis of the matrix used for making the composites, they are conventionally classified into four categories, viz. polymer matrix composite, metal matrix composite, ceramic matrix composite and carbon matrix composite or carbon-carbon composites. Polymer matrix composite can be classified into thermoplastic matrix, thermoset matrix and rubber or elastomer matrix [5]. Depending on the shape of reinforcement, composites are divided into particulate composite, chopped or short-fibre composites, long-fibre composites and laminar composite. On the basis of size of reinforcement, it divided into three categories namely macrocomposite, microcomposite and nanocomposites. In macrocomposite the reinforcement sizes are in the range of mm or cm. Stone or steel rod-reinforced concrete is a classic example of macrocomposite. Such composites are examples of the oldest man-made composite. When the reinforment size is in micron scale, then the related composites are called microcomposite. Particulate and fibre-reinforced polymer composites, which have formed the basis of multicrores industries today are examples of microcomposites. When the reinforcement size is reduced to nanoscale (1–100 nm) at least in one dimension, then they are called nanocomposites. The driving force towards reduction of reinforcement sizes is the increase in surface area, which offers better scope to manipulate the interaction of the reinforcement with matrix and design the interface to achieve desired properties. On the basis of application, composites can be divided into structural composites, functional composites and smart composites. Structural composite is again divided into two categories: sandwiched composite and laminated composite. The main focus of this book is on polymer matrix composites, which will be elaborated in subsequent sections.

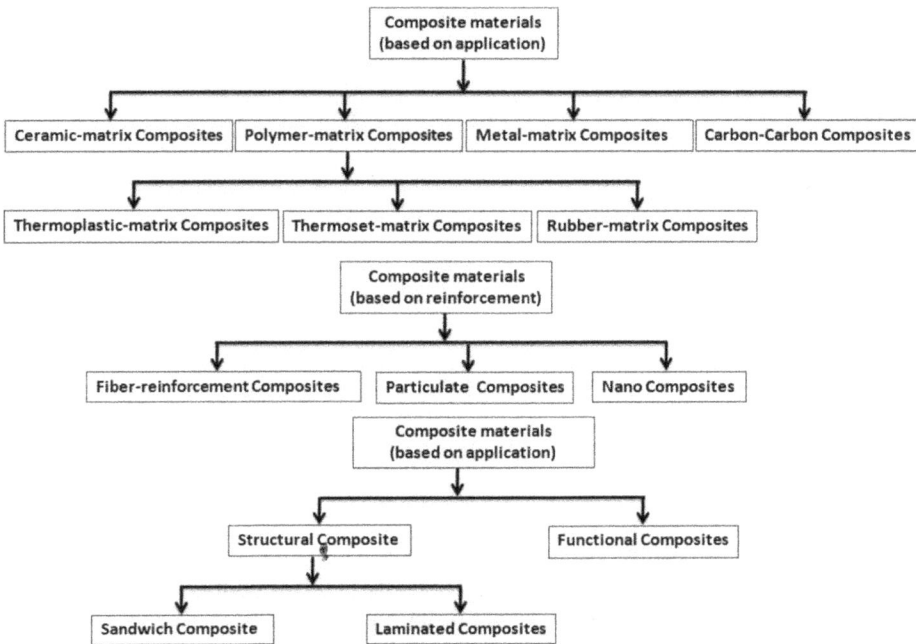

Fig. 1.1: Classification of composite materials.

1.1.2 Constituents of polymer composites

Polymer based composites are prepared by using two major constituents, viz. the polymer-based matrix (thermoplastic, thermoset or rubber) and the reinforcement. Unlike the polymer blends in which the properties of constituent polymers are lost, in polymer composites both the polymer matrix and reinforcing filler retain their identities. As a whole, the composite displays properties that can never be achieved with either of the polymer matrix or the reinforcing filler. The fibres are usually much stronger and harder than the matrix and predominantly responsible for load-bearing capacity of composites. The function of a matrix resin is to keep the fibres in the proper location and orientation. A minimum separation distance between the fibres is to be maintained; otherwise, mutual abrasion of fibres may take place during the deformation of the composites. When a load is applied into the composite, it is distributed into the fibres through the matrix. Therefore, toughness property of the composites depends on the matrix. By using a ductile matrix it is possible to develop a tough composite even by using very brittle fibres. However, translation of matrix toughness into the toughness of composite depends on various factors and various complex energy dissipating mechanisms are involved. Therefore, it is very difficult to predict the toughness of a composite using modelling and simulation.

1.2 Polymer matrices

The matrices used for polymer matrix composites are of three types, viz. thermoplastic resin, thermosetting resins and rubber or elastomer. The composites related to these three types of matrices, i.e. thermoplastic composites, thermoset composites and rubber composites will be elaborated in Chapters 2, 3 and 4, respectively. The corresponding matrices will be covered in the respective chapter. In the present chapter, some new materials like interpenetrating polymer network (IPN), reprocessable thermosets and vitrimers will be discussed.

1.2.1 Interpenetrating polymer network (IPN)

IPNs are formed when a crosslinkable polymer is crosslinked in the immediate presence of another polymer. If the other polymer is a thermoplastic and not capable of crosslinking, then the resulting IPN is called a semi-IPN. If both the polymers are capable of undergoing crosslinking reaction, then the generated IPN is called full IPN. From this we can understand that IPN offers a wide scope for modification to produce material with tunable and multifunctional properties by changing the synthetic variables available at our disposal, such as the type of the polymer pair, sequence of polymerization, rate of polymerization, ratio of the polymers and their compatibility, mechanism and degree of crosslinking, among others [6–10]. Topological features of IPN are quite different than observed in polymer bends or graft copolymers, in the sense that the constituent polymers in IPNs are held together by physical entanglements or secondary forces of interactions without any covalent bonding [11]. IPNs exhibit very broad viscoelastic transition region and the temperature range of the glass transition encompasses the relaxation or glass transition temperature of the individual polymers with a connected interphase. Due to this remarkable feature of viscoelastic transition in a very broad frequency range initial studies were focused on damping of sound waves [12, 13]. However, in recent years, IPNs have been viewed for many potential applications such as membranes, drug delivery, templates for porous networks, double network hydrogels [14, 15], etc.

Ratna et al. [16, 17] have reported semi-IPN of poly(ethylene oxide) (PEO) and polymethacrylates. Methyl methacrylate (MMA) and 2-hydroxyethyl methacrylate (HEMA) were used separately as a monomer for the synthesis of IPNs. The IPNs were synthesized by polymerizing the methacrylate monomer in the immediate presence of PEO, using benzoyl peroxide as an initiator and triethylene glycol dimethacrylate (TEGDM) as a crosslinker. The IPNs made using MMA and HEMA are designated as PEO/x-PMMA and PEO/x-PHEMA, respectively. PEO is miscible with the methacrylate monomers (MMA or HEMA) and TEGDM mixture at 60 °C. However, with the advancement of the free radical-initiated crosslinking reaction of methacrylate/TEGDM system, the combinatorial entropy of mixing decreases. As a result, the crystalline part of PEO undergoes reaction-induced phase separation leading to the formation of a two-phase microstructure, and the amorphous part of PEO remains dissolved with the methacrylate network. Thus, the methacrylate network forms the continuous phase and the intermingled crystalline phase of PEO forms the dispersed phase, as we can see from scanning electron microscopy (SEM) photographs shown in Fig. 1.2. The scheme for synthesis of IPN and DSC plot of PEO, PMMA and the IPN are also presented in Fig. 1.2. We can see that PEO shows a crystalline melting peak at 65 °C and PMMA shows glass transition at 100 °C, whereas IPN shows both the melting peak and glass transition although the melting peak and glass transition is shifted to lower temperature due to phase mixing to a certain extent. Such IPNs exhibit shape memory property because the crystalline PEO component acts as the switching phase due to the reversible process of melting and crystallization.

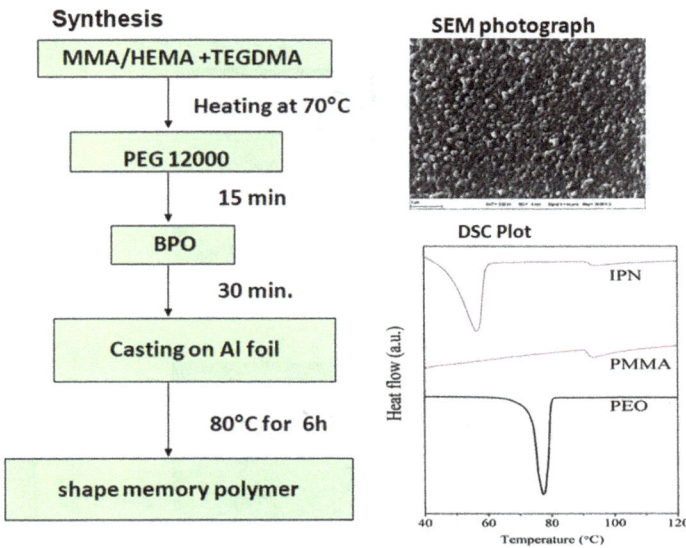

Fig. 1.2: The scheme for synthesis of IPN, DSC plots of PEO, PMMA and the IPN and SEM photographs of fracture surfaces of IPN. Reprinted with permission from Debdatta Ratna and J. Karger Kocsis Polymer 52 (2011) 1063. © 2011 Elsevier Publishers.

Such IPNs exhibit shapes memory property because the crystalline PEO component acts as the switching phase due to the reversible process of melting and crystallization. If a material can be deformed and fixed to temporary shape, and if it recovers to the permanent shape on demand by using a stimulus, then the material is said to have shape memory property [18]. Figure 1.3 demonstrates the shape memory property showing the recovery of a temporary shape at different times. The IPN shows almost 100% shape recovery. Shape memory polymers have potential applications [19–22] in many fields such as biomedical implant, drug delivery, smart textile etc. If a big biomedical implant is to be inserted into a human body, it requires a major surgery. If the implant is made of shape memory polymers, then it can be given a temporary shape like a string and can be inserted into the body with the help of a laparoscope. The implant is expanded on demand into a permanent shape at body temperature. In the process, a major surgery can be avoided.

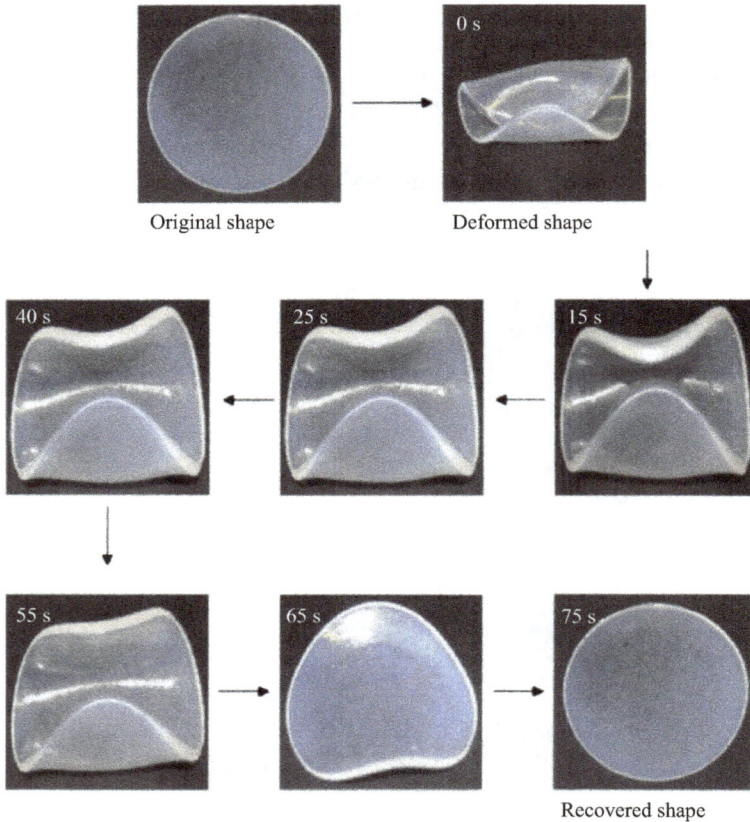

Fig. 1.3: Shape memory property of PEO/PMMA IPN showing the recovery of a temporary shape at different times. Reprinted with permission from Debdatta Ratna and J. Karger Kocsis Polymer 52 (2011) 1063. © 2011 Elsevier Publishers.

As mentioned above, when both the polymers used for synthesis of IPN are crosslinkable then it is called full IPN. Based on method of preparation IPN can be classified into two categories: sequential IPN and simultaneous IPN and latex IPN. Rubber-based IPNs are made either by the sequential method as discussed above or by latex polymerization. In sequential IPN, the second polymeric component is crosslinked after the completion of the crosslinking of the first one [23, 24]. The usual procedure is to first make a cured sheet of rubber by using a peroxide crosslinker such as dicumyl peroxide and swell the rubber sheet in a mixture of monomer, initiator plus crosslinker and allowed to crosslink. Note that sulphur, which is conventionally used to vulcanize the rubber, cannot be used for preparation of IPNs because it interferes the crosslinking reaction of second polymer. By varying the swelling time, the composition of the IPN can be adjusted and properties can be tailored for particular applications as discussed above. For the synthesis of latex IPN, rubber latex is swelled with a mixture of monomer and initiator and allowed to polymerize to prepare the IPN in the form of latex. This process is called the "monomer-flooded process." An alternate procedure can also be adopted by adding the monomer (for the second polymer) into the emulsion or dispersion at a slow and controlled rate in such a way that the rate of addition and the rate of polymerization are equal. This process is known as the monomer-starved process. It may be noted that the monomer used in this process must be water-soluble and one must use a water-soluble initiator like potassium or ammonium per sulphate. If we analyse in the perspective of application, then latex method certainly has an advantage because the latex can be directly used for coating application. Since there is no requirement for organic solvent the application process is environment-friendly. It may be noted that the need to reduce the volatile organic compound in various applications is not only expressed by acute government regulation but equally and persuasively by various social concerns for environment. However, preparation of IPN by latex method is associated with some issues. For example, there is the possibility of formation of graft copolymer during polymerization in the latex, which may adversely affect the final property of the IPN. Generally, the IPN prepared by latex method exhibits inferior thermal and water resistance compared to the same prepared by simultaneous method.

A number of IPNs based on rubber and polyacrylates have been reported [25–29]. The IPNs were prepared using a sequential method by swelling in different types of acrylate monomers. All the IPNs show broad glass transition, covering almost the entire range of T_gs of rubber and polyacrylates. However, the loss factor decreases as a result of IPN formation. For example, maximum value of loss factor for carboxylated nitrile rubber (XNBR) is 1.4, which reduces to 0.6 due to formation of IPN with poly(ethyl methacrylate) (PEMA) [29]. The behaviour can be explained thermodynamically by considering free energy of mixing of the two components. When free energy of mixing is highly negative, then phase mixing takes place leading to formation a single homogeneous phase, which shows a single narrow glass transition. This behaviour is observed in a compatible blend. When the free energy is highly positive, macrophase separation takes place, which shows two distinct T_g. Such behaviour is observed in an incompatible polymer blend. When the free energy of mixing is near zero, two phases are formed, namely component 1-rich phase and component 2-rich phase. When the phases are very small ($<1\ \mu$) as it happens in the case of an IPN, we get a single broad Tg due to formation of a microheterogeneous phases as shown in Fig. 1.4.

(A)

(B)

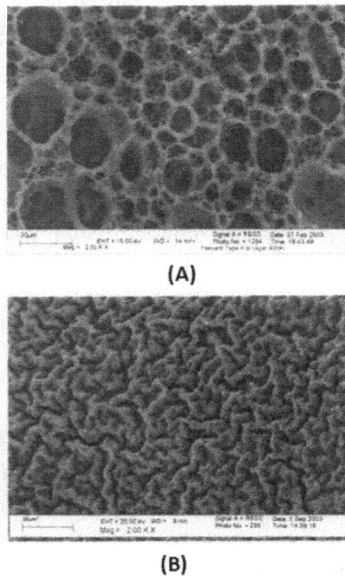

Fig. 1.4: SEM microphotographs of (A) X-NBR/poly(ethyl methacrylate) blend and (B) the corresponding IPN. Reprinted with permission from N. R. Manoj, R. D. Rout, P. Shivraman, D. Ratna, B.C. Chakraborty J. Appl. Polym Sci. 96 (2005) 4487. © 2005 John Wiley and Sons Publishers.

Unlike rubber-based IPNs, thermoset-based IPNs are generally prepared by a simultaneous method. In this method, IPNs are prepared by a process in which both the component networks are polymerized simultaneously. They are processed as a homogeneous solution containing the required crosslinker for network formation. The formation of both the networks takes place simultaneously. The polymerization mechanism for two polymers may be similar or different. One may face issues related to interference of one curing process with the other one. The major advantage of this type of IPNs is that they are well-suited for different types of coating applications, e.g. casting, brushing, spraying and injection moulding. However, as mentioned above, they have disadvantages in terms of achieving a homogeneous solution out of a mixture containing different resins and other ingredients; and curing reaction of one polymer may interfere with that of the other. Karger Kocsis and co-workers [30] reported simultaneous IPN of vinyl ester (VE) and epoxy resins (EP), in which EP undergoes crosslinking via polyaddition and homopolymerization, while VE resin gets crosslinked through free radical-induced copolymerization with styrene. IPNs having a broad range of VE/EP composition were prepared. Atomic force microscopy (AFM) indicates that strand width and mean roughness increase with an increasing concentration of the EP in the IPN. This result is substantiated by dynamic mechanical analysis indicating phase segregation and enhancement in toughness up to an optimum EP concentration in the IPN. Giavery et al. [31] reported an IPN binder formed by simultaneous polymerization of silicone and epoxide prepolymers and used for high-temperature (>400 °C) coating applications. When the IPN-based binder is applied on a metal surface and given thermal treatment at 450 °C, a ceramic layer comprising silica and fillers remains on the metal's surface and acts as a thermal barrier coating. The property can be improved by judiciously selecting the organic substituents of silicone or adding nanoparticles like graphene.

Simultaneous IPNs of epoxy and soybean oil-based polyurethane (PU) were also investigated [32]. Addition of PU converted the glassy epoxy into a rubbery elastomer resulting in a 13-fold increase in percentage of elongation at break as compared to the pure epoxy. Tg values of PU/epoxy IPNs are lower than those of the neat epoxy. Due to slower cure rate of epoxy compared to PU, some of the epoxide groups remain unreacted and act as plasticizers leading to a lowering of Tg of IPNs. When the reaction proceeds, Tg is increased due to formation of graft structure through the reaction of hydroxyl groups of epoxy with the isocyanate group of PU [33]. Yu et al. [34] incorporated PU into EP to prepare the IPNs and observed a significant improvement in mechanical properties and proposed for an application like water-lubricated bearing materials. In the case of vegetable oil (e.g. castor oil)-based PU, after curing reactions of castor oil with diisocyanate, the double bonds present along the fatty acid backbone can also participate in the radical polymerization, in the presence of vinyl or acrylic monomers leading to the formation of EP/PU IPNs. The IPNs based on castor oil-based polyurethane were copolymerized with different compounds derived from petroleum or renewable resources like nitrokonjac glucomannan, poly(ethylene oxide) (PEO), poly (hydroxyethyl methacrylate), benzyl starch, chitosan, nitrocellulose, etc. [35–40] to optimize the thermomechanical properties of the networks.

Meier et al. [41] reported a series IPNs based on novolac-type cyanate ester (CE) resin and phenyl-ethynyl-terminated imide (PETI), which exhibit thermal properties slightly higher than those of pure CE resin. However, no significant improvement in fracture toughness was observed, probably to the high crosslinking density as well as the existence of macroscopic phase separation. Such problems observed in an IPN system can be overcome [42] by generating chemical crosslink points between two networks, thus generating a class of IPNs called graft-IPNs. An example is the graft-IPNs based on a highly stiff copolymer phase, consisting of bisphenol-bis (2-hydroxy-3-methacryloxypropyl) ether (BisGMA) resin and two acrylic monomers: methyl methacrylate (MMA) and triethylene glycol dimethacrylate (TEDGMA), and a soft, rubbery polyurethane phase (PU). The crosslinking of the two networks is accomplished by polyaddition between the secondary (–OH) groups and the (NCO) groups of the isocyanate. It was observed that the miscibility of the different network and toughness properties of the resulting IPNs can be considerably enhanced by generating chemical crosslinks between the networks. Wen and co-workers [43] investigated IPNs based on highly soluble fluorinated ethynyl-terminated imide (FETI) oligomers and bisphenol A dicyanate ester (BADCy) through a solvent-free procedure. FETI oligomers were prepared via a conventional one-step method in m-cresol, using 4, 4'-(hexafluoro isopropylidene) diphthalic anhydride and 2, 2′-bis (trifluoromethyl) benzidine as the monomers and ethynylphthalic anhydride as the end-capper. The IPNs exhibited a glass transition temperature of > 300 °C and improved mechanical properties compared to pure BADCy. The IPNs could be potentially utilized for high-temperature adhesives and composite applications.

A series of IPNs prepared from glassy epoxy and crosslinked acrylates, wherein the former is crosslinked by a short chain diamine, i.e. ditheylene triamine (DETA), and the latter by diisocyantes, has been patented [44]. The crosslink density of DETA-cured epoxy is expected to be high and hence glassy, leading to IPNs with a narrow spectrum of segmental dynamics. Recently, Ratna and co-workers [45] reported sequential IPNs based on a rubbery epoxy and polymethyl methacrylate (PMMA). A polyether amine hardener (D-900) was used as hardener for epoxy, and azobisisobutyronitrile (AIBN) and tetraethyeneglycol methacrylate (TEGDM) are used as initiator and crosslinker for MMA, respectively. From the discussion mentioned above, it appears that rubber-based IPNs are prepared by the sequential method and thermoset-based IPNs reported so far are simultaneous types. The probable reason for this might be the high crosslink density of thermoset, i.e. low molecular weight between crosslinks (Mc), which makes it difficult to achieve the required level of swelling for an IPN synthesis. On the other hand, rubbers being lightly crosslinked allowed sufficient swelling of monomer due to the presence of high free volume.

Selection of rubbery epoxy network, in the present case, makes it possible to synthesize the IPNs using sequential method. Schematic of the sequential IPN formation steps is presented in Fig. 1.5. The details of the preparation of epoxy/PMMA IPNs using the sequential method are described below. Initially, cured epoxy network was made by mixing difunctional epoxy (LY556) and polyether amine hardener (D-900) in a ratio of 20:15 (w/w). In order to remove the air bubbles, the mixture was degassed under vacuum for 20 – 30 min. The mixture was then cast in a Teflon mould and cured at 80 °C for 5 – 6 h. The cured epoxy network was immersed in an MMA/AIBN/TEGDM solution for swelling to take place. The MMA was allowed to soak in the epoxy sheet till the desired weight gain was achieved. The swollen film of cured epoxy was kept 10 min for equilibrium and heated at 80 °C for 4 to 6 h and 100 °C for 2 h. Thus, the polymerization and crosslinking of MMA takes place in the existing cured epoxy network forming an IPN. The resultant film was hardened and kept in vacuum for the complete removal of the unreacted MMA monomer.

The IPNs were characterized by Positron annihilation lifetime spectroscopy (PALS) analysis to study the morphology of the networks at the molecular level. The details of PALS analysis are provided in Chapter 6. It was observed that the free volume decreases with increasing MMA concentration, which is attributed to the interpenetration of stiffer PMMA chains into the flexible epoxy matrix. However, interpenetration lead to generation of small free volume at the interface. Thus, with increase in PMMA concentration, the size of the smaller defects at the interfacial regions decreases while its concentration increases. The IPN shows minimum free volume at a PMMA concentra-

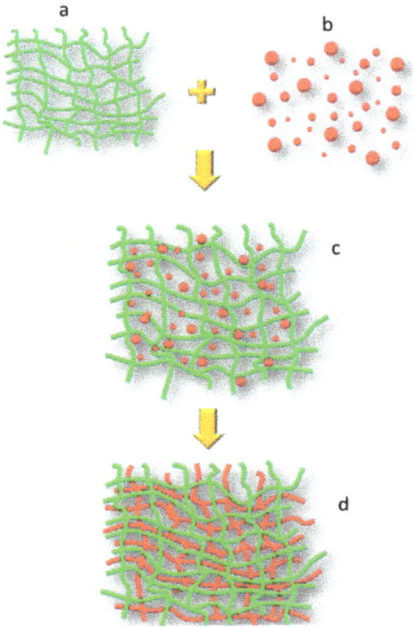

Fig. 1.5: Schematic of the sequential IPN formation: (a) preformed polyetherdiamine cured epoxy film, (b) MMA monomer, (c) swelling of the epoxy network by MMA and (d) interpenetration of polymer chains during polymerization and crosslinking of MMA leading to epoxy-PMMA sequential IPN. Reprinted with permission from Debdatta Ratna and J. Karger Kocsis Polymer 52 (2011) 1063. © 2011 Elsevier Publishers.

tion of 20 wt% and increases with further increase in PMMA concentration. This observation is quite different from the studies on styrene-butadiene rubber (SBR)/PMMA IPNs reported by James et al. [46] in which the free volume was reported to be monotonically increased up to 40 wt% of PMMA. This difference in behaviour can be explained by considering the inherently lower free volume of the elastomeric epoxy network compared to that of SBR. All the IPNs exhibit broad tan δ vs. temperature peak, which when deconvoluted appeared to be a combination of four peaks, as we can see in Fig. 1.6. The four peaks are attributed to the existence of four different phases in the IPNs, namely epoxy phase, PMMA phase, epoxy-rich phase and PMMA-rich phase. However, there is a shift of tan δ peak temperature of epoxy and PMMA. In the case of epoxy, it shifts to the higher side, whereas for PMMA it shifts to lower side. For example in the IPN, the tan δ peak temperature for epoxy phase is 18–20 °C, whereas pure epoxy network shows the same at 0 °C. As we can see, the Tg of epoxy used in the present case is much lower than the conventional amine or anhydride-cured epoxy network (Tg > 60 °C). This can be attributed to the use of polyether amine hardener, which enhances the flexibility of the network by introducing ether linkage in the network and reducing the crosslink density because of the presence of polyether chain. The IPNs have potential applications in the field of vibration damping and shape memory applications.

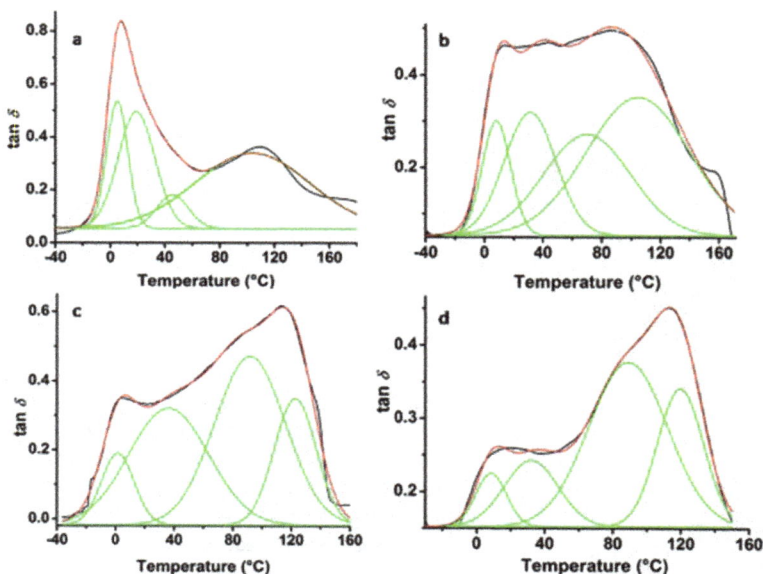

Fig. 1.6: Curve fitting results of tan δ vs temperature profiles of the IPNs into component Gaussian peak shapes. The black profiles correspond to the experimental result and the red curves are the simulated results: (a) EPMA90:10, (b) EPM80:20, (c) EPMA75:25, and (d) EPMA70:30. Reprinted with permission from Debdatta Ratna, Vishal G. Dalvi, Srikanth Billa, Sandeep K. Sharma, Sangram K. Rath, Kathi Sudarshan, and Pradeep K. Pujar ACS Appl. Polym. Mater. 3 (2021) 5073 – 5086. © 2021 ACS Publishing Company.

1.2.2 Ionomeric elastomers

1.2.2.1 Covalently crosslinked elastomer

Elastomers are basically lightly crosslinked rubber. Elastomers are a class of polymers that exhibit rubber-like elasticity and deformability. They are characterized by their ability to return to their original shape after being stretched or deformed, unlike thermoplastics, which can deform permanently under stress. Elastomers can be found in a wide variety of products and applications, from everyday consumer products like shoe soles and rubber bands to more specialized applications in the automotive, aerospace, and medical industries. Elastomers are typically formed through the polymerization of monomers that contain unsaturated bonds, such as butadiene, isoprene, and chloroprene. These unsaturated bonds are used for crosslinking by sulphur or peroxide. These crosslinks forms in moderate concentration give elastomers their rubber-like properties. It may be noted that if they are highly crosslinked they will turn hard and brittle.

Elastomers can be further classified into two categories, namely natural and synthetic. For example, natural rubber is an elastomer that is derived from the latex of certain trees and contains predominantly cis-1,4-polyisoprene chains. Some common synthetic elastomers include styrene-butadiene rubber (SBR), neoprene (poly chloro-prene), and nitrile rubber (NBR). Both natural and synthetic rubber are crosslinked by using sulphur and the process is known as vulcanization to improve their mechanical properties. The vulcanization process was first discovered by Charles Goodyear in the mid-nineteenth century when he accidentally spilled a mixture of natural rubber and sulphur on a hot stove, resulting in the creation of a strong and elastic material. Since then, vulcanization has become an essential process for the production of a wide range of rubber products. Vulcanization transforms a plastic rubber compound into a highly elastic product by forming a three-dimensional crosslinked network structure in the rubber matrix. The vulcanization proceeds through the abstraction of alpha hydrogen leading to the formation of sulphur-sulphur (S-S) bonds [47] as shown in Fig. 1.7. The degree of crosslinking and the resulting mechanical properties of the elastomer can be controlled by adjusting the temperature, time and amount of vulcanizing agent used.

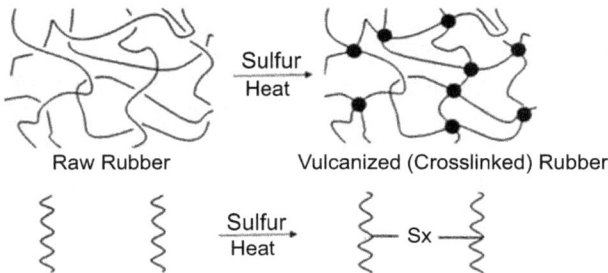

Fig. 1.7: Schematic representation of the vulcanization of rubber.

Vulcanization of rubbers by sulphur alone is an extremely slow and inefficient process. The chemical reaction between sulphur and the rubber occurs mainly at the unsaturation sites and each crosslink requires 40 to 55 sulphur atoms in the absence of an accelerator. The process takes around 6 h at 140 °C for completion, which is uneconomical by any production standards. The vulcanizates thus produced are extremely prone to oxidative degradation and do not possess adequate mechanical properties for practical rubber applications. These limitations were overcome through the invention of accelerators, which subsequently became a part of rubber-compounding formulations. A typical recipe of rubber compound comprises 0.5–3 parts per hundred grams of rubber (phr) sulphur, 0.1 to 4 phr accelerator, 3 to 10 phr Zinc oxide and 1 to 4 phr stearic acid. Some of the common accelerators are N-Cyclohexyl-2-benzothiazole Sulphenamide (CBS), bis(2-benzothiazole)disulphide (MBTS), 2-mercaptobenzothiazole (MBT) and zinc salt of mercaptobenzothiazole (ZMBT), tetramethyl thiuram monosulfide (TMTM), and tetramethyl thiuram disulphide (TMTD) and dipentamethylene thiuram tetrasulphide (DPTT) etc.

1.2.2.2 Ionic elastomer

Ionic elastomers are a class of elastomeric materials [48, 49] that possess mobile ionic groups within their polymer chains, which can facilitate the transport of ions and contribute to unique properties such as self-healing and actuation. One common method of crosslinking elastomeric ionomers is through the use of multifunctional crosslinking agents. These agents typically contain two or more functional groups, such as epoxide or carboxyl groups, that can react with ionomer polymer chains to form covalent bonds. The resulting crosslinked structure is highly resistant to deformation and has improved mechanical properties. Another method of crosslinking elastomeric ionomers is through ion-induced crosslinking. In this method, the ionic groups within the ionomer polymer chains can interact with ions in the surrounding environment, leading to the formation of crosslinks. For example, in a poly(acrylic acid)-based ionomer, the carboxylate groups can interact with metal ions to form crosslinks within the polymer matrix. Carboxylated nitrile butadiene ionic elastomer (XNBR) is a type of ionic elastomer that is made by copolymerizing nitrile butadiene rubber (NBR) with anionic monomers, such as acrylic acid. Due to the presence of carboxyl group, it can be ionically cured using metal oxides like zinc oxide, magnesium oxide and also can be covalently cured by using sulphur or peroxide due to the presence of unsaturation in the chemical structure as shown in Fig. 1.8. Because of the dual crosslinking, XNBR has two transitions, namely α and α1 at −16 °C and 45 °C respectively. Therefore, it offers superior acoustic property compared to conventional rubber.

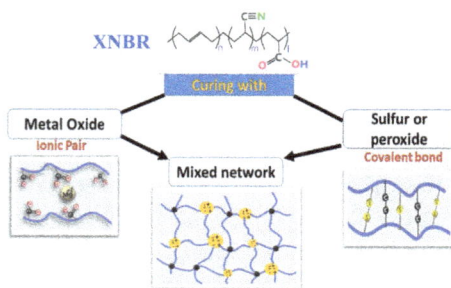

Fig. 1.8: Schematic representation of vulcanization chemistry of XNBR.

Prakash et al. [50] prepared a series of ionically (I-X series) and mixed system (covalent and ionic)-cured XNBR networks, (M-X series) with and without stearic acid (SA) as the processing aid in the curative system. Small angle X-ray (SAX) analysis shows a single ionomeric peak for the network processed without SA, implying a liquid-like ordering of ionic aggregate in the elastomer matrix. On the other hand, the network processed with SA shows multiple higher-order reflections as observed for a block copolymer in their SAXS profiles, suggestive of lamellar morphology resting due to self assembling of the ionic aggregates. The schematic of the ionic phase morphology XNBR networks processed with and without SA is shown in Fig. 1.9.

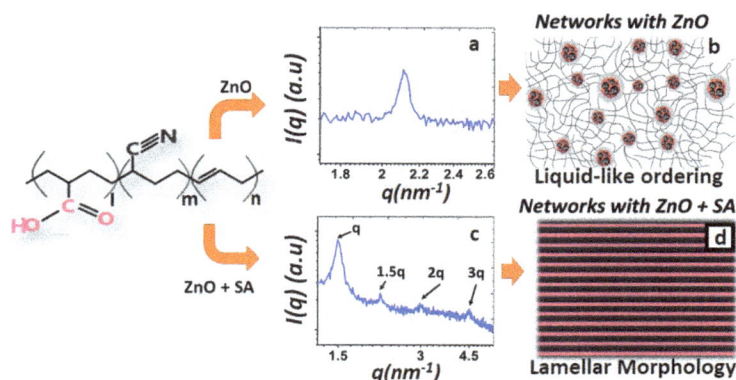

Fig. 1.9: Schematic of the ionic phase morphology XNBR networks processed with and without SA: networks processed without SA show a liquid-like ordering of ionic domains dispersed in the hydrocarbon XNBR matrix as evidenced from a single ionomeric peak in SAXS profile (a), which is schematically shown in (b). Lamellar morphology of the ionic domains is observed for the networks processed with SA, as evidenced from the multiple higher order reflections seen from SAXS profiles (c). Schematic of the lamellar morphology is shown in (d). Reprinted with permission from Vislavath, P., Billa, S., Bahadur, J., Sudarshan, K., Patro, T. U., Rath, S. K., & Ratna, D. Macromolecules,55 (2022) 6739–6749. © 2022 ACS Publishing Company.

1.2.2.3 Reprocessable thermoset

Today polymer composite industries are dominated by thermosetting resins. The composites used for aircraft application are predominantly carbon or glass epoxy prepreg. It is not only the composite industries; if we look at the paint & rubber industries, thermosetting polymers have become an indispensable part. This is because of the obvious advantages of thermosetting polymers, which a thermoplastic cannot offer. These advantages include ease of processing, better dimensional stability and environmental stress cracking resistance of the cured network. The properties of thermoplastic-based product may change even due to presence of solvent, which may initiate formation of a crack. However, thermosets being in the 3D network form in the cured state, their performance is not significantly affected due to presence of traces of solvent. However, the network structure, although provide lot of advantages, has a downside in terms of reprocessability. Thermosets cannot be reprocessed or recycled because of the formation of irreversible linking by covalent bonds. Note that in thermoplastic, the polymer chains are held together by weak interactions like van der Waals; therefore, they can be deformed reversibly above their glass-transition temperature or their crystalline melting and can be easily processed by extrusion, injection moulding and welding, like a thermoplastic.

Hence, after the life of a thermoset-based product is over, it will simply go to a landfill and cause environmental hazard because they cannot be destroyed biologically. If the materials are burnt, they will produce carbon dioxide and other toxic gases leading to environmental pollution and greenhouse effect. It may be noted that solid waste management issue has become alarming and has posed various social concerns for the environment. Hence, the issues need to be addressed very seriously. Thus, there is a need to develop a new generation of materials like thermoplastics that can be reprocessed, yet still retain the beneficial properties of a crosslinked thermoset material. In the last few years, a lot of research activities have been initiated all over the world with the objective of developing reprocessable thermoset, which will be reviewed in subsequent sections.

Introduction of exchangeable chemical bonds into thermoset network can generate plasticity in crosslinked polymer networks due to the formation of dynamic crosslinks. Dynamic crosslinks can also be supramolecular adaptable networks comprising noncovalent interactions such as hydrogen bonding, host-guest interactions, or metal-ligand coordination [51, 52], etc. If chemical crosslinks can be efficiently exchanged between different positions of the organic polymer chains, macroscopic flow can be achieved without causing structural damage or permanent loss of material properties. Polymer networks containing such exchangeable bonds are also known as covalent adaptable networks (CANs). CANs may be further classified into two groups depending on their exchange mechanism. The first group of CANs makes use of a dissociative crosslink exchange mechanism. In this exchange, chemical bonds are first broken and then formed again at another place. The second group of CANs makes use of associative bond exchanges between polymer chains in which the original crosslink is only broken when a new covalent bond to another position has been formed. In other words, a pendent reactive group within the network undergoes a substitution reaction with an existing crosslink.

Sumerlin and co-workers [53] published an elegant review highlighting the most enticing methods of accessing CANs and summarized the fundamental properties and behaviour of these materials to provide both new and experienced researchers in the field with a thorough sense of the state of the art as well as an unambiguous understanding of the characteristics of each class of CANs. As the author mentioned, the rapid pace of recent developments has obscured the previously distinct line between thermoplastics and thermosets, and there have also emerged challenges and misconceptions in distinguishing dissociative and associative mechanisms of bond exchange in CANs. The dissociative and associative bond exchange pathways reported for various CANs are depicted [53] in Fig. 1.10. Reprocessable thermosets have been reported using a number of thermoreversible crosslinking chemistries as shown in Fig. 1.11 such as Diels-Alder (DA) cycloadditions [54–58] triazolinedione Alder-ene reactions [59, 60] hindered urea exchange [61–63] oxime-enabled transcarbamoylation [64] 1,2,3-triazolium [65, 66] and anilinium [67] transalkylation, boronate ester [68] and aminal bond [69] exchange, thiol-Michael [70] and aza-Michael [71] reactions, radical oligosulphide exchange [72, 73], ring-opening metathesis polymerization (ROMP) [74]

Dissociative Exchange

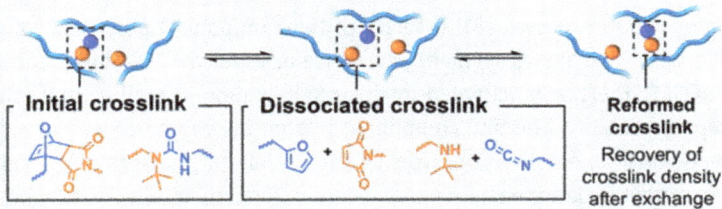

Initial crosslink — Dissociated crosslink — Reformed crosslink

Recovery of crosslink density after exchange

Associative Exchange

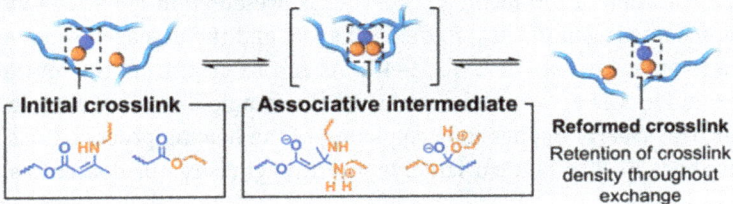

Initial crosslink — Associative intermediate — Reformed crosslink

Retention of crosslink density throughout exchange

Fig. 1.10: Depiction of dissociative and associative bond exchange pathways for covalent adaptable networks (CANs). Reprinted with permission from Georg M. Scheutz, Jacob J Lessard, Michael B. Sims, and Brent S. Sumerlin J. Am. Chem. Soc. 41 (2019) 16181–16196. © 2019 ACS Publishing Company.

Dissociative Exchange of Dynamic Crosslinks

1) Reversible pericyclic reactions

1a) Furan-maleimide Diels-Alder adducts

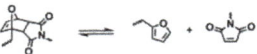

1b) TAD-indole Alder-ene adducts

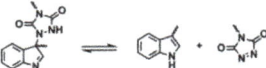

2) Urethane/urea dissociation

2a) Oxime-blocked isocyanates

2b) Hindered ureas

3) Nucleophilic transalkylation

3a) Alkyl 1,2,3-triazolium salts

3b) Alkyl anilinium salts

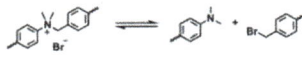

4) Aminal transamination

5) Ring-opening/closing metathesis

6) Stable free radical exchange

6a) TEMPO and TEMPS

6b) Biaryl radicals

7) Michael adduct exchange

X = S or NH

Fig. 1.11: Overview of crosslinking chemistries used for dissociative covalent adaptable networks. See refs 57–58 (1a), refs 59–60 (1b), ref 64 (2a), refs 61–63 (2b), refs 65–66 (3a), ref 67 (3b), ref 69 (4), ref 40 (5), refs 77–78 (6a), ref 79 (6b) and refs 70–71. Reprinted with permission from Georg M. Scheutz, Jacob J Lessard, Michael B. Sims, and Brent S. Sumerlin J. Am. Chem. Soc. 41 (2019) 16181–16196. © 2019 ACS Publishing Company.

and persistent radical approaches using nitroxide [75, 76], thionitroxide [77, 78] or biaryl radicals [79, 80].

Very recently, our group reported [81] a series of two component poly(urethane-urea) (PUU) networks with varied hard segment (HS) concentrations that exhibit quadruple shape memory (QSM), thermally activated intrinsic self-healing as well as multiple cycles of thermal reprocessability. The first component was an isocyanate-capped bi-soft segment blend of polybudadiene diol and polypropylene glycol and the second component consisted of an aromatic diamine chain extender solubilized in an oligomeric polyoxypropylene triol. The PUU networks showed thermally activated intrinsic self-healing property. The demonstration of self healing properties is presented in Fig. 1.12. A cut up to half of the thickness is generated into the PUU sheet and the damaged samples are subjected to a temperature of 120 °C for 1–4 h. As can be seen from the optical microscope images in Fig. 1.12 a, the cut marks for all the sample disappear after the thermal treatment. This clearly implies the completion of the healing process. It also may be noted that about 70–80% of the original tensile strength is retained after heal-

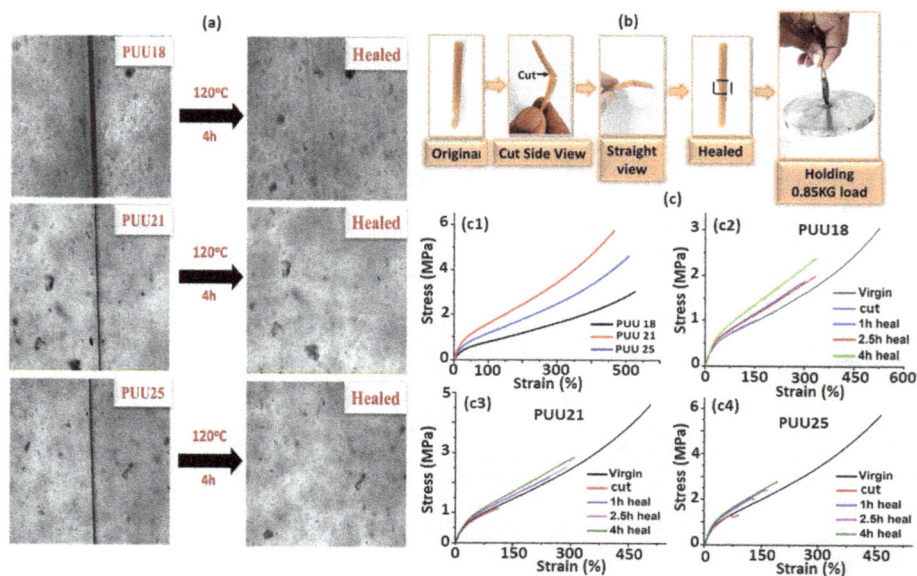

Fig. 1.12: Self-healing results of the PUU networks: (a) optical microscope images of the PUU networks showing the deep cut marks in the left panel and the disappearance of the marks upon thermal treatment at 120 for 4 h, (b) pictorial representation of the self healing process in which a strip of PUU sample is cut, healed at 120 °C for h, and then it is able to sustain a load of 0.84 kg after self-healing and (c) comparative stress-strain profiles of as-cast, cut and self-healed samples as a function of thermal treatment time at 120 °C: c1-pristine as cast samples; c2: PUU-18; c3-PUU-21 and c4-PUU-25. Reprinted with permission from Srikanth Billa, Prakash Vislavath, Jitendra Bahadur, Sangram K Rath, Debdatta Ratna, Manoj N R, Bikash C. Chakraborty, ACS Appl. Polym. Mater., 5 (2023) 3079–3095. © 2023 ACS Publishing Company.

ing. The high mobility of the hard segment at 120 °C, which is higher than the glass transition temperature of PUU, and considerable H-bonding and dynamic reversible cleavage of urea and urethane linkages (as confirmed by time-dependent FTIR) are the possible origins of self-healing property.

1.2.2.4 Vitrimer

The term "vitrimer" was coined by Leibler and colleagues in 2011 [82]. Vitrimers are materials that can be reprocessed or recycled like thermoplastics, and at the same time they offer all the useful thermomechanical properties and environmental stress cracking resistance like thermosets. In the previous section we have discussed about dissociative and associative CANs. Now the question is "should we call all CAN-containing materials as vitrimer?" The answer is no; only associative CAN-containing materials are known as vitrimer. Associative CANs-based mechanism is known as exchange reaction in which the original crosslinks are only broken when new ones are formed. These associative CANs have excellent solubility because their crosslink density is fixed and they retain their three-dimensional structure during rearrangement. Due to the acceleration of associative exchange reactions upon heating, these CANs that rely on exchange reactions exhibit an Arrhenius-like viscosity dependence as opposed to a pronounced viscosity decrease for normal thermoset. Due to vitreous silica's insolubility and gradual thermal viscosity behaviour, these materials are referred to as vitrimers by Leibler and co-workers [82, 83] who first demonstrated the unique properties by adding a transesterification catalyst to polyester-based epoxy networks. The associated behaviour is described as malleability. Since then, several dynamic chemistries, such as transamination, trans-alkylation, siloxane equilibration, disulfide metathesis, olefin metathesis, and dioxaborolane metathesis have been investigated to explore possible vitrimer materials [84–94].

Wu et al. [95] had developed a fully bio-based and recyclable vitrimer from epoxidized soybean oil and glycyrric acid (obtained from liquorice roots) using triazabicyclodecene (TBD) as a catalyst. The topological rearrangements of the network take place upon reaction of ester bonds with free –OH groups in the network (at 200 °C/15 min) and generate a fresh –OH group. With this mechanism, healing and recycling of the networks were achieved with efficiency more than 90%. Shape fixity and shape recovery efficiencies are also found to be decent with a three-cycle average of 92% and 94% respectively. Similar kind of work has been reported by Liu et al. [96] who developed eugenol-derived vitrimer networks. The networks are made by reacting epoxy-functionalized eugenol with succinic anhydride in the presence of Zn(acac)2 catalyst. The resultant networks are self-healed by heating at 190 °C/h, which recovers more than 90% of the mechanical strength of polymer. Dual shape memory of the networks was achieved by change in topology of the network segments at 160 °C. Hayashiand Katayama [97] reported an epoxy-based vitrimer using thiol click chemistry using commercially available epoxy. The molecular design for the synthesis of the vitrimer is shown in Fig. 1.13. The vitrimer is highly transparent and exhibits other attributes like room temperature ductility and shape memory property.

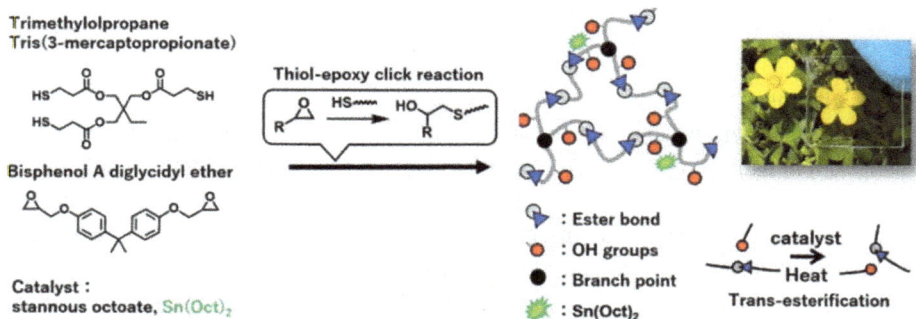

Fig. 1.13: Molecular design of the present vitrimer. The transparency of the vitrimer film is also provided. Reprinted with permission from Hayashi, M.; Katayama, ACS Appl. Polym. Mater. 2020, 2 (6), 2452–2457. © 2020 ACS Publishing Company.

Sheng Wang et al. [98] investigated the use of metal coordination to improve the properties of polyimine vitrimers. Addition of metal complexes such as Mg^{2+}, Fe^{3+}, and Cu^{2+} to the network of polyimine vitrimer, improved its thermal and mechanical properties, resistance to creep, resistance to solvents and chemical stability. The modified vitrimers retained their ability to be reprocessed and recycled. This research is a valuable contribution to the development of advanced materials that can be used in many different ways. Mao Chen et al. [99] developed a novel epoxy vitrimer with rapid stress relaxation and moderate temperature malleability, which was synthesized using two different exchange reactions of disulphide metathesis and carboxylate transesterification as shown in Fig. 1.14.

This is the first paper to describe the synthesis of vitrimer networks with dual exchange reactions, which they refer to as "dual dynamic vitrimers". They created the dual dynamic vitrimer using a classic epoxy chemical reaction of the diglycidyl ether of bisphenol A (DGEBA) and 4, 4'- dithiodibutyric acid (DTDA) at 180 °C for 4 h with triazobicyclodecene (TBD) as a transesterification catalyst. Once hydroxyl groups are formed as a result of the ring-opening reaction of epoxy and acid, they continue to react with epoxy or acid, resulting in polymer chain branching and network formation. When the epoxy network is heated, the polymer chains can be rearranged via simultaneous disulphide metathesis and carboxylate transesterification. The resulting vitrimer shows faster exchange reactions and more efficient stress relaxation and low T_m due to the dual exchangeable crosslinks of disulphide bonds and carboxylate ester bonds.

There is growing interest in the development of vitrimer composites for applications such as crack repairing and correction of moulding defects. Vitrimer-based com-

(a) Dual dynamic vitrimer:

(b) Network rearrangement and exchange reactions:

Disulfide bonds

(1) (2)

Ester bonds

Dual dynamic vitrimer

(1) Disulfide metathesis:

(2) Transesterification:

(c) Solvent resistance:

RT → Tricholorobenzene → 180°C

0.2cm   0.2cm

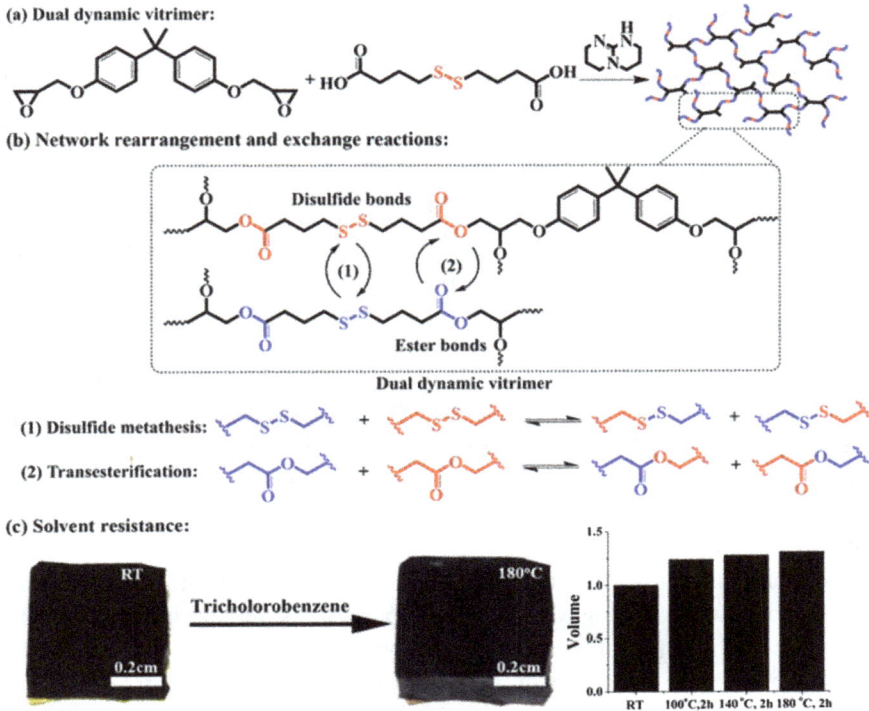

Fig. 1.14: (a) Synthesis and crosslinked structure of the dual dynamic vitrimer. (b) Network rearrangement of the dual dynamic vitrimer and the simultaneous exchange reactions of disulphide metathesis and carboxylate transesterification. (c) Swelling experiment of the dual dynamic vitrimer in trichlorobenzene from 100 to 180 °C. Reprinted with permission from M. Chen, L. Zhou, Y. Wu, X. Zhao, and Y. Zhang ACS Macro Lett. 8 (2019) 255–260. © 2019 ACS Publishing Company.

posites will have numerous advantages including reprocessability, reparability, and recyclability. Applying different types of process and process parameters for composite manufacturing and their optimization is essential for potential industrial prospects. However, successful introduction of vitrimer properties into a well-known resin may induce considerable modifications of the intrinsic parameters such as viscosity, gel time and curing procedure. It is important to choose a chemistry that works well with the curing chemistry and is based on monomers that are readily available commercially. This ensures that the dynamic chemistry can be easily integrated into the manufacturing process and will not cause any negative effects on the composite material's properties. The fibres or particles may positively or negatively interact with the dynamic chemistry and this should be carefully taken into consideration. Despite the challenges, vitrimer composites represent an important alternative toward durable and recyclable thermoset composites with expected bright future for industrial application.

1.3 Reinforcement

The reinforcement can be in the form particulate, whisker or fibre. For a particular fibre, again, there is wide variety in terms of nature and orientation. The reinforcing fibres can be classified into three categories, namely: i) inorganic fibre like glass, carbon, basalt, silicon nitride, etc. ii) polymeric fibre, e.g. Kevlar, spectra, etc. and iii) natural fibre, e.g. jute, coir, bamboo, etc. The fibres, as they are, if used for composite fabrication, do not result good mechanical properties because of poor interfacial adhesion. As is mentioned earlier, design of interface is very important for achieving desirable mechanical properties of a composite. The strength of interfacial bonding is generally quantified as inter laminar shear strength (ILSS). ILSS value of 30 MPa or above is required for a composite to be used for structural application. It may be noted that very high ILSS value is detrimental with respect to the impact properties of the composite. Therefore, the interface has to be designed optimally to achieve both the desirable mechanical and impact property or damage tolerance. Therefore virgin fibres are not used for composite fabrication. Fibre surfaces are to be treated judiciously depending upon the chemistry of fibre material. For example, glass fibres are treated with organisilane coating, which contains some function groups that can form chemical bonding or H-bonding with the matrix polymer. A number of oxidative and non-oxidative methods [100–104] are used to modify carbon fibre and polymeric fibres to improve their adhesion with polymeric matrices. For each fibre, there are again different variations possible, like unidirectional fibre, fibre mat, 2D fabric, 3D fabric, etc. Zhang and co-workers [105] reported various reinforcements in order to improve the out-of-plane properties of polymer composites, namely 3D orthogonal preform, 3D through-the-thickness preform, 3D angle interlock preform and 3D orthogonal circular perform [105, 106] as shown in Fig. 1.15.

Wide varieties of composites can be made using the same matrix, which may differ in respect of type of fibre, amount of fibre, fibre length, fibre orientation, etc. The properties of a composite can also tailored by fibre hybridization as discussed above. On the basis of length, fibres are classified in two categories, namely short fibre and long continuous fibres. Continuous fibre-reinforced composites contain reinforcements having lengths much greater than their cross-sectional dimensions. A composite is considered to be discontinuous or short fibre composite if its properties can be varied by changing the length of the fibre. On the other hand, when the length of the fibre is such that any further increase in length does not result in any enhancement in mechanical properties of the composite, then it is called continuous fibre- reinforced composite. Most continuous fibre-reinforced composites contain fibres, which are comparable in length to the overall dimension of the composite part. Thus, continuous fibre-reinforced composites offer superior mechanical properties compared to short fibre-based composites. Hence short fibre composites are used for secondary structural applications whereas continuous fibre composites are used in primary structural applications and considered high-performance engineering materials.

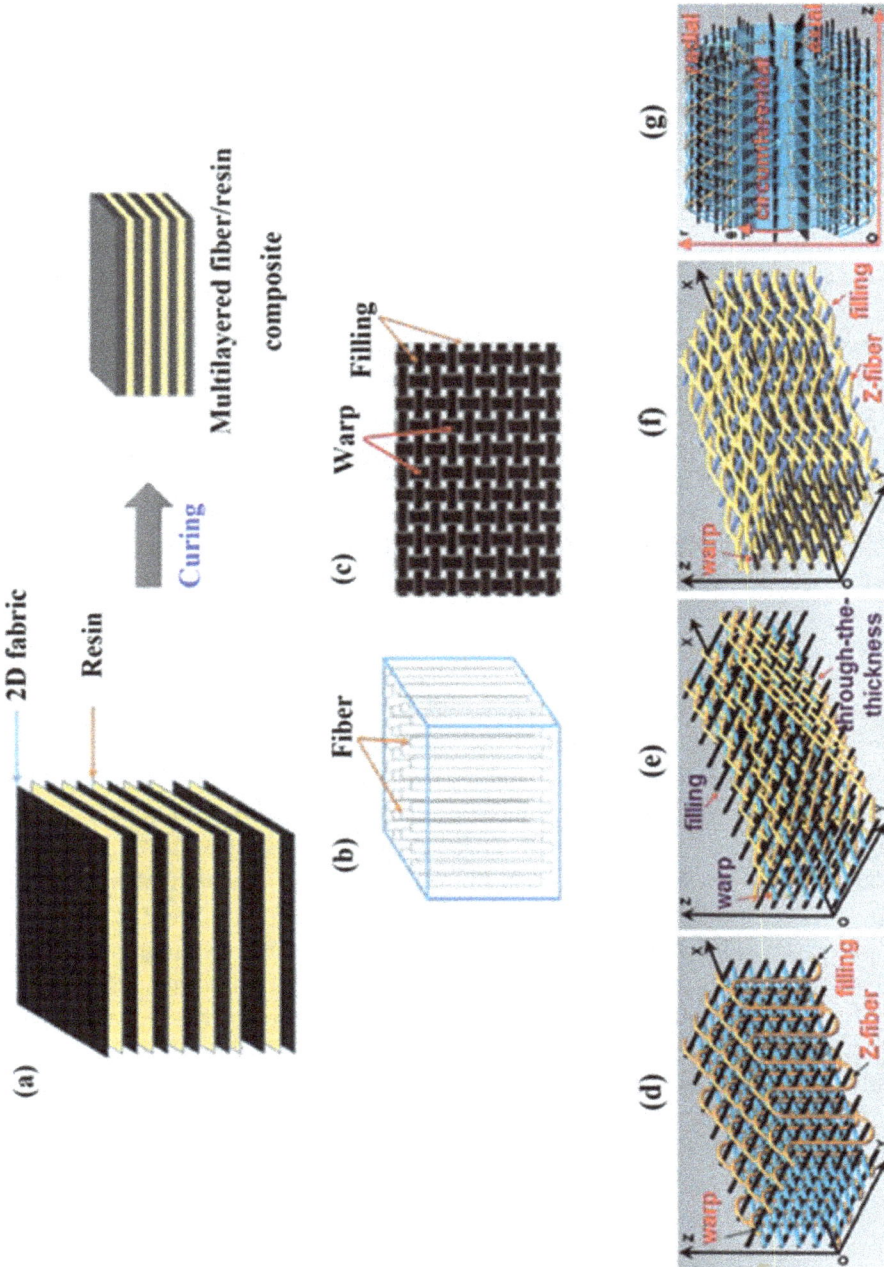

Fig. 1.15: Various forms of textile structural composites. a) Fibre laminate; b) non-woven mat; c) unidirectional fibres; d) non-crimp stitched fabric [106]; e) carbon woven fabrics; f) 3D orthogonal

1.3.1 Natural fibres

The inorganic and polymeric fibres as discussed above are not biodegradable. When the life of products made with such fibre is over, they simply go to landfill and contribute towards waste management issue. Burning of such fibre produces carbon dioxide and other toxic gases and creates greenhouse effect and environmental pollution. It may be noted that solid waste management is being viewed as a big challenge in times to come. The need has been expressed not only by government regulation but also various social concerns for environment. Therefore, for sustainable product development, it is necessary to use fibres that are biodegradable and produced from renewable resources. With this objective in mind, various natural fibres have been investigated with the intension to use them as substitutes for inorganic or polymeric fibres. In the last two decades, considerable works have been done on natural fibre composites, which will be reviewed in subsequent sections.

Natural fibres can be divided into three categories [108, 109] depending on their source of extraction: animal fibres (silk), vegetable fibres (abaca) and mineral fibres (asbestos) as shown in Fig. 1.16. Plant fibres have been extensively investigated for composite applications. Such fibres are composed of cellulose, hemicelluloses, lignin, pectin, wax, ash, and moisture. Precisely, the cellulose content in any natural fibre governs the mechanical properties of the fibre, as it controls the cell geometry condition, whereas the lignin content in natural fibres is responsible for water uptake into the core. Therefore, it is necessary to know their composition [110, 111]. Some of these fibres, which have gained popularity for their potential to use in fibre-reinforced polymer composites, are abaca, jute, sisal, kenaf, coconut and bamboo [112]. Table 1.1 presents [111–117] the average chemical composition of popular natural fibres. Table 1.2 shows [111] the mechanical and physical properties of the fibres. The mechanical properties of composites depend on the nature fibre, matrix, ILSS, fibre dispersion and fibre orientation [112]. The properties of some selected composites are presented [113, 115–117–123] in Tab. 1.3.

As we can see from the table natural fibre-based composites exhibit inferior mechanical properties compared to inorganic or polymeric fibre-based composites. Natural fibres are generally hydrophilic in nature because of the presence of many hydroxyl groups. When such fibres are used to reinforce polymer matrices, which are mostly hydrophobic in nature, the resulting composites show poor interfacial bonding. As a solution to this problem, natural fibres are modified with various pretreatment processes, which can enhance fibre-matrix adhesion, i.e. ILSS. Alkali treatment is one of the most important methods used to clean and modify the fibre surface

Fig. 1.15 (continued)
preform and g) 3D angle interlock preform [107] Reprinted with permission from Jianbin Li, Zhifang Zhang, Jiyang Fu, Zhihong Liang, and Karthik Ram Ramakrishnan Nanotechnology Reviews 10 (2021) 1438–1468. © 2021 De Gruyter Publishing Company.

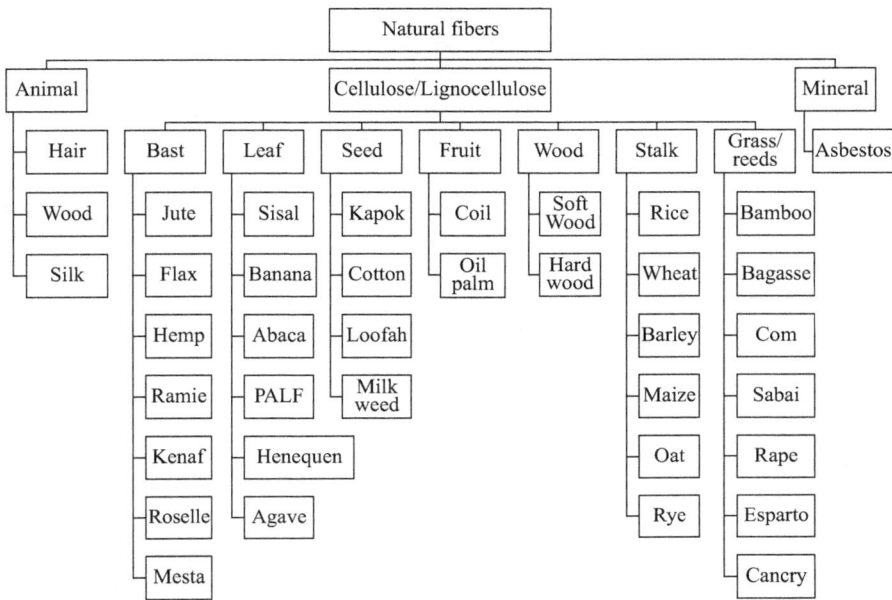

Fig. 1.16: Classification of NF [109]. Reprinted with permission from M. Alhijazi, Q. Zeeshan, Z. Qin, Babak Safaei, and M. Asmael, Nanotechnology Reviews 9 (2020) 853–875. © 2020 De Gruyter Publishing Company.

Tab. 1.1: Average chemical composition of natural fibres. Reprinted with permission from Agnivesh Kumar Sinha, Harendra K. Narang and Somnath Bhattacharya,J. Polym. Eng. 37 (2017) 879–895. © 2017 De Gruyter Publishing Company.

Type of fibre	Cellulose	Hemicelluloses	Lignin	Pectin	Wax	Ash	Moisture	Others	References
Abaca	56–64	25–29	11–14	–	–	–	–	–	[113]
Jute	64.4	12	0.2	11.8	0.5	0.5–2.1	10	–	[114]
Sisal	65.8	12	0.8	9.9	1.2	0.3	10	–	[114]
Kenaf	44.4	–	20.1	–	–	4.6	–	–	[115]
Coconut	37–43	24–28	26–28	–	–	–	–	7	[116]
Bamboo	78.83	–	1.15	–	–	–	–	–	[117]

Tab. 1.2: Mechanical and physical properties of natural fibres. Reprinted with permission from Agnivesh Kumar Sinha, Harendra K. Narang and Somnath Bhattacharya, J. Polym. Eng. 37 (2017) 879–895. © 2017 De Gruyter Publishing Company.

Type of fibre	Diameter (um)	Density (g/cm^3)	Tensile strength (MPa)	Young's modulus (GPa)	References
Abaca	250–300	1.5	717	18.6	[113]
Jute	250–2500	1.3–1.49	393–800	13–26.5	[118]
Sisal	2.5–300	1.41	350–370	12.8	[119]
Kenaf	83.5	1.2	282.6	7.13	[115, 120]
Coconut	396.98	1.2	140–225	3.–5.	[116, 121, 122]
Bamboo	–	1.2–1.5	500–575	27–40	[123]

Tab. 1.3: Mechanical properties of various NFCs. Reprinted with permission from Agnivesh Kumar Sinha, Harendra K. Narang and Somnath Bhattacharya, J. Polym. Eng. 37 (2017) 879–895. © 2017 De Gruyter Publishing Company.

Fibre type	Matrix	Fibre loading (WT.%)	Treatment	Tensile strength (MPa)	Flexural strength (MPa)	Impact strength
Jute/GF bidirectional woven mat	Epoxy	–	–	62.99	–	–
Jute/glass fibre	Polyester	–	–	229.54	–	–
Jute-sisal / glass fibre	Polyester	–	–	200	–	–
Oil palm – jute (4:1)	Epoxy	40	–	25.3	–	–
Oil palm-jute (1:1)	Epoxy	40	–	28.3	–	–
Oil palm-jute (1:4)	Epoxy	40	–	37.9	–	–
Jute	Epoxy	40	–	45.5	–	–
Jute (longitudinally laminated)	Epoxy	–	–	112.69	138.94	–
Jute (transverse laminated)	Epoxy	–	–	11.06	18.24	–
Jute	PP	–	Benzene/ethanol v:v,2:1	–	–	–
Bidirectional jute	Epoxy	48	–	110	55	4.9 J
Jute	UP	–	–	23	–	–
Jute-banana (100/0)wt. ratio	Epoxy	–	–	16.62	57.22	13.44 J
Jute-banana (75/25)wt. ratio	Epoxy	–	–	17.89	58.6	15.81 J

Tab. 1.3 (continued)

Fibre type	Matrix	Fibre loading (WT.%)	Treatment	Tensile strength (MPa)	Flexural strength (MPa)	Impact strength
Jute-banana (50/50)wt. ratio	Epoxy	–	–	18.96	59.8	18.32 J
Jute-banana (25/75)wt. ratio	Epoxy	–	–	18.25	59.3	17.89 J
Pultruded jute/glass fibre (50/50) by volume	UP	–	–	266.22	343.32	–
Woven jute (at low frequency)	HDPE	–	Oxygen plasma Treatment (90 W)	26.3	32.5	–
Woven jute (at radiofrequency)	HDPE	–	Oxygen plasma Treatment (90 W)	41.5	54.9	–
Abaca	PLA	–	–	74	124	5.3 kJ/m^2
Abaca	PP	10	Benzene diazonium salt	30	54	44 J/m
Abaca	PP	25	Benzene diazonium salt	28	55	52 J/m
Abaca	PP	–	Fungamix (enzyme)	45	69	5.2 mJ/m^2
Intralayer abaca-jute-glass fibre	Epoxy	–	–	42.4	3.06	4.01 J
Abaca/glass fibre	Epoxy	–	–	44.5	12.5	16 J
Abaca-jute/glass fibre	Epoxy	–	–	57	12.1	12 J
Coconut sheath fibre	Epoxy	–	5% NaOH 1 h	48.35	64.6	4.51 J/mm
Coconut sheath fibre	Epoxy	–	5% NaOH 1 h	58.6	76.8	5.61 J/mm
Coconut spathe: kenaf bast fibre (2.5:7.5)	Epoxy	10	5% NaOH 4 h	30	35	2 J
Coconut spathe: kenaf bast fibre (5.5)	Epoxy	10	5% NaOH 4 h	15	25	1.6 J
Coconut spathe: kenaf bast fibre (7.5:2.5)	Epoxy	10	5% NaOH 4 h	25	28	2.5 J
Coconut spathe	Epoxy	10	5% NaOH 4 h	20	30	4 J
Kenaf	Epoxy	–	–	58	–	–
Kenaf	Epoxy	–	–	124	–	–
Kenaf	Epoxy	–	–	164	–	–

Tab. 1.3 (continued)

Fibre type	Matrix	Fibre loading (WT.%)	Treatment	Tensile strength (MPa)	Flexural strength (MPa)	Impact strength
Woven kenaf (K) aramid (A) four-layered (A/K/A/K)	Epoxy	–	6% NaOH 3 h	123	64.7	–
Random oriented kenaf mat	Epoxy	–	6% NaOH 48 h	31.3	56.4	–
Random oriented kenaf mat	Epoxy	–	6% NaOH 48 h	42.5	81.4	–
Unidirectional kenaf	Epoxy	–	6% NaOH 48 h	95.4	123.2	–
Unidirectional kenaf	Epoxy	–	6% NaOH 48 h	106.1	177.6	–
Sisal-banana	Epoxy	–	–	25	62	–
Sisal	Epoxy	–	–	23	61	–
Sisal unidirectional at 90°	Epoxy	12	5% NaOH 4 h	56.5	371.33	1.3 kJ/m^2
Sisal unidirectional at 0°, 90°,0°	Epoxy	12	5% NaOH 4 h	16	96.38	1.35 kJ/m^2
Sisal	Epoxy	15	–	66.74	204.3	–
Sisal	Epoxy	20	–	87.54	167.3	–
Sisal	Epoxy	25	–	74.89	235.3	–
Sisal	Epoxy	30	–	132.73	288.6	–
Sisal mat	Epoxy	30	–	89.3	152.12	–
Bamboo	Polyester	–	6% NaOH	21	44.2	–
Bamboo cellulosic fibre	PLA	–	5% NaOH	56	–	4.2 kJ/m^2
Bamboo	Epoxy	57	–	392	–	–
Unidirectional bamboo fibre (longitudinal)	Epoxy	57	–	–	226	–
Unidirectional bamboo fibre (transverse)	Epoxy	57	–	–	11.89	–

to lower surface tension and enhance interfacial adhesion between natural fibre and polymeric matrix [124].

Another disadvantage of natural-fibre-reinforced thermosetting PMCs is their poor toughness. This issue can also be successfully resolved by chemical treatment of natural fibre [125] such as alkali treatment, silane treatment, acrylic acid (AA) treatment and benzylation. The fact can be elaborated taking bagasse fibre as an example. Bagasse is the by-product of sugarcane industry. It is a fibrous matter that remains after the extraction of sugarcane juice. The inner (pith) and outer (rind) part of the bagasse fibre show different properties [126]. Incorporation of 15 vol % of outer part of bagasse-fibre in unsaturated polyester resin resulted an increase in flexural strength to 30 MPa from 13 MPa for the unfilled resin. When the same volume of inner part of bagasse-fibre was incorporated into the polyester resin, the increase in flexural strength was found to be for 22.4 MPa. This has been explained by considering the fact that the outer part of the fibre, i.e. pith contains small fibres consisting mainly of sucrose whereas the rind contains longer and finer fibres arranged throughout the stem and bound together by lignin and hemicelluloses, providing better strength.

Cao et al. [127] reported that when bagasse fibres polyester treated with 1% alkali (NaOH) solution were used to make composites using unsaturated polyester matrix, an improvement by 13% in tensile strength (TS), 14% in flexural strength and 30% in impact strength of the related composites were observed. Scanning electron microscopy (SEM) analysis shows better surface modification and improved fibre-matrix adhesion of alkali-treated fibres. Similar observation was reported by Goulart et al. [128] who have investigated the effect on the mechanical properties of sugarcane-bagasse-fibre-reinforced polypropylene (PP) composites due to the chemical treatment of the fibres. The fibres were pre-treated with 10% sulphuric acid solution and delignified with 1% NaOH solution. An improvement in the TS by 16%, flexural modulus by 51% and impact strength by 45% compared to pure polymers was observed. The reason for the improvement of properties is the pre-treatment process carried out on the fibres, which removes considerable amount of moisture and other materials such as cellulose lignin, hemicelluloses and lignin from fibres, thus increasing fibre-matrix adhesion.

Recently, Lala et al. [129] reviewed natural fibre-based composites using both plant and animal based natural fibres. Plant-based natural fibres and composites have been elaborated in previous sections. Biocomposites in the true sense can be made by using natural fibre as well as biodegradable polymers like poly(lactic acid) (PLA) and poly(3 hydroxybutyrate-co-hydroxyvalerate). Such composites are also covered by Lala et al. [129] in their work. The properties of various biofibre-based composites are presented in Tab. 1.4. Although mechanical properties of natural fibre-based composites have been improved by adopting various chemical modifications, not much success has been so far achieved in enhancing the thermal properties of such composites. It may be noted that natural fibres are inferior to inorganic fibres in term of their thermal properties. A lot of focused research and development activities are to be initiated with the objective of making a breakthrough in this direction.

Tab. 1.4: Mechanical properties of composites. Reprinted with permission from Sumit Das Lala , Ashish B. Deoghare and Sushovan Chatterjee, Sci Eng Compos Mater 25 (2018) 1039–1058. © 2018 De Gruyter Publishing Company.

S.no	Reinforcements		Mechanical properties							
		Tensile strength	Elastic modulus	Flexural strength	Flexural modulus	Storage modulus	Impact strength	Yield strength	Hardness
1	Sugar beet pulp	37.5 ± 0.5MPa	–	–	–	–	–	–	–
2	Bagasse	–	–	30MPa (outer) 22.4MPa (inner)	–	2750 MPa (untreated fibres) 2900 MPa (NaOH-treated) 2780 MPa (acrylic acid-treated)	–	–	–
3	Hemp	30MPa 59.3MPa with PLA matrix	3.5GPa(for poly(3-hydroxybutyrate-co-hydroxy valerate) matrix) 6 GPa for copolyester amide matrix	124.2 MPa with PLA matrix	–	–	26.3 kJ/m² PLA matrix	–	–
4	Ramie	59.3 MPa 61.4 MPa, for 1,6-hexane Diisocyanate (HDI) 62.0 MPa for 4,40-diphenyl methane diisocyanate (MDI) 60.1 MPa for isophorone diisocyanate (IPDI)	–	136.8 MPa	–	–	–	–	–

5	Kenaf	223 MPa	–	254 MPa	–	–	–	–
6	Abaca	–	–	40 MPa	2240 MPa	–	–	–
7	Beech fibres	65 MPa	5049 MPa	–	–	–	–	–
8	Hildegardia populifolia	–	–	–	–	–	30.4 MPa (treated) 32.4 MPa (untreated) fibres at 20% fibre weight 35.2 MPa (treated) 34.5 MPa (untreated) fibres at 40% fibre weight	–
9	Corn husk	24.3MPa	–	–	–	–	–	3.02kP/mm^2
10	Roselle and sisal	32.4 MPa (10 wt.%)41.7 MPa (20 wt.%) and 48.8 MPa (30 wt.%) 50 mm fibre length, respectively 48.1MPa (10 wt%), 50.9 MPa (20 wt%), 58.7 MPa (20 wt%), 58.7 MPa 30 wt% at 100 mm fibre length	–	–	–	1.41 kJ/m^2 at 150 mm fibre length	–	–
11	Jowar	302 MPa	6.99 GPa	–	–	–	–	–

(continued)

Tab. 1.4 (continued)

S.no	Reinforcements	Mechanical properties							
		Tensile strength	Elastic modulus	Flexural strength	Flexural modulus	Storage modulus	Impact strength	Yield strength	Hardness
12	Pine fibre	–	2613 MPa	43.8 MPa	2943 MPa	–	–	–	–
13	Sisal fibres	17.5 MPa	2.39 GPa	36.8 MPa	–	–	5.6 kJ/m^2	–	–
14	Pineapple	37.28 MPa	687.02 MPa	–	–	–	–	–	–
15	Banana/sisal hybrid fibre	–	–	–	–	2818MPa	–	–	–
16	Sugar palm fibre	19.3 MPa	1706 MPa	–	–	–	–	–	–
17	Argan-nut shell	27.17 MPa	–	–	–	–	–	–	–
18	Rubber seed shell	17.29 (60–80 mesh) 16.07 (80–100 mesh) 17.04 MPa (100–120 mesh)	–	–	–	3819 MPa	–	–	–
19	Almond shell	26 MPa (with coupling agent), 27 MPa (without coupling agent), 28.14 MPa (binary untreated) & 29 MPa (ternary untreated)	–	–	–	–	–	–	–

20	Fish water soluble protein (FWSP)	5.33 ± 0.95 MPa for 5 FWSP 11.88 ± 1.82 MPa for 10 FWSP	–	–	–	–	–
21	Bamboo fibre	32 MPa (without compatibilizer) 36 MPa (addition of compatibilizer)	5 GPa (without compatibilizer) 6 GPa (addition of compatibilizer)	–	–	–	–
22	Acetylated cellulose nanocrystals	–	1502 MPa	–	–	44 MPa	–
23	Micro fibrillated cellulose (MFC)	–	687 MPa for 5 wt% MFC, 1033 MPa for 10 wt% MFC, 3898 MPa for 50 wt% MFC	–	–	–	–
24	Microcrystalline cellulose (MCC)	106 MPa	7.6 GPa	–	5275 MPa at 35% fibre wt	–	–
25	Cellulose nanocrystal (CNC)	–	2192 MPa at 1.5 wt% CNC	–	–	37 MPa at 1.5 wt% CNC	–
26	Lignocellulosic fibres	460 MPa	28 GPa	–	–	–	–

(continued)

Tab. 1.4 (continued)

S.no	Reinforcements	Mechanical properties							
		Tensile strength	Elastic modulus	Flexural strength	Flexural modulus	Storage modulus	Impact strength	Yield strength	Hardness
27	Cellulose and cellulignin fibres from sugarcane bagasse	20.1–26.3 MPa (in between)	–	–	–	–	–	–	–
28	Bo-cell fibres	910 MPa	23 GPa	–	–	–	–	–	–
29	Polylactic acid	40.1 (±2.4) MPa	1093.0 (± 7.0) MPa	–	–	–	–	–	–
30	Chitosan	26.37 MPa (for non-treated fibres) 35.40 MPa (for treated fibres)	1513.48 MPa (for non-treated fibres) 2296.94 MPa (for treated fibres)	–	–	–	48.5 J/m (for non-treated fibres) 65.8 J/m (for non-treated fibres)	–	–

1.4 Strength and toughness of composites

Polymer matrix composites can be either homogeneous or anisotropic in physical properties. For example, rubber with fine pigment loading can behave as a homogeneous mass, with same strength in all directions, but the same rubber with short fibre will show anisotropy in strength, and the extent of anisotropy would depend on orientation. However, the anisotropy depends on the size and critical aspect ratio of the non-spherical particle. For example, a polymer with nanoparticle of graphene, which has the maximum dimension as 500 nm, would behave almost as an isotropic composite, although the aspect ratio could be more than 300. For micron-sized long or continuous fibre and other non-spherical fillers, the anisotropy is predominant.

Toughness, defined as the ability to absorb strain energy till failure, is decided by the strength as well as flexibility of the polymer and also the reinforcing ability of the dispersed phase. The forces of interaction among the polymer molecules and the filler and clustering of the filler particles play important role in the toughness of a composite. It is generally observed that the presence of hydrogen bonding, polar groups and even a long chain of -CH_2 (forming a flexible spring) would result in high toughness compared to a stiff polymer such as cured epoxy thermoset. Fracture toughness, on the other hand, is the ability to resist the growth of a critical crack in the composite. The arresting of a critical crack growth can be done in a stiff matrix by chemical modification (flexibilizing) or even physically adding a rubbery material.

Toughness of a composite is decided by the matrix as well as type of inclusion. Certain inclusions reinforce the polymers in all mechanical properties, such as carbon black, carbon nanotubes, nanosilica, etc. Toughness can be much compromised in case of long or continuous brittle fibre inclusion compared to an isotropic composite. The toughness is also predominantly dependent on dispersion and distribution of fillers. A non-uniformity in dispersion may lead to considerable difference in shear stress in two neighbouring phases of a microregion of the filler-matrix blend, and the failure can occur prematurely than for a uniform dispersion of the same filler.

1.4.1 Tensile cracking

The tensile force acting on a composite results in different strain in the matrix, the interface of filler and matrix, the solitary filler particle and the cluster of filler. The interface decides the extent of shear stress for debonding of the filler from the matrix, and the difference in strain for the filler is almost negligible compared to that in the matrix. Hence, the transfer of stress from either phase is very minimal, when the modulus of the filler is very high compared to that of the matrix. Tensile cracking could occur due to several reasons, viz. breaking of interface after a critical shear stress, braking of cluster due to the tensile stress exceeding the interparticle cohesion, or matrix failure mainly due to defects such as critical cracks, or split in the composite due to interlaminar shear failure (in case of long/continuous fibre-reinforced composites).

Analytical studies to define each mode of tensile cracking is difficult due to superimposing events, and more so because the matrix as such is non-linear in high strain regime. There is also a difference in transverse cracking of the matrix in off-axis tensile loading between prepreg- based or dry fabric-based multidirectional fibre laying. Transverse cracking for prepreg-based composites is more likely because the waviness of the dry fabric (multidirectional) results in better off-axis strength. The high waviness of the layers introduced in the fibre bundles in the out-of-plane direction improves the interlaminar fracture toughness and reduces the important in-plane properties compared to prepreg laminates. As an alternative to conventional prepreg laminates, the use of weaving, stitching knitting, braiding, etc. with resin infiltration helps overcome such disadvantages. However, matrix failure due to defects and the cluster cracking can be neglected as the dominant cause, since it is generally the interface that fails first, and therefore, for fibre-reinforced polymer matrix composites, interlaminar shear failure or split, as it is defined, is common [130–133]. Patel and Wass [133] studied interlaminar splitting in carbon fibre-epoxy composite having multidirectional fibre orientation, using multiscale modelling of progressive damage analysis tool validated by open-hole tension experiments. Wang et al. [134] studied the thread pull-off failure of carbon fibre-epoxy composites under tensile load. The authors experimentally determined the pull-off strength, conducted an FEM-based simulation to calculate matrix tensile damage distribution under loads of different percentages and observed the damaged specimens by ultrasonic imaging. Schiffer et al. [135] used an underwater tensile loading simulation device to explore the response of laminates under high-strength underwater tensile loads and established a theoretical analysis model for the dynamic response of composite laminates.

The studies indicate that the tensile stress in a composite is non-linear, and the composite shows plastic deformation near the matrix failure load, and the final composite failure in tension is due to fibre cracking. The off-axis tension failure is more likely to occur earlier during a tension experiment. The failure process is the usual progressive damage phenomenon. In the process of tensile cracking of a composite laminate, the earliest damage is expected to the fibre or particle-matrix interface failure due to excessive shearing or tension beyond the critical interface shear stress. Intermediate failures can occur among the twisted bundled fibres due to poor interface in fibre-based composites and breakdown of clusters of particulate fillers (especially nanofillers) irreversibly due to continuous tensile loading. The failure sequence can, therefore be summarized as: (1) transverse matrix cracking due to manufacturing defects also in long fibre-composites, (2) delamination/filler-matrix interface failure and (3) fibre cracking or declustering of particle agglomerates.

The development of each crack has two stages: initiation and propagation. The initiation is usually from a micro-crack present as a defect in the matrix or interface and is independent of mesoscale geometry. The propagation is due to attainment of the critical tension level, beyond which the crack growth takes place and the phenomenon is explained according to linear elastic fracture mechanics (LEFM).The probability of transverse crack initiation is defined by The Weibull initiation stress distribution as [136]:

$$P_{in} = 1 - \exp\left[-\frac{V}{V_0}\left(\frac{\sigma_T}{\sigma_{in0}}\right)^m\right] \tag{1.1}$$

where P_{in} is the probability of crack initiation when the element's transverse tensile stress is σ_T, m and σ_{in0} are the shape and the scale parameters obtained in tests with reference specimens of the element volume V_0 and V is the volume of the considered element. The probability of crack initiation at a specific stress P_{in} (σ_{T0}) is defined as the ratio of the number of elements with initiated cracks to the number of elements, M in the layer:

$$P_{in}(\sigma_{T0}) = M_{cr}/M \tag{1.2}$$

and

$$M = L/t_k \tag{1.3}$$

where L is the length of the composite layer containing M elements, and M_{cr} number of initial cracks, while t_k is the ply thickness. The length of one element is taken as equal to the ply thickness and the stresses of a crack do not interfere with that of the neighbouring crack.

Under the cyclic tensile loading, eq. (1.1) is modified as:

$$P_{in} = \frac{\rho_k\left(\sigma_T^{fat}, N\right)}{\rho_{k\,max}} = 1 - \exp\left[-\frac{V}{V_0}N^n\left(\frac{\sigma_T^{fat}}{\sigma_{00}}\right)^m\right] \tag{1.4}$$

where N is the number of loading cycles and σ_T^{fat} is the cyclic stress, ρ_k is the crack density in a damaged k-th layer and the suffix "max" indicates maximum value.

The experimental determination of the cracks in a composite laminate is best done by the method of acoustic emission. An acoustic signal, however small in intensity, is generated as the crack propagates and the energy released is higher than crack initiation. The frequency of such acoustic signal varies from 10 to 100 kHz. A sensitive acoustic transducer of low mass is attached to the test specimen and the acoustic signal and the strain are simultaneously recorded with time to obtain the crack density evolution as a function of strain attained at the instant.

X-ray computed tomography (XCT) is also more or less accurate and applicable for any complex shape of many heterogeneous items for determination of internal design and defects. XCT illuminates the object layer-wise and the image formation is dependent on the relative absorption coefficient of the particular portion of the object (say, defect or damage) acquired from different angles of illumination. The image contrast relies on differences in the attenuation of X-rays through the portion of the object. However, in case of some composites, the poor phase contrast and long acquisition times among other drawbacks are limiting the applications of the XCT to the study of composites.

Optical microscopy can be used off-line, taking the sample out of the machine. It is easy and quite fast, but difficult to detect any edge crack without loading (initial). There can be a solution to this by printing the edge crack on a rubber film under a very low force and the film can be observed after the test in the microscope. This method can eliminate the removal of the specimen from the test set-up. However, suitability of any method would depend on the specific composite laminate.

1.4.2 Interlaminar stresses

Long or continuous fibres and woven fabrics are most commonly used for production of conventional polymer matrix composites. These composites are much stronger than short fibre composites by dough moulding compounds and particulate composites. The continuous or long fibre/fabric composites are thus used commonly as structural elements that replace the metals. A typical carbon fabric composite may have a Young's modulus of 50 GPa, while that of a DMC composite of short carbon fibre would be 22 GPa. Composites of long fibres and fabrics are made as layers of resin-wet fibre/fabric assembly in either one (0 degree) or multi-direction (0, 90, + 45, − 45 etc.). Manufacturing processes such as hand lay-up, resin transfer, vacuum-assisted resin transfer, resin infusion, prepreg method, etc. are widely used to obtain high-fibre composites with maximum mechanical strength. Unidirectional composites have very high strength and modulus in the direction of laying but are quite weak in of-axis load, because in the fibre direction, the maximum load is borne by the fibre and in transverse direction, the load is borne by the resin. In multidirectional composite, the anisotropy in behaviour is much reduced and the strength is almost similar in all directions for equal lay-up in several cross directions (0,90, + 45, − 45). Figure 1.17 shows typical unidirectional and multidirectional composites.

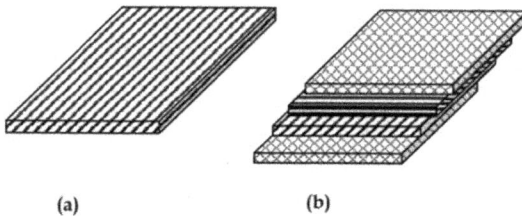

(a)   (b)

Fig. 1.17: (a) unidirectional and (b) multidirectional fibre-based composite.

In all the above arrangements, the subsequent layers are bonded by the interfacial force between the fibre and the resin. In case of glass fibres, "sizing" is used for a fabric, such as aminopropyltriethoxysilane (APTES) or methacryloxypropyltrimethoxysilane (MoPTMS), or glycidoxy- propyltriethoxysilane (GPTES), applied as a coating on glass fibre for enhancing the bond strength of the interface between the fibre and the resin [137]. The interface is therefore a vulnerable site for composite failure. If the fibres are assumed to be of circular cross section, then the individual fibre-resin interface may experience stresses in radial, axial and circumferential directions. The radial stress can be simulated with a direct pull-off analysis in tension, but the axial and circumferential detachment is almost like a lap shear failure.

Figure 1.18 shows the stresses in axial, circumferential and radial direction that can arise due to straining a composite in tension. The stresses at the interface in axial direction and circumferential direction are mainly shear; therefore, in general, the upper limit of the stress just before failure is less than the tensile strength of the resin and far less than the tensile strength of the common reinforcing fibres such as glass, Kevlar or carbon.

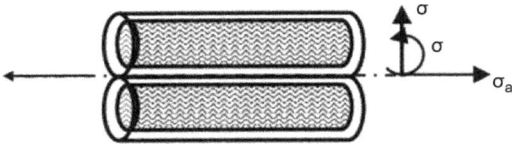

Fig. 1.18: Fibre-resin interface and three-directional stresses at the interface in a composite.

In automated fibre placement process (AFP), gaps and overlaps parallel to the fibre direction can be introduced between the adjoining tapes. These gaps and overlaps can cause a reduction in strength compared with pristine conditions. Li et al. [138] modelled the gaps and overlaps in a composite in the direction of fibre lay-up in FEM and analysed the intra-ply and inter-ply stresses to capture the influence of splitting and delamination. Out-of-plane waviness and ply thickness variations caused by gaps and overlaps were modelled. The FEM models thus designed were used to predict the reduction of strength as a function of the magnitude and type of the defects. The pristine composite tensile strength was about 740 MPa. The defects (gaps and overlaps) were created across the whole width of the specimen with the defect size varying between 0 and 4 mm, the percentage area of the defects varying between 0 and 5.7%, and the failure initiation in tensile load was at about 670 MPa. In general, with various combinations of staggering of ply and defects, the average reduction in tensile strength due to the defects was about 10%.

To test the interlaminar stresses to failure, there must be a special experimental arrangement for tension, bending and shear mode. As described in various international test standards, the most used methods are: (1) double cantilever beam tension (ASTM D5528 Standard), (2) sliding Shear and (3) three-point bending for interlaminar shear (ILSS) tests. The DCB specimens are prepared with a non-adhesive insert on the midplane of the laminate and the width and thickness measurements at midpoint are done as described in the method. The standard specimen length is at least 125 mm (5.0 in), the standard width is 20 to 25 mm (0.8 to 1.0 in) and the standard laminate thickness is 3 to 5 mm (0.12 to 0.2 in). The sample loading in each case is shown in Fig. 1.19. In the three-point bending test, the ILSS is calculated as:

$$ILSS = \frac{3P}{4bh}$$

(1.5)

where P is the load, b and h are width and thickness, respectively, of the composite specimen.

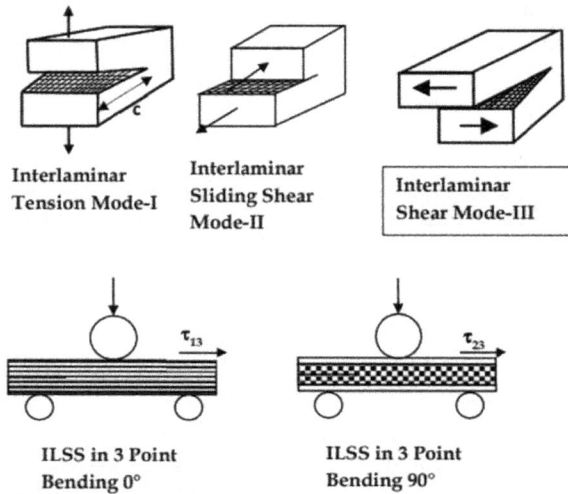

Interlaminar Tension Mode-I

Interlaminar Sliding Shear Mode-II

Interlaminar Shear Mode-III

ILSS in 3 Point Bending 0°

ILSS in 3 Point Bending 90°

Fig. 1.19: Different modes of interlaminar stress testing.

1.4.3 Prediction of strength and toughness

The strength of a polymer matrix composite can be defined in many ways, involving parameters of deformation vectors, stress tensors, inherent properties termed as hardness, elastic modulus, creep, relaxation, dynamic complex stiffness, etc. and are estimated according to the end application. Two examples can explain the differences in the concept of strength, such as a high-pressure hydraulic ball valve and an engine mount, both made of composite. The first one needs high ultimate strength to withstand hoop stress, while the second must have good complex stiffness to damp undue vibration. Therefore, prediction of composite strength depends on the property to be considered in a specific stress environment. In a simple consideration, tensile strength estimation or Young's modulus prediction can be done with composites of different reinforcements. Details about the prediction of strength and elastic modulus and effect of shape, size and stiffness of the reinforcements on mechanical properties of elastomer and thermoplastics are already discussed in Chapters 2 and 4. However, there is no direct predictive method for composite fracture toughness. However, the toughness in terms of dynamic mechanical properties can be predicted to some extent, which in turn can be applicable to fracture toughness, since viscoelastic properties are responsible also for regulating the strain energy. In a dynamic mechanical study, the loss of energy in each cycle of straining is calculated as an imaginary part of the input energy. The loss occurs due to the viscoelastic damping by the matrix polymer. Therefore, it prevents failure due to brittle fracture. For example, the dynamic strain energy of different carbon black-filled NBR-PVC based composites is shown in Fig. 1.20, calculated from dynamic mechanical response at 0.05% strain and 1 Hz frequency. The strain energy increases with the carbon black loading, indicating the increase in resisting brittle failure or crack propagation.

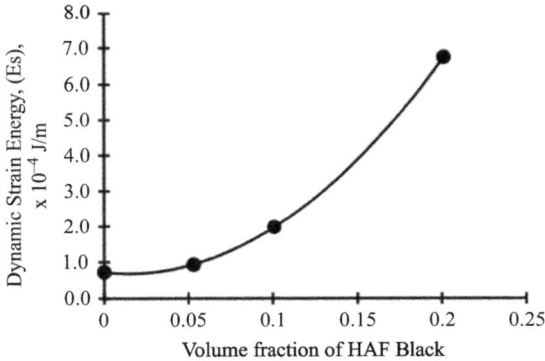

Fig. 1.20: Dynamic strain energy of NBR-PVC- HAF black composites.

The ability to withstand an impact or brittle fracture is related by a viscoelastic loss energy per cycle of stressing per unit volume of the sample as:

$$W_L = \pi E'' \varepsilon_0^2 \qquad (1.6)$$

In a composite, tensile strength, dynamic modulus, etc. cannot be predicted by simple law of mixing, as understood by micromechanical theories, and hence the loss too cannot be predicted. However, since the strain energy absorption by the fillers is almost nil and occurs only due to the polymer matrix, the modified energy loss can be written as [139]:

$$W_L = \pi E_M'' \varepsilon_M^2 (1 - \varphi_e) \qquad (1.7)$$

where the subscript M refers to the viscoelastic matrix. The quantity $(1 - \varphi e)$ is the volume fraction of the matrix material capable of dissipating energy.

The Kerner equation [140] pertaining to modulus reinforcement in simple tension is:

$$\frac{E_C'}{E_M'} = \frac{\{G_f'\varphi_f/[(7-5v)G_M' + (8-10v)G_f']\} + \{(1-\varphi_f)/[15(1-v)]\}}{\{G_M'\varphi_f/[(7-5v)G_M' + (8-10v)G_f']\} + \{(1-\varphi_f)/[15(1-v)]\}} \tag{1.8}$$

The elastic tensile modulus E' and shear modulus G' here are real parts of the corresponding complex modulus in dynamic stressing of the composite. However, the same could be applied for quasi-static experimental results too. The symbol v represents Poisson's ratio, taken as 0.5 for elastomer composites.

The equation above does not show direct dependence of composite modulus on the filler content unlike many subsequent equations, although semi-empirical in nature.

The storage modulus and loss modulus of a filled polymer are estimated by the modified Kerner equation on reinforcement of a polymer by rigid, spherical particles of very high elastic modulus compared to the polymer and assuming that the shear modulus of the polymer is approximately one-third of the tensile modulus (Poisson's Ratio = 0.5) [141–143]. Following are the modified Kerner equations suggested by Ziegel and Rommanov [141]:

$$\frac{E_c'}{E_M'} = \frac{1 + 1.5\varphi_f B}{1 - \varphi_f B} \tag{1.9}$$

$$\frac{E_M''}{E_c''} = 1 - \varphi_f\left(1 + \frac{\Delta R}{R}\right)^3 \tag{1.10}$$

$$\tan\delta_c = \tan\delta_M/\left(1 + 1.5\varphi_f B\right) \tag{1.11}$$

where,

$$B = \left(1 + \frac{\Delta R}{R}\right)^3 \tag{1.12}$$

B is an interaction parameter and defines the interaction at the interface of the filler and polymer. Higher value of B is attributed to better interaction, φ_f is the volume fraction of the filler and suffixes M and c indicate unfilled and filled polymer, respectively. The ratio $(\Delta R/R)$ represents the relative increase in the size of a filler (assumed spherical) in the matrix polymer due to the adhered polymer to the filler particle. If $\Delta R = 0$, there is no interaction, while for any finite values of ΔR, there is interaction between the filler and the polymer [141]. Figure 1.21 shows the interaction parameters of nanocomposites of NBR-PVC blend with graphene nanoplates at various frequencies of dynamic strain. The real and imaginary modulus were obtained from a DMA experiment in frequency scale at 30 °C. Higher filler-polymer interaction was noticed for lower volume fraction of graphene because at higher loading, filler-filler interaction becomes more dominant. Additionally, there is hardly any difference in interaction with change in frequency.

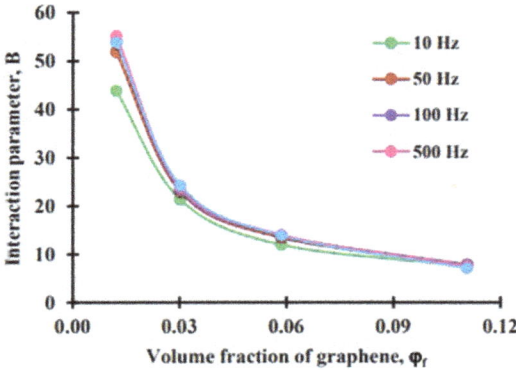

Fig. 1.21: Interaction parameter for NBR-PVC-graphene nanocomposite at various frequencies of dynamic loading.

There is a difference in the volume fraction φ_e in eq. (1.7) and φ_f in all subsequent equations. If it is accepted that due to the filler-polymer interaction, the true volume fraction includes the adhered polymer material, the volume fraction φ_e would be related to φ_f by the following equation:

$$\varphi_e = \varphi_f(1+\Delta R/R_0)^3 \tag{1.13}$$

where the term $\Delta R/Ro$ is the relative increase in particle diameter for spheres and can be estimated from the eq. (1.10).

A general theory of strength of anisotropic materials was given by Tsai and Wu [144] way back in 1971, which is still used because of excellent agreement with experimental findings for continuous fibre composites. The authors suggested that the basic assumption of strength criterion is that there exists a failure surface in the stress-space as:

$$f(\sigma_k) = F_i\sigma_i + F_{ij}\sigma_{ij} = 1 \tag{1.14}$$

where i, j, k are = 1,2,.6 and F_i and F_{ij} represent strength tensors of the second and fourth rank, respectively. The linear terms in σ_i takes into account the internal stresses and the quadratic terms in $\sigma_i\sigma_j$ define an ellipsoid in the stress space. The authors ignored higher-order terms for practical consideration. The failure takes place due to all the type of stresses in a composite, but with different weightage.

It is difficult to predict the failure stress because of the inhomogeneous morphology of the composite and it is difficult to exactly find out by experiment, where the failure actually initiated – the matrix, or the fibre-matrix interface or in the fibre. The failure stress level and the failure mechanism in these three phases are definitely different. The difficulty in predicting constituent failure is in part due to the lack of information about the correlation between macro (ply) stress/strain and micro stress/strain. Subsequently, there are many modifications of this original work such as by Chang and Chang [145], who considered several possibilities of failures: matrix cracking, fibre-matrix shearing and fibre breakage, also considered a material degradation model. Ha et al. [146] introduced a set of equations to calculate micro stresses from macro stresses and used micromechanics of failure (MMF) for predicting fibre-matrix interface failure, matrix failure and fibre failure for continuous fibre-reinforced composites.

Accelerated test methods developed along with MMF models can best describe ultimate strength and failure criteria of a composite, which in any form is far from homogeneous material, except for small amounts of inclusion of micron-size spherical particles, without agglomeration. More detail on degradation of composites due to ageing is discussed in Chapter-7.

Determination of interaction parameters, Krenchel orientation factors for modified mixing rules, Halpin-Tsai model or many such micromechanical models can then be applied with reasonable accuracy. The theoretical and semi-empirical analysis of the polymer nanocomposites with plate-type filler such as nanoclay or graphene showed that the nature of the constitutive equations are only valid with different coefficients and indices of volume fractions of the fillers, as shown by Murali et al. [147] Nguyen and co-workers [148] attempted to predict microstructure and mechanical properties of long- glass fibre (13 mm length, 17.4 µm diameter) 40% loaded in polypropylene injection- moulded thermoplastic composite, but opined that to match the experimental results, accuracy in fibre length distribution and fibre orientation distribution measurement/prediction is essential. The authors defined number average length ($\bar{L}_n$) and weight average length ($\bar{L}_w$) of fibres and Weibull distribution function to match the experimental fibre length distribution. Haneefa et al. [149], however, found good agreement with theoretical models for prediction of tensile strength and elastic modulus for composite of hybrid short fibre consisting of treated banana and glass (both 6 mm chopped) in polystyrene matrix.

Huang et al. [150] suggested that in any mechanics of continuum media, the stress of any point in a material is defined as an average value of an infinitesimal element containing the point as:

$$\sigma_i = \frac{1}{V^f} \left(\int_{vf} \tilde{\sigma}_i dV \right) \tag{1.15}$$

The resulting stress, σ_i is a point-wise value. V_f is the volume of the representative volume element (RVE). For a composite, however, one cannot take an infinitesimal element, because both the fibre and matrix must form one element. Thus, a composite stress is defined as a homogenized one satisfying eq. (1.15), as long as only two constituents are involved, the fibre and the matrix:

$$\sigma_i = \frac{1}{V^f}\left(\int_{vf} \tilde{\sigma}_i dV\right) \equiv V^f \sigma_i^f + V_m \sigma_i^m \tag{1.16}$$

The subscript m stands for the matrix polymer. The bigger the volume element, the more will be the error in strength prediction compared to the experimental result. The smallest volume element in a short-fibre composite should be a single fibre in a volume of matrix, which is adhered to the fibre and extended to such a distance from the fibre surface that the outer surface of the volume element is the same in strength as the bulk matrix, since there could be a positive or negative gradient of adhesion strength from the surface of the fibre towards the bulk matrix. However, since fibre strength is always much higher than the matrix, the failure is essentially at the matrix or matrix-fibre interface depending on the interaction of the fibre and matrix. Therefore, for a true value of stress in the composite, one must have a bridging equation, as:

$$\{\sigma_i^m\} = [a_{ij}]\{\sigma_j^f\} \tag{1.17}$$

where $[a_{ij}]$ is a bridging tensor, and $i, j = 1, 2, \ldots 6$. The tensor consists of longitudinal modulus, shear modulus of the matrix and for the fibre, the tensor values are for longitudinal, transverse and in plane shear modes, Poisson's ratio of both and compliance tensors. The homogenized stresses in the matrix and the fibre are given by:

$$\{\sigma_i^m\} = [a_{ij}]\left(V_f[I] + V_m[a_{ij}]\right)^{-1}\{\sigma_j\} \tag{1.18}$$

$$\{\sigma_i^f\} = \left(V_f[I] + V_m[a_{ij}]\right)^{-1}\{\sigma_j\} \tag{1.19}$$

where [I] is a unit tensor and σ_j is the external stress applied. The stresses in the fibre are uniform, but in the matrix, there can be different stress concentrations, and therefore, the authors determined the true stresses in the matrix by multiplying the stress concentration factors, which were elaborately described [150]. The results of their study showed fairly good agreement with experimental data on tensile strength of unidirectionally aligned short-fibre composite and the bridging model was seen to be good for glass fibre epoxy than carbon fibre-epoxy at $V_f = 0.55$ for both and aspect ratio of both the fibres as 455. For randomly oriented fibre, the bridging model is essential for better prediction of strength.

Barnett and co-workers [151] studied discontinuous carbon fibre-loaded composites of (1) polyphenylene sulphide (PPS), (2) acrylonitrile butadiene styrene (ABS) and (3) epoxy thermoset composite. Stochastic microstructure was considered for prediction of strength and modulus. The fibre used was of two varieties, recycled and virgin, in the form of non-woven mat and the composites were fabricated by compression moulding. The average lengths were about 30 and 40 mm for the virgin and recycled carbon fibres, while the diameters were similar, about 7.9 and 7.3 μm, respectively. The volume fraction and void volumes in the composites were determined by optical microscopy, image analysis and a MATLAB programme for pixel counting. The fibre orientation in layers was determined by X-ray tomography. The Cox shear lag model was used to predict the stiffness in the fibre direction:

$$E_l = E_{lf} V_f \eta_L + E_m V_m \tag{1.20}$$

where

$$\eta_L = 1 - \frac{\tanh(ns)}{ns} \tag{1.21}$$

$$s = \frac{L_f}{2r_f} \tag{1.22}$$

$$n = \sqrt{2 \frac{G_m}{E_{lf} \log(r_m/r_f)}} \tag{1.23}$$

where V_f is the fibre volume fraction, V_m is the matrix volume fraction, E_{lf} is the longitudinal modulus of the fibre, E_m and G_m are the longitudinal and the shear modulus of the matrix, related through Poisson's ratio (v_m) as:

$$G_m = \frac{E_m}{2(1 + v_m)} \tag{1.24}$$

L_f and r_f are the length and radius of the fibre, r_m is the radius of the matrix layer around the fibre (Fig. 1.18) and is given by:

$$r_m = \frac{r_f}{\sqrt{V_f}} \tag{1.25}$$

The Poisson's ratio of each digital lamina was calculated using the rule-of-mixtures:

$$v_{12} = v_{12f} V_f + v_m V_m \tag{1.26}$$

The failure stresses in longitudinal, transverse and in pane shear mode was calculated as:

$$\sigma_1 = \frac{E_1 \sigma_f}{E_{1f}} \frac{1}{1 - \mathrm{sech}\left(\frac{L_{fc}}{2} \sqrt{\frac{E_m}{E_{1f} r_f (r_m - r_f)(1 + v_m)}}\right)} \tag{1.27}$$

$$\sigma_2 = \sigma_m \left(1 - 2\sqrt{\frac{V_f}{\pi}}\right) \tag{1.28}$$

$$\tau_{12} = \frac{\sigma_m}{\sqrt{3}} \tag{1.29}$$

The failure stress depends on the critical fibre length L_{fc}, given by:

$$L_{fc} = \frac{r_f \sigma_f}{\tau_i} \tag{1.30}$$

where τ_i is the fibre-matrix interfacial shear strength.

The authors [151] reported quite good agreement with the findings of lamina microstructure determined by X-ray tomography and optical microscopy. Consequently, the mathematical calculation of strength, longitudinal, transverse, shear modulus and strain to failure of the composites were also in good agreement with the experimental results.

Estimation of sustainable strength and failure stress can be an important aspect when a composite is subjected to continuous vibration, at random or specified frequency. In that case, prediction of the strength under fatigue is more complicated. Long-term creep prediction can be one method to study the long-term fatigue. Generally, a set of limited time experiment at different temperature is converted to a time-scale master curve, generated by time-temperature superposition principle, for example using WLF equation, to shift the isotherms to time scale by shift factors. However, if it is assumed that the time-temperature transformation process can predict a long-term creep of a composite, then the same shift factors can be used to predict long-term fatigue from a limited time experiment at different temperatures, just like a creep study. At each fatigue cycle, the time to failure is a deciding parameter. In order to calculate the shift factors at each temperature in creep experiment, it is best to resort to graphical shifting process, since in that case, there would be no assumptions of numerical coefficients or WLF constants. However, the long-term creep determination has another complication of physical ageing, which shifts the relaxation and retardation time of the matrix polymer, especially when the glass transition of the polymer is above the application temperature. The details of the physical ageing and corresponding determination of relaxation time and creep compliance are discussed separately in the Chapter 7.

1.5 Fracture mechanics

Fracture in composites is caused by sudden stresses or by a constant loading, when there is a crack growth in the material to such an extent that the material fails catastrophically. When a crack reaches a certain critical length, it can propagate catastrophically through the structure, even though the gross stress is much less than would normally cause yield or failure in a tensile specimen. The crack growth may be initiated due to a small crack with sharp tip, wherein the stress concentration far exceeds that in the bulk, and the growth of the crack becomes fast enough to fracture the item suddenly. The stress concentration depends on the crack tip radius; hence sharp tips would be more prone to grow and fracture. However, even a round hole is considered as a crack with blunt tip. The energy required for growth of a crack is the summation of the energy required for overcoming the surface energy of the atomic layer to create a new surface and the energy required to overcome the cohesive energy of the material. For a ductile material, the energy required to overcome the viscoelastic energy is dominant.

The growth of a micro-crack to a critical crack length under a stress depends on the ductility of the material, because of large deformability without losing integrity. Under increasing stretching, the crack growth is also arrested to a large extent due to the high strain energy absorbing capacity of the material. This is a basic principle of designing polymeric composite backing of bullet-proof armour panels. A typical example is ultrahigh molecular weight polyethylene. The arresting of crack growth can be achieved by a multi-phase matrix where the dispersed phase is made up of flexible particles finely distributed in the continuous phase, so that the flexible particles arrest the crack tip from further propagation by absorbing the strain energy. A typical example is a blend of epoxy and carboxyl-terminated polybutadiene.

The definition of toughness may be the ability to absorb strain energy under quasi-static and dynamic stress while fracture toughness can be defined as the ability to arrest a fracture or crack from further propagation. Toughness is generally quantified by one of the three quantities, namely the area under the stress-strain plot (quasi-static experiment), or by the area under the loss modulus (vs. temperature or frequency) plot or by the Izod or Charpy impact energy.

1.5.1 Fracture toughness

The term is used to quantitatively determine the ability of the material to arrest a crack so that the material does not fail catastrophically under a mechanical stress. The definition does not include the condition of any crack generation, but assumes that a crack already exists in the material. For fracture toughness, the extent of critical stress intensity factor and strain energy of fracture are computed to define the fracture toughness. The quantification of stress intensity factor and the strain energy release rates are described in brief in the section below.

The stress intensity approach was developed by Inglis [152], way back in 1913. He stated that the local stress at the corner or a hole in a plate could be many times higher than the average applied stress. He showed that the degree of magnification of a local stress depends on the radius of curvature of the hole, higher stress for smaller radius. Therefore, he defined a Stress Intensity Factor $(SIF)_k$, for quantification of the local stress. For an elliptical hole in a plate, Inglis proposed following expression:

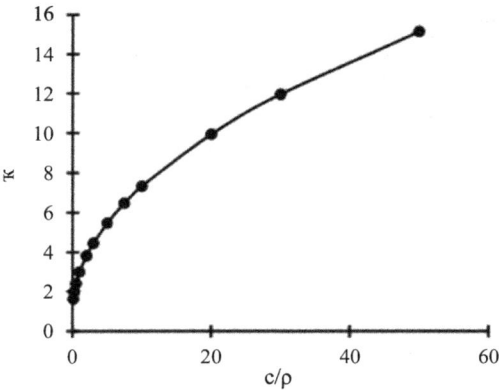

Fig. 1.22: Stress intensity factor for varying ratios of c/ρ in eq. (1.6).

$$\kappa = 1 + 2\sqrt{\frac{c}{\rho}} \qquad (1.31)$$

where c is the radius of the hole and ρ is the curvature at the tip of the hole where the local stress is estimated.

Figure 1.22 explains the extent of stress concentration at various c/ρ ratios. Obviously, for a circular hole, $c/\rho = 1$ and hence, the factor $_k = 3$.

Irwin [153] showed that the stress field $\sigma(r,\theta)$ in the vicinity of an infinitely sharp crack tip could be described mathematically by:

$$\sigma_x = \frac{K_I}{\sqrt{2\pi r}} \cos\frac{\theta}{2}\left(1 - \sin\frac{\theta}{2}\sin\frac{3\theta}{2}\right) \qquad (1.32)$$

The first term on the right-hand side describes the magnitude of the stress, while the terms associated with θ describe the distribution of the stress. The constant K_1 is termed as Stress Intensity factor and is defined as:

$$K_1 = \sigma_a Y\sqrt{\pi c} \qquad (1.33)$$

The term Y is a geometric factor, σ_a is the externally applied stress and c is the half crack length. For a particular crack system, π and Y are constants, so the stress intensity factor indicates that the magnitude of the stress at position (r,θ) depends only on the external stress applied and the square root of the crack length.

At large distances from the crack tip, these relations cease to apply and the stresses approach their far-field values that would be obtained where the crack is not present. The K_I in eq. (1.33) is a very important parameter known as The Stress Intensity Factor (SIF). The factor K_I contains the dependence on applied stress σ_a, the crack length a, and the specimen geometry. The K_I factor gives the overall intensity of the stress distribution.

The equation shows three factors that, taken together, depict the stress state near the crack tip: the denominator factor $(2\pi r)^{-1/2}$ shows the singular nature of the stress distribution; σ approaches infinity as the crack tip is approached, with $a/r^{1/2}$ dependency. The angular dependence is separable as another factor; e.g.

$$f(x) = \cos\frac{\theta}{2}\left(1 - \sin\frac{\theta}{2}\sin\frac{3\theta}{2}\right) \tag{1.34}$$

For the specific case of a central crack of width $2a$ or an edge crack of length $2a$ in a large sheet,

$$K_I = \sigma_\infty\sqrt{\pi a} \tag{1.35}$$

and for an edge crack of length a in a large sheet,

$$K_I = 1.12\sigma_\infty\sqrt{\pi a} \tag{1.36}$$

These stress intensity factors are used in design and analysis under a consideration that the material can withstand crack tip stresses up to a critical value of stress intensity, termed K_{Ic}, beyond which the crack propagates rapidly. This critical stress intensity factor is then a measure of material toughness. The failure stress σ_f is then related to the crack length a and the fracture toughness by:

$$\sigma_f = \frac{K_{1c}}{a\sqrt{\pi a}} \tag{1.37}$$

where a is a geometrical parameter equal to 1 for edge crack and generally unity in many situations. Expressions for a are tabulated for a wide variety of specimen and crack geometries, and specialty finite element methods are used to compute it for new situations.

Strain energy concept was developed by Griffith [154], who proposed that the reduction in strain energy due to the formation of a crack must be equal to or greater than the increase in surface energy required by the new crack faces. Using an energy balance method, he arrived at an equation for the stress to failure relating to the strain energy as:

$$\sigma_f = \sqrt{\frac{EG_c}{\pi a}} \tag{1.38}$$

where E is the Young's modulus and G_c is the Critical Strain Energy Release Rate, and "a" is the size of the flaw. This expression describes the interrelation between three important aspects of the fracture process: the material property as G_c, the stress level σ_f and the size, a, of the flaw. In a design situation, one might choose a value of a based on the smallest crack that could be easily detected. Then for a given material with its associated value of G_c, the safe level of stress σ_f could be determined. The structure would then be sized so as to keep the working stress comfortably below this critical value.

1.5.1.1 Compliance calibration
A number of means are available by which the material property G_c can be measured. One of these is known as compliance calibration, which employs the concept of compliance as a ratio of deformation to applied load:

$$C = \frac{\delta}{P} \tag{1.39}$$

The total strain energy U can be written in terms of this compliance as:

$$U = \frac{1}{2}P\delta = \frac{1}{2}CP^2 \tag{1.40}$$

$$G = \frac{\partial U}{\partial a} = \frac{1}{2}P^2\frac{\partial C}{\partial a} \tag{1.41}$$

If the compliance C is plotted against corresponding flaw size a, then the slope at the point of a is s, taken to calculate the strain energy release rate G according to eq. (1.41). Therefore, at the critical crack length a_c, the slope is taken to calculate the Critical Strain Energy Release Rate G_C.

$$G_C = \frac{1}{2}P_C^2\frac{\partial C}{\partial a}\bigg|_{a=a_C} \tag{1.42}$$

The mode-I (refer Fig. 1.19) delamination energy (G_{IC}) of the composite specimens are calculated using the double cantilever beam (DCB) tests according to the ASTM D 5528 standard [155]. The general formula from linear elastic fracture mechanics is used to evaluate G_{IC} as follows [156]:

$$G_{IC} = \frac{P^2}{2b}\frac{\partial C}{\partial a} = \frac{P\delta}{2bC}\frac{\partial C}{\partial a} \tag{1.43}$$

The term δ is the crack opening displacement (COD) and a is the crack length. The crosshead displacement of the DCB test is taken as COD of the specimen. There is a small correction, though not very necessary for practical applications, about the crack length, because of slight deformation around the crack tip. The crack length can be modified as $a+\Delta$, and it can be determined by plotting $C^{1/3}$ against crack extension [155–157]. Finally, the modified G_{IC} is calculated as:

$$G_{IC} = \frac{3P\delta}{2b(a+\Delta)} \tag{1.44}$$

For Mode-II delamination test according to ILSS as shown in Fig. 1.19 with end-notched flexure (ENF) specimen, the G_{IIC} is calculated as [158]:

$$G_{IIC} = \frac{9P\delta a^2}{2b(2L^3 + 3a^3)} \tag{1.45}$$

Here, $2L$ is the length of the beam in between the two end supports and a is the crack length measured from the end support on the beam.

1.5.1.2 Example

A typical compliance calibration curve (tensile cracking) is shown in Fig. 1.23 for an FRP composite of E-glass fabric and epoxy having Young's modulus of 22 GPa and tensile strength of 350 MPa, maximum % elongation at break is 2.5%. The dimension for the Mode-I fracture toughness study was 180 mm × 12.5 mm × 3.5 mm and an initial crack length was 50 mm. The critical crack length was 68 mm ($\delta = 18$ mm) and the corresponding slope $(\delta C/\delta a)$ was calculated at this crack length and corresponding G_{IC} was found to be 365 J/m^2 using the eq. (1.43).

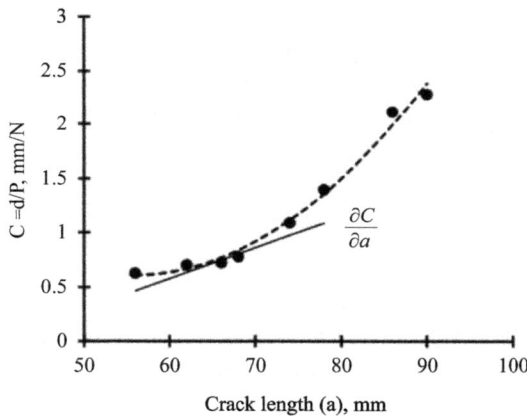

Fig. 1.23: Compliance calibration for Mode-I fracture toughness determination of an FRP composite.

1.5.1.3 Mode-II fracture toughness: End Notch Flexure

The Mode-II fracture energy release rate G_{II} can be determined without monitoring the crack length "a" during the experiment, which is somewhat difficult, following the equation [159]:

$$G_{II} = \frac{9P^2 a_e^2}{16B^2 b^3 E_f} \tag{1.46}$$

Effective flexural modulus E_f is given by:

$$E_f = \frac{3a_0^3 + 2L^3}{8Bb^3} \left(C_0 - \frac{3L}{10BbG_{LT}} \right)^{-1} \tag{1.47}$$

where
P = load, C_0 = initial compliance, a_0 = initial crack length, L = half span length, B = specimen width, b = specimen height, G_{LT} = Shear modulus of specimen and

$$a_e = \left[\frac{C_c}{C_{0c}} a_0^3 + \frac{2}{3} \left(\frac{C_c}{C_{0c}} - 1 \right) L^3 \right]^{1/3} \tag{1.48}$$

where

$$C_c = C - \frac{3L}{10BbG_{LT}}; \quad C_{0c} = C_0 - \frac{3L}{10BbG_{LT}} \tag{1.49}$$

The specimen compliance C can be given following the Timoshenko beam theory [160, 161]:

$$C = \frac{3a^3 + 2L^3}{8Bb^3E_L} + \frac{3L}{10BbG_{LT}} \tag{1.50}$$

where E_L is the longitudinal elastic Modulus.

1.5.1.4 Example

Consider one epoxy-chopped carbon fibre composite laminate with following properties and sample dimension for ENF test to determine the Mode-II fracture energy release rate G_{II}:

E_L: 50,000 N/mm^2, G_{LT}: 5000 N/mm^2, L: 50 mm, B: 25 mm, **b**: 5 mm, a_0: 25 mm, C_0: 3.269 × 10^{-4} mm/N.

The load-deflection curve in ENF test is given below as Fig. 1.24. The plot shows that the composite is almost linear in flexure deformation. The absence of viscoelastic effect is negligible due to the rigid thermoset and high Young's modulus of the carbon fibres. The sample fractured to failure at 650 N, corresponding to 6.5 mm deflection as seen in the plot.

All the data on properties, sample initial crack length and load-deflection values are arbitrary, taken as an example only. For subsequent calculations, eqs. (1.46) to (1.50) were used here.

Figure 1.25 shows the compliance at progressive crack lengths during the ENF test, calculated using eq. (1.50) and not by simply taking the ratio of deflection to load (δ/P). The compliance increases with crack propagation, indicating the loss of elastic modulus as the crack increases. The nature of the plot is a second-order polynomial, but with a very small non-linearity. The fracture energy release rate G_{II} for Mode-II fracture is shown in Fig. 1.26 at progressing crack length. The fracture energy is obvi-

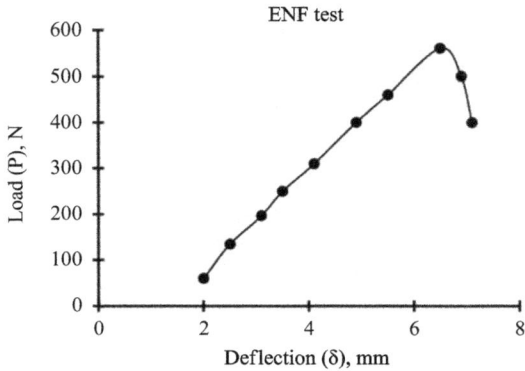

Fig. 1.24: Load-deflection curve for ENF test (all data are arbitrary, only used as example).

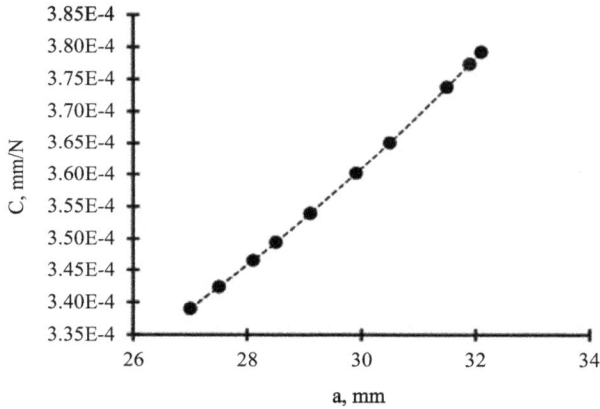

Fig. 1.25: Compliance calculated using eq. (1.50).

ously low at initial stage of deformation, but non-linearly increases with deflection, till the failure. The critical energy release rate G_{IIC} is 2205 J/m^2 in the present example.

1.5.1.5 Toughening methods

A large number of studies have been done on fracture toughness determination of composites based on thermosets and engineering plastics reinforced with various fibres of glass, carbon, Kevlar, SiC, boron nitride, etc., particulate mineral fillers, nano-fillers of various aspect ratios and those which have unique reinforcing ability. The fracture toughness arises basically from the ability to arrest the crack opening and propagating in the continuous phase (matrix). That is exactly the reason for many research studies on toughening of composites by means of:

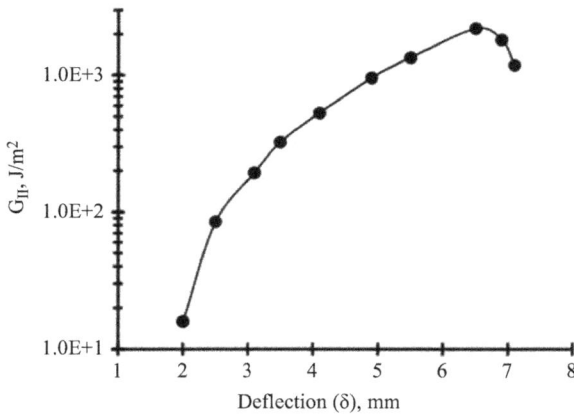

Fig. 1.26: Fracture energy release rate for the composite Mode-II ENF test.

(1) physical blends of a rubber or a viscous oligomer in the thermoset matrix, such as nitrile rubber, polyurethane, epoxidized natural rubber, pitch or tar, low molecular weight resol, etc.

(2) nanofillers such as fumed silica, nano titanium dioxide, carbon nanofibres, nanotubes, graphene oxide, reduced graphene oxide, asbestos, etc.

(3) organically modified layered silicate clays such as cloisite and montmorillonite, which can be intercalated or exfoliated by the low molecular weight uncured resin upon physical mixing.

(4) chemically bonding with hyperbranched polymers with selected functional groups.

(5) chemically bonding a soft moiety such ascarboxyl or amin terminated butadiene-nitrile rubber (CTBN/ATBN), amine-terminated polysulphones, acrylic oligomers, carboxyl-terminated polyethylene glycol adipate (CTPEGA), etc.

(6) using long chain reactants such as long-chain epoxy resin, or long-chain poly (ether) amine, or both.

Way back in 1976, Bolger [160] reviewed different studies on toughening of epoxy thermoset by addition of nitrile rubber and other elastomers, polyurethanes, glass fibres, asbestos and other fillers, and also chain-extending latent curing agents. The author also presented experimental data on a series of flexible anhydride-cured epoxy and it was possible to incorporate toughness as well as good hydrolytic and thermal stability by using flexible polyesters and avoiding use of excessive amines for curing. Bucknall and Gilbert [161] and Hourston and Lane [162], studied blending of polyetherimide in tetra-functional epoxy for toughening. Ratna et al. developed carboxyl-terminated poly(2-ethyl hexyl acrylate) (CTPEHA) liquid oligomer [163], amine-terminated polysulphones [164], and carboxyl-terminated polyethylene glycol adipate (CTPEGA) [165, 166] for chemical modification of epoxy resin as well as a flexible, damping-type toughened epoxy using a series of long chain polyetheramine (Jeffamine) as curing agent [167] and also studied the effect of organically modified nanoclay (amine modified Cloisite Na+clay)-filled flexible epoxy-based composite [168] for dynamic mechanical loss, which indicates toughness. Zavareh and Vahdat [169] have shown that only 2% bitumen addition improved the impact energy and fracture toughness by three times, while the mechanical strength, elongation and elastic modulus remained unchanged. Unnikrishnan and Thachil [170] reviewed several toughening methods for epoxy thermosets using various functional group-terminated butadiene-acrylonitrile (low molecular weight nitrile rubber), such as carboxyl (CTBN), amine (ATBN), epoxy (ETBN), phenol (PTBN), hydroxyl (HTBN) and mercaptan (MTBN). The authors dwelt extensively on phase morphology, miscibility, mechanical properties and fracture toughness properties of the unmodified and modified epoxy thermoset, in particular, CTBN modified ones, and tabulated the fracture toughness in terms of critical stress intensity factor (K_{IC}) and Mode-I critical fracture energy release rate (G_{IC}). It was shown that the fracture toughness improved from the pure epoxy thermoset $K_{IC} = 0.8$ M/Nm$^{3/2}$ and $G_{IC} = 0.3$ kJ/m^2 to about 2M/Nm$^{3/2}$ and 2.1 kJ/m^2, respectively when CTBN or ATBN was used as toughening agents to the extent of 10–15% by weight.

Ratna et al. [171] studied impact energy of clay-reinforced epoxy nanocomposite and shown that 5% addition of octadecylammonium salt of montmorillonite clay (Nanocore I 30E) could improve impact energy from 0.7 kJ/m to about 1.1 kJ/m. Seyhan et al. [172] investigated Mode-I and Mode-II interlaminar fracture toughness, and interlaminar shear strength of E-glass non-crimp fabric/multiwall carbon nanotube-modified polymer matrix composites. The MWCNT was amino-functionalized and added to the vinyl ester-polyester hybrid resin to the extent of only 1% by weight. The mixing of MWCNT in the base epoxy oligomer was done in a triple roll mill, which is usually used for making solvent-free paints and putty type pastes. The compliance C was calculated using the following equation for a three-point bending ENF test:

$$C = \frac{2L^3 + 3a^3}{8E_1bh^3} \tag{1.51}$$

C is also the ratio of the beam deflection at the load point to the corresponding applied load ($C = \delta/P$). Mode-II critical fracture energy release rate is calculated according to eq. (1.45).

The authors reported that there was no significant difference in onset fracture energy in Mode-I. The propagation Mode-I critical fracture energy was highest at 20–30 mm increase in crack length ($\Delta a = 0.02$ to 0.03 m) and lower thereafter for the base laminate without MWCNT, but the fracture toughness has comparatively lower value all through for any Δa, but almost monotonically increased. However, the Mode-II fracture toughness has shown about 11% higher value for MWCNT-modified laminates compared to the base laminate.

1.5.1.6 Mode-I and III fracture with crack orientation

Alsaadi and Erkliğ [156] studied S-glass fibre-epoxy composite laminates filled with sodium tetraborate decahydrate ($Na_2B_4O_7$, $10H_2O$), commonly known as borax and silicon carbide (SiC) particles separately. The particles were maximum 35 mm size and were added at 5, 10, 15 and 20%. The composite laminates were produced by adding borax or SiC particles in epoxy resin. It was observed that the borax improved mechanical properties of the laminates better than SiC, and 10% borax addition showed maximum improvement while 15% SiC showed best result among all SiC contents. In fact, flexural strength and modulus were improved to maximum extent for 10% borax among all the studied laminates. However, Mode-I fracture toughness showed 46% increase for 15% borax-filled laminate at the onset and 34% at the propagation, while the corresponding value was 37% and 23% for 15% addition of SiC. The interesting part of the study was observation of fibre pull-out for the base laminate, while the particles had higher interaction with the resin, showing no debonding at particle-resin interface, and decrease in properties and toughness for high loading was simply because of agglomeration.

In an attempt to use unconventional wastes for improvement of polymeric composites, Aslaadi and co-workers [173] used sewage sludge ash (SSA), from domestic and industrial sewage systems, which contains mainly phosphate and silicates of calcium, magnesium, aluminium, and traces of sodium, manganese, iron, zinc, etc., and fly ash (FA), which is a waste from industry such as thermal power plants, contains mainly silicates of iron, aluminium, calcium, potassium, magnesium and traces of sodium and sulphates. The effectiveness of these two fillers were compared with SiC filler. All the fillers were of maximum 35 μm size. These were used as fillers in an unsaturated polyester resin, which was cured with the catalyst methyl ethyl ketone peroxide (MEKP). The authors carried out extensive mechanical and fracture toughness studies covering the following tests: (1) tensile test, (2) single end notch bend test (SENB) for mixed Mode-I/III fracture toughness, (3) SENB with different angles of the notch as shown in Fig. 1.27, the crack inclination angles of $\theta = 30°$, $45°$, $75°$ and $90°$ to the beam axis. The beam dimensions are also shown in the figure.

The tensile strength of all the filled composites were almost the same as the pristine polyester thermoset up to 10% fillers (about 48–51 MPa), and reduced to about 40 MPa for 20% loading, irrespective of the filler type. However, the Young's modulus increased significantly and uniformly from 2100 MPa to about 2900 MPa for SSA, 3350 MPa for FA and 3530 for SiC at 20% addition. The Poisson's ratio also decreased from 0.36 to 0.31/0.32 for 20% filler loading. The change in these properties indicated that the effect of the fillers so chosen were to increase the hardness without any improvement in strength. Similarly, three-point bending test (flexural test) for un-notched samples indicated slight increase in fracture load from 3040N to about 3300N at 5% loading of SSA and FA filler, but not for SiC filler. The fracture load decreased steadily with further addition of the fillers, from 3040N to about 2670N for SSA, 2750 for FA fillers, and lowest load of 2420N was observed for SiC filler, all at 20% loading in the thermoset. The results indicated that flexural strength of the filled thermoset was bet-

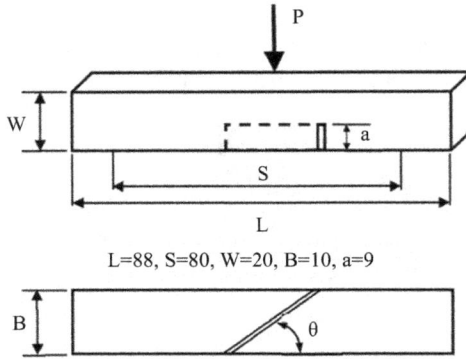

L=88, S=80, W=20, B=10, a=9

Fig. 1.27: Sample dimensions and initial notch for SENB test.

ter for SSA and FA filers compared to the conventional SiC. The reason for reduction of strength at high loading (>10%) is due to agglomeration. The authors observed the size of agglomerates to be on average 41, 71 and 101 μm for 20% FA, SSA and SiC-loaded thermoset, respectively.

The critical stress intensity factor for Mode-I was calculated by using initial notch depth method from the following equations:

$$K_A = \left(\frac{P}{BW^{1/2}}\right) f\left(\frac{a}{W}\right) \tag{1.52}$$

where K_A is termed as apparent stress intensity factor and $f(a/W)$ is a correction factor given by [174]:

$$f\left(\frac{a}{W}\right) = \frac{3sA^{1/2}}{2W} \frac{1.99 - A(1-A)(2.15 - 3.93\,A + 2.7\,A^2)}{(1+2A)(1-A)^{3/2}} \tag{1.53}$$

where $\qquad A = \dfrac{a}{W}$

The Mode-I and Mode-III critical stress intensity factors were determined from K_A as [173]:

$$K_{IC} = K_A \sin^2\theta \tag{1.54}$$

$$K_{IIIC} = K_A \sin\theta\cos\theta \tag{1.55}$$

The corresponding critical strain energy release rates were calculated as:

$$G_{IC} = \frac{(1-v^2)}{E} K_{IC}^2 \tag{1.56}$$

$$G_{IIIC} = \frac{(1+v)}{E} K_{IIIC}^2 \tag{1.57}$$

where v is the Poisson's ratio and E is the Young's modulus.

The results of the study showed that the K_{IC} of all the composites increased with angle of notch, attaining a maximum beyond 80° and K_{IIIC} values are highest at about 30–40° and decrease with notch angle and is practically zero at 90° crack orientation angle. Further, K_{IC} values are much higher than K_{IIIC} values for al composites, due to the mode of deformation imposed. A typical example of dependence of K_{IC} and K_{IIIC} on the notch (initial crack) angle for sewage slag ash (SSA)-filled polyester thermoset as studied by the authors is represented here as Fig. 1.28. It is seen that the fracture toughness, both in Mode-I and Mode-III deformations, has improved significantly on inclusion of 5% SSA filler, also for higher loading, but with less effect.

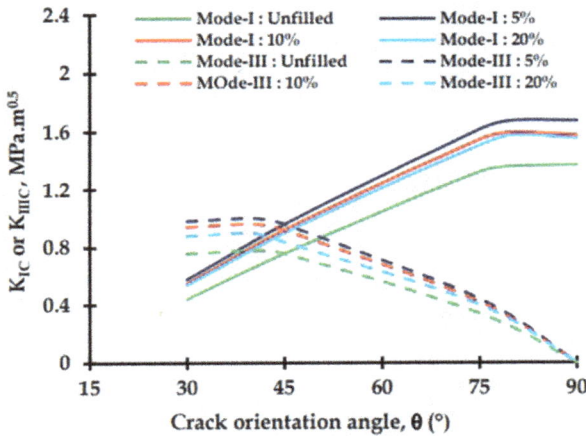

Fig. 1.28: Mode-I and Mode-III critical stress intensity factors for polyester thermoset-SSA particulate composites: dependence on crack orientation angle (refer Fig. 1.11). Reprinted with permission from Alsaadi, M., Erkliğ A., Bulut, M. Sci Eng Compos Mater 25 (2018) 679–687. © 2018, De Gruyter Publishing Company.

However, the critical fracture energy release rates G_{IC} and G_{IIIC} of unfilled polyester thermoset was higher than most filled composites, except for the 5% filler loading, and for SSA filler, the values of both G_{IC} and G_{IIIC} for unfilled thermoset were same as those for 5% loading as taken from Ref [173]. in Fig. 1.29. The reason for higher value for unfilled thermoset could be lower value of the Young's modulus compared to composites, since it was seen that the enhancement in modulus was quite significant (about 40% to 60%).

Ngo et al. [175] studied a large number of epoxy-clay (Cloisite 30B) nanocomposites for fracture toughness in terms of critical stress intensity factor (K_{IC}) and critical strain energy release rate (G_{IC}). Epoxy resin cured with Jeffamine D-2000 is a rubbery material with a glass transition temperature $T_g = -46.3$ °C, while the material prepared with Jeffamine D-230 is a glassy solid with a higher T_g of 86.8°. A glassy solid with $T_g = 150.4$ °C was obtained on curing with BF_3-MEA via polymerization of epoxy and hy-

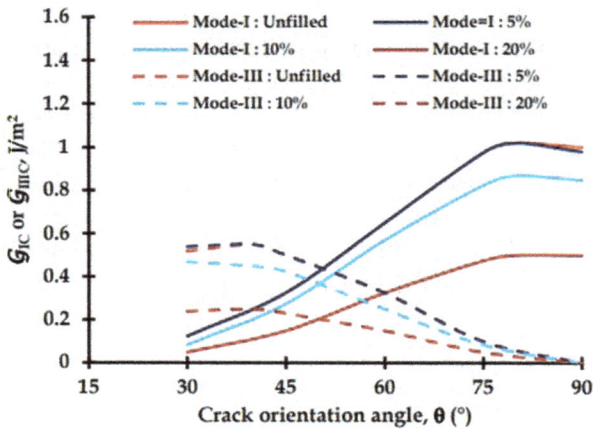

Fig. 1.29: Mode-I and Mode-III critical strain energy release rates for polyester thermoset-SSA particulate composites: dependence on crack orientation angle (refer Fig. 1.11). Reprinted with permission from Alsaadi, M., Erkliğ A., Bulut, M. Sci Eng Compos Mater 25 (2018) 679–687. © 2018, De Gruyter Publishing Company.

droxyl groups. The authors observed that both K_{IC} and G_{IC} decreased sharply with increasing size of clay agglomerates, especially if the size is above 1 micron. At 2% clay loading, the fracture toughness was maximum. The study showed that the enhancement of fracture toughness in J2000 cured epoxy was highest (K_{IC} 4.8 times, G_{IC} 9 times), for curing with BF_3 (2.1 times and 4.6 times) and least for J230-cured epoxy (47% and 40% increase), all with 6% clay loading in each, in comparison to the corresponding pristine epoxy.

Sharuddin [176] studied the mechanical properties of polyester-based nanocomposites using halloysite nanotubes (HNT) in small quantities (approximately 1%) and also halloysite-graphene hybrid nanocomposite for comparison. The dynamic viscoelastic properties and glass transition improved significantly with incorporation of 1% HNT, and water absorption studies showed plasticizing effect, reducing the T_g by about 2–3 °C. Seawater absorption was also seen to be substantially reduced on 1% HNT addition, but the T_g decreased more for higher (>0.75%) HNT in seawater immersion test, possibly due to the interaction of water with the polar organic modifier of the HNT. The degradation of dynamic viscoelastic properties on seawater immersion was significant, reducing the loss modulus (E'') value as well as the temperature of E''_{max}. A single edge notch three-point bending (SEN-TPB) test was used to investigate the Mode-I fracture toughness K_{IC} of the nanocomposite, following the equation:

$$K_{IC} = \frac{P_{max}f\left(\dfrac{a}{W}\right)}{BW^{1/2}} \tag{1.58}$$

where

$$f\left(\frac{a}{W}\right) = \frac{\left(2+\dfrac{a}{W}\right)\left\{0.0866 + 4.64\left(\dfrac{a}{W}\right) - 13.32\left(\dfrac{a}{W}\right)^2 + 14.72\left(\dfrac{a}{W}\right)^3 - 5.6\left(\dfrac{a}{W}\right)^4\right\}}{\left(1-\dfrac{a}{W}\right)^{3/2}} \qquad (1.59)$$

where a = crack length (between 0.45 W and 0.55 W), W = width of the sample, B = thickness of the sample, P_{max} = maximum load to fracture.

The fracture toughness of individual nanocomposites of HNT and multilayer graphene (MLG) on 0.1%addition had shown maximum improvement compared to the pristine polyester thermoset, while the hybrid nanofiller composite (0.5% each filler) was having less effect. Secondly, the K_{IC} value increased to maximum extent (100%) for MLG composite, but less so for HNT or hybrid composites.

1.5.2 Environmental stress cracking (ESC)

Crack development and catastrophic failure are very common for polymer matrix composites in the presence of a hostile environment, such as a reactive chemical fluid, a solvent or any penetrator moiety. A seemingly not-so-harmful environment alone or a seemingly safe stress alone might not cause a failure of the composite, but in combination, these might cause crack propagation leading to failure. The effect of the environment may be in two ways, either physical or chemical. In the case of physical effect, mass transfer of a gas or liquid molecule in a composite can occur in two steps, the first one being sorption, meaning entry into the composite via the exposed surface and subsequently diffusion, which is a random distribution of the molecules inside the composite, driven by concentration gradient. The result would be plasticization, and weakening of the secondary valence bonds between the polymeric molecules (van der Waals) such as simple dispersive forces, polar attraction, induced dipoles and hydrogen bonds, thus increased separation of the polymer chains creating larger void volume. Secondly, the penetrant can weaken the interface of the reinforcing material (fibre or particulates) and the polymer matrix. Chemical effect is the reaction with the polymer, the easiest example being hydrolysis by moisture, acid or alkali. There are many more examples depending on the environment in which the composites are used.

However, it is very difficult to accommodate the two simultaneous parameters in a mathematical model described so far simply because the molecular level phenomenon involved in the effect of environment – either rate of weakening of the secondary valence forces due to mass transfer or the deterioration rate due to chemical reaction cannot be coupled with a mechanical phenomenon of crack propagation. Only through experimental methods of simultaneous determination of the two phenomena it may be possible, such as, seawater ingress with Mode-I cracking experiment on a FRP used in underwater application at regular interval of exposure. This method of experiment

may lead to a large number of studies on composite-environment-stress combinations. The ESC study being based on experimentation, there are many reports on such studies, especially with polyethylene products. A few international standards of ESC tests are:
(1) ISO 4599 – Bent Strip Method
(2) ISO 6252–Constant Tensile Stress method
(3) ASTM D 3681 – Fibreglass (glass-reinforced thermostetting resin) Pipe in a deflected condition

1.5.2.1 Factors affecting ESC

In a composite, the polymer material is the most vulnerable for ESC. A polymer of low molecular weight has lower interchain bond strength and more end groups and hence, is more prone to failure due to interaction with the penetrant molecules, by both physical and chemical process. Higher the molecular weight, greater would be the intermolecular attraction and chain entanglement and hence more resistive to the penetrant, hence less vulnerable to ESC. However, a rigid chain segment may cause easier crack propagation under even a small interaction with the penetrant. That is exactly why semi-crystalline polymers are more investigated than other polymers.

The second factor is compatibility of the solute-solvent pair, defined by solubility parameter:

$$\delta = \sqrt{(CED)} = \sqrt{\frac{H_V - RT}{V_m}} \tag{1.60}$$

where δ = Hildebrand Solubility Parameter, CED = Cohesive Energy Density, H_v = heat of vaporization, V_m = molar volume, R = universal gas constant and T = temperature (in K).

If the solubility parameter of a polymer and the solvent is similar then the solvent will be compatible with the polymer, and hence will be more affected by diffusion and absorption. However, in order to accommodate polar force, induced dipoles and hydrogen bonding in the polymer molecule, the total solubility parameter is expressed as:

$$\delta^2 = \delta_D^2 + \delta_P^2 + \delta_H^2 \tag{1.61}$$

where D stands for dispersive, P is for polar forces and H is for hydrogen bonds. These are termed as Hansen solubility parameters and similarity of these individual parameters of a solvent and polymer decides the solubility.

In addition, the difference in the solubility parameters is a measure of extent of interaction among the polymer and the solvent. The quantification of the difference or similarity in these parameters of a solute and solvent is done by defining a distance of solubility parameters as:

$$R_a^2 = 4(\delta_{D2} - \delta_{D1})^2 + (\delta_{P2} - \delta_{P1})^2 + (\delta_{H2} - \delta_{H1})^2 \qquad (1.62)$$

Here, 4 stands for constant representing solubility data as a sphere encompassing good solvent.

The relative energy difference is defined as the ratio of the R_a to the radius of interaction sphere in Hansen space calculated for each solute-solvent system to understand the solubility of a solvent in a polymer. If the relative energy difference is zero, then the solvent can be perfectly miscible, and a value greater than 1 indicates less likely to be compatible.

The determination of solubility parameter of the polymer is sufficient, since to use the composite is not a practical proposition due to possible disintegration of the filler-polymer interface. The solubility parameter of a solid polymer is determined by swelling experiment in various solvents, graded as polarity and solubility parameter. The solubility parameter of the solvent in which the polymer swelled to maximum extent, is considered as the solubility parameter of the polymer [177].

Most common applications of composites are in atmospheric environment with humidity and temperature. Effect of water ingress and deterioration in strength depends on surface tension of water and the polymer. For most engineering thermoplastics, the surface tension is far less than that of water (72.5 dynes/cm) [178], and hence the wetting of the polymer surface by water is poor. In case of polar polymers such as polylactic acid, the wettability is much better. When a composite is to be immersed in water continuously, the effect of water on the ESC is much higher than that of atmospheric moisture, where the water is in gaseous state, although the mobility in gas phase is much higher than the liquid phase, where the water molecules are in association or cluster due to hydrogen bonding. According to the mass transfer principles, the absorption and diffusion of moisture in a polymer exposed to humid atmosphere is dependent of the relative humidity of the air.

For effect of chemicals in degradation of properties of plastics and elastomers, PDL (Plastic Design Library) table of rating of resistance to chemicals is mostly referred as a first information to begin with a study on ESC. The chemical resistance is graded according to the extent of effect it could have on the test material [178]. There are similar rating tables available from commercial manufacturers of plastic items.

Overall general conclusion [179] on the chemical resistance and ESC of polymers can be:
(i) Glassy amorphous thermoplastics/elastomers, without exception, offer poor resistance to dynamic fatigue and corrosive chemicals.
(ii) Glassy amorphous thermoplastics/elastomers are poor in resistance to a wide range of organic fluids under stress.

1.5.2.2 ESC of composites

Chen and co-workers [180] studied GRP based on epoxy resin with and without epoxidized natural rubber (ENR) as toughening agent and short glass fibre (6 mm length) to the extent of 12% for ESC. The glass fibre was used as such and also as treated with 2% solution of a silane coupling agent. The experiment was done in flexure mode (3 point bending) in air (as natural environment), dilute aqueous NaOH and HCl media. The water absorption testing was performed to study the percentage of water absorbed by the GRPs with different proportions of short glass fibre. The observation made in the study was that the surface treatment of the fibre together with the toughening of the epoxy with ENR had a profound effect on improvement in environmental stress cracking resistance (ESCR). Secondly, there was not much difference in the ESCR in different media for the treated fibre-toughened epoxy GRP, with slight lower values of ESCR for alkaline and acidic media, which showed excellent ESCR for the treated short fibre-toughened GRP.

Compared to glass fibre, aramid fibre is more ductile and is expected to improve fracture toughness and environmental stress cracking resistance in an FRP. Imielińska [181] studied epoxy glass composite (GRP) and epoxy-glass-aramid hybrid fibre composite (A-GFRP) for ESC in water under constant stress for three months and observed the crack growth and also in air medium for comparison. Both the fibres were in the form of fabric and the volume fraction was approximately 40% in all the composites. The water uptake of (A-GFRP) composite is higher than glass fibre GRP in general. However, the hybrid composite was reported to be relatively immune to the environmental effects in crack propagation in the time span of the study, since aramid was less prone to corrosion by water, which prevented fibre fracture and resulted in complex failure mechanism. The crack rate in air and water for the GP was the same under the identical tensile load, which is obvious due to the time-dependent surface absorption process. Subsequently, the crack rate in water increased significantly with time. The stress intensity factor of the GRP in water reduced by 50% compared to that in air. Transverse fracture of the glass fabric and fibre pull-out was observed for the GRP in water under stress in three months of test.

With the advancement in environment-friendly composites for structural application, many natural fibres such as jute, kenaf, cotton, flax, banana, wood, etc. are being increasingly used. These fibres mostly consist of cellulose fibre and some hemicellulose, and are more ductile, but have lower strength compared to glass, Kevlar or carbon fibres. Due to the ductility, these fibres might be better in fracture toughness and environmental stress cracking resistance (ESCR). However, the inherent characteristics and shortcoming of natural fibres, like poor adhesion, high moisture absorption, poor wettability, etc. cause lower bonding with the polymer matrix. Therefore, modification or pre-treatment of natural fibres is done prior to composite preparation. Alkali treatment is mostly used to reduce hydrogen bonding, remove hemicellulose, lignin, impurity, etc., followed by drying to make the fibre surface rough, and for better interface bond with polymer matrix [182, 183].

Bonnia et al. [184] studied ESCR of kenaf fibre-loaded composite of unsaturated polyester resin cured with styrene as crosslinker in the presence of organo-cobalt compound as catalyst. The authors used 3% liquid natural rubber as toughening agent for the polyester thermoset. The addition of the toughening agent improved flexural and impact strength by 66–70%. The ESC study was done in 1% acidic, 1% alkaline and air medium for maximum 300s with three-point bending mode. Critical stress intensity factor K_{IC} was improved to some extent for the rubber toughened composites. However, 20% fibre content was optimum for highest fracture toughness. Similarly, ESCR was slightly better for rubber-toughened composite. The composite was more affected by acid than by alkaline medium or air. In fact, there was not much difference in alkaline and air medium in respect of resistance to crack.

References

[1]   Gujjala, R., Bandyopadhyay, S., eds. Toughened Composites: Micro and Macro Systems. CRC press, London, 12 Dec 2022.
[2]   Wang, R.-M., Zheng, S.-R., Zheng, Y.-P. G. Polymer Matrix Composites and Technology. Elsevier, London, 14-Jul-2011.
[3]   Middleton, D. H., ed. Composite Materials in Aircraft Structure. Longman, New York, 1990.
[4]   Bakker, A. A., Jones, R., Callinan, R. J. Compos. Struct. 1985, 15, 154.
[5]   Jang, B. Z. Sci. Eng. Compos. Mater. 1991, 2, 29.
[6]   Lipatov, Y. S., Sergeeva, L. M. Russ. Chem. Rev. 1976, 45, 63–74.
[7]   Roland, C. M., Kobayashi, S., Mullen, K., eds. Encyclopedia of Polymeric Nanomaterials. Springer Berlin Heidelberg, Berlin and Heidelberg, Germany, 2015, pp. 1004–1011.
[8]   Hourston, D. J., Mccluskey, J. A. J. Appl. Polym. Sci. 1986, 31, 645–655.
[9]   Narine, S. S., Kong, X., Bouzidi, L., Sporns, P. I. Elastomers. J. Am. Oil Chem. Soc. 2007, 84, 55–63.
[10]   Madbouly, S. A. Polym. Test. 2020, 89, 106586.
[11]   Silverstein, M. S. Polymer. 2020, 207, 122929–122943.
[12]   Sperling, L. H. An introduction to Polymer Networks and IPNs. Plenum Press, New York, 1981, Chapter 1, pp. 1–10.
[13]   Klempner, D., Frisch, K. C., eds. Introduction to Physical Polymer Science. Technomic Publishing Co., Lancaster, PA, Vol. I–III, 1989, pp. 1–15.
[14]   Lohani, A., Singh, G., Bhattacharya, S. S., Verma, A. J. Drug Deliv. 2014, 2014, 583612. Chao Meng, Sheng Huang, Dongmei Han, Shan Ren, Shuanjin Wang, and Min Xiao Advances in Polymer Technology 23 (2020) 1–8.
[15]   Ratna, D., Karger Kocsis, J. Polymer. 2011, 52, 1063.
[16]   Jagtap, S., Dalvi, V., Shankar, K., Ratna, D. Polym. Int. 2019, 68, 8812–8817.
[17]   Han, J., Fei, G., Li, G., Xia, H. Macromol. Chem. Phys. 2013, 214, 1195.
[18]   Ratna, D., Karger-Kocsis, J. J. Mater. Sci. 2008, 43, 254–269.
[19]   Metzger, M. F., Wilson, T. S., Schumann, D., Mathew, D. L., Maitland, D. J. Biomed. Microdevices. 2002, 4, 89.
[20]   Wache, H. M., Tartakowska, D. J., Hentrich, A., Wagner, M. H. J. Mater. Sci. Mater. Med. 2003, 14, 109.
[21]   Gall, K., Mikulas, M., Munsi, N., Tupper, M. J. Intell. Mater. Syst. Struct. 2000, 11, 877.
[22]   Enomoto, M., Suehiro, K. Text Res. J. 1997, 67, 601.

[23] Roland, C. M. Interpenetrating Polymer Networks (IPN): Structure and Mechanical Behavior. In Kobayashi, S., Müllen, K., eds. Encyclopedia of Polymeric Nanomaterials. Berlin, Heidelberg, Springer, 2013.

[24] Gong, J. P., Katsuyama, Y., Kurokawa, T., Osada, Y. Adv. Mater. 2003, 15, 1155–1158.

[25] Samui, A. B., Suryavansi, U. G., Patri, M., Chakraborty, B. C., Deb, P. C. J. Appl. Polym. Sci. 1998, 68, 275.

[26] Mathew, A., Chakraborty, B. C., Deb, P. C. J. Appl. Polym. Sci. 1997, 65, 549.

[27] Mathew, A., Deb, P. C. J. Appl. Polym. Sci. 1994, 53, 1104.

[28] Samui, A. B., dalvi, V. G., Chandrasekhar, L., Patri, M., Chakraborty, B. C. J. Appl. Polym. Sci. 2006, 99, 2542.

[29] Manoj, N. R., Rout, R. D., Shivraman, P., Ratna, D., Chakraborty, B. C. J. Appl. Polym. Sci. 2005, 96, 4487.

[30] Gryshchuk, O., Karger-kocsis, J. J. Nanosci. Nanotechnol. 2006, 6, 345.

[31] Giaveri, S., Gronchi, P., Barzoni, A. Coatings. 2017, 7, 213.

[32] Jin, H., Zhang, Y., Wang, C., Sun, Y., Yuan, Z., Pan, Y., Xie, H., Cheng, R. J. Therm. Anal. Calorim. 2014, 117, 773.

[33] Zhang, Y., Hourston, D. J. J. Appl. Polym. Sci. 1998, 69, 271.

[34] Yu, P., Li, G., Zhang, L., Zhao, F., Chen, S., Dmitriev, A. I., Zhang, G. Tribol. Int. 2019, 131, 454.

[35] Gao, S., Zhang, L., Huang, Q. Polymer networks. J. Appl. Polym. Sci. 2003, 90, 1948.

[36] Lu, K. T., Liu, C. T., Lee, H. L. J. Appl. Polym. Sci. 2003, 89, 2157.

[37] Huang, J., Zhang, L. J. Appl. Polym. Sci. 2002, 86, 1799.

[38] Zhang, L., Huang, J. J. Appl. Polym. Sci. 2001, 81, 3251.

[39] Cao, X., Zhang, L. Biomacromolecules. 2005, 6, 671.

[40] Yeganeh, H., Hojati-Talemi, P. Polym. Degrad. Stab. 2007, 92, 480.

[41] Meier, C., Parlevliet, P., Dring, M. High Perform. Polym. 2017, 29, 556.

[42] Ballestero, R., Sundaram, B. M., Tippur, H. V., Express, M. L. Polym. Lett. 2016, 10, 204.

[43] Wen, Y., Yan, J., liu, J., Wang, Z. Express Polym. Lett. 2017, 11, 963.

[44] Patent No. CN110964282 A (China) EP/PMMA-IPN ring crush enhancer and preparation method thereof:// w.w.w. webofscience.com/wos/alldb/full-record/DIIDW:2020314970.

[45] Ratna, D., Dalvi, V., Billa, S., Sharma, S. K., Rath, S. K., Sudarshan, K., Pujari, P. K. ACS Appl. Polym. Mater. 2021, 3, 5073–5086.

[46] James, J., Thomas, G. V., Akhil, P. M., Nambissan, G., Nandakumar, K., Thomes, S. Phys. Chem. Chem. Phys. 2020, 22, 18169–18182.

[47] Eisenberg, A., Kim, J., . S. Introduction to Ionomers. John Wiley & Sons, New York, 1988.

[48] Grady, B. P. Polym. Eng. Sci. 2008, 48, 1029–1051.

[49] Basu, D., Das, A., Stockelhuber, K. W., Jehnichen, D., Formanek, P., Sarlin, E., Vuorinen, J., Heinrich, G. Macromolecules. 2014, 47, 3436–3450.

[50] Vislavath, P., Billa, S., Bahadur, J., Sudarshan, K., Patro, T. U., Rath, S. K., Ratna, D. Macromolecules. 2022, 55, 6739–6749.

[51] Krishnakumar, B., Sanka, R. V. S. P., Binder, W. H., Parthasarthy, V., Rana, S., Karak, N. Chem. Eng. J. 2020, 385, 123820.

[52] Yue, L., Guo, H., Kennedy, A., Patel, A., Gong, X., Ju, T., Gray, T., Manas-Zloczower, I. ACS Macro. Lett. 2020, 9, 836–842.

[53] Scheutz, G. M., Lessard, J. J., Sims, M. B., Sumerlin, B. S. J. Am. Chem. Soc. 2019, 41, 16181–16196.

[54] Chen, X., Dam, M. A., Ono, K., Mal, A., Shen, H., Nutt, S. R., Sheran, K., Wudl, F. A. Science. 2002, 295, 1698–1702.

[55] Oehlenschlaeger, K. K., Mueller, J. O., Brandt, J., Hilf, S., Lederer, A., Wilhelm, M., Graf, R., Coote, M. L., Schmidt, F. G., Barner-Kowollik, C. Adv. Mater. 2014, 26, 3561–3566.

[56] Reutenauer, P., Buhler, E., Boul, P. J., Candau, S. J., Lehn, J. M. Chem. Eur. J. 2009, 15, 1893–1900.

[57] Bergman, S. D., Wudl, F. Mendable polymers. J. Mater. Chem. 2008, 18, 41–62.

[58] Adzima, B. J., Aguirre, H. A., Kloxin, C. J., Scott, T. F., Bowman, C. N. Macromolecules. 2008, 41, 9112–9117.

[59] Billiet, S., De Bruycker, K., Driessen, F., Goossens, H., Van Speybroeck, V., Winne, J. M., Du Prez, F. E. Nat. Chem. 2014, 6, 815.

[60] Van Herck, N., Du Prez, F. E. Macromolecules. 2018, 51, 3405–3414.

[61] Ying, H., Zhang, Y., Cheng, J. Nat. Commun. 2014, 5, 3218.

[62] Zhang, Y., Ying, H., Hart, K. R., Wu, Y., Hsu, A. J., Coppola, A. M., Kim, T. A., Yang, K., Sottos, N. R., White, S. R., Cheng, J. Adv. Mater. 2016, 28, 7646–7651.

[63] Zhang, L., Rowan, S. J. Macromolecules. 2017, 50, 5051–5060.

[64] Liu, W.-X., Zhang, C., Zhang, H., Zhao, N., Yu, Z.-X., Xu, J. Chem. Soc. 2017, 139, 8678–8684.

[65] Obadia, M. M., Mudraboyina, B. P., Serghei, A., Montarnal, D., Drockenmuller, E. Chem. Soc. 2015, 137, 6078–6083.

[66] Obadia, M. M., Jourdain, A., Cassagnau, P., Montarnal, D., Drockenmuller, E. Adv. Funct. Mater. 2017, 27.

[67] Chakma, P., Digby, Z. A., Shulman, M. P., Kuhn, L. R., Morley, C. N., Sparks, J. L. ACS Macro. Lett. 2019, 95–100.

[68] Cash, J. J., Kubo, T., Bapat, A. P., Sumerlin, B. S. Macromolecules. 2015, 48, 2098–2106.

[69] Chao, A., Zhang, D. Macromolecules. 2019, 52, 495–503.

[70] Zhang, B., Digby, Z. A., Flum, J. A., Chakma, P., Saul, J. M., Sparks, J. L., Konkolewicz, D. Macromolecules. 2016, 49, 6871–6878.

[71] Baruah, R., Kumar, A., Ujjwal, R. R., Kedia, S., Ranjan, A., Ojha, U. Recyclable Macromol. 2016, 49, 7814–7824.

[72] Griebel, J. J., Nguyen, N. A., Astashkin, A. V., Glass, R. S., Mackay, M. E., Char, K., Pyun, J. ACS Macro. Lett. 2014, 3, 1258–1261.

[73] Kleine, T. S., Nguyen, N. A., Anderson, L. E., Namnabat, S., LaVilla, E. A., Showghi, S. A., Dirlam, P. T., Arrington, C. B., Manchester, M. S., Schwiegerling, J., Glass, R. S., Char, K., Norwood, R. A., Mackay, M. E., Pyun, J. ACS Macro. Lett. 2016, 5, 1152–1156.

[74] Liu, H., Nelson, A. Z., Ren, Y., Yang, K., Ewoldt, R. H., Moore, J. S. ACS Macro. Lett. 2018, 7, 933–937.

[75] Amamoto, Y., Kikuchi, M., Masunaga, H., Sasaki, S., Otsuka, H., Takahara, A. Macromolecules. 2009, 42, 8733–8738.

[76] Jin, K., Li, L., Torkelson, J. M. Adv. Mater. 2016, 28, 6746–6750.

[77] Takahashi, A., Goseki, R., Ito, K., Otsuka, H. ACS Macro. Lett. 2017, 1280–1284.

[78] Takahashi, A., Goseki, R., Otsuka, H. Chem. Int. Ed. 2017, 56, 2016–2021.

[79] Ping, Z. Z., Zhi, R. M., Qiu, Z. M. Funct. Mater. 2018, 28, 1706050.

[80] Zhang, C., Liu, Z., Shi, Z., Yin, J., Tian, M. ACS Macro. Lett. 2018, 7, 1371–1375.

[81] Billa, S., Vislavath, P., Bahadur, J., Rath, S. K., Ratna, D., Manoj, N. R., Chakraborty, B. C. ACS Appl. Polym. Mater. 2023, 5, 3079–3095.

[82] Montarnal, D., Mathieu, C., François, T., Leibler, L. Science. 2011, 334, 965–968.

[83] Capelot, M., Montarnal, D., Tournilhac, F., Leibler, L. J.Am. Chem. Soc. 2012, 134, 7664–7667.

[84] Yang, Y., Pei, Z., Zhang, X., Tao, L., Wei, Y., Ji, Y. Chem. Sci. 2014, 5, 3486–3492.

[85] Brutman, J. P., Delgado, P. A., Hillmyer, M. A. ACS Macro. Lett. 2014, 3, 607–610.

[86] Altuna, F. I., Pettarin, V., Williams, R. Green Chem. 2013, 15, 3360–3366.

[87] Denissen, W., Rivero, G., Nicolÿa, R., Leibler, L., Winne, J. M., Du Prez, F. E. Adv. Funct. Mater. 2015, 25, 2451–2457.

[88] Obadia, M. M., Mudraboyina, B. P., Serghei, A., Montarnal, D., Drockenmuller, E. J. Am. Chem. Soc. 2015, 137, 6078–6083.

[89] Lu, Y.-X., Tournilhac, F., Leibler, L., Guan, Z. J. Am. Chem.Soc. 2012, 134, 8424–8427.

[90] Pepels, M., Filot, I., Klumperman, B., Goossens, H. Polym. Chem. 2013, 4, 4955–4965.

[91] Taynton, P., Yu, K., Shoemaker, R. K., Jin, Y., Qi, H. J., Zhang, W. Adv. Mater. 2014, 26, 3938–3942.

[92]   Krishnakumar, B., Sanka, R. V. S. P., Binder, W. H., Parthasarthy, V., Rana, S., Karak, N. Chem. Eng. J. 2020, 385, 123820.
[93]   Yue, L., Guo, H., Kennedy, A., Patel, A., Gong, X., Ju, T., Gray, T., Manas-Zloczower, I. ACS Macro. Lett. 2020, 9, 836–842.
[94]   Rottger, M., Domenech, T., Weegen, R., Breuillac, A., Nicolay, R., Leibler, L. Science. 2017, 356, 62–65.
[95]   Wu, J., Yu, X., Zhang, H., Guo, J., Hu, J., Li, M. ACS Sustain. Chem. Eng. 2020, 8, 6479–6487.
[96]   Liu, T., Hao, C., Wang, L., Li, Y., Liu, W., Xin, J., Zhang, J. Macromolecules. 2017, 50, 8588–8597.
[97]   Hayashi, M., Katayama, A. ACS Appl. Polym. Mater. 2020, 2(6), 2452–2457.
[98]   Wang, S., Ma, S., Li, Q., Xu, X., Wang, B., Huang, K., liu, Y., Zhu, J. Maromolecules. 2020, 53, 2919–2931.
[99]   Chen, M., Zhou, L., Wu, Y., Zhao, X., Zhang, Y. ACS Macro. Lett. 2019, 8, 255–260.
[100]   Drzal, L. T., Rich, M., Lloyd, P. J. Adhes. 1983, 16, 1.
[101]   Hook, K. J., Agrawal, R. K., Drzal, L. T. J. Adhes. 1990, 32, 157.
[102]   Zhang, H., Zhang, Z., Claudia, B. Compos. Sci. Technol. 2004, 64, 2021.
[103]   Yang, Y., He, F., Wang, M., Zhang, B. J. Mater. Sci. 1998, 33, 3651.
[104]   Ratna, D., Kushwaha, R., Dalvi, V., Manoj, N. R., Samui, A. B., Chakraborty, B. C. J. Adhes. Sci. Technol. 2008, 22(1), 93.
[105]   Jianbin, L., Zhang, Z., Jiyang, F., Liang, Z., Ramakrishnan, K. R. Nanotechnol. Rev. 2021, 10, 1438–1468.
[106]   Bilisik, K., Syduzzaman, M. Polym. Compos. 2021, 42, 1670–1697.
[107]   Bilisik, K. Text Res. J. 2017, 87, 2275.
[108]   Alhijazi, M., Zeeshan, Q., Qin, Z., Safaei, B., Asmael, M. Nanotechnol. Rev. 2020, 9, 853–875.
[109]   Kozlowski, R., Wladyka-Przybylak, M. Uses of Natural Fiber Reinforced Plastics. Natural Fibers, Plastics and Composites. Springer, Boston, MA, 2004, pp. 249–274.
[110]   Faruk, O., Bledzki, A. K., Fink, H., Sain, M. Prog. Polym. Sci. 2012, 37, 1552–1596.
[111]   Kumar Sinha, A., Narang, H. K., Bhattacharya, S. J. Polym. Eng. 2017, 37, 879–895.
[112]   Pickering, K. L., Efendy, M. G. A., Le, T. M. Compos. A Appl. Sci. Manuf. 2016, 83, 98–112.
[113]   Cai, M., Takagi, H., Nakagaito, A. N., Li, Y., Waterhouse, G. I. N. Compos. Part A Appl. Sci. Manuf. 2016, 90, 589–597.
[114]   Bledzki, A. K., Gassan, J. Progr. Polym. Sci. 1999, 24, 221–274.
[115]   Fiore, V., Di Bella, G., Valenza, A. Compos. Part B Eng. 2015, 68, 14–21.
[116]   Zakikhani, P., Zahari, R., Sultan, M. T. H., Majid, D. L. Mater. Des. 2014, 63, 820–828.
[117]   Vijaya, R. B., Manickavasagam, V. M., Elanchezhian, C., Vinodh, K. C., Karthik, S., Saravanan, K. Mater. Des. 2014, 60, 643–652.
[118]   Ramesh, M., Palanikumar, K., Reddy, K. H. Compos. Part B Eng. 2013, 48, 1–9.
[119]   Asim, M., Jawaid, M., Abdan, K., Ishak, M. R. J. Bionic Eng. 2016, 13, 426–435.
[120]   Brahmakumar, M., Pavithran, C., Pillai, R. M. Compos. Sci. Technol. 2005, 65, 563–569.
[121]   Justiz-Smith, N. G., Virgo, G. J., Buchanan, V. E. Mater. Charact. 2008, 59, 1273–1278.
[122]   Abdul Khalil, H. P. S., Bhat, I. U. H., Jawaid, M., Zaidon, A., Hermawan, D., Hadi, Y. S. Mater. Des. 2012, 42, 353–368.
[123]   Bachtiar, D., Sapuan, S. M., Hamdan, M. M. Mater. Des. 2008, 29, 1285–1290.
[124]   Liu, Q., Hughes, M. Compos. Part A Appl. Sci. Manuf. 2008, 39, 1644–1652.
[125]   Nirmal, U., Hasim, J., Megat, M. M. H. Tribol. Intl. 2015, 83, 77–104.
[126]   Lee, S. C., Mariatti, M. Mater. Lett. 2008, 62, 2253–2256.
[127]   Cao, Y., Shibata, S., Fukumoto, I. Compos. Part A Appl. Sci. Manuf. 2006, 37, 423–429.
[128]   Goulart, S. A. S., Oliveira, T. A., Teixeira, A., Mileo, P. C., Mulinari, D. R. Procedia Eng. 2011, 10, 2034–2039.
[129]   Das Lala, S., Deoghare, A. B., Chatterjee, S. Sci. Eng. Compos. Mater. 2018, 25, 1039–1058.
[130]   Kashtalyan, M., Soutis, C. Prog. Aerosp. Sci. 2005, 41, 152–173.

[131] Prabhakar, P., Waas, A. M. Compos. Struct. 2013, 98, 85–92.

[132] Pineda, E. J., Bednarcyk, B. A., Waas, A. M., Arnold, S. M. Int. J. Solids Struct. 2013, 50, 1203–1216.

[133] Patel, D. K., Waas, A. M. Compos. Part C. 2020, 2, 100016.

[134] Wang, J., Chen, L., Shen, W., Zhu, L. Polymers. 2022, 14, 2318.

[135] Schiffer, A., Tagarielli, V. L. Int. J. Impact Eng. 2014, 70, 1–13.

[136] Kahla, H. B. Micro-cracking and delamination of composite laminates under tensile quasistatic and cyclic loading. Ph.D. Thesis, Division of Materials Science, Department of Engineering Sciences and Mathematics, Luleå University of Technology, SE-971 87. Luleå, Sweden, 2019 Ch.1, p.8.

[137] Thomason, J. A review of the analysis and characterisation of polymeric glass fibre sizings. Polym. Test. 2020, 85, Article 106421.

[138] Li, X., Hallett, S. R., Wisnom, M. R. Sci. Eng. Compos. Mater. 2015, 22, 115–129.

[139] Ziegel, K. D., Romanov, A. I. Inorganic fillers. J. Appl. Polym. Sci. 1973, 17, 1119.

[140] Kerner, E. H. Proc. Phys. Soc B. 1956, 69, 808–813.

[141] Ziegel, K. D. J. Colloid Interface Sci. 1969, 29, 72–80.

[142] Rooj, S., Das, A., Stöckelhuber, K. W., Reuter, U., Heinrich, G. Macromol. Mater. Eng. 2012, 297, 369–383.

[143] Tsai, S. W., Wu, E. M. J. Compos. Mater. 1971, 5, 58–80.

[144] Chang, F. K., Chang, K. Y. J. Compos. Mater. 1987, 21, 834–855.

[145] Ha, S. K., Jin, K. K., Huang, Y. J. Compos. Mater. 2008, 42, 1873–1895.

[146] Murali, D. M., Chakraborty, B. C., Begum, S. S. Iran Polym. J. 2022, 31, 1129–1145.

[147] Nguyen, B. N., Bapanapalli, S. K., Holbery, J. D., Smith, M. T., Kunc, V., Frame, B. J., Phelps, J. H., Tucker, C. L. J. Compos. Mater. 2008, 42, 1003–1029.

[148] Haneefa, A., Bindu, P., Aravind, I., Thomas, S. J. Compos. Mater. 2008, 42, 1471–1489.

[149] Huang, Z. M., Guo, W. J., Huang, H. B., Zhang, C. C. Materials. 2021, 14, 2708.

[150] Barnett, P. R., Young, S. A., Patel, N. J., Penumadu, D. Compos. Sci. Technol. 2021, 211, 108857.

[151] Inglis, C. E. Trans. Inst. Nav. Archit. London. 1913, 55, 219–230.

[152] Irwin, G. R. J. Appl. Mech. 1957, 24, 361–364.

[153] Griffith, A. A. Philos. Trans. R. Soc. London Ser A. 1920, 221, 163–198.

[154] ASTM Standard D 5528-94a. Test Method for Mode-I Interlaminar Fracture Toughness of Unidirectional Fiber-Reinforced Polymer Matrix Composites. American Society for Testing and Materials, West Conshohocken, PA, 2001.

[155] Srivastava, V. K., Hogg, P. J. J. Mater. Sci. 1998, 33, 1119–1128.

[156] Alsaadi, M., Erkliğ, A. Arab. J. Sci. Eng. 2017, 42, 4759–4769.

[157] Seyhan, A., Tanoglu, M., Schulte, K. Eng. Fract. Mech. 2008, 75, 5151–5162.

[158] de moura, M. F. S. F., Dourado, N., Morais, J. J. L., Pereira, F. A. M. Fatigue Fract. Eng. Mater. Struct. 2010, 34, 149–158.

[159] de Moura, M. F. S. F., Silva, M. A. L., de Morais, A. B., Morais, J. J. L. Eng. Fract. Mech. 2006, 73, 978–993.

[160] Bolger, J. C. SAE Trans. 1976, 85, 1065–1072.

[161] Bucknall, C. B., Gilbert, A. H. Polymer. 1989, 30, 213–217.

[162] Hourston, D. J., Lane, J. M. Polymer. 1992, 33, 1379–1383.

[163] Ratna, D., Banthia, A. K., Deb, P. C. J. Appl. Polym. Sci. 2001, 80, 1792–1801.

[164] Ratna, D., Patri, M., Chakraborty, B. C., Deb, P. C. J. Appl. Polym. Sci. 1997, 65, 901–907.

[165] Ratna, D., Samui, A. B., Chakraborty, B. C. Flexibility improvement of epoxy resin by chemical modification. Polym. Int. 2004, 53, 1882–1887.

[166] Murali, M., Ratna, D., Samui, A. B., Chakraborty, B. C. J. Appl. Polym. Sci. 2007, 103, 1723–1730.

[167] Ratna, D., Manoj, N. R., Chandrasekhar, L., Chakraborty, B. C. Polym. Adv. Technol. 2004, 15, 583–586.

[168] Jagtap, S. B., Rao, V. S., Ratna, D. J. Reinf. Plast. Compos. 2013, 32, 183–196.

[169] Zavareh, S., Vahdat, G. J. Reinf. Plast. Compos. 2012, 31, 247–258.

[170] Unnikrishnan, K. P., Thachil, E. T. Des. Monomers Polym. 2006, 9, 129–152.

[171] Ratna, D., Manoj, N. R., Varley, R., Raman, R. K. S., Simon, G. P. Polym. Int. 2003, 52, 1403–1407.

[172] Seyhan, A. T., Tanoglu, M. Eng. Fract. Mech. 2008, 75, 5151–5162.

[173] Alsaadi, M., Erkliğ, A., Bulut, M. Sci. Eng. Compos. Mater. 2018, 25, 679–687.

[174] Moghadam, B., Taheri, F. J. Strain Anal. Engng. Design. 2013, 48, 245–257.

[175] Ngo, T. D., Ton-That, M. T., Hoa, S. V., Cole, K. C. Sci. Eng. Compos. Mater. 2010, 17, 19–30.

[176] Saharudin, M. Mechanical properties of polyester nano-composites exposed to liquid media. Ph. D. thesis, Faculty of Engineering and Environment, Northumbria University, 2017, Accessed April 1, 2023 at http://nrl.northumbria.ac.uk/policies.html.

[177] Mathew, A., Chakraborty, B. C., Deb, P. C. J. Appl. Polym. Sci. 1994, 53, 1107.

[178] Ebnesajjad, S. Effect of Chemicals on Plastics. In Baur, E., Ruhrberg, K., Woishnis, W., eds. Chemical Resistance of Commodity Thermoplastics -A volume in Plastics Design Library. Elsevier, 2016, Ch.2, xI.

[179] Chandrasekaran, C. Anticorrosive Rubber Lining – A volume in Plastics Design Library. Elsevier, 2017, Ch.23, pp. 191–197.

[180] Chen, R. S., Muhammad, Y. H., Ahmad, S. Polym. Test. 2021, 96, Article No. 107088.

[181] Imielińska, K. Kompozyty (Composites). 2006, 6, 4.

[182] Asim, M., Jawaid, M., Abdan, K., Ishak, M. R. J. Bionic Eng. 2016, 13, 426–435.

[183] Fiore, V., Di Bella, G., Valenza, A. Compos. B Eng. 2015, 68, 14–21.

[184] Bonnia, N. N., Ahmad, S. H., Zainol, I., Mamun, A. A., Beg, M. D. H., Bledzki, A. K. EXPRESS Polym. Lett. 2010, 4, 55–61.

Chapter 2
Thermoplastic matrix composites (TMC)

Abbreviations

Abbreviation	Description
PE	Polyethylene
PP	Polypropylene
TPU	Thermoplastic polyurethane
PMMA	Polymethyl methacrylate
PES	Polyether sulphone
PEEK	Polyether ether ketone
PTFE	Polytetrafluoroethylene
PET	Polyethylene terephthalate
PC	Polycarbonate
PVC	Polyvinyl chloride
PPO	Poly(phenylene oxide)
CPVC	Critical pigment volume concentration
CB	Carbon black
HDPE	High density polyethylene
CSA	Composite sphere assemblage
PTC	Positive temperature coefficient
SG	Synthetic graphite
PVDF-CIP	Polyvinylidene fluoride-carbonyl iron powder
TPU	Thermoplastic polyurethane
TPE	Thermoplastic elastomer
GNP	Graphene nanoplates
OMMT	Organically modified montmorillonite
MA	Maleic anhydride
WLF	William-Landel-Ferry
ISB	Injection stretch blow
LDH	Layered double hydroxide
PHA	Polyhydroxyalkanoates
PLA	Polylactic acid

2.1 Thermoplastic polymer

The earliest linear polymer identified was the naturally occurring cellulose in 1838 by the French chemist Anselme Payen, who isolated it from plant matter and determined its chemical formula. Subsequently, the first thermoplastic "Celluloid", which is nitro-cellulose, was made by Hyatt manufacturing Co. in 1870. Other modifications such as methyl cellulose, carboxyl methyl cellulose, hydroxyl propyl cellulose, etc. were subsequently developed and commercialised.

https://doi.org/10.1515/9783110781571-002

However, following the famous invention of Ziegler-Natta catalyst in the mid-twentieth century of the synthesis process of alkene polymerisation, a new, vast area of thermoplastics were opened with innumerable varieties of homopolymers, copolymers and derivatives. Subsequently, in the 1930s, Wallace Carothers and other researchers of Du Pont invented a step-growth polymerization process to synthesize polyester, which is a thermoplastic. Today, most common thermoplastics such as Nylon, polyethylene (PE), polypropylene (PP), polyvinyl chloride (PVC), polycarbonate (PC), polyethylene terephthalate (PET), polytetrafluoroethylene (PTFE), polyether ether ketone (PEEK), polyether sulphone (PES), Polymethyl methacrylate (PMMA), thermoplastic polyurethane (TPU), etc. are considered essential materials for consumer goods and many other industrial applications. A few commonly used thermoplastics with their physical properties are listed in Tab. 2.1.

The usefulness of a thermoplastic is based on its chemical structure, molecular weight, polydispersity and stereo specific configuration, since these characteristics decide all other properties such as ease of processing, crystallinity, mechanical strength, toughness, electrical and thermal properties. A large number of thermoplastics are referred to as "Engineering Thermoplastics" because of their high mechanical strength, toughness and thermal stability. Most engineering thermoplastics have high Young's modulus, about 2–5 GPa at ambient temperature, and high thermal stability. One advantage of using thermoplastics for large-scale commodity is that these are recyclable as these can be moulded repeatedly and hence do not require incineration or other waste disposal methods, which increase the atmospheric pollution. However, excessive use of PE, in particular, is causing waste disposal problems due to the techno-economic constraints of quality of reprocessed items and the corresponding cost. A typical example is the reprocessing cost of very thin PE bags and wrappers. Most thermoplastics are not bio-compatible or bio-degradable, for example PE, PP, PS, PVC, PEEK, PES, etc. The present awareness in environment has seen more application of bio-compatible and bio-degradable polymers such as starch, polylactic acid and their derivatives in packaging, the food processing industry and in medical fields.

Being linear polymers, one disadvantage of thermoplastics for structural application is the higher creep and stress relaxation compared to thermosets like epoxy, phenolics, etc. Secondly, the rheology of thermoplastic melts is very complex and can be varying for the same material, according to the molecular weight and polydispersity. Due to the complexity of the chain configuration, some more complications are encountered such as the effects of normal force, time-dependent behaviour, etc., which complicate the process technology. The polarity, hydrogen bonding and crystallinity also aggravate the rheological complexity. The other disadvantage is that the fabrication of items with thermoplastics is energy-intensive due to the high temperature requirement for the melting and moulding process. Added to this, there is criticality in pressure distribution for accurate mould flow. This necessitates the use of relatively expensive and efficient machineries such as extruder, injection moulding, thermoforming, blow moulding systems, etc.

Nevertheless, consumption of thermoplastics has increased from about 270 million metric tons in 2010 to about 400 million metric tons in 2020 and growing at the rate of almost 6–8% every 5 years, with PE and PP being the major contributors.

Tab. 2.1: Common thermoplastics.

Thermoplastc	Glass transition temperature (T_g), °C	Density, g/cc	Molecular weight range (g/mol)
Poly(ethylene)	−25 (max.)	0.85–0.92	30,000–600,000
Poly(propylene)	−34 to −10	0.85–0.95	67,000–250,000
Poly(vinyl chloride)	85	1.385	99,000–230,000
Poly(methyl methacrylate)	114	1.17	250,000–600,000
Poly(styrene)	100	1.05	100,000–400,000
Poly(lactic acid)	60	1.26	120,000–150,000
Polyamide (Nylon 6 and Nylon 66)	50–80	1.15	10,000–20,000
Poly(ethylene terephthalate) (PET)	80	1.38	8,000–31,000
Polycarbonate (PC)	147	1.2	50,000–300,000
Poly(ether sulphone) (PES)	225	1.33	50,000–60,000
Poly(ether ether ketone) (PEEK)	145	1.24	30,000–100,000
Poly(tetrafluoroethylene) (PTFE)	−110 (DMA), −50 (DSC & Calc.), +130 (DMA), +110 (Dilatometry),	2.20	100,000–5,000,000
Poly(phenylene oxide) (PPO)	215	1.04	15,000=164,000

2.2 Particulate thermoplastic composite

Transparent thermoplastics are very common in the commodity market. However, in many applications, thermoplastics are filled with various types of additives such as colouring agents, pigments, particulate fillers, etc. for reinforcement, aesthetics and cost reduction. For example, reinforcement of polypropylene by carbon black improves both strength and UV stability. The effect of a particulate filler reinforcement for a thermoplastic depends on the elastic modulus of the plastic and the filler, the polarity of the plastic and the filler, particle size and aspect ratio of the filler and its agglomeration due to the filler-filler interaction, and obviously the extent of filler loading in the composite. The critical pigment volume concentration (CPVC) can be determined for a polymer-filler system by measuring its mechanical properties at various filler loading levels.

Most engineering thermoplastics have reasonably high Young's modulus and hence, the effective transfer of force from the filler to the matrix is expected to be much better than of rubber. Common reinforcing particulate filler is carbon black (CB) in a variety of forms as described in Chapter-4. Because of the small size and the presence of some organic groups on the surface of CB, it enhances the mechanical strength to a great extent. It is highly compatible with all types of thermoplastics. One other most important effect is that the resistance against UV degradation of plastics is largely improved by the addition of carbon black. In fact, CB-reinforced PE and PP pipes and vessels can be used in open sunlight for several years. Certain conducting CB can be used for making conducting thermoplastic composites for use as antistatic application. Liang and Yang [1] used CB of particle size of about 18 microns in high-density PE (HDPE) and observed the increase in tensile fracture strength by about 100% at 8% loading of CB, while the tensile strength did not improve at all. The elongation-at-break improved by about 12% at 5% CB and reduced on further addition. The Flexural modulus monotonically improved linearly with CB content and at 8%, the improvement was about 12%. Similarly, the flexural strength improved by about 10%. Huang and Wu [2] studied the effect of CB on the processing parameters of mixing, improvement of mechanical and electrical properties of EVA-CB composites, based on three types of EVA filled with 2.5 to 30% CB of surface area 1475 m^2/g. The work energy of mixing 30% CB was 80 times higher than that of unfilled EVA of two varieties. The tensile strength reduced by 9–28% in two types of EVA but increased by 25% in the third variety. The elongation-at-break drastically reduced for all three EVA. The electrical volume resistivity reduced drastically from 10^{15} ohm-cm to 10^5 ohm-cm for 25% CB-filled EVA of all varieties, with the threshold at 6–10% CB.

Gill et al. [3] reported a composite filled with hybrid CB-Nickel particles, where the matrix was a blend of PP with 5% PP-graft-maleic anhydride for imparting some polarity to enable better dispersion of the filler mixture, as confirmed by SEM micrograph. Both the tensile strength and modulus attained a maximum at 2% filler loading, reducing on further addition (up to 5%). The non-linearity in modulus with respect to filler loading cannot be explained by a simple mixing rule. Electrical conductivity also had a percolation threshold at 2.5% hybrid filler. The optimum hybrid filler loading was thus reported as 2.5%. Ding et al. reported a PP-CB composite [4] where the electrical percolation threshold was at 3% CB (by volume). A silane-coupling agent was used to treat the CB for better dispersion in PP, and the authors reported significant (20–30%) improvement in the tensile properties and impact energy using this method of reinforcement.

Graphite is another form of flake-type carbon filler that is used for imparting electrical conductivity to thermoplastic composites. Thermal and electrical conductivity, thermal diffusivity of high-density polyethylene (HDPE) and polystyrene/graphite composites were investigated by Krupa and Chodak [5]. The authors observed different percolation thresholds for HDPE composites, with two types of graphite differing in morphology. Young's modulus, thermal and electrical properties showed significant improvement on graphite loading at 0.3 volume fraction (30 volume %), while relative elongation decreased drastically. Altay et al. [6] reported a composite of polyamide 4,6 with synthetic graphite-graphene nanoplatelet hybrid filler and observed synergistic effect on the mechanical and thermal properties of the composite. Addition of only 5% graphene in 40% graphite-loaded composite enhanced the thermal conductivity 1.5 times and the tensile modulus by 1.65 times. However, the ultimate tensile strength with and without graphene addition in 40% graphite-loaded PA did not change, and was reported to reduce by 43% compared to pristine polyamide.

Several minerals and glass beads are also reported to be used in thermoplastics for the enhancement of strength [7–13].

Metal powder-filled electrically conducting thermoplastic composites are reported in some literatures. Krupa et al. [14] studied HDPE-nickel (special grade-branched structured nickel particle) composite to improve the electrical conductivity and also adhesion with metal surfaces. The authors reported the electrical percolation threshold to be at 8 volume% nickel in the composite. The hydrophilicity of the composite was observed to be increased significantly (contact angle 80° compared to 93° of pristine HDPE), but the adhesive bond strength for aluminium foil improved to 1.7 times the bond strength with pristine HDPE. Mamunya et al. [15] used nickel and copper particles as fillers in PVC, and Boudenne et al. [16] used copper particle in PP for the enhancement of thermal and electrical conductivities.

2.2.1 Effect of filler: mechanical property

The effect of rigid particle inclusion on the tensile strength can be expressed by the Nicolais and Narkis [11] equation as:

$$\sigma_c = \sigma_m \left(1 - 1.21\phi_f^{2/3}\right) \tag{2.1}$$

where σ_c and σ_m represent the tensile strengths of the composite and the matrix, respectively, and ϕ_f represents the volume fraction of the rigid filler.

Whereas, Liang and Li [12, 13] introduced a new concept of the interfacial adhesion angle (θ), and deduced the following modified equation for tensile strength:

$$\sigma_c = \sigma_m \left(1 - 1.21\sin^2\theta\phi_f^{2/3}\right) \tag{2.2}$$

If the angle is 90°, then eq. (2.2) becomes identical to eq. (2.1).

According to these equations, the strength reduces on the addition of the rigid particulate filler, instead of reinforcing. The improvement in the Young's modulus is, on the other hand, observed in all such reinforcements, particularly, where the particles interact with the polymer, due to some functionality or polarity.

Liang and Yang [1] proposed the following relationship for flexural strength and modulus for the composite of HDPE, based on carbon black and mica reinforcement:

$$\sigma_{Fl} = \beta_0 + \beta_1\phi_f \tag{2.3}$$

$$E_{Fl} = \alpha_0 + \alpha_1\phi_f \tag{2.4}$$

β_0 and α_0 are the values of the pristine polymer and ϕ_f is the weight fraction of the filler in the composite.

The impact strength was seen to be decreasing and the authors have shown that it is best described by the binomial equation:

$$\sigma_{Im} = B_0 + B_1\phi_f + B_2\phi_f^2 \tag{2.5}$$

B_0 being the impact strength of the unfilled HDPE. All the coefficients in eqs. (2.3) to (2.5) are dependent on the type of polymer and filler and on the polymer-filler and filler-filler interactions. It is therefore best to determine these coefficients for a pair of filler-polymer by experimentation. The authors reported a negative value of B_1 and a small positive value of B_2 in eq. (2.5) for the HDPE-CB composite.

A generalized equation of the filler effect on polymeric composite is given by [17]:

$$\frac{E_C}{E_p} = \frac{1 + AB\phi_f}{1 - B\phi_f} \tag{2.6}$$

where

$$B = \frac{(E_f/E_p) - 1}{(E_f/E_p) + A} \tag{2.7}$$

The constant A is a material-dependent term, involving the Poisson's ratio and the aspect ratio of the filler, and is related to the generalized Einstein coefficient, k.

Finally, the shear modulus ratio of the composite to the polymer matrix as given by Nielsen [17] is:

$$\frac{E_C}{E_p} = \frac{1 + (k-1)B\phi_f}{1 - B\phi_f} \tag{2.8}$$

For, the low volume fraction of the filler (filler-filler interaction neglected) and the ratio of filler to polymer modulus are very high.

Nielsen also suggested the value of k for different Poisson's ratios of the filler, taking the Poisson's ratio of the polymer as 0.5. However, the assumption may not be applicable for thermoplastics because the Poisson's ratio of most engineering thermoplastic is much less than 0.5.

Krupa et al. [14] took the value of A as 6.5 for HDPE-nickel powder composite and estimated the composite modulus to be in fairly accurate agreement with the experimental result.

The model by Gold as shown in Chapter-4, eq. (4.17) without considering agglomeration and eq. (4.18) for agglomerated filler, for example carbon black in a rubber matrix, may also be used for thermoplastic-particulate composites with some errors.

Ogorckiewicz and Weidman [9] reported modelling and experiments with a polypropylene-glass spheres composite, and suggested some models in series and in parallel and in combinations of stress components in the prism-in-prism, cube-in-cube, etc. form of unit cells. Two types of equations yielded close results of composite Young's modulus in tension.

With the series stress model:

$$E_C = \frac{E_f E_p}{\phi_f E_p + \left(1 - \phi_f\right) E_f} \tag{2.9}$$

Cube-in cube model:

$$E_C = \frac{E_p\left\{ E_f\left(1 - \phi_f^{1/3} + \phi_f\right) + E_p\left(\phi_f^{1/3} - \phi_f\right)\right\}}{E_f\left(1 - \phi_f^{1/3}\right) + E_p\phi_f^{1/3}} \tag{2.10}$$

Bourkas et al. [18] derived composite elastic modulus using a cube-in -cube model where the filler particle is a considered as a small cube and the load components in the matrix-filler unit cell are in various series-parallel combinations. The authors assumed perfect filler-matrix interactions and no agglomeration. Further, the uniaxial stresses have no other components in any other direction.

The longitudinal modulus was derived as:

$$E_C = E_p\left(1 + \frac{\phi_f}{1/(m-1) + \phi_f^{1/3} - \phi_f^{2/3}}\right) \tag{2.11}$$

where m = ratio of the longitudinal modulus of filler to polymer, (E_f/E_p).

The corresponding shear modulus was given by:

$$G_C = G_p\left(1 + \frac{\phi_f}{1/(\rho-1) + \phi_f^{1/3} - \phi_f^{2/3}}\right) \tag{2.12}$$

where ρ = ratio of the shear modulus of filler to polymer, (G_f/G_p).

A simplified model for spherical inclusion as a filled polymer composite is given by Kerner-Nielsen's equation as follows [19, 20]:

$$E_C = E_p\left(1 + \frac{\phi_f}{1 - \phi_f}\left[\frac{15\left(1 - \nu_p\right)}{8 - 10\nu_p}\right]\right) \tag{2.13}$$

where ν_p is the Poisson's ratio of the polymer.

Ramsteiner and Theysohn carried out experiments on the tensile behaviour of PP with different types of filler in which small glass beads of diameter less than 15 μm were used as spherical fillers. They observed that the cube-in-cube model was in good agreement with the experimental data and the Kerner-Nielsen model was in close approximation.

2.2.2 Effect of filler: thermal conductivity

The Hashin-Shtrikman model [14, 21, 22] predicts a lower bound in thermal conductivity of a filled thermoplastic:

$$\lambda_c = \lambda_m + \frac{\phi_f}{\dfrac{1}{\lambda_f - \lambda_m} + \dfrac{1 - \phi_f}{3\lambda_m}} \tag{2.14}$$

Here, ϕ represents the volume faction of the filler and λ represents the thermal conductivity. The above expression considers a uniform particle distribution, therefore, the percolation due to the contact of the conducting filler particles is not taken into account. The equation is thus valid only for up to 10–12% filler (by volume). Budiansky [23] proposed the following expression for N number of fillers:

$$\sum_{i=1}^{N} \phi_i \left[\frac{2}{3} + \frac{1}{3} \left(\frac{\lambda_i}{\lambda_c} \right) \right]^{-1} = 1 \tag{2.15}$$

The Lewis-Nielsen model is defined by following equations for various shapes of fillers:

$$\lambda = \lambda_m \frac{1 + AB\phi_f}{1 - B\psi\phi_f} \tag{2.16}$$

where

$$\psi = 1 + \frac{(1 - \phi_{max})\phi_f}{\phi_{max}^2}, \quad B = \frac{\dfrac{\lambda_f}{\lambda_m} - 1}{\dfrac{\lambda_f}{\lambda_m} + A} \tag{2.17}$$

'A' is a parameter that depends on the shape of the particles and ϕ_{max} is the maximum packing fraction. Different values A and ϕ_{max} are reported in the literature [14, 24] for various shapes (spheres, fibres, flakes, irregular particles and various packing geometries; e.g., hexagonal, face and body centred cubic, simple cubic, random). Krupa et al. [14] found fitting values as A = 5.5 ± 0.7 and ϕ_{max} = 0.6 (approximately) for the composite they studied.

2.2.3 Example

Let us take an example of an antistatic thermoplastic shield to be designed with HDPE of melt-flow index (MFI) of about 6 g/10 min, a 99.3% pure graphite non-spherical powder of average particle size of 2.6 μm, thermal conductivity of the particle as

159.6 W/m.K. and that of HDPE as 0.53 W/m.K. The Young's modulus in tension is 1.07 GPa for HDPE and 20GPa for graphite powder and the corresponding densities are 0.97 and 2.25 g/ml. The Poisson's ratio of HDPE is taken as 0.40. The loading of graphite was up to 30% by volume of the composite. The predicted Young's modulus and thermal conductivity of this composite with respect to the filler volume fraction is to be calculated with the help of the various models described above, for comparison.

Figure 2.1 shows the composite Tensile Modulus calculated for HDPE-Graphite particulate composite using the mixing rule (eq. 2.4), generalized model by Nielsen (eq. 2.6), Ogorckiewicz and Weidman (O&W) model on series stress (eq. 2.9) and cube-in-cube model (eq. 2.10), Bourkas model (eq. 2.11) and Kerner-Nielsen model (eq. 2.13). The prediction by the mixing rule showed highest reinforcement, while the series stress model showed a negative growth with increase in the volume fraction of graphite. However, except the mixing rule and the series stress, all the other four composite models gave somewhat close results.

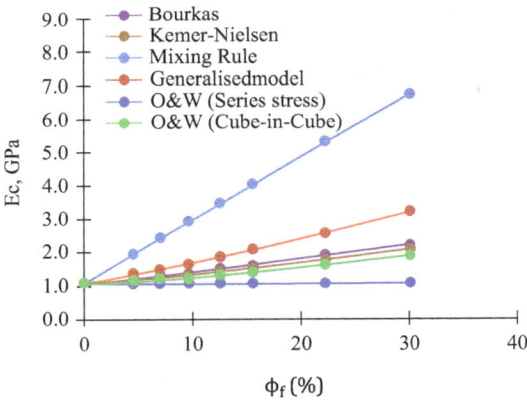

Fig. 2.1: Tensile modulus of HDPE-Graphite composite: theoretical models.

Figure 2.2 shows the composite conductivity, calculated using the models by Hashin-Shtrikman (eq. 2.14), Budiansky (eq. 2.15) taking the binary phase, and by Lewis-Nielsen (eqs. 2.16 and 2.17). The specific gravities of HDPE and graphite were taken as 0.97 and 2.25, respectively. Graphite volume fraction was taken from 0 to 0.30 (50% HDPE and 50% Graphite by weight). For the Lewis-Nielsen model, the values of A and ϕ_{max} values taken are 6 and 0.6, respectively.

2.2.4 Electrical properties

One of the important applications of particulate thermoplastic composites is in the field of electrically conducting material, which is a structural component in large elec-

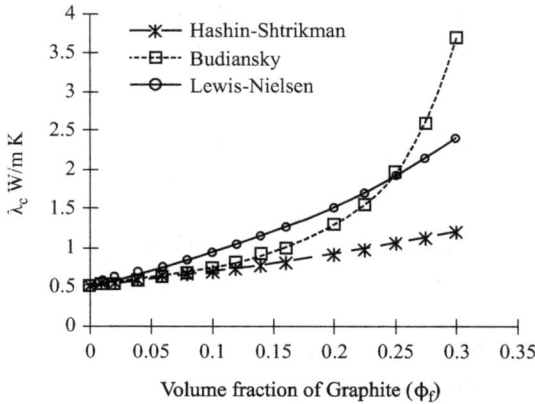

Fig. 2.2: Thermal conductivity of HDPE-Graphite composite: theoretical models.

trical power systems, such as in EMI shielding, antistatic sheets and floor mats, radar absorbing sheets, etc. The particulate fillers, especially metal powders such as carbonyl iron, nickel, silver, conducting carbon, etc. are widely used to impart electrical conductivity, similar to a semiconductor. All thermoplastics are basically electrical insulators, having electrical resistivity in the range of 10^{12} to 10^{16} Ohm-cm [2, 14, 15, 24, 25].

In this part of discussion, the so-called conducting polymers, having linear chain with conjugate double bonds, are not included as these polymers do not have other structural properties like a typical engineering thermoplastic. Some examples are polyacetylene, polythiophene, polyaniline, polypyrrole, etc. Even those conducting polymers have electrical conductivity that is much lower than typical semiconductors, and need to be doped with ionic molecules so that electron hopping is possible from one chain to the other, which is actually the cause of electronic conductivity of these polymers. Typical examples are polyaniline doped with acids such as H_2SO_4, $HClO_4$, HNO_3 HI, HCl, HBR, etc. [26, 27] These doped conducting polymers can have conductivity as high as 0.10 to 10^2 s/cm (undoped about 10^{-6} to 10^{-4} s/cm) and are best used in organic coating (paint) as a conducting material dispersed in a resin matrix, such as acrylic or polyurethanes.

The mathematical models describing the effect of conducting filler inclusion on electrical conductivity is almost similar to thermal conductivity models. Hashin and Shtrikman [21] derived the following equation for the prediction of an articulate composite conductivity, based on the composite sphere assemblage (CSA) model.

$$\mu_c = \mu_p + \cfrac{\phi_f}{\cfrac{1}{\mu_f - \mu_p} + \cfrac{1 - \phi_f}{3\mu_p}} \qquad (2.17)$$

The same can be applicable to magnetic permeability, electrical permittivity, thermal conductivity, diffusivity, etc.

However, the increase in conductivity with conducting filler loading is not linear, and has a percolation threshold concentration of the filler at which the conductivity sharply increases, similar to the trend of thermal conductivity. When the volume fraction of the conducting filler reaches the maximum packing, known as the percolation threshold, the conductivity abruptly increases by several folds and thereafter reaches the maximum conductivity of the composite. Below the threshold, the conductivity is very low, almost equal to or slightly higher than that of the polymer, and increases slowly. Above the threshold also, the increase is slow till the maximum value.

Huang and Wu [2] studied the electrical conductivity of three types of ethylene-vinyl acetate (EVA) copolymer filled with carbon black and reported a decrease in volume resistivity by a ratio of 10^{12} for all the three EVAs at a CB loading of 25% (by wt). The percolation threshold concentration was 5–10% CB, depending on the type of EVA. Beyond the percolation threshold, the decrease in resistivity was slow, as expected.

Krupa and Chodák [5] studied polystyrene (PS) and HDPE filled with two types of graphite particles of average size 15 to 40 μm, and evaluated the various physical properties. The conductivity ratio of the composite to polymer (σ_c/σ_p) of PS showed 100 times higher than that of HDPE composite with similar graphite particles size at similar CB loading. Another observation was that the smaller particles showed better reinforcement in terms of electrical conduction. The percolation threshold was almost the same for the HDPE and PS filled with the particles of the same size. Krupa et al. [14] studied HDPE-nickel (average particle size: 2–20 μm) composite for electrical conductivity and found the percolation concentration range of 0.03 to 0.10 volume fraction (mid value = 0.08) of the nickel filler. The conductivity of the composite was enhanced compared to HDPE by a factor of 10^{22} for nickel volume fraction of 0.30.

Percolation is defined as a point of inflection where the eq. (2.17) is discontinuous. An empirical equation to define this point is suggested as [5, 28]:

$$\log(\sigma_C/\sigma_m) = B\left(1 - \exp[-a\phi_f]\right)^n \tag{2.18}$$

where B, a and n are empirical constants and can be determined through experimental result. σ is the conductivity, c and m denote the composite and matrix, respectively, and ϕ_f is the volume fraction of the filler. The coefficient B is approximately given by:

$$B \cong \log(\sigma_{C\max}/\sigma_m) \tag{2.19}$$

where σ_{Cmax} is the maximum conductivity attained for the composite.

The authors also suggested the percolation concentration from the following equation:

$$\phi_i = \phi_c = \ln(n)/a \tag{2.20}$$

where ϕ_i is the volume fraction of the filler at the point of inflection, defined as the percolation threshold. The authors [5] summarized the sets of values of B, a and n and also calculated the percolation threshold for each polymer-graphite combination from experimental results.

The percolation threshold volume fraction of the conductive site is related to the maximum packing factor (F) as:

$$\phi_C = X_C F \tag{2.21}$$

where X_C is a percolation probability factor and the values of X_C and F are such that the ϕ_C is = 0.16 in evenly distributed filler with the most probable packing in a polymer matrix [15, 29].

The packing factor is most important for a filler-polymer system as it is the packing limit of the filler that is equal to the highest possible filler volume fraction at a particular type of packing. It decides the conducting connectivity through a continuous path and the maximum reinforcement by filler. The packing factor can be calculated as:

$$F = \frac{V_f}{V_f + V_p} \tag{2.22}$$

where V_f is the volume occupied by the filler at the highest possible filler volume fraction and V_p is the volume occupied by the polymer in the space among the filler particles [15].

For spherical particulate fillers with the most probable packing and uniformly dispersed in a polymer matrix, F = 0.64. Beyond the percolation concentration of the filler, the conductivity changes with the filler volume fraction as [15]:

$$\sigma_C = (\phi - \phi_c)^t \tag{2.23}$$

where ϕ_c is the filler volume fraction at the percolation threshold, $\phi > \phi_c$ and t is the critical exponent = 1.6 to 1.9. The experimental value can be different due to the filler–polymer interaction, clustering, uniformity in distribution, particle packing type, etc. The above equation can also be written in a normalized form:

$$(\sigma - \sigma_c)/(\sigma_m - \sigma_c) = [(\phi - \phi_c)/(F - \phi_c)]^t \qquad \text{for } \phi > \phi_c \tag{2.24}$$

where σ_m is the maximum conductivity of the composite.

Mamunya et al. [15] stated that for a randomly distributed spherical conducting filler in a non-conducting matrix, F = 0.64, X_C = 0.25 and ϕ_f = 0.16. The authors used carbonyl nickel (10 μm) and copper (100 μm) particles as conducting fillers in PVC and obtained the percolation threshold in the range of 4 to 8.5%, depending on the filler type, and the critical exponent (t) was also much higher than the usual range 1.6–1.9.

At the threshold point, the conductivity abruptly increased by approximately 10^{17} folds compared to the pristine polymer (PVC).

Ding et al. [4] studied the effect of a highly conducting carbon black filler (resistivity approximately 0.8 Ohm-cm) for a PP-CB composite on the change in conductivity and on the positive temperature coefficient (PTC) effect. The commercially obtained carbon black was surface treated with a silane-coupling agent and heat treated for improvement in dispersion in PP. The authors observed a 6-fold decrease in electrical volume resistivity at the percolation threshold filler volume fraction of 0.03. The authors also reported a positive temperature coefficient (PTC) effect of the CB-filled PP near about 145–150 °C (onset of increase in resistivity) and found maximum resistivity at about 165–176 °C (melting region), depending on the carbon black content in the composite. The addition of the treated carbon black also improved the mechanical properties.

Altay et al. [6] developed a particulate composite of polyamide 4, 6 (PA46) filled with synthetic graphite (SG) of 50μm average particle size. There was a 10^5-fold increase in the electrical conductivity when the SG content increased from 30% to 40%, and the overall increase in conductivity was 10^{11}-folds, compared to pristine PA46. However, addition of 5% graphene nanoplatelets (10–20 nm) improved the conductivity by another 10^2-folds. However, the composites showed reduction in mechanical strength, but increase in elastic modulus, indicating harder and lower flexibility.

Joseph et al. [25] developed a composite EMI shield with polyvinylidene fluoride-carbonyl iron powder (PVDF-CIP), with maximum 50% volume of the filler. CIP is known to be highly conducting material, being a small particle (1–3 μm), while PVDF is a flexible thermoplastic, having excellent dielectric properties, excellent atmospheric and thermal stability. The electrical conductivity of the composite with the highest filler content was maximum about 10–11 S/cm, and the EMI shielding was observed to be more due to the absorption of electromagnetic wave energy rather than by reflection.

However, some thermoplastic composites are designed to improve the electrical insulation property along with dielectric strength and frequency-dependent dielectric properties for many industrial applications. Insulation properties of non-polar thermoplastics are good in general, but the ease of processing and tailoring the properties are not as good as those of thermoplastic polyurethane. Thermoplastic polyurethane (TPUs) based on polyether polyol and aromatic diisocyanate are highly stable and are commercially available in a wide range of hardness, viscoelastic properties and offering resistance to moisture and to many chemicals.

Ramraj [30] used PTFE powder (300–600 μm size) as a filler in thermoplastic polyurethane (TPU). The PTFE was used for a maximum of 10 wt% in the composite and a 10-fold increase in volume resistivity and substantial improvement in dielectric strength were observed. However, there was a decrease in mechanical strength and increase in flexibility.

He at al. [31] developed a filler of core-shell structure using a polydopamine-treated 40–100 nm sized titanium carbide particles (PDA-TiC) and loaded in a TPU for improvement of dielectric properties, specifically frequency-dependent dielectric constant (μ') and dielectric loss factor (tan δ) in the KHz to MHz range. The dielectric constant and the loss factor both improved substantially on adding the filler system, and the increase was monotonic with the doping of TiC. The effect is reduced as the frequency is increased. The reason for the increase of complex dielectric property at lower frequencies was suggested by the authors as the polarization effect of the interface between PDA and TiC, the disruption of hydrogen bond among the chains of TPU due to the presence of TiC and the corresponding increase in the dipole polarization of the TPU chains. However, relaxation of dipoles is a strongly frequency-dependent phenomenon. Polarization effects are not observed at a higher frequency due to the mismatch in the relaxation frequency of the dipoles with that of the dynamic electric field. The dielectric strength improved to some extent for PDA-TiC filled composite up to a certain doping content and beyond that, the strength gradually decreased with dopant content. The dielectric strength of the TPU with undoped TiC was less than that of pristine TPU.

2.3 Short-fibre composite

Short fibres play an important role in reinforcing thermoplastics for the improvement in their overall mechanical strength, thermal and electrical properties and also barrier properties. The advantages of short fibres are the ease of mixing and processing, higher reinforcing effect compared to particulate fillers and near-homogeneous composite with almost isotropic properties. The methods of composite manufacturing can be hot compression moulding of the polymer and the chopped fibre or hot pressing of the chopped fibre-polymer prepreg, extrusion and injection moulding. Chopped glass fibres are most common among short fibres, although carbon and Kevlar short fibres are also used in some thermoplastic composites. There are some exceptional short fibres such as PEEK fibres in thermoplastic matrix, for example, single-polymer composite. Generally, the length of such short fibres is 2–6 mm and their cross sections are in microns, similar to long fibres. In fact, short fibres or chopped strands are produced by chopping the long fibres.

Apart from filler–polymer interaction, the reinforcement by a short fibre depends on the aspect ratio (length-to-diameter), and below a critical aspect ratio, the reinforcement is almost similar to the agglomerated particulate filler. Similarly, the viscoelastic damping by a shot-fibre composite is more if the fibre is beyond a critical length so that friction damping and longitudinal-to-shear mode conversion under vibration significantly contribute to more damping than particulate-filled thermoplastics. The advantages of the effect of enhanced localized loss of shear strain energy and frictional damping is neither possible by particulate fillers nor by the continuous fibres. The other advantage is the less possibility of agglomeration (bunching) and entanglement. The size range of short fibres as such is quite big enough for inter-particle physical bonding, and short enough for any entanglement. This facilitates almost homogeneous dispersion and distribution of the short fibres in the polymer matrix.

2.3.1 Mathematical models

There are various mathematical attempts to analyze and describe the properties of a short fibre thermoplastic composite. The most discussed model, way back in 1938, was given by Guthand Gold [32], while predicting the composite's viscosity, and Guth used the same expression for carbon black particles forming agglomerates in a rod-shape, cube, etc., similar to a short cylindrical inclusion.

However, there is a critical length of the short fibre at which the reinforcement is optimum. The shear stress at the end of a fibre is zero and gradually increases with distance along the surface as the load is transferred to the fibre from the matrix. Maximum reinforcement is achieved when the fibre is long enough for complete stress transfer. The minimum length required for this condition is given by the following equation by Rosen [33] using a shear-lag analysis. It was assumed that the matrix and the fibre are both elastic materials (no viscoelastic loss).

$$\frac{\delta}{d} = \left[\frac{1}{2} \frac{E_f}{G_m} \frac{1 - v_f^{1/2}}{v_f^{1/2}} \right]^{1/2} \tag{2.25}$$

There are classical rules in mixing theory for describing the strength of perfectly aligned short fibres in a composite [34]:

$$\sigma_c = \sigma_f V_f F(l/l_c) + \sigma'_m (1 - V_f) \tag{2.26}$$

where σ'_m is the stress on the matrix at the breaking strain of the fibre, l is the length and lc is the critical length of the short fibre. The function $F(l/lc)$ is ideally maximum 1 at $l = lc$, and it is the factor that defines the efficiency of the reinforcement when the length of the fibre is below the critical value. However, considering the higher stress concentration at the end of the fibre, the upper bound of F was found by the FEM method as 0.5, at a high fibre aspect ratio [35].

For non-aligned fibre orientation, the strength is described by:

$$\sigma_c = C\sigma_f V_f F(l/l_c) + \sigma'_m (1 - V_f) \tag{2.27}$$

The value of the coefficient C varies on orientation, for example, Cox [36] used the classical shear-lag analysis and estimated C = 1/3 for 2-D and 1/6 for 3D orientation of the fibre.

Halpin-Sai models [37–40] suggest the elastic modulus of short-fibre composites in oriented and transverse directions as:

$$E_{11} = E_m \left[\frac{1 + \xi \eta V_f}{1 - \eta V_f} \right] \tag{2.28}$$

where

$$\eta = \frac{E_f/E_m - 1}{E_f/E_m + \xi}, \qquad \xi = 2\left(\frac{l}{d} \right) \tag{2.29}$$

The transverse modulus E_{22} is the same, with a difference that $\xi = 2$ in that case [38]. A modified Halpin-Tsai equation is used for the overall composite modulus, combining longitudinal and transverse moduli, assuming uniform distribution of the short fibre [40]:

$$\frac{E_c}{E_m} = \frac{3}{8} \left(\frac{1 + \xi_f \eta_{fL} V_f}{1 - \eta_{fL} V_f} \right) + \frac{5}{8} \left(\frac{1 + 2\eta_{fT} V_f}{1 - \eta_{fT} V_f} \right) \tag{2.30}$$

Here, V_f is the volume fraction of the filler and E_m is the matrix elastic modulus, while ξ and η are defined as in eq. (2.29).

Load transfer models based on the shear lag theory [41, 42] are based on some assumptions such as the fibre is a circular cylinder, the fibre to matrix adhesion is perfect, the axial stress around it reduces with distance of the fibre surface from the matrix, and is equal to the general stress at a certain distance R from the fibre. Both the fibre and the matrix behave elastically with no loss; the axial stress at the fibre end is zero; and the fibre–fibre interaction is absent. Here, the fibre surrounded by a cylindrical matrix element becomes the composite element in such an analysis as shown in Fig. 2.3. Lacroix et al. derived that the maximum interfacial and axial stress values in a short-fibre polymer composite occur at the fibre end and at the centre of the fibre, respectively. Both fibres are fully elastic in nature.

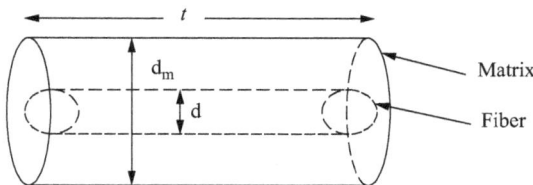

Fig. 2.3: A model element of a cylindrical fibre in a cylindrical matrix.

In the case of perfect adhesion between the fibre and the matrix, load transfer occurs in a purely elastic manner and the following stress distribution is obtained from the shear lag theory [42]:

Axial stress: $\sigma_t = E_t\varepsilon + a\sinh(2nx/d) + b\cosh(2nx/d)$ (2.31)

Shear stress: $\tau = -\dfrac{n}{2}[b\sinh(2nx/d) + a\cosh(2nx/d)]$ (2.32)

where

$$n^2 = \frac{E_m}{E_f(1+v_m)\ln(2R/d)}$$ (2.33)

The interface radial stress arising out of Poisson's contraction for perfect adhesion at the interface and when the matrix element diameter (d_m) is 100 times larger than fibre diameter (d), is given by:

$$\sigma_{Pois} = \frac{E_f E_m (v_f - v_m)}{E_f(1+v_m) + E_m(1+v_f)(1-2v_f)}\varepsilon$$ (2.34)

where v_f and v_m are the Poisson's ratios, E_f and E_m are elastic modulus of fibre and the matrix, respectively, ε is the axial strain and s is the aspect ratio = (l/d) of the fibre. The maximum axial and shear stress values are:

$$\sigma_{f,max} = E_f\varepsilon[1 - \mathrm{sech}(ns)]$$
$$\tau_{max} = \frac{1}{2}nE_f\varepsilon\tanh(ns)$$ (2.35)

Carrara and McGarry [43] derived the force in the matrix surrounding a discontinuous fibre at any distance x from the fibre end (x is in the perpendicular direction to the axial load) as:

$$P = E_f A_f\varepsilon\left[1 - \frac{\sinh\beta\left(\dfrac{l}{2} - x\right)}{\cosh\beta\left(\dfrac{l}{2}\right)}\right]$$ (2.36)

where

$$\beta = \left[\frac{H}{E_f A_f}\right]^{1/2} \quad \text{and} \quad H = \frac{2\pi G_m}{\ln(r_m/r_f)}$$ (2.37)

The shear stress parallel to the direction of fibre's length at any point on the interface is given by the authors as:

$$\tau_{int} = \sigma_c K_1\frac{\sinh\beta\left(\dfrac{l}{2} - x\right)}{\cosh\beta(l/2)}$$ (2.38)

where

$$K_1 = \frac{BA_f\left(E_f - E_m\right)}{2\pi r_f E_m} \tag{2.39}$$

σ_c is the uniform composite stress and ε is the composite strain applied in the axial direction.

For a randomly oriented short-fibre composite, Tsai-Pagano model [44, 45] may be more effectively used to predict the Young's modulus and shear modulus as also the tensile and shear strengths:

$$\text{Young's Modulus:} \quad E_c = \frac{3}{8}\left[V_f E_f + \left(1 - V_f\right)E_m\right] + \frac{5}{8}\frac{E_f E_m}{V_f E_m + \left(1 - V_f\right)E_f} \tag{2.40}$$

$$\text{Tensile strength:} \quad \sigma_c = \frac{3}{8}\left[V_f \sigma_f + \left(1 - V_f\right)\sigma_m\right] + \frac{5}{8}\frac{\sigma_f \sigma_m}{V_f \sigma_m + \left(1 - V_f\right)\sigma_f} \tag{2.41}$$

$$\text{Shear Modulus:} \quad G_c = \frac{3}{8}\left[V_f G_f + \left(1 - V_f\right)G_m\right] + \frac{5}{8}\frac{G_f G_m}{V_f G_m + \left(1 - V_f\right)G_f} \tag{2.42}$$

The first term on the right hand side is the Voigt model, assuming the fibre is oriented along the direction of the applied force and the second term is the Reuss model, assuming the fibre's alignment is in a direction that is transversal to the force.

2.3.2 Example

Let us take an example of a graphite fibre-loaded thermoplastic elastomer (TPE) having Young's modulus = 1.242×10^9 N/m^2 and shear modulus of 4.43×10^8 N/m^2, Poisson's ratio of 0.40, ultimate tensile strength of 5.6×10^7 N/m^2 and elongation-at-break of about 123%. The graphite fibre is 0.5 μm in diameter and four samples of varying average lengths –7.5 μm, 15 mm, 25 mm and 50 μm are considered. The morphology of one sample by SEM is shown in the Fig. 2.4. The average aspect ratio, defined as length-to-diameter is therefore 15, 30, 50 and 100. The Young's modulus of the graphite along the axis is about 1.42×10^{11} N/m^2.

The axial modulus of the composite for perfect alignment of the fibre along the axis of the load application is calculated using eq. (2.28), together with eq. (2.29), for fibre volume fractions of 0.0422 to 0.166 (10 phr to 45 phr) and the ratio of the Young's modulus of the composite to the matrix is shown in Fig. 2.5. The transverse modulus for this condition was also calculated and plotted in the figure.

The transverse modulus is the same for all aspect ratios of the fibre, obviously because the value of ξ is 2 and hence η is also constant at 0.974. The example shows that the modulus of the composite increases with aspect ratio. However, the increase in not linear for any fixed fibre content. Figure 2.6 shows the modulus ratio E_c/E_m for two volume fractions: 0.081 (20 phr) and 0.165 (45 phr) of graphite fibre against fibre

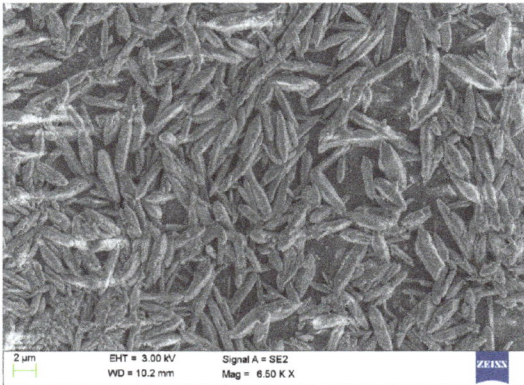

Fig. 2.4: SEM micrograph of a sample of graphite.

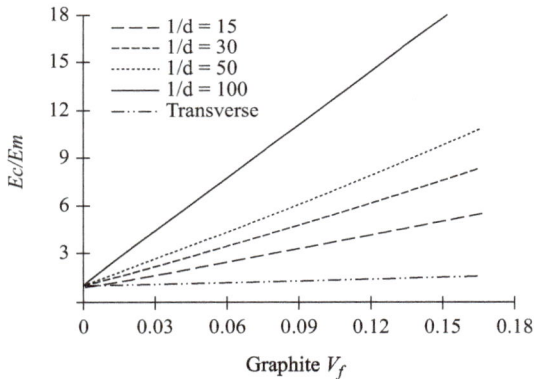

Fig. 2.5: Axial Young's modulus of TPE/Graphite fibre composites at different aspect ratios of the graphite fiber.

aspect ratio (l/d), theoretically calculated taking the matrix and graphite fibre properties. The optimum aspect ratio is near 400 as seen in the figure, for both the volume fractions. This value is for an ideal reinforcement, assuming no filler–filler interaction and the interface of the filler–matrix is perfect.

In a real situation, these are not realized because the small particles tend to agglomerate, and more so when they have polar groups or when there is a possibility of hydrogen bonding; the interface of the filer–matrix is never perfect because at a critical shear stress, there will be a slip at the interface due to the large difference in the shear modulus of the filler and the matrix. If the interactive bond is a physical bond, then this slip is a reversible phenomenon.

When the fibres are not perfectly aligned and are uniformly distributed, the reinforcement is reduced. This is demonstrated in Fig. 2.7 by calculating the modulus ratio

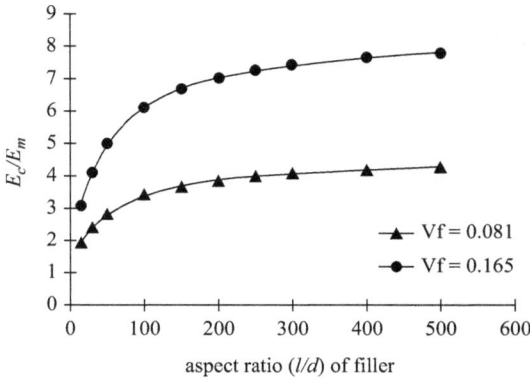

Fig. 2.6: Ratio of composite's to matrix's modulus at different aspect ratios of graphite fibre (theoretical calculation as example).

by eq. (2.30). The effective reinforcement is obviously much lower than the perfect alignment as shown in Fig. 2.5 above. In this case, the transverse stress is calculated taking the ξ value as 2 and the corresponding η is calculated using the axial Young's modulus of the graphite fibre.

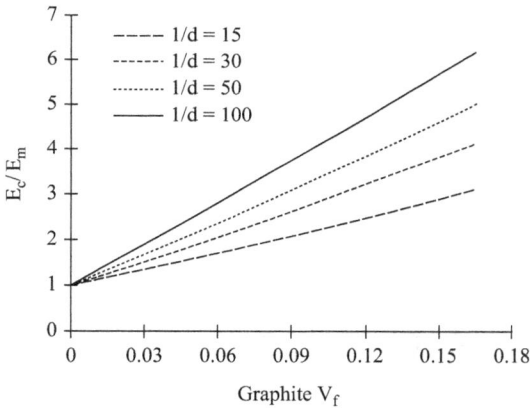

Fig. 2.7: Reinforcement by randomly oriented graphite fibre in the thermoplastic elastomer.

A comparison of the Guth equation for the composite modulus for a non-spherical shape of the filler such as rod-like agglomerate of carbon black particle filler in a rubber, and the modified Halpin-Tsai equation (eq. 2.30) is done for two aspect ratios. The Guth equation used was [46]:

$$E_C = E_m \left[1 + 0.67\phi . V_f + 1.62\phi^2 V_f^2\right] \tag{2.43}$$

where ϕ is the aspect ratio of the filler. Figure 2.8 shows the comparison of these two theories for the same thermoplastic elastomer filled with the graphite filler.

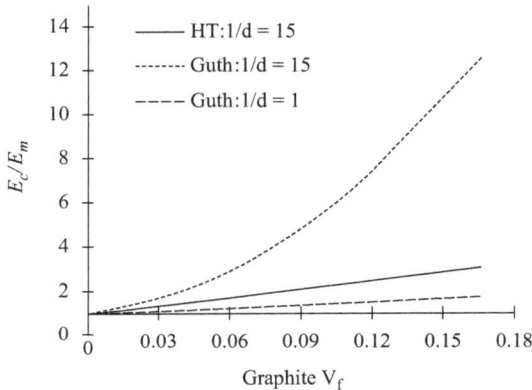

Fig. 2.8: Reinforcement by randomly oriented graphite fibre in the thermoplastic elastomer: Comparison of the modified Halpin-Tsai with the Guth model.

It is seen that the semi-empirical Guth equation yields quite a different prediction, a much higher reinforcement for the same aspect ratio of 15 compared to the modified Halpin-Tsai model for random orientation of the filler (eq. 2.30). This is obvious because the semi-empirical equation was developed for the carbon black reinforcement of elastomers.

It is well known that carbon black reinforces a polymer more than any other graphite because carbon black, as a result of its manufacturing process, contains some organic functional groups that enhance the intermolecular interaction with the polymers, while graphite is a much purer form of carbon and therefore does not reinforce to that extent as carbon black. Hence, the prediction of carbon black-based composite is higher and also non-linear, which fairly agrees with experiments on rubber reinforcement with carbon black having some rod-shaped agglomerate. Secondly, it is seen from the figure that a spherical filler ($l/d = 1$) does not augment mechanical modulus to the extent a fibre or plate-type filler ($l/d \gg 1$) can.

Unal and Mimaroglu [47] experimentally measured the tensile and flexural properties and Izod impact energy of short-fibre composites of polyamide (Nylon 6) filled with a mineral fibre (RF-MF) and E-glass fibre (GF), having average diameters of 5.5 μm and 10.5 μm and lengths of 150–200 μm and 3000–4500 μm, respectively. The Young's modulus in both tensile and flexure modes were seen to be much higher for glass-fibre reinforcement, almost 4 times the pristine PA at 30% GF content by weight as against 2 times by the rock fibre at the same filler content. A comparison of the mathematical model, as per eq. 2.30 (random orientation of short fibber), with the experimental result by the authors is done and the results are shown in Fig. 2.9. For the purpose of the calculations, the Elastic modulus of the RF-MF was taken as 70GPa (average value), as determined for similar mineral fibres by Bombac et al. [48] and that for short-GF (E-glass) as 80 GPa [49]. The volume fraction of each filler in each composite was calculated, taking the weight percent of the respective short-fibre and PA, 6 and the densities of each component from ref [45]. The tensile strength was increased from 69 MPa for pristine PA 6 to about 165 MPa for 30 wt% short GF, but only up to 81 MPa for the same loading of the mineral fibre. The stiffness (Young's modulus) of the short fibres effectively improved at the expense of the % elongation at break, which was seen to be drastically reduced from 105% for pristine PA 6 to about 3.5% and 17% for even 10 wt% loading of short GF and mineral fibre, respectively. The loss of flexibility is one disadvantage of brittle fibres. A combination of these two types of fibres also resulted in similar mechanical properties, and much closer to the properties of the glass-fibre composite.

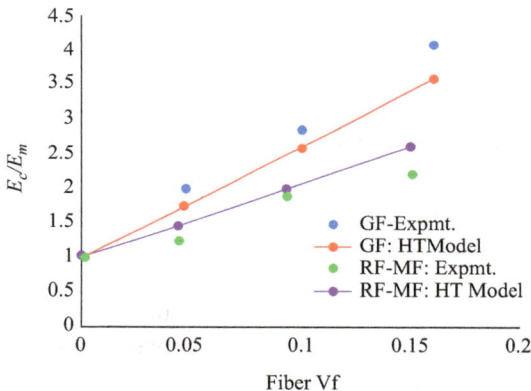

Fig. 2.9: Experimental and Halpin Tsai prediction of the Young's modulus of the two types of short-fibre-filled Polyamide (PA 6) composites, Reprinted with permission from Unal, H., Mimaroglu, Sci. Eng. Compos. Mater. 21 (2014), 323–328. © 2014, De Gruyter Publishing Company.

Lou et al. [50] reported an interesting method of improvement of short-glass-fibre PA 66 composite in respect of dimensional stability and properties, upon injection moulding, using the supporting mould by three routes of hydrothermal conditioning treatment, including immersion in a boiling water bath and drying at an elevated temperature. The dimensional accuracy was improved after the hydrothermal treatment. More noticeable improvement was for mechanical strength in tension and bending, as these improved by about 140% and 160%, respectively. The authors explained that the hydrothermal conditioning reduced the residual stress in the composite that is caused due to moulding.

Zheng and co-workers [51] reported improvement in the hardness and tribological properties of different commercial PTFE-short-glass-fibre composites with molybdenum sulphide (as additive for reduction of coefficient of friction). Li and others [52] also studied PEEK-chopped glass-fibre composite for possible improvements of tribological properties. The authors used a commercial product of 30 wt% short-glass-fibre-loaded PEEK and evaluated the tensile, flexural and tribological properties by a specifically designed ball-on-disc wear test apparatus. The tensile and flexural strengths improved to 1.65 times of the unfilled poly(ether ether ketone) (PEEK), but the flexural modulus improved better (2.4 times) compared to the tensile modulus (1.75 times).The coefficient of friction increased by 50% for the composite compared to that of unfilled PEEK at lower load, but the difference substantially reduced when a higher load was applied for the wear test. However, the wear substantially reduced for the composite, by about 65% at low load, and by about 40% at the highest load applied.

Ramsaroop et al. [53] studied fracture toughness of PP-chopped glass fibre (l = 15 mm, d = 10 μm) composite by the determination of Mode 1 stress intensity factor K1, according to ASTM D 5045. The authors compared the fracture toughness of chopped glass-fibre composite with MMT nanoclay composite and woven glass-fibre composites and found that the chopped glass-fibre composite showed the highest fracture toughness among these three types. The critical stress intensity factor for chopped glass fibre-PP composite was about 4 times that of pristine PP, and the failure load was also more than 4 times that for PP. The results show that the short fibre incorporation at 0.3 weight fraction can reinforce a polymer more effectively than nanoclay or continuous random fibre. Yuan and others [54] studied the properties of extruded pipes made with a commercial short glass fibre-PP composite (30% SGF by weight) blended with HDPE. The mass ratio was HDPE/SGFPP = 60/40. The composite pipes were extruded using a shearing-drawing two-dimensional compound stress-field pipe-extrusion device. The axial strength was not much improved compared to that of the conventional composite pipe, but the circumferential (circular) strength improved substantially (about 37%). Ari and co-workers [55] compared the properties of PP-based composites filled with chopped glass fibre, carbon fibre and aramid fibre, each 6 mm in length, and the maximum volume fraction was 0.30. The authors used a twin screw extruder to make the composites. The tensile strength of the carbon fibre was about 18% higher than that of glass fibre and 42% higher than that of aramid fibre. However, the authors observed the maximum strength by chopped glass-fibre inclusion at 30% by volume, although the values are almost similar for all the three chopped fibres at the same loading. The fibre efficiency was best for aramid, possibly due to more compatibility. The flexural strength of glass-fibre composite was remarkably higher than of other composites. However, the impact force was least for chopped glass-fibre composite and highest for chopped carbon-fibre composite at all volume fractions.

Kabiri and others [56] developed the PP-short-chopped glass-fibre composite (PP-SCGF) for potential use as bone fracture fixation plates. The process of making the composite plate was by hot pressing of the PP-SCGF in a mould. The Young's modulus in tension and flexure and the strengths and impact energy of such a composite was seen to be meeting the desired requirement. The tensile modulus was 3 times the unfilled PP and the corresponding strength was 1.5 times that of unfilled PP when the SCGF was 15 volume% in the composite.

2.3.3 Natural short-fibre composites

Natural fibres such as fibres of flax, jute, hemp, coir, sisal, bamboo, rice husk, etc. are being used in untreated and treated forms for the development of more environmentally acceptable and biocompatible/degradable or recyclable thermoplastic composites. For example, Kuruvilla and Mattoso [57] discussed the different composites with sisal fibre reinforcement way back in 1998. Recently, Rabbi et al. [58] reviewed research works pertaining to natural fibre-based thermoplastic and thermoset composites filled with both long and short natural fibres. For short natural fibres, various aspects of the injection moulding process for thermoplastic composites were discussed. An example of a completely biocompatible/degradable composite is a short sisal fibre used as filler in starch-based polymer by Alvarez et al. [59, 60] However, synthetic polymers are also being used, such as in PP by Joseph et al. [61] and PLA by Samouh et al. [62].

Natural fibres are polar, due to the presence of -OH groups and therefore, hydrophilic, while many host polymers are hydrophobic such as PE,PP, PVC, PLA, etc. Therefore, the interface of the fibre and the matrix is poor if the fibre is not treated specially for composite application. Secondly, natural fibres are not as thermally stable as synthetic fibres. Hence, natural fibres are often treated to improve their reinforcing ability, compatibility and thermal stability. Most common is alkali treatment [59], which removes the lignin, wax, and oil on the surface of natural fibres and increases the roughness of the surface of fibres. This ensures a better interlocking of the fibres with the polymer matrix. Sun and Mingming [63] studied the effect of sol–gel modification of sisal fibre on PP-based composite's properties. Apart from chemical treatment, physical treatments are also done, for example, using a coupling agent for the improvement of compatibility. Sood and Dwivedi [64] reviewed the various treatments on natural fibres being reported in literature since 2000. The authors gave an approximate analysis of the composition of the different natural fibres. Cellulose is the main component while other ingredients, in small quantities, are lignin, hemicellulose, pectin and wax. Among the fibres, the most available are jute, wood fibrils, cotton, hemp and sisal. In Asian coastal areas, coconut coir is another short fibre, which is a potential reinforcing filler. Hassani and co-workers studied the injection moulding process with the prediction of mould filling of PP-coir short-fibre composite [65] using the melt-flow process, applying the incompressible Navier-Stokes equation to study the flow front, and claimed to be in good agreement with the experimental results of moulding. In a separate article, the authors also predicted the mechanical properties of the composite [45]. The authors determined the gradual decrease in fibre length by grinding, sieving, extrusion with PP and injection moulding of the composite to obtain an almost homogeneous length of 2.25 mm and an average diameter of 290 μm. The tensile strength did not improve much on 10 wt% coir addition, but the Young's modulus was remarkably increased by about 16%, compared to unfilled PP. Further, it was observed from the predictive analysis that the self-consistent model and the Tsai-Pagano model were more suitable than other models for mechanical properties such as Young's modulus, tensile strength, etc. Adeniyi et al. [66] published a review of coir-filled polymeric composites.

2.3.4 Bio-degradable composites

Research on natural fibre-based bio-compatible and bio-degradable composites is ever increasing as an obvious necessity. Till now, the strength and thermal properties of such composites are not in any scale near to those of synthetic non-degradable thermoplastic-based composites reinforced with glass, carbon, aramid, etc. Therefore, many researchers are working on the development of synthetic polymer-natural fibre combinations to obtain more acceptable alternatives. PE, PP and poly(butyleneadipate-co-terephthalate), which are non-degradable are used for making natural fibre composites, where at least, on an average, 20% strength and elastic modulus enhancement is possible, by a simple theoretical calculation using the Halpin Tsai or Tsai-Pagano models. Some natural fibres have tensile strength ranging from 400 MPa to 1000 MPa such as cotton, hemp, flax, jute, sisal, pineapple, abaca, ramie and banana, with Young's modulus of 30 to 80 GPa. Jute, hemp, sisal and pineapple fibres have both maximum strength and modulus. The problem of waste disposal of synthetic polymer- short-fibre thermoplastic composites can be reduced by using hybrid short fibres like a combination of glass-sisal, glass-jute, etc., with an added advantage of higher strength in both tensile and flexure modes. At present, bio-degradable thermoplastic composites are being used in some sectors such as components in automobiles, furniture, packaging, household small items and fancy artefacts, but in limited scale. Considering the recyclability/reprocessability of thermoplastic short-fibre composites, the use of many natural polymers and fibres has the disadvantage of stability during melt processing, lower strength and modulus, and durability in various application environments. Fully bio-degradable composites made from poly(lactic acid) (PLA), modified starch, cellulose derivatives, poly(3-hydroxyalkanoate), poly(3-hydroxybutyrate),poly(3-hydroxybutirate-3-hydroxyvalerate), polycaprolactone (PCL), etc. have the advantage of disposal by bio-degradation, and are not dependent on recycling by melt processing such as injection moulding or extrusion. End-of-life options for bio-compatible/degradable thermoplastic-composites are (1) aerobic digestion (domestic and industrial composting), (2) anaerobic digestion (to biogas), (3) mechanical recycling, (4) chemical recycling, (5) incineration, (6) land filling, (7) metabolization by soil or with organism. Recycling at industrial scale has not been taken up till the present day due to purely economic considerations since the scale of bio-composite production has not been reached for cost effective recycling. For example, the cost of a polyethylene bag is much less than a PLA-based bag; hence the production scale of PLA is far less than that of polyethylene, consequently, recycling of PLA composite is not economically viable.

2.4 Nanocomposite

Nanocomposites of thermoplastics have been, of late, the most researched materials. The high reinforcement ability of nano particles has opened a new generation of high-performance thermoplastic composites. The reinforcement ability being directly proportional to the surface area of a filler, nano fillers are very effective even in very minute quantity, commonly 0.1 to 5%. This results in substantial reduction in the weight of the composite and hence saves transportation energy. On a simple analysis, work/energy of moving an object is directly proportional to mass, and hence even 10% reduction in mass is equivalent to a 10% reduction in energy consumption. Because of the high physical force of attraction with any molecule, there is an obvious advantage of a higher reinforcement effect compared to micron-sized particles but the disadvantage is also of a higher extent of agglomeration. The high agglomeration of nano particles leads to difficulty in dispersion in thermoplastics because the melt viscosity of the thermoplastics is very high, and is non-linear with shear rate. A typical apparent viscosity of molten PP is in the range of 50 to 280 Pa.s at the corresponding shear rates of 600 to 30/s at 190 °C and the viscosity-shear rate relationship is in logarithmic form. This implies that to reduce the viscosity by about six folds, the shear rate has to be increased by 20 times. Therefore, to obtain homogeneity in dispersion and distribution of the nano particles in thermoplastics, high shear force in a mixing machine, such as an internal mixer or extruder is needed. The best way of mixing nano particles in a thermoplastic is by using a twin-screw extruder, either as single or multiple pass, as needed. The morphology of the nanocomposite is extremely important to achieve the maximum benefit of the nano filler since all nano filers are used in small quantities. Secondly, depending on the polarity of the nano filler, the filler–matrix interaction as well as filler–filler interaction, simultaneously affect the reinforcing ability and agglomeration. Depending on the type of nano filler, the physical and, in some cases, the chemical properties of the thermoplastics change with filler content. With a spherical nano particle such as nano titanium dioxide or nanoceramic powder, the increase in mechanical strength of a polyolefin improves almost linearly, but with a rod-like or plate type nano filler, the strength increases non-linearly. The reinforcement depends much on the shape, in addition to size. For example, a carbon nanotube with a diameter of 20 nm and length of 1 μm has an aspect ratio of 50 (l/d), but for a length of 2 μm, the aspect ratio is 100, and hence the effective composite modulus would be enhanced accordingly. Unlike elastomers, which are soft materials, thermoplastics have high elastic modulus, commonly 1–4 GPa, and hence the load transfer between the filler and the matrix is better in thermoplastic nanocomposites. The remarkable advantage of an efficient reinforcement is reported in innumerable research work on thermoplastic nanocomposites. A simple analysis of how the surface area of a filler increases with the reduction of particle size was shown by Su and others [67] for spherical, cylindrical and plate-type fillers:

$$\text{Spherical: } S = 4\pi r^2 \left[\frac{Va}{(4/3)\pi r^3} \right] = \frac{3aV}{r} \tag{2.44}$$

$$\text{Cylindrical: } S = 2\pi rh \left[\frac{Va}{\pi r^2 h} \right] = \frac{2a}{r} V \tag{2.45}$$

$$\text{Plate type: } S = 2r^2 \left[\frac{Va}{r^2 t} \right] = 2\frac{a}{t} V \tag{2.46}$$

where r is the radius in (2.44) and (2.45), but lateral dimension in (2.46), V is the volume of a cube containing N number of particles, a is the fraction of the volume occupied by the particles and t is the thickness of the plates in (2.46). The inter-particle space for spherical particles in a cube of volume V (in a composite) having a lateral dimension L is also given by:

$$\Delta L = r \left(\sqrt[3]{\frac{4\pi}{3a}} - 2 \right) \tag{2.47}$$

Let us consider two composites, one with 20 nm radius (r) and the other with 20 μm radius (r) spherical filler in an identical volume (V) of 50 μm³ of both the composites. The total surface area of the fillers in each composite was calculated using eq. (2.44) for volume fractions (a) from 0.01 to 0.5 for each filler. Figure 2.10 shows the variation of the surface area (S) of nano-sized sphere compared to microspheres at various volume fractions of the fillers. The total surface area of nanospheres is exactly 1000 times that of microspheres. This demonstrates the scale of reinforcement by a nano filler compared to micron-sized fillers at the same volume occupied in a composite. However, the extent of reinforcement enhancement is never realized due to more agglomeration as the particle size decreases. The theoretical inter-particle distance calculated using eq. (2.47) is shown in Fig. 2.11 for this example.

The reinforcement by a nano particle cannot be expressed by a simple mixing rule because of the very small size compared to the size of the test sample. As discussed earlier, there is a critical length of the fibre at which the load transfer is optimum. Therefore, a length efficiency factor is introduced in the mixing rule for the determination of composite elastic modulus:

$$E_C = \left(\eta_l E_f - E_m \right) V_f + E_m \tag{2.48}$$

where η_l is the Krenchel length efficiency factor [68] and is less than 1, below a critical length of the fibre. The above expression is valid for fibres perfectly aligned in the direction of the applied force. The factor η_l is dependent on the aspect ratio (l/d) and modulus ratio (E_m/E_f) as well as the volume fraction (V_f) of the filler:

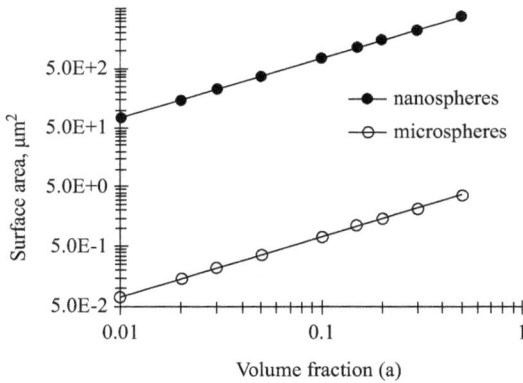

Fig. 2.10: Surface area of nano- and micron-sized fillers.

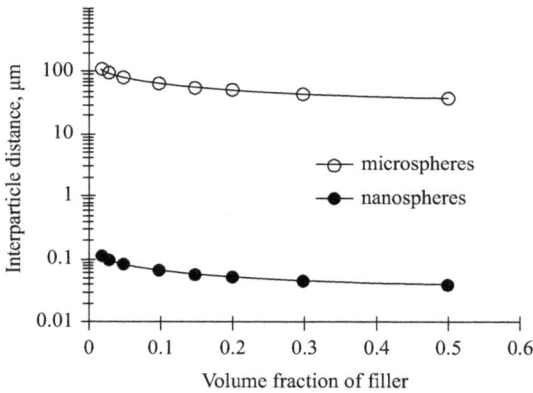

Fig. 2.11: Inter-particle distance for nano- and micron-sized fillers at various volume fractions.

$$\eta_l = 1 - \frac{\tanh[a(l/d)]}{[a(l/d)]} \tag{2.49}$$

and a is given by:

$$a = \sqrt{\frac{-3E_m}{2E_f \ln\left(V_f\right)}} \tag{2.50}$$

The Krenchel factor increases with the volume fraction of the filler. However, the orientation of such tiny particles can be in different directions in one plane or randomly in three dimensions. An orientation factor is therefore introduced into the eq. (2.48) to account for the reduced reinforcing effect of the filler [69]:

$$E_C = \left(\eta_0 \eta_l E_f - E_m\right) V_f + E_m \qquad (2.51)$$

where η_0 is estimated [69] as 3/8 for 2D orientation (all fibres in one plane) and 1/5 for random orientation in 3D.

Considering a nanocomposite of PP of Young's modulus 1.2 GPa, filled with an MWCNT sample of average diameter of 20 nm and length of 500 nm and a Young's modulus of 500 GPa, the Young's modulus of the composite was calculated taking eq. (2.48) along with eq. (2.49) and eq. (2.50) and also taking eq. (2.51) for both 2D and 3D orientations. The variation of the composite modulus with the volume fraction of the multi wall carbon nanotube (MWNCT) is shown in Fig. 2.12.The reinforcing effect is drastically reduced due to the non-alignment of the filler as seen in the figure. The composite with 3D oriented filler showed least reinforcement, as expected. Since the filler modulus is much higher than the polymer, the ideal mixing rule predicts a very high composite modulus due to the assumption of perfect load transfer between the fibre and polymer with prefect alignment and orientation, while the values calculated using the effective factors of length and orientation leads to approximate closeness to real examples [67–69].

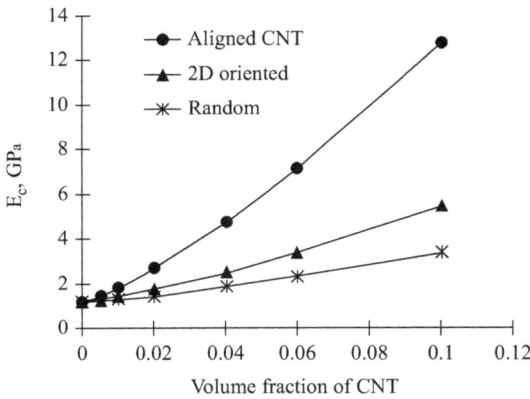

Fig. 2.12: Composite Young's modulus for various alignments.

As an example, Mohan and Kanny [70] developed a composite fibre of PP with Cloisite 20A nanoclay (up to 5 wt%) using extrusion. Since the fibre is aligned in the axial direction of the extruder, it can be assumed that the alignment of most of the fibre is along the length of the fibre. The Young's modulus in tension was measured by the authors and is taken for comparison with the theoretical modulus as expressed by eq. (2.48), which assumes orientation factor as 1 and with the simple mixing rule as eq. (2.52), which assumes both the length factor and orientation factors as 1.

$$E_C = (E_f - E_m)V_f + E_m \qquad (2.52)$$

The composite modulus is in close agreement with eq. (2.52) when the filler modulus is not very different from that of the matrix. In the present case, the Young's modulus of PP is 1.25 GPa and that of Cloisite 20A is about 50 GPa, and densities are 0.95 and 1.77 g/cc, respectively. The aspect ratio (l/d) of Cloisite was taken as 60 (l = 6 μm, d = 100 nm). The experimental and the calculated values using eqs. (2.48) and (2.52) are presented in Fig. 2.13. The theoretical values are in close agreement with the experimental values in this case, which are better for taking the length efficiency factor (Krenchel factor) by eq. (2.48).

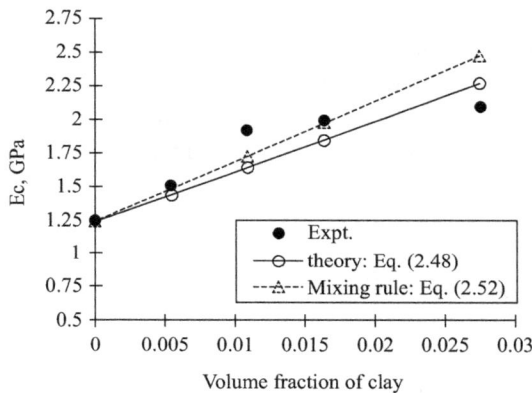

Fig. 2.13: PP-nanoclay Composite Young's modulus for various volume fractions of clay: experimental result and aligned clay particles eq. (2.46), Reprinted with permission from T.P. Mohan and K. Kanny, J. Polym. Eng 35 (2015) 773–784. © 2015, De Gruyter Publishing Company.

Deviations from established models are also reported in literature. In a study by Ivanova and Kotsilkova [71] on two types of polymer nanocomposites using PLA-reinforced with MWCNT and graphene nanoplatelets (GNP), the dynamic shear modulus at 200 °C was determined for each composite and found to follow a logarithmic dependence of the modulus with log volume fraction. The aspect ratio of the GNP was about 240 and that of MWCNT was nearly 1000. The values of dynamic shear modulus at a very low frequency (0.1 rad/s = 0.016 Hz) was only 1.5 Pa for pristine PLA and increased by 5 orders of magnitude from pristine PLA to 12 wt% MWCNT and 4 orders for 12 wt% GNP. Figure 2.14 shows the composite dynamic shear modulus of PLA-MWCNT composites with varying volume fractions. The density of PLA was taken as 1.24 g/cc, that of CNT as 2.1 g/cc and the maximum volume fraction of MWCNT was 0.071, corresponding to 12 wt%. The equation fitting to the composite modulus is given by:

$$\log\left(G'_C - G'_m\right) = 3.1456 \log\left(V_f\right) + 9.156 \qquad (2.53)$$

We can see a good agreement with the experimental data ($R^2 = 0.988$) by the authors [71]. The equation shows that the reinforcement by MWCNT to PLA is non-linear and increases sharply as the volume fraction of the filler crosses 1%.

Similarly, the composite dynamic shear modulus for PLA-GNP is also non-linear with the filler volume fraction and sharply increases beyond 2% filler volume. Figure 2.15 shows the variation of dynamic shear modulus at 0.017 Hz for PLA-GNP nanocomposite with filler volume fraction. The density of the GNP was taken as 2.2 and the maximum volume % was 0.071, corresponding to 12 wt%.

The fitting equation for GNP-reinforced PLA composite is obtained as:

$$\log\left(G'_C\right) = 64.7 V_f + 0.176 \qquad (2.54)$$

with a very good agreement ($R^2 = 0.997$) with the experimental data of the authors [71].

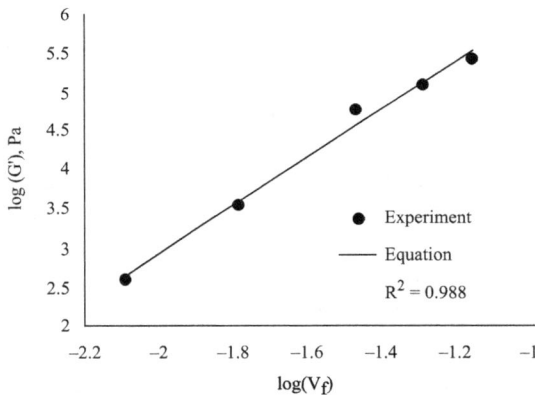

Fig. 2.14: Dynamic shear modulus of PLA-MWCNT composite at varying filler volume fractions, Reprinted with permission from Ivanova, R., Kotsilkova, R. Rheological, Appl. Rheol. 28 (2018) 54014. © 2018, De Gruyter Publishing Company.

However, the modulus enhancement by GNP is one order lower than that by MWCNT, which is expected since CNT is known to be more efficient in reinforcement due to the much larger aspect ratio and correspondingly large specific surface area, which results in additional force of attraction.

2.4.1 Thermoplastic elastomer nanocomposite

Thermoplastic elastomers (TPE) are a special class of thermoplastics that show mechanical properties nearer to a vulcanized elastomer but are mostly linear polymers.

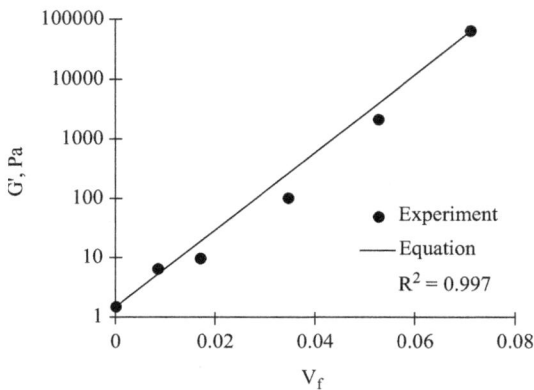

Fig. 2.15: Dynamic shear modulus of PLA-GNP composite at varying filler volume fractions, Reprinted with permission from, Ivanova, R., Kotsilkova, R. Rheological, Application. Appl. Rheol. 28 (2018) 54014. © 2018, De Gruyter Publishing Company.

Many thermoplastic elastomers are made by blending an engineering thermoplastic having high stiffness with a highly flexible linear polymer. Typical example is PC-Copoly(ether-ester) blend[72, 73]. A special class of thermoplastic elastomer is thermoplastic polyurethane (TPU), which is blended with a variety of linear polymers such as ABS, PEO, PBT, PMMA, PLA, PTFE, PP, PC, etc. [74]. A special thermoplastic elastomer blend was developed by Manoj et al. with 50/50 PVC-NBR blend and TPU [75]. Wide use of TPU compared to other similar soft thermoplastics is due to the enormous possibilities of molecular size and structural variation using varying sizes and chemistry of polyols, diisocyanates and NCO:OH ratios. A TPU hardness can range from 20 to 90 Shore "A", which is not possible by any other thermoplastic material.

In the electrical insulation industry, thermoplastic elastomer-based composites are third-generation insulators having good electrical resistance, adequate flexibility and strength. Very common TPEs in this industry are based on PE and PP with EPDM, TPU and styrene-ethylene-butylene-styrene (SEBS) triblock copolymer. An example of thermoplastic elastomer-nanocomposite is PP/EPDM nanocomposite filled with hybrid fillers (1.5 vol% ammonium polyphosphate (APP) and 2 vol% organoclay) for good electrical insulation and dielectric breakdown strength, low dielectric constant and low dielectric loss [76]. A very comprehensive review of thermoplastic elastomer nanocomposites for electrical application is published by Ismail and Mustapha [77]. Other few examples of thermoplastic elastomer nanocomposites are given in reference [78–81].

The advantage of TPEs and, rather, more with TPU is the low energy requirement in manufacture and processing to composite formation because of the low melting pint and low melt viscosity. In TPU, the nanocomposite formation can take place during the urethane formation as such in a mould or in melting TPU granules like other thermoplastics. However, blending of engineering thermoplastics with low molecular weight soft Copoly (ester-ether) [82] and TPU requires high temperature for melt processing of nanocomposites because of the high melting temperature of the engineering thermoplastic.

2.5 Rheological properties

Flow behaviour of the composites of thermoplastic melts is widely researched for the industrial application to predict the operating parameters for the perfect item fabrication by mould filling, extrusion, blow moulding, injection moulding, thermoforming and many such processes. In an industrial scale of production, the rate of production with the desired quality requires very precise rheological analysis. On the other hand, solid thermoplastic composites show non-linear deformation behaviour under an external force, static or time-dependent (oscillatory). This complex elastic–inelastic response is another area of rheological study of thermoplastics. The precise determination of elastic and inelastic energies can be achieved by a dynamic force with varying frequency of the oscillatory force. In addition, the temperature of melt processing could change the rheological behaviour of the same thermoplastic composite. Thermoplastic nanocomposite processing requires detailed analysis of the effects of the filler geometry, size, filler–polymer interaction, filler–filler interaction and the extent of filling. Processing of the nanocomposite of a thermoplastic is best done by a melt blending in a twin-screw extruder. The advantage of such a mixing method is the complex shearing in the extruder and the repeatability of operation in a short time. However, different thermoplastic melts behave differently under similar melt-flow conditions.

2.5.1 Thermoplastic melts

Rheological behaviours of thermoplastic melts can be classified into two main categories, time-independent and time-dependent. In the case of time-independent fluids, the viscosity changes with shear rate, but does not change with time at a fixed shear rate, while for a time-dependent fluid, the viscosity changes with time. In the time-independent category, different types of fluids show reduction of viscosity with shear rate in various patterns. The most common behaviour can be defined by the Power Law or Ostwald-de Waele relationship [83],:

$$\sigma = k \left(\frac{dy}{dt} \right)^n \tag{2.55}$$

and the apparent viscosity is defined by instantaneous viscosity, which changes with the shear rate. For a pseudoplastic, the viscosity gradually decreases with the shear rate and the value of "n" is positive, less than 1, while for a dilatant fluid, the viscosity increases with shear rate and "n" is greater than 1. The apparent viscosity can be expressed as:

$$\eta_{app} = \frac{\sigma}{d\gamma/dt} = k\left(\frac{d\gamma}{dt}\right)^{n-1} \tag{2.56}$$

Figure 2.16 shows a typical apparent viscosity at varying shear rates for a dilatant and pseudoplastic fluid as an example.

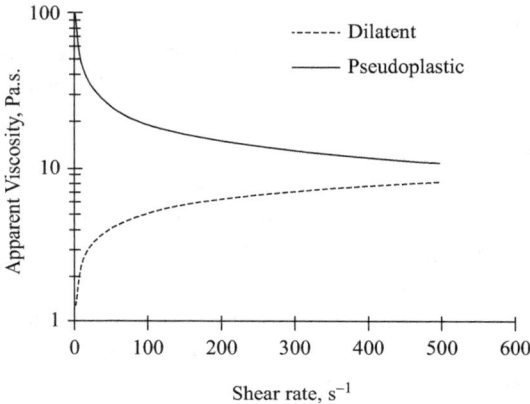

Fig. 2.16: Example of dilatant and pseudoplastic fluids. Numerical values are arbitrary, chosen only for example.

Bingham Model: The fluid has a threshold stress value, below which the flow does not take place, and beyond that, the behaviour can be predicted by a simple Newtonian law.

$$\sigma = \sigma_0 + k\frac{d\gamma}{dt}, \quad \text{for} \quad \sigma > \sigma_0$$

$$\frac{d\gamma}{dt} = 0, \qquad \text{for} \quad \sigma \leq \sigma_0 \tag{2.57}$$

However, many thermoplastic composites show different non-Newtonian characteristics other than the above, and some of the established predictive constitutive equations relating the shear flow parameters (rate, stress and apparent viscosity) are given below:

Herschel-Bulkley Model: It is a modification of the Bingham model where the shear rate has an index n:

$$\sigma = \sigma_0 + k\left(\frac{d\gamma}{dt}\right)^n, \quad \text{for} \quad \sigma > \sigma_0$$

$$\frac{d\gamma}{dt} = 0, \qquad\qquad \text{for} \quad \sigma \leq \sigma_0$$

(2.58)

Casson Model: The Casson model is a modification of the Bingham plastic model and shows non-linearity beyond the threshold stress as follows:

$$\sqrt{\sigma} = \sqrt{\sigma_0} + \sqrt{k\frac{d\gamma}{dt}}, \quad \text{for} \quad \sigma > \sigma_0$$

$$\frac{d\gamma}{dt} = 0, \qquad\qquad \text{for} \quad \sigma \leq \sigma_0$$

(2.59)

Cross Model: It is a general shear-thinning fluid model where the apparent viscosity is constant at low shear rate and is termed as a "zero-shear viscosity" and beyond a certain shear rate, the fluid shows shear-thinning effect like a pseudoplastic, but with a finite constant viscosity at infinite shear rate. The model equation is given as:

$$\frac{\eta_0 - \eta_{app}}{\eta_0 - \eta_\infty} = \frac{1}{1 + k(\dot{\gamma})^m}$$

(2.60)

and

$$\tau = \eta_\infty (d\gamma/dt)^n + \frac{(\eta_0 - \eta_\infty)}{1 + a(d\gamma/dt)^n}(d\gamma/dt)$$

(2.61)

Carreau Model: It is also a generalised fluid model, where the relaxation time of the polymer chain segments is taken to accommodate the viscosity expression:

$$\frac{\eta_{app} - \eta_\infty}{\eta_0 - \eta_\infty} = \left[1 + \tau\left(\frac{d\gamma}{dt}\right)^2\right]^{\frac{n-1}{2}}$$

(2.62)

where τ is the relaxation time of the segmental mobility.

Ellis Model: It is an inverse form of the above models described so far:

$$\eta_{app} = \frac{\eta_0}{1 + (\sigma/\sigma_{1/2})^{a-1}}$$

(2.63)

where η_0 is Zero-Shear viscosity and a is found out from the best fit of data on shear stress vs viscosity. $\sigma_{1/2}$ is taken as the stress at which the apparent viscosity has dropped to half the Newtonian value. For shear-thinning effect, a must be greater than 1. It can be readily seen that at very high value of $\sigma_{1/2}$, the model becomes a Newtonian fluid model and at $(\sigma/\sigma_{1/2}) \gg 1$, the model becomes a Power Law.

The above expressions are for the fluid behaviour where, on a constant shear force, the apparent viscosity does not change with the time of measurement. However, there are fluids, where the viscosity changes with time at a constant shear, for example a *Thixotropic* fluid, which shows a decreasing viscosity with time and a *Rheopectic* fluid, which shows an increasing viscosity with time. Therefore, a thixotropic fluid thins down with time while the rheopectic fluid thickens with time. The first category is very useful for thickness building of a paint coating in a single coating operation and widely used in paint industry, while the rheopectic fluids are best utilized in smart shock and vibration damping application. These fluids have the characteristics of building reversible physical structure under a constant shear force with time, either breaking for thixotropy or forming for rheopecty behaviour. On withdrawal of the shear force, fluid viscosity is restored with time to zero-shear value.

The other important molecular response to resist the flow of a long chain polymer melt is the elastic force. The resistance to movement is dependent on the length of the chain, secondary valence forces among the chains and the extent of entanglement. This resistance has both elastic and shear stress components in the bulk melt. The total pressure difference required for the flow to occur is the sum of the elastic and shear stresses. The effect of the elastic force is to retract the molecules to coil up again, on withdrawal of the force, with a time-dependent relaxation, and results in a *normal force* that acts in the transverse direction of the flow, while the shear force is in the direction of the flow. Since a minimum molecular weight, *Mc*, is necessary for segmental entanglements to influence bulk viscosity appreciably, a higher minimum molecular weight (Mc)e, (probably > *2Mc)* is expected for such interactions to produce an effective elastic network.

The processes of this uncoiling/recoiling lead to multiple phenomena that are usually called *elastic effects*. Extrudate swell, sharkskin, melt fracture, Weissenberg Effect (rod climbing), draw down and frozen-in orientation are some of the important elastic effects. The viscosity, calculated from the extensional strain, is termed as *Elongational Viscosity*, also known as *Extensional Viscosity*; it is a coefficient of viscosity when applied stress is an extensional stress. This parameter is often used for characterizing polymer melts. It can be measured by a rheometer that applies an extensional force. Acoustic Rheometer is an example of such an instrument

$$\eta_e = \frac{\sigma_n}{d\varepsilon/dt} \tag{2.64}$$

where σ_n is the normal stress and ε is the extensional strain.

Elongational viscosity is expressed in a melt flow as:

$$\eta_e = \frac{3(n+1)\Delta P_e}{8(d\varepsilon/dt)} \tag{2.65}$$

where

$$\frac{d\varepsilon}{dt} = \frac{4\eta_s(dy/dt)}{3(n+1)\Delta P_e} \tag{2.66}$$

However, the elongational viscosity can be related to shear viscosity by a relaxation time (dynamic force), as given by Liang and Zhong for single axial extension and steady state flow [84]:

$$\eta_e(\dot{\gamma}) = 2\eta_s(\dot{\gamma}) \bigg/ \left[1 - \frac{2}{\sqrt{3}}\lambda(\dot{\gamma})\dot{\gamma}\right] + \eta_s(\dot{\gamma}) \bigg/ \left[1 + \frac{1}{\sqrt{3}}\lambda(\dot{\gamma})\dot{\gamma}\right] \tag{2.67}$$

where $\dot{\gamma}$ is the shear strain rate and λ is the relaxation time. Since a Newtonian fluid has a very small relaxation time compared to the shear strain rate in practical extruder operations, the above equation can be approximately:

$$\eta_e = 3\eta_s \text{ (Trouton ratio)} \tag{2.68}$$

Zhao and Zhong [85] derived a modified expression of eq. (2.67) taking into consideration three material parameters k_1, k_2 and k_3:

$$\eta_e(\dot{\varepsilon}) = \left[2\sqrt{\frac{k_2}{k_1\dot{\varepsilon}^2}\eta_0 + \frac{1}{4\dot{\varepsilon}^2}\left(\frac{k_2}{k_1}\right)^2} - \frac{k_2}{2k_1\dot{\varepsilon}^2}\right] \bigg/ \left[1 - \frac{2}{\sqrt{3}}\lambda_1\dot{\varepsilon}/(1 + k_3\dot{\varepsilon})\right]$$
$$+ \left[\sqrt{\frac{k_2}{k_1\dot{\varepsilon}^2}\eta_0 + \frac{1}{4\dot{\varepsilon}^2}\left(\frac{k_2}{k_1}\right)^2} - \frac{k_2}{2k_1\dot{\varepsilon}^2}\right] \bigg/ \left[1 - \frac{1}{\sqrt{3}}\lambda_1\dot{\varepsilon}/(1 + k_3\dot{\varepsilon})\right] \tag{2.69}$$

where η_0 is the zero-shear viscosity, k_2/k_1 has a unit of viscosity, k_3 is a dimensionless coefficient and λ_1 is the relaxation time.

As the relaxation of chain segments are time-dependent, the elastic behaviour increases with the increase in the residence time of the melt at a molten state. For example, Schreiber et al. [86] has shown that the die swell (ratio of the outer diameter of the extrudate polymer to the die's internal diameter) increases form 1.22 to 1.41 for a residence time of 0.5 h and 2 h at an extrusion temperature of 190 °C for polyethylene.

2.5.2 Rheology of thermoplastic composites

Thermoplastic composites are more complicated in rheological characteristics, which demand additional analytical considerations apart from the above classical rheology of non-Newtonian equations of flow. Adding fillers to a neat polymer melt changes its rheology, influencing both the way the melt processes and the properties of the ultimate product. Key factors are filler size and shape, filler concentration, and the extent of any interactions among the particles. The consequences of adding fillers are an increase in melt viscosity and a decrease in die swell. Moreover, particle interactions increase the non-Newtonian range and cause it to occur at a lower shear rate than for the unfilled polymer melt. Filled polymers have a higher viscosity at low shear rates, and yielding may occur with increased filler concentration. In addition, the temperature of melt processing is very important for analysis and control of the rheological behaviour in respect of operating parameters such as threshold pressure requirement, L/D, of the extruder barrel, die swell, elongational stress, etc.

2.5.3 Mathematical models

There are many relationship expressions involving the Normalized Rheological Function. Γ (= suspension viscosity/solvent viscosity) and the filler volume fraction, ξ_g. The equations assume that rigid fillers are suspended in the bulk amorphous phase. Some of the early developments are listed by Lamberti et al. [87].

The earliest theory of viscosity of suspensions was given by Einstein [88] as:

$$\Gamma = 1 + \frac{5}{2}\xi_g \tag{2.70}$$

The above linear equation assumes rigid, non-interacting spherical fillers in low concentration (below 2.5%) in a Newtonian Fluid. Batchelor [89] modified the above using a second-order polynomial expression, taking into account the interactive nature of the spherical, rigid filler, also at low concentration:

$$\Gamma = 1 + \frac{5}{2}\xi_g + 6.2\xi_g^{\,2} \tag{2.71}$$

Mooney [90] gave an expression for the viscosity of a suspension with higher concentration of rigid, spherical fillers. The author modelled the rise in viscosity exponentially with filler volume fraction:

$$\Gamma = \exp\left[\left(\frac{5}{2}\right)\frac{\xi_g}{1 - (\xi_g/\xi_{g\,max})}\right] \tag{2.72}$$

where ξ_{gmax} is the maximum filler volume fraction occupied by the spheres in the suspension.

Ball and Richmond [91] proposed an expression of concentrated rigid spherical inclusion as:

$$\Gamma = \left(1 - \frac{\xi_g}{\xi_{g\,max}}\right)^{-\frac{5}{2}\xi_{g\,max}} \tag{2.73}$$

An almost similar expression was proposed by Krieger & Dougherty [92] for concentrated suspension of a rigid particle, with the intrinsic viscosity [η] of the unfilled fluid as the negative index, which signifies the effect of hydrodynamic volume of the unfilled melt/solution:

$$\Gamma = \left(1 - \frac{\xi_g}{\xi_{g\,max}}\right)^{-\xi_{g\,max}[\eta]} \tag{2.74}$$

Kitano et al. [93] gave a similar expression involving a shape factor of the fillers to accommodate non-spherical shapes in a polymer melt:

$$\Gamma = \left(1 - \frac{\xi_g}{A}\right)^{-2} \tag{2.75}$$

A = 0.68 for smooth spheres.

Graham [94] proposed the following equation for viscosity of concentrated suspension of spherical rigid fillers:

$$\Gamma = 1 + \frac{5}{2}\xi_g + \frac{9}{4}\left[\frac{1}{\psi(1 + 0.5\psi)(1 + \psi)^2}\right] \tag{2.76}$$

where ψ is given by:

$$\psi = 2\left[\left(1 - \sqrt[3]{\xi_g/\xi_{g\,max}}\right)\right] / \left(\sqrt[3]{\xi_g/\xi_{g\,max}}\right) \tag{2.77}$$

Frankel and Activos [95] suggested the following viscosity expression for concentrated suspension of rigid spheres:

$$\Gamma = \left(\frac{9}{8}\right)\frac{\left(\xi_g/\xi_{g\,max}\right)^{1/3}}{1 - \left(\xi_g/\xi_{g\,max}\right)^{1/3}} \tag{2.78}$$

Tanner [96] proposed a second-order polynomial expression, similar to eq. (2.71) by Batchelor, with the coefficients as fitting parameters for any filled polymer solution or melt:

$$\Gamma = 1 + a_1\xi_g + a_2\xi_g^2 \tag{2.79}$$

2.5.4 Example

Let us consider the rheological properties of a nanocomposite of polylactic acid and graphene nanoplates studied by Evanova and Kotsilkova [71] for steady-state viscosity under varying shear rates. A TEM image of the PLA-6%GNP is shown in Fig. 2.17. The GNP filler is thin platelets and some clustering is seen. The image is a cross-section of the filament of the composite. The GNP used by the authors had average size of 5–7 μm, thickness 30 nm and aspect ratio of approximately 240. A series of compositions with 0, 1.5, 3, 6, 9 and 12% GNP was made and melt viscosities were measured at shear rates from 0.05 to 500/s at 200 °C. The data on shear rate vs. apparent viscosity was taken for each GNP content, at a constant 0.05/s shear rate. The above equations were examined to observe the closeness to the experimental data. For calculation, ξ_g is taken from data in Fig. 2.15 and ξ_{gmax} was taken as the highest content (12 wt% = 0.071 Volume fraction). Four models by Einstein, Mooney, Graham and Frankel & Activos are used for the comparison. Figure 2.18 shows the experimental [71] and the calculated data using the four models.

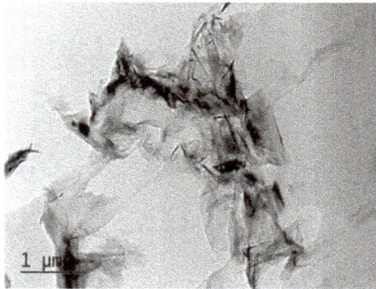

Fig. 2.17: TEM image of PLA-GNP nanocomposite with 6 wt% GNP, Reprinted with permission from, Ivanova, R., Kotsilkova, R. Rheological, Application. Appl. Rheol. 28 (2018) 54014. © 2018, De Gruyter Publishing Company.

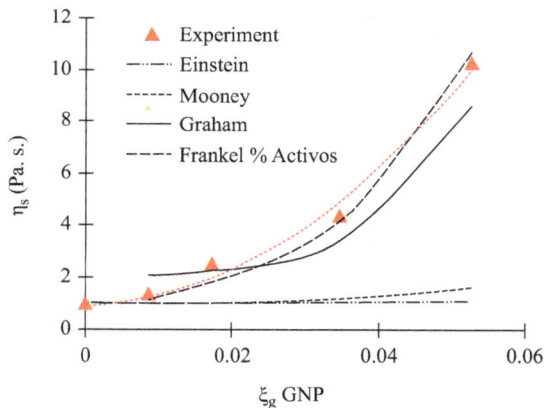

Fig. 2.18: Apparent viscosity of GNP-filled PLA nanocomposite melts as a function of GNP weight fraction at 200 °C and at a constant shear rate of 0.05 s-1. Red dotted line is a second order polynomial fit as per Tanner model, Reprinted with permission from, Ivanova, R.,Kotsilkova, R. Rheological, Application. Appl. Rheol. 28 (2018) 54014. © 2018, De Gruyter Publishing Company.

From the Fig. 2.18, it is seen that the models by Graham (eq. (2.76) with eq. (2.77)) and Frankel & Activos (eq. 2.77) are somewhat close to the experimental data. In the plot, the data for 12 wt% GNP composite is avoided since it is taken as the maximum filler content (ξ_{gmax}). A polynomial fit as Tanner model was used to obtain the coefficients a_1 and a_2 (eq. 2.79). The fitting equation is given below:

$$\Gamma = 1 - 1.455\xi_g + 1132\xi_g^2 \quad (R^2 = 0.988) \tag{2.80}$$

The coefficient of the first-order is quite small compared to the second, and the negative value also indicates a slow increase on the addition of graphene nanoplates at lower concentration levels. There is a rapid rise in the composite viscosity beyond 5% and for the model calculations too. It was seen from the study that the Newtonian range for the pristine PLA melt is extended till 10/s shear rate, which can be termed as the *Zero-Shear Viscosity*. This is also true for lower GNP content (below 6 wt%). This is reflected in the Fig. 2.18 since the reinforcing effect of GNP is significant only beyond 6 wt% in the nanocomposite.

Nanoclay is another non-carbon reinforcing filler, which is used to make polymer-clay nanocomposites. The melt processing of the thermoplastic clay nanocomposite is very common, as it is efficient in distributing and dispersing nanoclay in the polymer and shows better intercalation/exfoliation. The rheological properties of polymer–clay nanocomposite is therefore of great interest. Chafidz and co-workers [97] studied the melt rheology of PP–clay nanocomposite. PP with 0, 5, 10, and 15 wt% organically modified montmorillonite clay (OMMT) was melt-compounded and to study the effect of reprocessing on the properties of the nanocomposites, the melt compounding process was carried for two cycles. The PP-OMMT master batch containing 50% OMMT was a commercial product and the four experimental composites were made using necessary additional PP, along with maleic anhydride (MA) as the compatibilizer. The authors measured the dynamic complex viscosity (η^*) using a parallel disc rheometer (25 mm dia, 2 mm gap) at a constant temperature (260 °C) and oscillation stress (500 Pa), with an angular frequency range of 0.1–628.3 rad/s. The trend of complex viscosity with increasing angular frequency (a measure of shear rate) showed that (1) The zero-shear viscosity range (Newtonian range) reduces with increasing OMMT loading and is almost nil for 15% OMMT composite, (2) The second cycle melt processing marginally improved the viscosity compared to first cycle, and the trend of improvement is similar. Figure 2.19 shows the experimental plots as reported by the authors.

The experimental result also revealed that the sharp rise in viscosity, beyond 10% OMMT, is only observed at lower frequency (lower shear rate). The effect of OMMT volume fraction on the logarithm of complex viscosity at 0.1, 10 and 100 rad/s angular frequency is plotted in Fig. 2.20. The viscosity is normalized by the viscosity of the pristine PP melt and is used as Γ^*, and the volume fraction (ξ_g) is calculated taking the density of OMMT as 2.5 g/cc and for PP as 0.95 g/cc. The OMMT volume fractions are 0.019, 0.037 and 0.054 for 5, 10 and 15 wt%, respectively.

The change in log (complex viscosity) at 100 rad/s is linear with filler volume fraction, but follows a second-order polynomial against filler volume fraction for lower angular frequencies of 0.10 and 10 rad/s. The fitting equations are listed in Tab. 2.2 below.

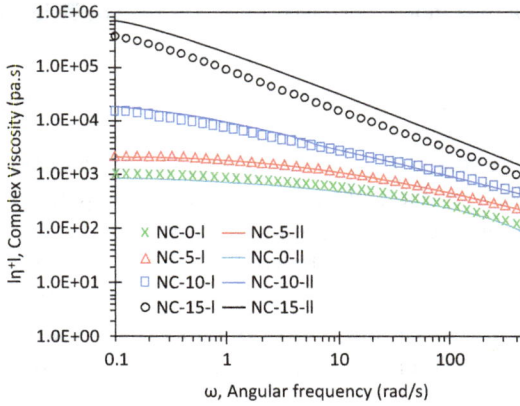

Fig. 2.19: Complex viscosity at varying angular frequencies for PP-OMMT composites for two cycles of processing. (Symbols: Cycle I, lines: Cycle II), Reprinted with permission from, Chafidz, A., Ma'mun, S., Fardhyanti, D. S., Kustiningsih, I., Hidayat, IOP Conf. Ser. Mater. Sci. Eng. 543 (2019) 012036. © 2019, IOP Publishing Company.

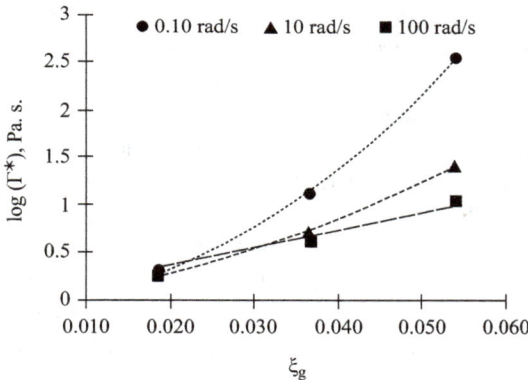

Fig. 2.20: Complex viscosity of the PP-OMMT clay nanocomposite at various filler contents at different angular frequencies. Reprinted with permission from, Chafidz, A., Ma'mun, S., Fardhyanti, D. S., Kustiningsih, I., Hidayat, IOP Conf. Ser. Mater. Sci. Eng. 543 (2019) 012036. © 2019, IOP Publishing Company.

The logarithm of complex viscosity indicates that the effect of the filler is quite significant in enhancing the melt viscosity and hence also the dynamic mechanical properties. Further, the polynomial fit indicates much more enhancement with increasing filler loading. The data on pristine PP is eliminated for fitting the equations since Γ^* is 1, and hence $\log(\Gamma^*) = 0$ in that case. The fitting empirical equations are modifications of the Tanner model, as $\log(\Gamma^*)$ is used here instead of Γ^*. The dispersion and distribution of OMMT is observed to be uniform but the authors could not

Tab. 2.2: Dependence of complex viscosity on OMMT volume fraction in PP-OMMT composite melts.

ω, rad/s	Fitting equation	R^2
0.10	$\log(\Gamma^*) = 900\xi_g^2 - 1.57\xi_g$	1
10	$\log(\Gamma^*) = 356\xi_g^2 + 6.95\xi_g$	0.996
100	$\log(\Gamma^*) = 18.24\xi_g$	0.991

conclude about the extent of exfoliation or intercalation of the clay layers. Figure 2.21 shows the SEM morphology of 5% OMMT-filled composite after the first and second cycles. The platelets of clay are seen as bright thin plates in the micrograph.

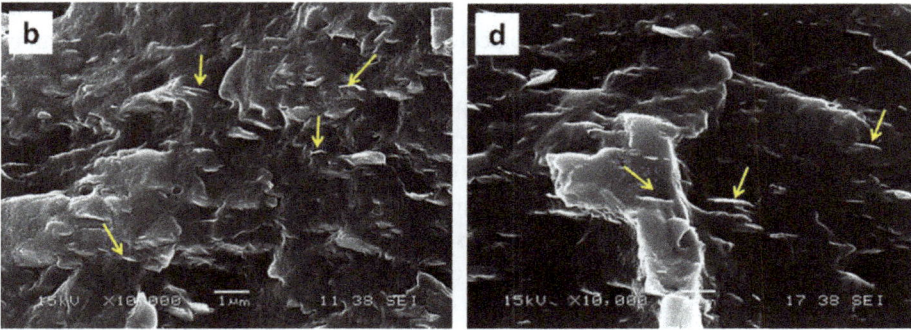

Fig. 2.21: SEM morphology of the fractured surface of 5% OMMT-filled PP composite. Reprinted with permission from, Chafidz, A., Ma'mun, S., Fardhyanti, D. S., Kustiningsih, I., Hidayat, IOP Conf. Ser. Mater. Sci. Eng. 543 (2019) 012036. © 2019, IOP Publishing Company.

In both the above examples, it is observed from the reinforcement effect of two completely different nano fillers that the deviation from classical models is an indication of the nano effect, small size, increased surface area of contact with the matrix per unit mass and very high aspect ratio, all of which results in higher forces of attraction between a filler particle and the polymer matrix. There was a quantum jump of apparent viscosity at low shear rates (zero-shear viscosity) for about 10 wt% fillers in the thermoplastic composite melts. However, enhancement of melt viscosity may not be observed in all types of thermoplastic nanocomposite processing. For instance, Cho and Paul [98] studied melt mixing and rheology of Nylon 6 (polycaprolactam, also known as PA6) filled with Na+ MMT (Cloisite Na), OMMT and short glass fibre. At 5% loading of all fillers, it was seen that the apparent viscosity of PA6-OMMT nanocomposite is much less than the pristine PA6 throughout the experimental shear range. The OMMT was well dispersed and partially exfoliated as reported by the authors. PA6 is a polyamide, partially crystalline, highly polar, with hydrogen bonding and all the amide groups are in same direction in the chain. As the polymer segments entered the clay gallery, both the polar attraction and the hydrogen bonding are partly disrupted in the molten state, thus reducing the inter-chain interaction and the melt viscosity, although the mechanical properties of the solid composite PA6-OMMT was better than that of the pristine polymer.

2.6 Processing

Processing of thermoplastic is a well-established industrial practice. Processing of thermoplastic-based composites is no exception. Addition of reinforcing or inert fillers could change the rheological properties due to the variation in thermal diffusivity, elastic forces (elongational viscosity) and shear forces (shear viscosity), compared to the neat polymer. Discussions in the previous para brought out some numerical examples of wide variation in the rheological parameters of some thermoplastic composites. However, the processing principles are almost unaltered for the composites, except for a few changes in the case of special fillers such as nanoparticles. Due to their very tiny size, the nanoparticle dispersion in a thermoplastic melt is much more difficult than micron-sized fillers, irrespective of the shape/aspect ratio. The first consideration in a proper unit operation is homogenizing the solid particles in the molten state of the polymer. Therefore, melt mixing is done to incorporate the filler uniformly under severe deformation of the pellets. Subsequent melting under shear force and heat allow the dispersion of the solid filler in the melt, finally forming homogeneous composite pellets, on cooling. This process can be effectively carried out in a single- or twin-screw extruder with proper barrel length, temperature zones, motor power and attachments for pellet production at the die outlet. The composite pellets thus produced can be processed to make moulded items by several methods such as injection moulding, blow moulding, coaxial extrusion, filament making, compression moulding, thermoforming, etc. according to the end product. Some important processes for thermoplastic composite are described here.

2.6.1 Extrusion

The basic function of the extruder is to convey the solid polymer and the filler from the feed section to the forward sections towards the exit die. A sketch of the barrel-and-screw geometry is shown below as Fig.2.22. The polymer pellets are heated up in the first zone due to the high frictional force in the annular space of the barrel and then in the rotating screw. Then, as the mass moves to the higher temperature zone, the polymer melts and dispersion of the filler takes place in the molten mass. The advantage of a twin-screw is to enhance the mixing process by additional shearing between the two adjacent screws. The homogenizing may require more than one cycle of extrusion. The use of twin-screw reduces the number of repeat cycles.

The study by Chafidz et al. [97] showed that two cycles of extrusion mixing of PA6-Clay and PP-Glass fibre in a twin-screw extruder was marginally beneficial, and only one cycle can be adopted with fair homogeneity. For composite processing, the residence time of the mix has to be such that the mixing is uniform. This is achieved by a lower rotational speed of the screw and by taking a higher L/D ratio of the extruder screw compared to those in the case of pure polymer processing. The complete melt mixing of the polymer-filler is ensured by the longer residence time and by setting the end zone temperature somewhat above the melting point of the polymer for reducing the viscosity and the normal force. For example, the melting point of polypropylene (atactic) is about 160–170 °C, while the extruder processing of neat PP is at 190–210 °C, but for PP-based composites with glass, mineral or talc filler, the processing temperature is higher, about 230 to 260 °C. One of the measuring parameters is the volumetric flow rate of extrusion, which is the resultant flow due to drag, and opposed by the viscous force (pressure).

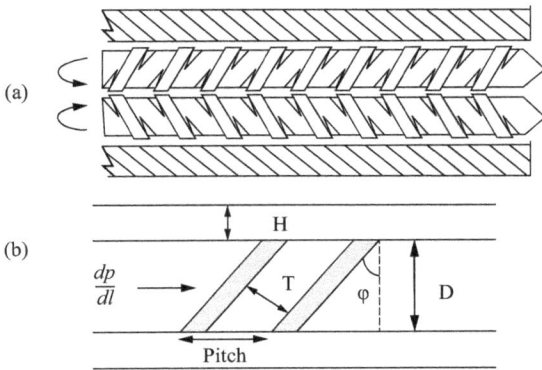

Fig. 2.22: (a) Twin-screw Extruder, with contra-rotating screws and (b) barrel-and-screw geometry in a single-screw.

The expressions for drag flow, pressure flow and the resultant volumetric flow rate are:

$$\text{Drag Flow: } Q_d = \frac{1}{2}\pi^2 D^2 NH \sin\varphi \cos\varphi \tag{2.81}$$

$$\text{Pressure Flow: } Q_p = \frac{1}{\eta}\left(\frac{dP}{dl}\right)\left[\frac{\pi DH^3 \sin^2\varphi}{12}\right] \tag{2.82}$$

$$\text{Total Flow: } Q = Q_d - Q_p = \frac{1}{2}\pi^2 D^2 NH \sin\varphi \cos\varphi - \frac{1}{\eta}\left(\frac{dP}{dl}\right)\left[\frac{\pi DH^3 \sin^2\varphi}{12}\right] \tag{2.83}$$

where D is the diameter of the screw, H is the annular space between the barrel and the screw, φ is the angle of the screw helix, N is the rotation of the screw (cycles per second) and dP/dl is the pressure per unit length in the flow direction (refer Fig.2.22). The expression does not consider wall-slip.

The term η is assumed to be Newtonian viscosity in this expression, which means that it is independent on the rotaional speed of the screw. However, in the case of non-Newtonian fluids such as polymer melts, it can be apparent viscosity, and can be calculated at equivalent shear rates from a previous viscometric experiment. For example, the relation between N and shear rate in an extruder is given by:

$$\frac{d\gamma}{dt} = \frac{\pi DN}{H} \tag{2.84}$$

The factor ½ in the drag-flow expression should be replaced by f, which can be experimentally determined for individual composite melt processing.

When the pressure is very low, only the drag flow is significant, and the flow rate is maximum through the die. This condition is stated as the characteristic, Q_{max}, of the extruder:

$$Q = Q_{max} = \frac{1}{2}\pi^2 D^2 NH \sin\varphi \cos\varphi \tag{2.85}$$

When the pressure is very high, the flow stops, and the condition gives the maximum pressure, P_{max}, extruder characteristics:

$$Q = 0, \quad \frac{1}{2}\pi^2 D^2 NH \sin\varphi \cos\varphi = \frac{1}{\eta}\left(\frac{P}{l}\right)\left[\frac{\pi DH^3 \sin^2\varphi}{12}\right] \tag{2.86}$$

$$P_{max} = \frac{6\pi DNl\eta}{H^2 \tan\varphi}$$

From the knowledge of conduit flow of a fluid, it can be said that the output flow rate and the corresponding pressure are definitely linearly related. Hence,

$$Q = kP \tag{2.87}$$

From the equations of momentum balance, we get

$$k = \frac{\pi R^4}{8\eta L} \tag{2.88}$$

The plot of output flow rate against the corresponding pressure is the characteristic plot for the extruder, and is shown as Fig. 2.23.

The melting capacity of a screw increases at a slower rate compared to the screw speed, mainly because the residence time of the solid bed inside the screw decreases as the screw speed increases. The melting capacity eventually becomes the limiting factor

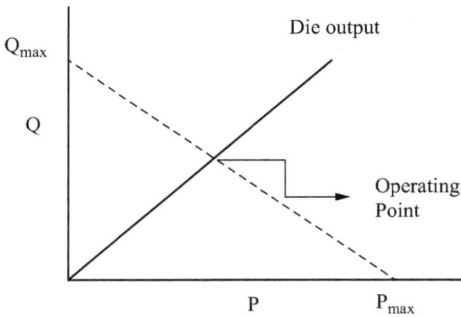

Fig. 2.23: Extruder characteristics: Output volumetric flow rate vs. pressure.

in the output rate as the screw speed increases. Some pellets in the solid bed eventually become incompletely molten at the end of the screw at high screw speeds, resulting in poor melt quality. Most high-performance screws are designed to eliminate incompletely molten pellets and to increase the melting capacity. Due to the high viscosity of thermoplastics composite melts, a large amount of heat is generated within the melt by dissipating the mechanical energy of the drive motor. Since viscosity is the result of friction between molecules, the mechanical energy converts into thermal energy. The melt transfers heat into the solid bed, melting the solid bed at the melt film–solid bed interface. This melting mechanism is termed as *dissipative melting*. Preferred conditions for a high dissipative melting capacity of a screw at a given screw speed are:

- High barrel temperature, well above the melting point of the polymer
- Large solid bed–barrel contact area
- Small channel width
- Tight flight clearance
- High feed temperature
- High polymer viscosity for large amount of heat generation
- Low melting point
- Low heat capacity for melting

When the solid reduces in the bed, the bed as such becomes unstable, known as solid bed breakup, and the solids get mixed up with the melt over an approximate L/D = 20–25. The solids subsequently melts due to the conduction of heat from the surrounding hot molten mass. This phenomenon is termed as *conduction melting*. Preferred conditions for fast conduction melting are:

- Small pellet size with a large surface area
- Good mixing of the melt and the solid pieces
- High feed temperature
- Low melting point
- Low heat capacity for melting

Generally, the viscosity in processing is reduced by employing a temperature higher than the melting point. Since polymers are generally bad conductors of heat, the power requirement for raising the temperature is substantial, even considering the initial frictional heat generated in the feed zone. However, when a carbonaceous filler like carbon black, CNT, graphite, graphene, etc. is used for making the composite, the thermal diffusivity in the mix drastically increases, which facilitates uniformity in temperature and also aids in reducing the thermal requirement and melt viscosity. Minerals, glass fibre, etc., on the contrary, take more heat as these are bad conductors, and the extruder temperature has to be increased to more than that for carbon-based composites.

Nevertheless, the estimation of viscosity at any temperature can be done by the Arrhenius or William-Landel-Ferry (WLF) equation as shown below:

$$\text{Arrhenius: } \eta = \eta_0 \exp\left(\frac{E}{RT}\right) \tag{2.89}$$

$$\text{WLF: } \frac{\eta(T)}{\eta(T_r)} = \log a_T = \frac{-C_1(T - T_r)}{C_2 + (T - T_r)} \tag{2.90}$$

The values of C_1 and C_2 are 8.86 and 101.6, respectively.

From eq. (2.85), the Activation energy can be calculated from the slope of a straight line plot of $\ln(\eta)$ against $1/T$. These two equations are useful to predict the viscosity at any temperature, once the shift factor (ratio of viscosity at a temperature and that at the reference temperature) is known.

2.6.2 Blow moulding

Blow moulding is a manufacturing process by which hollow plastic parts are formed. It is a well-known method of making glass vessels. In general, there are three main types of blow moulding:
(1) Extrusion Blow Moulding,
(2) Injection Blow Moulding, and
(3) Injection Stretch Blow Moulding.

The blow moulding process begins with the melting down of the plastic and forming it into a *Parison* or in the case of injection and injection stretch blow moulding (ISB), a *Preform*. The parison is a tube-like piece of plastic with a hole at one end through which compressed air can pass.

The parison is then clamped into a mould and air is blown into it. The air pressure then pushes the plastic out to match the mould. Once the plastic has cooled and hardened, the mould opens up and the part is ejected. The cost of blow-moulded parts is higher than that of injection–moulded parts but lower than that of rotational-moulded parts.

Conventional process of extrusion blow moulding can be batch, continuous or in-termittent process. Continuous process is fast and is used to determine the thickness of the moulded object. It will swell when it comes out of the mould. When the molten object emerges from the die, it swells due to the recovery of elastic deformation. Im-portant dimensions in blow moulding are shown as a sketch in Fig. 2.24.

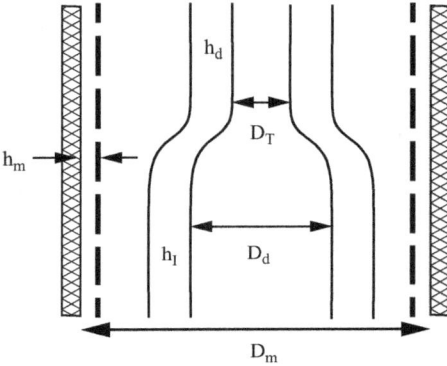

Fig. 2.24: Sketch of a Blow moulding system.

It can be shown that the following relationship applies:

$$B_{SH} = B_{ST}^2 \qquad (2.91)$$

where
B_{SH} – Swelling of the thickness = h_i/h_d
B_{ST} – Swelling of the diameter = D_i/D_d
h_i = Swelling occurs in Parison
h_d = Swelling occurs in Die
D_i = Diameter of Parison
D_d = Diameter of Die
and

$$h_i = h_d \left(\frac{D_i}{D_d}\right)^2 \qquad (2.92)$$

Now consider the situation that the parison is inflated to fill a cylinder die of diameter D_m as in Fig. 2.24. Assuming constant volume and neglecting draw-down effect,

$$\pi D_i h_i = \pi D_m h_m, \quad \therefore h_m = \frac{D_i}{D_m} h_i$$

From eqs. (2.91) and (2.92),

$$h_m = \frac{D_i}{D_m} \left(h_d B_{ST}^2 \right)$$

$$\because D_i = D_d B_{ST}, \quad \therefore h_m = B_{ST}^3 h_d \frac{D_d}{D_m} \tag{2.93}$$

The most common method of extrusion blow moulding is based on the principle of downward extrusion of a tube of molten polymer, known as parison. Two mould halves close around the tube, which is then blown to the shape of the mould by compressed air. In the case of bottles it is possible to blow upside down and inflating it by introducing compressed air through a spigot that is inserted into the neck of the bottle. In order to achieve a high rate of extrusion, the moulding time must be at a minimum. It is important to control the thickness to avoid wastage. In order to obtain a strong item, the parison should not have die lines.

The rheological parameters are temperature–viscosity relationship, shear-induced crystallisation, die swell, draw-down characteristics, flow behaviour in the extruder die, melt fracture and/or shark skin. Study of elongational viscosity, molecular weight distribution, and effect of filler are also important to control the die swell and the subsequent thickness of the blown item.

Moulding time is minimized by lowering the operational temperature, but viscosity is high and thickness may not be optimum at low temperatures. Moreover, elastic effects are high at a low temperature and could cause melt fracture or shark skin. Crystalline polymers when used at a temperatures of a few degrees higher than the melting point may show shear-induced crystallization, thus making the operation difficult.

Die swell is very important in blow moulding since the blown plastic items are generally thin. The slight swelling may change the thickness to unacceptable quality, and in large production volumes, such as mineral water bottles, the economy is greatly affected. Die swell increases with shear rate (up to the critical shear rate), and with MW and polydispersity index (MWD). However, the swell decreases with temperature and with the length of the die. In the downward extrusion of the parison, an uneven thickness is caused by the opposing effects of die swell and draw down.

2.6.3 Fibre spinning

Use of thermoplastic nanocomposite fibres is a new area; most probably, the earliest repot was in 1993 by Kojima et al. [99, 100] on melt-spinning of Nylon 6-clay nanocomposite. Fibres with colouring pigments that are conventionally used are not reinforcing, and do not qualify for a high-performance fibre in critical applications such as in Defence, aerospace and medical fields. Exception is carbon black.

Fibre spinning of thermoplastic polymers is conventionally done using any of the three methods, viz., Melt spinning, Wet spinning and Dry spinning. Of these methods, melt spinning is the preferred method so far for thermoplastic nanocomposites.

Melt-spinning is simply the extrusion of the molten polymer composite through a spinneret plate with large number of holes for multiple fibre production. In the recent past, melt spinning of layered double hydroxide (LDH)/high density polyethylene (HDPE) nanocomposites was reported for the first time by Kutlu et al. [101]. Ivanova and Kotsilkova [71] developed filaments of 1.75 mm diameter with two types of PLA nanocomposites with MWCNT and Graphene nanoplates for fused deposition modelling using 3D printing. Redondo and co-workers [102] developed cellulose nanocrystal-filled polyurethane melt-spun fibres. Sandler and others [103] studied melt-spinning of Nylon 12-based nanocomposites with two nanocarbons, MWCNT and nanocarbon fibre.

A schematic presentation of the dry spinning process is shown in Fig. 2.25. In the dry spinning method, the polymer composite is in a solution form and the solvent evaporates after the filament leaves the spinneret hole. The filament is thus dried by hot air. Dry spinning of nanocomposite fibres is rare and no significant work has been done so far for dry spinning of nanocomposite fibres.

Wet spinning is a method of extrusion of the solution made by dissolving the thermoset in a solvent and extruded through the spinneret hole. The drawn fibre is passed through a solvent–non-solvent mixture to extract the solvent, thus forming the precipitate in the fibre form. Figure 2.26 shows a sketch of the wet-spinning process. In this process, the drawing speed is much less than melt spinning due to the lower strength of the polymer solution. During the spinning process, the filaments are first extruded through an air gap where the filaments undergo strains of 2 to 3 times, which produces a high degree of molecular orientation in the filaments. This air gap is of the order of one inch. It also allows the spinneret plate to be warm (100 °C) while the extraction bath can be cool (ca 15 °C). The hot filaments strike the cooling bath where the filaments are "quenched" and much of the orientation is locked in by the rapid cooling action.

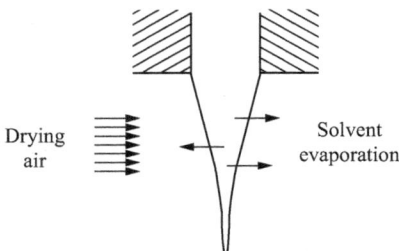

Drying air · Solvent evaporation

Fig. 2.25: Dry spinning.

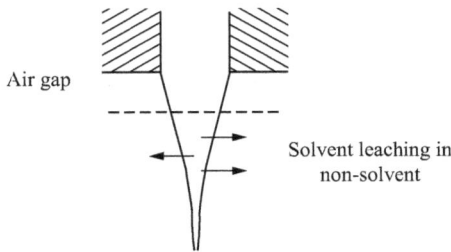

Fig. 2.26: Wet spinning.

Lu et al. [104] used wet spinning of conductive composite fibre made from surface-modified silver nanowire and thermoplastic polyurethane. The authors claimed high electrical conductivity of the fibres containing a maximum of 75% of nano silver wires by weight in the composite fibre and stretched to 200%. There is no significant progress in this area of nanocomposite fibre spinning technology so far.

2.6.4 Additive manufacturing

The conventional composite manufacturing processes, as discussed above, inherit a lot of defects that limit their use for many applications. Such defects primarily include delamination, interlaminar voids, improper orientation, variation in dimensions, etc. Mechanical properties of the parts produced by such processes also show a considerable deviation, which makes it very difficult to standardize the properties. Additive manufacturing of such components can be utilized to overcome the mentioned defects. Additive manufacturing (AM), commonly known as 3D printing, is defined as the process of layer-by-layer creation of three-dimensional physical model, directly from 3D-CAD data. AM technique is also known as rapid prototyping, automated fabrication, solid free-form fabrication, etc. 3D printing was first described by Charles W. Hull in the 1984 patented stereolithography additive manufacturing process. Earlier, AM was considered only for prototyping purpose.

The first step of 3D printing is to create a 3D model of the object to be printed. This model can be designed with a computer-aided design (CAD) software. The CAD file is then converted to a standard additive manufacturing file format i.e. standard tessellation language (STL) file. The file is then digitally sliced into layers. In the fourth step, the machine that is only controlled by a computer builds the model layer-by-layer, in other words, this step is basically providing printing parameters to slice the model and providing supports for overhang parts. The fifth and sixth steps are part building and post-processing (removal of supports, surface finishing), respectively. The layer thickness dictates the final quality and this depends on the machine and the process [105].

According to ISO/ASTM D52900-15 [106], additive manufacturing is categorized as material extrusion, vat photo polymerization, powder bed fusion, material jetting, binder jetting, sheet lamination and direct energy deposition. The use of the processes depends upon the material to be manufactured. It offers several technological advantages over conventional manufacturing processes. For example, 3D printing does not require any tooling, which is essential for conventional manufacturing techniques such as injection moulding. The requirement of tooling in widely used manufacturing techniques mentioned above can be a barrier to production due to the high cost. In the case of 3D printing, various parts of a component can be sent digitally and printed in homes or locations near to consumers. As a result, the necessity of transport and the related expenditure can be reduced. Thus, 3D printing is a more environmental friendly process compared to conventional manufacturing processes. Due to the process flexibility, additive manufacturing processes offer more design and material freedom than conventional composite manufacturing processes. Taking such advantage of the process accuracy, good dimensional tolerances and surface finish can be achieved. With only a smaller deviation in mechanical properties as compared to traditionally manufactured composites, additively manufactured composites can also be vastly used in many fields such as in aerospace, automotive, energy production, etc.

Additive manufacturing of composites can be achieved through several techniques such as stereolithography (SLA), selective laser sintering (SLS), direct energy deposition (DED), fused deposition modelling (FDM), etc. [107]. Of all these methods, FDM is the most investigated one and utilized for composite manufacturing. S. Scott Crump invented and patented the FDM process in 1989. FDM rapid prototyping, in general, consists of an extrusion system, which has the ability to precisely control where the extrudate is laid in three-dimensional space in order to build a prototype with complex geometry. Basically, layers are built by extruding a small bead of material in a particular lay-down pattern such that the layers are covered with the beads. After a layer is completed, the height of the extrusion head is increased and the subsequent layers are built to construct the part. The quality of the printed parts can be controlled by altering the printing parameters such as layer thickness, printing orientation, raster width, and raster angle and air gap. FDM printers also offer advantages, including low cost, high speed and simplicity. It allows the deposition of diverse materials simultaneously by using multiple extrusion nozzles. Since different materials can be loaded at a time, multifunctional parts with the desired composition can be easily made. The viscosity of molten material plays an important role in the FDP process. If the viscosity is very high it will be difficult to extrude the same. At the same time if the viscosity is very low, the structural support will be disturbed. Therefore, the optimum melt viscosity of the polymer has to be used for processing. The molten viscosity should be high enough to provide structural support and low enough to enable extrusion. Also, complete removal of the support structure used during printing may be difficult.

The process works on the principle of heating the filament material slightly above its melting temperature, allowing it to flow from the nozzle and re-solidification of the molten material due to cooling via convection. Figure 2.27 shows the schematic of an FDM process. The material spool rotates to feed the material to the nozzle using an extrusion mechanism. The heating element inside the nozzle head heats the material to slightly above its melting temperature. The molten material flows from the nozzle to the build platform. The print bed is allowed to move laterally to print the molten material within the layer. After the layer is completed, the nozzle head moves upward by an equivalent layer thickness and the next layer is deposited.

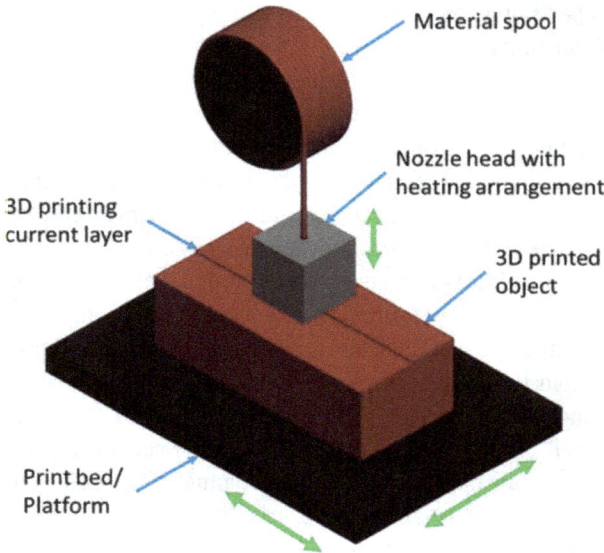

Fig. 2.27: Schematic of Fused Deposition Modelling.

In the FDP process, the fibre reinforcement can be achieved in multiple ways. Blending short fibres in polymer spool is one of the ways to obtain a fibre-reinforced composite material as shown in Fig. 2.28 (a). Mixing the continuous fibres with the matrix filament inside the nozzle head just prior to printing is also presented in Fig. 2.28 (b). However, an improper bonding between the fibre and the matrix material can cause defects in the printed part. A new technique has been developed which uses two nozzles to build the part, one each for the fibre and the matrix. Initially, the matrix layers are deposited on the build platform and then the fibre layers are deposited, which ensures the proper adhesion between them Fig. 2.28 (c). This technique has emerged as a prominent research domain due to its ability to 3D-print continuous fibre as well as deliver a good quality product.

Mechanical properties of traditionally manufactured composites are found to slightly vary from that manufactured with additive manufacturing. Current research focuses on the composite material using a variety of fibres and matrices, along with hybrid manufacturing processes to take advantage of the merits of both the conventional and additive manufacturing. New processes are being developed to eliminate

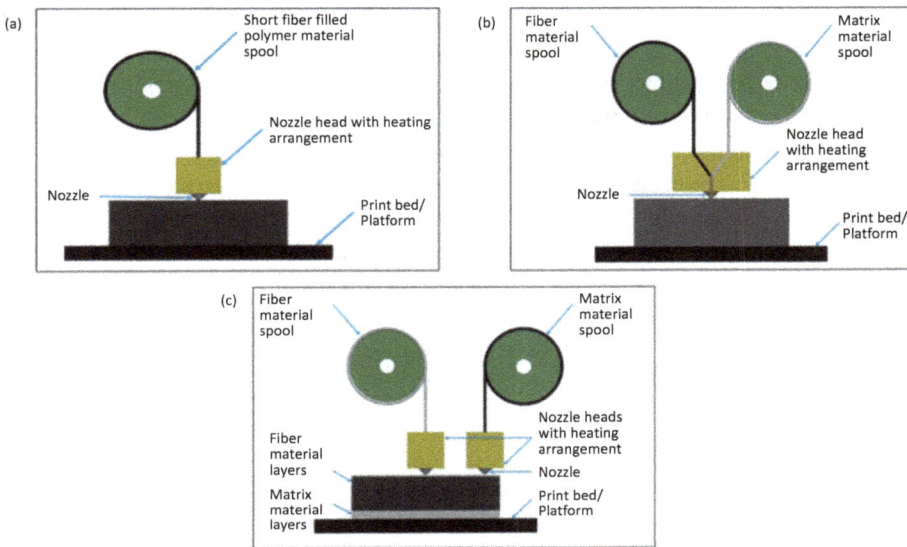

Fig. 2.28: Additive manufacturing of composite materials using FDM.

the fibre-matrix delamination issues and to improve the product quality in terms of mechanical properties and surface finish. Such processes can help to implement the additive technology for end-use products.

2.7 Conclusion

A very important advantage of thermoplastic-based composites is recyclability, thus reducing the solid and gaseous pollution and also carbon footprint when compared with composites of cross-linked elastomer and thermosets. Thermoplastics can have a wide range of hardness and flexibility to replace some applications where elastomers are being used. Particularly, thermoplastic polyurethane (TPU)-based nanocomposites are of great promise. As for engineering plastics, especially, PE, PP, PVC, acrylates, PEEK, PEKK, PC, ABS, etc., there is a wider opportunity to replace thermoset composites based on epoxy, phenolic, polyester vinyl ester, etc. Most of the desired properties, specific to an application, can be imparted to thermoplastics by blending, copolymerisation, nano inclusion and chemical modifications of thermoplastics. With the increasing use of polymer technology in the medical field and with increasing awareness of compatibility with nature, more bio-compatible and bio-degradable thermoplastic composites are being developed. Numerous studies are being reported in recent times on chemistry and technologies of thermoplastic composites, aiming applications in the medical, food packaging, and agriculture industries and for raw stock preservation. Polylactic acid

(PLA) and its derivatives, polyhydroxyalkanoates (PHA), and proteins (which are thermoplastic heteropolymers) and their composites are best examples. There are also new thoughts on bio-compatible reinforcements such as starch nano powders, cellulosic fibres from nature (plants, flowers, vegetation) like cotton, jute, flax fibres, wood shavings, saw dust and also natural basalt rock fibre, etc. The advantage of recycling can be more for polymer-nanocomposites over long- or short-fibre-reinforcement because of better accuracy in mould flow, blowing, extrusion and thermoforming. Overall, the prospects of reusable and bio-compatible thermoplastic nanocomposites with natural, biocompatible reinforcements are increasing in critical applications of medical and food processing industries.

References

[1]   Liang, J. Z., Yang, Q. Q. J. Reinf. Plast. Compos. 2009, 28, 295–304.
[2]   Huang, J. C., Wu, C. L. Adv. Polym. Technol. 2000, 19, 132–139.
[3]   Gill, Y. Q., Irfan, M. S., Khan, F. H., Saeed, F., Ehsan, H., Shakoor, A. Res. Express 2019, 6, 085343.
[4]   Ding, X., Wang, J., Zhang, S., Wang, J., Li, S. Polym. Bull. 2016, 73, 369–383.
[5]   Krupa, I., Chodák, I. Eur. Polym. J 2001, 37, 2159–2168.
[6]   Altay, L., Kızılkan, E., Seki, Y., İşbilir, A., Sarıkanat, M. Plast. Rubber Compos. 2021, 15, 1–12.
[7]   Alghamdi, M. N. Int. J. Eng. Res. Technol. (IJERT) 2016, 5, 633–637.
[8]   Helanto, K., Talja, R., Rojas, O. J. J. Polym. Eng. 2021, 41, 746–758.
[9]   Ogorckiewicz, R. M., Weidman, G. W. J. Mech. Eng. Sci. 1974, 16, 10–17.
[10] Agoudjil, B., Ibos, L., Majeste, J. C., Candau, Y., Mamunya, Y. P. Compos. A. 2008, 39, 342–351.
[11] Nicolais, L., Narkis, M. Polym. Eng. Sci. 1971, 11, 194–199.
[12] Liang, J. Z., Li, R. K. Y. Polym. Compos 1998, 19, 698–703.
[13] Liang, J. Z., Li, R. K. Y. Polym. Int 2000, 49, 170–174.
[14] Krupa, I., Cecen, V., Boudenne, A., Prokeš, J., Novak, I. Mater. Des. 2013, 51, 620–628.
[15] Mamunya, Y. P., Davydenko, V. V., Pissis, P., Lebedev, E. V. Eur. Polym. J. 2001, 38, 1887–1897.
[16] Boudenne, A., Ibos, L., Fois, M., Majeste´, J. C., Ge´hin, E. Compos. A. 2005, 36, 1545–1554.
[17] Nielsen, L. E. Generalized Equation for the Elastic Moduli of Composite Materials. J. Appl. Phys. 1970, 41, 4626–4627.
[18] Bourkas, G., Prassianakis, I., Kytopoulos, V., Sideridis, E., Younis, C. Adv. Mater. Sci. Eng. 2010, 1–13.
[19] Ramsteiner, F., Theysohn, R. On the tensile behavior of filled composites. Composites 1984, 5, 121–128.
[20] Nielsen, L. E. Mechanical Properties of Polymers. Reinholds Publishing Corp., Chapman & Hall, London, 1962.
[21] Hashin, Z., Shtrikman, S. J. Appl. Phys. 1962, 33, 3125–3131.
[22] Hashin, Z. J. Appl. Mech 1983, 50, 481–505.
[23] Budiansky, B. J. Compos. Mater. 1970, 4, 286–295.
[24] Bigg, D. M. Adv. Polym. Sci. 1995, 119, 1–30.
[25] Joseph, N., Sebastian, M. T. Mater. Lett. 2013, 90, 64–67.
[26] Catedral, M. D., Tapia, A. K. G., Sarmago, R. V., Tamayo, J. P., Del rosario, E. J. Sci. Diliman 2004, 16, 41–46.
[27] Hammo, S. M. Tikrit J. Pure Sci. 2012, 17, 76–78.
[28] Chodak, I., Krupa, I. J. Mater. Sci. Lett 1999, 18, 1457–1459.
[29] Scher, H., Zallen, R. J. Chem. Phys. 1970, 53, 3759–3761.

[30] Ramaraj, B. Polym. Plast. Technol. Eng 2007, 46, 575–578.
[31] He, X., Zhou, J., Jin, L., Long, X., Wu, H., Xu, L., Gong, Y., Zhou, W. Materials. 2020, 15, 3341.
[32] Guth, E., Gold, O. Phys. Rev 1938, 53, 322.
[33] Rosen, B. W. Tensile failure of fibrous composites. in Mechanics of Fiber Strengthening. Fiber Composite Materials, ASM Publication October 1960, Chapter 3.
[34] Termonia, Y. J. Polym. Sci. B: Polymer Physics 1994, 32, 969–979.
[35] Chen, P. E. Eng. Sci. 1971, 11, 51.
[36] Cox, H. L. J. Appl. Phys 1952, 3, 72–79.
[37] Halpin, J. C., Kardos, J. L. Polym. Eng. Sci. 1976, 16, 344–352.
[38] Halpin, J. C., Kardos, J. L. Polym. Eng. Sci. 1978, 18, 496–504.
[39] Osoka, E. C., Onukwuli, O. D. Int. J. Eng. Sci. Invent (IJESI) 2018, 7, 63–70.
[40] Luo, Z., Li, X., Shang, J., Zhu, H., Fang, D. Adv. Mech. Eng. 2018, 10, 1–13.
[41] Piggott, M. R. Load Bearing Fibre Composites. Pergamon, Oxford, UK, 1980, 83–89.
[42] Piggott, M. R. J. Mater. Sci 1978, 13, 1709–1716.
[43] Carrara, A. S., McGarry, F. J. J. Compos. Mater. 1968, 2, 222–243.
[44] Hassani, F.-Z. S. A., Ouarhim, W., Zari, N., Bensalah, M. O., Rodrigue, D., Bouhfid, R., Qaiss, A. Polym. Compos. 2019, 40, 1919–1929.
[45] Ariawan, D., Surojo, E., Triyono, J., Purbayanto, I. F., Pamungkas, A. F., Prabowo, A. R. Theor. Appl. Mech. Lett. 2020, 10, 57–65.
[46] Guth, E. Theory of Filler Reinforcement. J. Appl. Phys. 1945, 16, 20–27.
[47] Unal, H., Mimaroglu, A. Sci. Eng. Compos. Mater. 2014, 21, 323–328.
[48] Bombac, D., Lamut, M., Mrvar, P., Širok, B., Bizjan, B. Materials 2021, 14, 6108.
[49] E-Glass Fibre, Documents@AZoM.com. AZO Materials, Azonetwork UK Ltd. www.AZoM.com (accessed on 27 December, 2022).
[50] Lou, S., Zhang, H., Liu, F., Wang, J., Yin, W., Chen, Z., Su, C. Sci. Eng. Compos. Mater. 2021, 28, 466–477.
[51] Zheng, J., Zhou, H., Wan, Z., Sang, R. Adv. Mater. Res. 2012, 535–537, 144–148.
[52] Li, E. Z., Guo, W. L., Wang, H. D., Xu, B. S., Liu, X. T. Physics Procedia 2013, 50, 453–460.
[53] Ramsaroop, A., Kanny, K., Mohan, T. P. Mater. Sci. Appl. 2010, 1, 301–309.
[54] Yuan, Y., Liu, C., Huang, M. Materials 2019, 12, 1323.
[55] Ari, A., Bayram, A., Karahan, M., Karagöz, S. Polym. Polym. Compos. 2022, 30, 1–13.
[56] Kabiri, A., Liaghat, G., Alavi, F., Saidpour, H., Hedayati, S. K., Ansari, M., Chizari, M. J. Compos. Mater. 2020, 54, 4903–4919.
[57] Kuruvilla, J., Mattoso, L. H. C. Nat. Polym. Compos. 1998, 333.
[58] Rabbi, M. S., Islam, T., Islam, G. M. S. Int. J. Mech. Mater. Eng. 2021, 16.
[59] Alvarez, V. A., Ruscekaite, R. A., Vazquez, A. J. Comp. Mat. 2004, 37, 1575–1588.
[60] Alvarez, V., Vazquez, A., Bernal, C. J. Compos. Mater. 2006, 40, 21–35.
[61] Joseph, K., Thomas, S., Pavithran, C., Brahmakumar, M. J. Appl. Polym. Sci. 1993, 47, 1731–1739.
[62] Samouh, Z., Molnar, K., Boussu, F., Cherkaoui, O., El Moznine, R. Polym. Adv. Technol. 2019, 30, 529–537.
[63] Sun, Z., Mingming, W. Ind. Crop. Prod. 2019, 137, 89–97.
[64] Sood, M., Dwivedi, G. Egypt. J. Pet. 2018, 27, 775–783.
[65] Hassani, F. Z. S. A., Ouarhim, W., Zari, N., Bensalah, M. O., Rodrigue, D., Bouhfid, R., Qaiss, A. Polym. Compos. 2019, 40, 4042–4055.
[66] Adeniyi, A. G., Onifade, D. V., Ighalo, J. O., Adeoye, A. S. Compos. Part B. Eng. 2019, 176, 107305.
[67] Su, X., Wang, R., Li, X., Araby, S., Kuan, H.-C., Naeem, M., Ma, J. Nano. Mater. Sci. 2022, 4, 185–204.
[68] Li, Z., Young, R. J., Wilson, N. R., Kinloch, I. A., Vallés, C., Li, Z. Compos. Sci. Technol. 2016, 123, 125–133.
[69] Li, Z., Young, R. J., Kinloch, I. A., Wilson, N. R., Marsden, A. J., Raju, A. P. A. Carbon 2015, 88, 215–224.
[70] Mohan, T. P., Kanny, K. J. Polym. Eng. 2015, 35, 773–784.
[71] Ivanova, R., Kotsilkova, R. Appl. Rheol. 2018, 28, 54014.

[72]   Mishra, V. S., Khuswa, R., Chandrasekhar, L., Barman, S., Sivaraman, P., Manoj, N. R., Samui, A. B., Chakraborty, B. C. Polym. Test. 2004, 23, 527–532.
[73]   Gaztelumendi, M., Nazabal, J. J. Polym. Sci. Part B. Polym. Phys. 1995, 33, 603–610.
[74]   Ahad, N. A. InIOP Conference Series: Materials Science and Engineering 957 2020, 012045. IOP Publishing.
[75]   Manoj, N. R., Sivaraman, P., Praveen, S., Raut, R. D., Chandrasekhar, L., Chakraborty, B. C. J. Appl. Polym. Sci. 2005, 97, 1763–1770.
[76]   Hamzah, M. S., Mariatti, M., Kamarol, M. J. Polym. Mater 2016, 33, 605–613.
[77]   Ismail, N. H., Mustapha, M. Polym. Eng. Sci. 2018, 58, E36–E63.
[78]   Balaji, S., Balasubramanian, R., Fathima, R. M., Sarojadevi, M. Polym Eng. Sci. 2018, 58, 691–700.
[79]   Guo, J., Xu, Y., He, W., Wang, N., Tang, M., Chen, X., Hu, S., He, M., Qin, S. Polym. Eng.sci. 2018, 58, 752–758.
[80]   Tumnantong, D., Rempel, G. L., Prasassarakich, P. Polym. Eng. Sci. 2018, 58, 759–766.
[81]   Liang, N., Liu, W., Zuo, D., Peng, P., Qu, R., Chen, D., Zhang, H. Q. Polym. Eng. Sci. 2018, 85, 767–774.
[82]   Dupont Hytrel Thermoplastic Polyester Elastomer Product Reference Guide. http://www.dupont.com/products-and-services/plastics-polymers-resins/thermoplastics. (Accessed on 7 January, 2023).
[83]   Ostwald, W. Ueber die rechnerische Darstellung des Strukturgebietes der Viskosität. Kolloid-Zeitschrift 1929, 47, 176–187.
[84]   Liang, J. Z., Zhong, L. Polym. Test. 2010, 29, 972–976.
[85]   Liang, J.-Z., Zhong, L. J. Elastomers Plast. 2014, 46, 662–672.
[86]   Schreiber, H. P., Rudin, A., Bagley, E. B. J. Appl. Polym. Sci. 1965, 9, 887–892.
[87]   Lamberti, G., Peters, G. M. W., Titomanlio, G. Intern. Polymer Processing XXII 2007, 3, 303–310.
[88]   Einstein, A. Annalen der Physik 1906, 19, 289–306.
[89]   Batchelor, G. K. J. Fluid. Mech 1977, 83, 97–117.
[90]   Mooney, M. J. Colloid Sci. 1951, 6, 162–170.
[91]   Ball, R., Richmond, P. J. Phys. Chem. Liquids 1980, 9, 91–116.
[92]   Krieger, I. M., Dougherty, T. J. Trans. Soc. Rheol. 1959, 3, 137–152.
[93]   Kitano, T., Kataoka, T., Shirota, T. Rheol. Acta. 1981, 20, 207–209.
[94]   Graham, A. L. Appl. Sci. Res. 1981, 37, 275–286.
[95]   Frankel, N. A., Activos, A. Eng. Sci. 1967, 22, 847–853.
[96]   Tanner, R. L. J. Non-Newtonian. Fluid Mech. 2003, 112, 251–268.
[97]   Chafidz, A., Ma mun, S., Fardhyanti, D. S., Kustiningsih, I., Hidayat, A. IOP Conf. Ser. Mater. Sci. Eng. 2019, 543, 012036.
[98]   Cho, J. W., Paul, D. R. Polymer 2001, 42, 1083–1094.
[99]   Kojima, Y., Usuki, A., Kawasumi, M., Okada, A., Kurauchi, T., Kamigaito, O. J. Polym. Sci. Part A: Polym. Chem. 1993, 31, 983–986.
[100] Usuki, A., Kojima, Y., Kawasumi, M., Okada, A., Fukushima, Y., Kurauchi, T., Kamigaito, O. J. Mater Res. 1993, 8, 1179–1184.
[101] Kutlu, B., Mein, J., Leuteritz, A., Brünig, H., Heinrich, G. Polymer 2013, 54, 5712–5718.
[102] Redondo, A., Chatterjee, S., Brodard, P., Korley, L. T. J., Weder, C., Gunkel, I., Steiner, U. Polymers 2019, 11, 1912.
[103] Sandler, J. K. W., Pegel, S., Cadek, M., Gojny, F., Van es, M., Lohmar, J., Blau, W. J., Schulte, K., Windle, A. H., Shaffer, M. S. P. Polymer 2004, 45, 2001–2015.
[104] Lu, Y., Jiang, J., Yoon, S., Kim, K.-S., Kim, J.-H., Park, S., Kim, S.-H., Piao, L. ACS Appl. Mater. Interfaces 2018, 10, 2093–2104.
[105] Gibson, I., Rosen, D. W., Stucker, B. Additive Manufacturing Technologies. Springer, 2015.
[106] ISO/ASTM D52900-15, Additive manufacturing — General principles — Terminology Dec, 2015.
[107] Yakout, M., Elbestawi, M. A., Additive Manufacturing of Composite Materials: An Overview. VMPT, 2017, Montréal.

Chapter 3
Thermosetting polymer-based composites

Description of abbreviations

Abbreviation	Description
FRP	Fibre-reinforced plastic
PMC	Polymer matrix composite
MMC	Metal matrix composite
CMC	Ceramic matrix composite
RTM	Resin transfer moulding
VARTM	Vacuum-assisted resin transfer moulding
FRC	Fibre-reinforced composite
SRIM	So-called structural RIM
RFI	Resin film infusion
ASTM	American Society for Testing and Materials
SBS	Short beam shear
DMA	Dynamic mechanical analysis
TGAP	Trifunctional epoxy resin
GFRP	Glass fibre-reinforced plastic
UPE	Unsaturated polyester
GFRT	Glass fibre-reinforced thermoset
CFRT	Carbon fibre-reinforced thermoset
BFRT	Besalt fibre-reinforced thermoset composite
KFRT	Kevlar fibre-reinforced thermoset
RIM	Reaction injection moulding
CFRP	Carbon fibre-reinforced plastic
TGDDM	Tetraglycidylether of 4,4′-diaminodiphenyl methane
DGEBA	Diglycidyl ether of bisphenol-A
ILSS	Interlaminar shear stress
TGAP	Triglycidyl p-aminophenol
LCM	Liquid composites moulding
CFRC	Carbon fibre-reinforced composite
CNT	Carbon nanotube
CNF	Carbon nanofibre
CVD	Chemical vapour deposition
DCB	Double cantilever beam
CFRC	Carbon fibre-reinforced composite
CVD	Chemical vapour deposition
CSRP	Core-shell rubber particle
SEM	Scanning electron microscope
PEBA	Poly(ether-*block*-amide)
CTPEGA	Carboxyl-terminated poly(ethylene glycol)
DGEBA	Diglycidyl ether of bisphenol-A
HBP	Hyperbranched polymer
CSRP	Core-shell rubber particle
CF	Carbon fibre
RIFT	Resin infusion flexible tooling

https://doi.org/10.1515/9783110781571-003

3.1 Thermoset matrix

In previous chapter we have discussed composites based on thermoplastic matrix. Unlike thermoplastic, thermosetting resins do not soften on heating and stiffen on cooling repeatedly several times. Like rubbers they undergo some chemical curing on heating and form 3-dimensional (3-D) network resulting in conversion into an infusible and insoluble material. The difference between rubbers and thermosetting resin is that rubbers are generally high-molecular-weight polymer whereas thermosetting resins are mostly low-molecular-weight viscous liquid.

The composite industries are dominated by thermoset resins. The reason behind this popularity of thermoset resin for composite application is their historic availability with varying consistency like from viscous liquid to tack-free solid, relative ease of processing, lower cost of capital equipment for processing and low material cost. Since the thermosetting resins are available in oligomeric or monomeric low-viscosity liquid forms, they have excellent flow properties to facilitate resin impregnation of fibre bundles and proper wetting of the fibre surface by the resin. Thermoset-based products can be moulded, in general, at a much lower temperature and pressure compared to thermoplastics. By virtue of their crosslinked structure, thermoset composites offer better creep properties and environmental stress-cracking resistance compared to many thermoplastics, for example, polycarbonate [1–4]. The presence of the traces of solvent may cause catastrophic failure in case of thermoplastic whereas cured thermosetting polymers being in the 3D network state are not affected significantly due to the presence of solvent. Another important characteristic of a thermosetting resin is that their properties are not only dependent on the chemistry and molecular weight of the resin (as in the case of thermoplastic) but also largely dependent on the crosslink density of the resin network. A wide range of properties can be achieved by simply tailoring the crosslink density of a network without changing the chemical structure.

For detailed chemistry and properties of thermoset resins, the readers may consult the author's recent book published [5] in 2022. However, for the sake of the readers the thermosetting matrix will be elaborated briefly. The thermosetting resins can be classified into two categories, namely general-purpose resin, for example, unsaturated polyester (UPE), epoxy, vinyl ester and high-temperature resins, for example, phenolic, benzoxazine, polyimide, bismaleimide, cyanate ester and pthalonitrie. Looking at the properties, a resin can be selected for a particular application considering especially the service temperature in which the product will be used. It may be noted that thermal stability of reinforcing fillers is much higher compared to the resin system. Since thermal resistance of polymeric resin is much inferior to the reinforcing filer, the thermal performance of a composite material is mostly controlled by resin system used as a matrix. If service temperature (the maximum temperature at which a material has to work) is higher than the glass transition temperature of the resin, the related composite will lose its dimensional stability whatever be the strength of reinforcing fibre used for making the composite. Once the resin loses its dimensional stability the related composite cannot be used anymore for load-bearing applications. The resin should be selected in such a way that the glass transition temperature of the resin should be at least 50 °C higher than the service temperature. For an application, it may so happen that a material has to work partly at a low temperature and partly at a high temperature. In that case, the higher service temperature has to be considered for the selection of resin system for fabrication of composite for such applications. However, for other application this condition may not hold good. For example, in case of sealant application wherein the resin has to be rubbery in nature, the glass transition temperature of the resin should be well below the service temperature. If the service temperature is less than the glass transition temperature of the resin, the material will be hard and brittle at the service temperature and will not be suitable for sealing applications. For adhesive and coating applications the glass transition temperature should be close to the service temperature. In the subsequent sections, various thermosetting resins will be briefly discussed under the heading of general-purpose and high-temperature resins.

3.1.1 General-purpose resins

3.1.1.1 Unsaturated polyester resin

Unsaturated polyester resin (UPE) resins are a mixture of unsaturated polyester, styrene and an inhibitor [6–9]. Polyesters are a type of condensation polymer, which are made by reacting a diacid or dianhydride with a dihydroxy compound (diols). Unsaturated polyester can be made by using acid monomer containing unsaturation in its structure (namely maleic anhydride or fumeric acid) in addition to saturated acid monomers like phthalic anhydride. The most common diols used for unsaturated polyester synthesis are ethylene glycol or propylene glycol. The unsaturation generated in the structure due to incorporation of maleic anhydride is utilized for curing of the resin through free radical mechanism. Styrene is added up to a concentration of 40 wt% to act as a reactive diluent to reduce viscosity of the resins. Styrene can also be polymerized by free radical mechanism. The inhibitor is used to enhance the storage life of the resin. When the resin is mixed with a peroxide initiator and activator (cobalt octoate/naphthenate), the free radicals are formed. The free radicals produced at the initial stage are consumed by the inhibitor due to higher stability of inhibitor radicals. Once the inhibitor molecules are depleted, the free radicals, produced from the initiator, initiate the polymerization of the polyesters and styrene; both the intermolecular and intramolecular crosslinking of polyester with or without involving styrene. Chain branching on the polyester molecules and homopolymerization of styrene may also take place.

Organic peroxides, azine and azo compounds are used as to generate free radicals. UPE industry is dominated by the use of peroxides namely methyl ethyl ketone peroxide (MEKP), benzoyl peroxide (BPO) and *t*-butyl peroxide (TBO). MEKP decomposes at room temperature whereas BPO and TBO require higher temperatures 70 and 140 °C, respectively, to decompose. It is necessary to use some accelerators which accelerate the decomposition of peroxides in order to ensure room temperature or low-temperature curing at a reasonable rate. Generally, transition metal oxides (e.g. cobalt naphthenate and cobalt octoate) are used as accelerators for curing UPE resins. The free radicals initiate the chain reaction which propagates through the unsaturated sites of the polyester and monomer leading to the formation of a network structure, which is insoluble and infusible.

3.1.1.2 Epoxy resins

Epoxy resins constitute a class of thermosetting polymers characterized by the presence of two or more oxirane rings or epoxy groups within their molecular structure [3, 4]. The resins form network on curing through the reaction of epoxide groups with a suitable curing agent. The most common epoxy resin is diglycidylether of bisphenol-A (DGEBA), which is prepared by the reaction of epichlorohydrin (ECD) and bisphenol-A (BPA). Today, a wide variety of epoxy resins of varying consistency and physicochemical properties are available in the market after the first production started simultaneously in Europe and in the United States in the early 1940s. It is important for a material scientist to judiciously select the resin considering the nature of fibre, sizing and the processing technique to be used [10–21]. The common types of epoxy resins and curing agents that are used [22] for commercial application and research and development are presented in Tab. 3.1.

Tab. 3.1: The epoxy resins and curing agents commonly used in researches and productions.

Resin epoxy	Curing agent	Application	References
DGEBA	Cycloaliphatic diamine, bis *p*-amino cyclo-hexyl methane	CF sizing	[10, 11]
Phenyl glycidyl ether	Cyclohexane dicarboxylic anhydride	Research of carbon fibre-reinforced plastic (CFRP) composite	[12]
Tetraglycidyl diaminodiphenyl methane (TGDDM)	Diaminodiphenyl sulfone (DDS)	Prepreg CFRP	[13]
DGEBA	Aliphatic polyethertriamine	CFRP sheet	[14]
DGEBA	Anhydride	CF sizing	[15]
DGEBA and diglycidyl ether of bisphenol-F (DGEBF)	Trioxatridecane diamine	Adhesive	[16]
DGEBA	Mixing of diethylenetriamine and triethylenetetramine	Civil construction	[17]

Tab. 3.1 (continued)

Resin epoxy	Curing agent	Application	References
Diglycidyl ester of hexahydrophthalic acid and 3,4-epoxycyclohexyl methyl-3,4 epoxycyclohexane carboxylate	Hexahydrophthalic anhydride and methylhexahydrophthalic anhydride	CRFP composite	[18]
TGDDM and novolac resin	DDS	Modified by novolac resin	[19]
DGEBA, merk Araldite GY-250 with solvent of triglycidylether of trimethylolpropane (TGETMP)	90 wt% isopharone diamines (vestamin IPD) and 10 wt% trimethyl hexamethylene diamine (vestamin TMD)	CFRP composite	[20]
4,5-Epoxy cyclohexane 1,2-dicarboxylate diglycidyl	Polyamine and acid anhydride	Prepreg CFRP	[21]

A difunctional epoxy resin, that is, diglycidyl ether of bisphenol-A (reactive functionality = 2) is mostly used for composite applications as a matrix material. Multifunctional epoxies like trifunctional, tetrafunctional or novolac epoxy resins are recommended for applications where service temperature is higher than 100 °C (e.g. aerospace). Amines and anhydrides are used as curing agents. In case of amine, the epoxy group first undergoes ring-opening reaction with primary amine to produce and secondary amine and hydroxyl group. Then, the secondary amine reacts with other epoxide groups producing additional hydroxyl groups and tertiary amines as shown in Fig. 3.1. This reaction continues until all active groups of hardener and/or epoxy resins have been completely consumed. However, the reaction becomes diffusion-controlled and extremely slow once the glass transition temperature of the network crosses the cure temperature. This is called vitrification. Post-curing is necessary to expedite the further reaction. Although a single activation energy and heat of reaction are experimentally obtained for both the steps, reactivities of primary and secondary amino groups may be different. The hydroxyl groups generated during the cure can also react with the epoxy ring, forming ether bonds (etherification). The etherification reaction competes with the amine-epoxy cure when the reactivity of the amine is low or when there is an excess of epoxy groups. The tendency of etherification reaction to take place depends on the temperature and basicity of the diamines and increases with the concentration of epoxy in the mixture.

Let us now discuss the curing reaction in the presence of anhydride hardener. Because the anhydride group cannot react directly with the epoxy group, the anhydride ring first initiates a reaction by binding to the hydroxyl (OH) group existing in the system to form a monoester containing a carboxylate group. Furthermore, the carboxylate group undergoes esterification reaction with an epoxy group to produce an ester and hydroxyl bond. The reaction scheme is shown in Fig. 3.1. The curing reaction

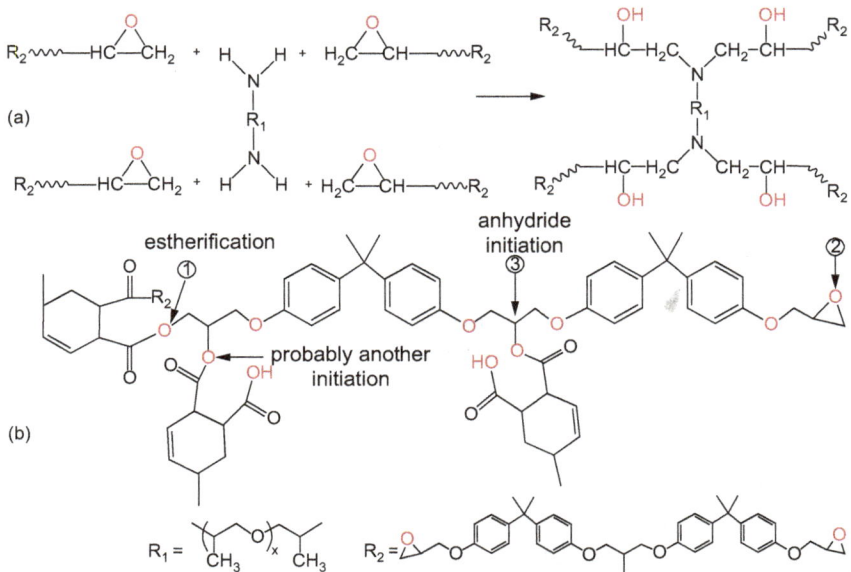

Fig. 3.1: Network scheme generated by the reaction between epoxy DGEBA: (a) polyetheramine hardener and (b) anhydride hardener [23].

with anhydride generally requires high temperature (200 °C) and the cure temperature can be reduced by using a suitable catalyst. Unlike in the amine-cured system, post-curing does not have much effect on the final structure of the epoxy resins [23]. The epoxy group can also react with the OH group in the system forming ether bonds (the reaction is called etherification), and if the OH group comes from a resin backbone, a homopolymerization reaction will occur. Esterification and homopolymerization reactions are likely to occur with an increase of the curing temperature, either with an amine curing agent or anhydride.

3.1.1.3 Vinyl ester resin

VE resin was first produced commercially by Shell Chemical Co. in 1965 under the trade name "Epocryl". Subsequently Dow Chemical Co. introduced a similar series of resins under the name "Derakane". VE resins are prepared by reacting an epoxy resin (difunctional or multifunctional) with an unsaturated carboxylic acid like acrylic acid or methacrylic acid [24, 25]. For example, reaction between 1 mol of diglycidyl ether of bisphenol-A and 2 mol of methacrylic acid can be used to synthesize a simple form of vinyl ester. Synthesis of a VE resin is usually catalysed by tertiary amines, phosphines and alkalies or onium salts. The reaction is carried out in a temperature range of 90–120 °C and taking the reactants in a stoichiometric ratio. The complete esterification of the epoxy resin, in practice, is difficult to achieve because of the gelation of the product before an acid value below 10 is attained. Therefore, an excess of epoxide resin is always employed in the esterification reactions. In order to stop the polymerization of methacrylic acid an inhibitor such as hydroquinone is used. A wide variety of VE resin can be synthesized by changing the nature of base commercial epoxy resins namely halogenated or phosphorous-containing resins and unsaturated acids (e.g. acrylic, methacrylic, crotonic and cinnamic).

Like UPE resin, it is necessary to add reactive diluents like monofunctional vinyl compounds or acrylates to overcome the increased processing time and to reduce the high risks of inhomogeneity and consequent defect generation in the specimens [26]. The diluents are required for better control over the process and to modify the rheological and thermomechanical properties of the related VE resins.

3.1.2 High-temperature resin

3.1.2.1 Phenolic resin

Phenolic resin or phenol formaldehyde (PF) resin is the first man-made polymer material produced commercially. The concept of producing high-molecular-weight substances from the oligomeric product was first discovered by Leo H. Baekeland in 1909 [27]. Although nature has long demonstrated the synthesis of polymers like protein, carbohydrate and natural rubber and modification of natural polymers using sulphur, that is, vulcanization or curing of natural rubber, was discovered by Goodyear in 1839, the macromolecular hypothesis was profounded by Staudinger in 1920 [28] or a base catalyst and the process initially adopted by Barkeland has been perfected to higher and higher level of sophistication and reached the modern phenolic resin technology. When phenol is used in excess the product is called novolac and when formaldehyde is used in excess the product is called resole. Resole is self-curable and does not need any additional curing agent. However, novolac requires extra formaldehyde for curing to take place and generally cured with 10–15% of hexamethylene tetramine (HMTA), a condensation product of formaldehyde and ammonia [29]. Under the curing condition, HTMA decomposes to generate formaldehyde which converts novolac to a crosslinked network structure. In the case of novolac it is necessary to add plasticizers like methyl benzoate (non-reactive) or furfuryl alcohol (reactive) to achieve uniform and complete curing required to ensure high modulus and performance at elevated temperatures [30].

Phenolic resins by virtue of its aromatic backbone display a number of useful properties namely high mechanical strength, long-term thermal and mechanical stability, excellent fire, smoke and low toxicity characteristics and excellent thermal insulating capability. The smoke density is 40 times higher than epoxy and vinyl ester resins. The cured phenolic resin exhibit T_g more than 150 °C and therefore suitable for high-temperature applications. Novolac and resole-based phenolic resins and their composites are extensively used in aerospace industries due to good heat and fire resistance and excellent ablative properties. The main attraction of such phenolic resin is low cost in addition to the properties mentioned above. However, several technological issues are associated with the processing of such resins and the related composites. This is because of the cure chemistry involving condensation mechanism with the evolution of volatile gases, which initiates the formation of microvoids leading to the substantial deterioration of deterioration of mechanical properties of the related components. Moreover, acidic catalyst used to cure the resin initiates corrosion in the metallic structure. Most importantly the phenolic resin networks are highly brittle, which limits their use for many applications. As a solution to these problems of conventional phenolic resins, addition-cure phenolic resins have been developed [31–37] by using various approaches like (i) incorporation of thermally stable addition-curable groups onto a novolac backbone; (ii) structural modification (transformation) involving phenolic hydroxyl groups; (iii) curing of novolac by suitable curatives through addition reactions of OH groups and (iv) reactive blending of structurally modified phenolic resin with a functional reactant.

3.1.2.2 Benzoxazine resin

Benzoxazine monomers are characterized by the presence of oxazine rings, that is, heterocyclic six-membered rings with oxygen and nitrogen atoms and attached to a benzene ring. Holly and Cope [38] first reported the synthesis through Mannich condensations of phenols with formaldehyde and amines. The work related to the possibility of commercial applications for the preparation of phenolic materials with improved properties was first patented by Schreiber [39]. Polybenzoxazines are prepared by reacting a phenol with formaldehyde and an aromatic amine and are cured by thermal ring-opening polymerization [40, 41]. Aminomethylol group is formed as an intermediate due to reaction between the amine and formaldehyde which further reacts to form 1,3,5-triaza-like structure. Subsequent reaction with phenol and formaldehyde leads to the formation of oxazine rings. The resin can polymerize via a thermally induced ring-opening reaction via ortho attack to form a phenolic structure characterized by a Mannich base bridge (CH2-NR-CH2). If we compare with novolac or resole-based phenolic resin, then we can see the presence of Mannich base bridge (CH_2–NR–CH_2) in benzoxazine resin instead of the methylene bridge structure associated with traditional phenolic resins [42, 43]. The better impact properties of banzoxazine compared conventional phenolic can be explained considering the change in chemical structure. The synthetic chemistry offers wide scope to play with chemical structure using different types of phenol and amine to synthesize a variety of benzoxazine polymers. The polybenzoxazine polymers derived from monofunctional monomers (containing one oxazine ring) forms linear polymers, whereas di- or polyfunctional benzoxazine monomers (having two or more oxazine rings) display crosslinking networks which are very useful for commercial applications. Benzoxazines are known for their excellent thermal and dielectric properties, low water absorption and surface energy, zero shrinkage and structure design flexibility.

Recently Yan et al. [44] developed bio-based resin which is curable completely below 200 °C. They used bio-based phloretic acid (PHA), *p*-coumaric acid (pCOA) and ferulic acid (FEA) containing a carboxylic group as a phenol source and furfurylamine (FA) as an amine source are used to prepare three full bio-based benzoxazines (PHA-fa, pCOA-fa and FEA-fa). Synthetic routes of PHA-fa, pCOA-fa and FEA-fa monomers are shown in Fig. 3.2. Note that other biobased enzoxazine resins exhibit lower thermal stability and inferior low-temperature curing performance compared to the resin made from fossil feedstock. The curing mechanisms of PHA-fa, pCOA-fa and FEA-fa monomers are shown in Fig. 3.3. The mechanism includes ring opening, electrophilic substitution, structural rearrangement and crosslinking of the double bond. On heating, the oxazine ring is first opened to generate a Schiff base structure, and then the Schiff base structure attacks the phenol ortho-position (aryl attack), O atom in the oxazine ring (O-attack) or the furan ring (furan ring attack). Meanwhile, Mannich-base phenoxy-type polybenzoxazine was formed by insertion of the monomers via the oxygen of the oxazine ring. Finally phenoxy-type polybenzoxazine rearranges into phenolic-type polybenzoxazine. The presence of double bond creates the possibility of formation of another crosslinked structure formed by the double bond for poly(pCOA-fa) and poly(FEA-fa) resins. Apart from crosslinkable benzoxazine resins, main-chain benzoxazines are also developed [45–48]. The main-chain benzoxazine resins behave like a thermoplastic and can be processed as self-supporting films, which can used for various applications.

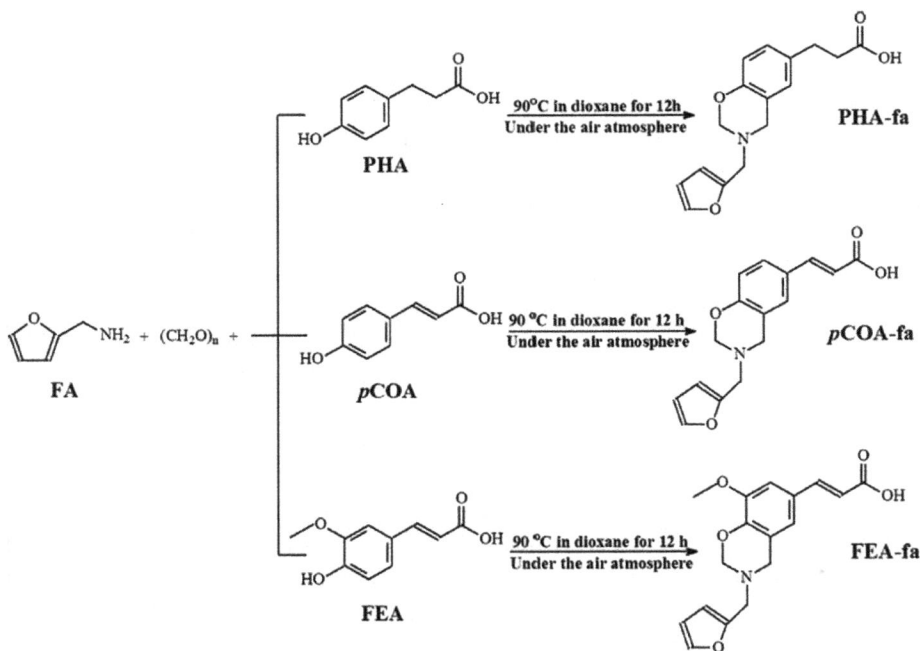

Fig. 3.2: Synthetic routes of PHA-fa, pCOA-fa and FEA-fa monomers, Yan, Zuomin Zhan (reprinted with permission from Yan, Zuomin Zhan, Huaqing Wang, Jie Cheng and Zhengping Fang, *ACS Appl. Polym. Mater.* 3 (2021) 3392−3401. © 2021, ACS Publishing Company).

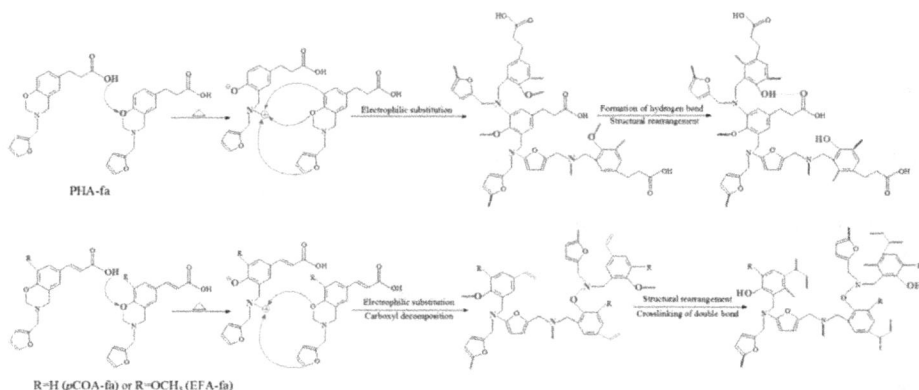

Fig. 3.3: Curing mechanisms of PHA-fa, pCOA-fa and FEA-fa monomers (reprinted with permission from Yan, Zuomin Zhan, Huaqing Wang, Jie Cheng and Zhengping Fang, *ACS Appl. Polym. Mater.* 3 (2021) 3392–3401. © 2021, ACS Publishing Company).

3.1.2.3 Polyimides

Polyimides are another class high-temperature resins, which were made commercial available by DuPont in the early 1960s in the trade name of Kapton, Vespel, etc. Polyimides are generally made by reacting a dianhydride with an aromatic amine. A wide variety of polyimides can be made by using different types of anhydrides and amine compound. The first step of the reaction is the formation of a polyamic acid intermediate in an organic solvent like dimethylacetamide or *N*-methyl pyrrolidone. The intermediate formation requires 6–20 h and the reaction time depends on reactivity of the precursors. The polyamic acid intermediate undergoes further polymerization under heat treatment to produce polyimide.

Polyimides are polymers having ring structure with –CO–NR–CO– backbone [49–52] and by virtue of their aromatic backbone display outstanding thermal stability. Because of their excellent thermal stability, superior mechanical strength and good chemical resistance properties, aromatic PIs are widely used in various industrial applications. However, such outstanding properties of PIs are also associated with poor processability due to their stiff chain structures and strong interchain interactions. When flexible linkages are introduced into the structure the glass transition temperature tends to decrease leading to a sacrifice in thermal stability. Han et al. [53] reported a series of homo- and copolyimides based on mixed thioetherdiphthalic anhydride (TDPA). TDPA is an aromatic bridged anhydride and available as a mixture of three isomeric form as shown in Fig. 3.4. The chemistry of synthesis of polyimides with TDA and different kinds of aromatic diamines is shown in Fig. 3.5. The idea is to introduce the flexible thioether moiety into the network structure to improve the flexibility and easy melt processability. The copolyimides from 1,3-phenylenediamine (m-PDA) and 3,4'-oxydianiline (3,4'-ODA) were obtained with higher T_g and lower melt viscosity. Addition polyimides can synthesized in a similar way as discussed above for condensation polyimide using an aromatic compound with the desired unsaturated group. By changing the unsaturated functional groups, for example, vinyl, nadic, acetylene and phenyl ethynyl, different types of addition polyimides can be synthesized.

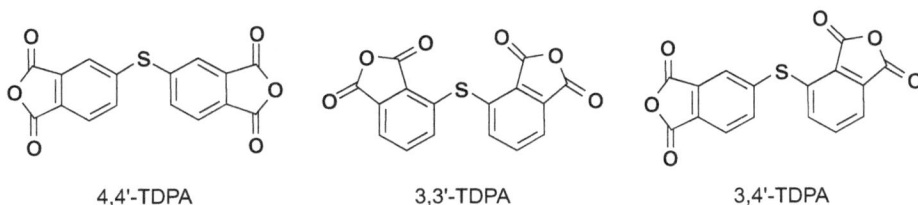

4,4'-TDPA   **3,3'-TDPA**   **3,4'-TDPA**

Fig. 3.4: The structures of thioetherdiphthalic anhydride isomer (reprinted with permission from Y. Han, X. Z. Fang, X. X. Zuo, *eXPRESS Polymer Letters* 4 (2010) 712–722. © 2010 Online Publishing).

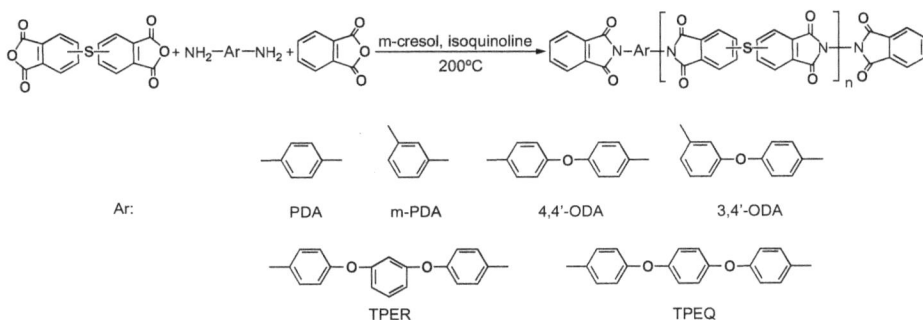

Ar: PDA   m-PDA   4,4'-ODA   3,4'-ODA

TPER   TPEQ

Fig. 3.5: Synthesis of polyimides derived from mixed-TDPA via one-step polymerization (reprinted with permission from Y. Han, X. Z. Fang, X. X. Zuo, *eXPRESS Polymer Letters* 4 (2010) 712–722. © 2010 Online Publishing).

When a cured polyimide product is processed through such polyamic acid, the presence of traces of high boiling solvent causes a significant reduction of the properties of the polyimide networks. As a solution to this problem in situ polymerization of monomeric reactants (PMR) resins have been developed in which half ester of a tetracarboxylic acid like 3,3',4,4'-benzophenone tetracarboxylic acid anhydride (BTDA) is used in place of dianhydride and mixed with an amine using low boiling solvent like methanol or ethanol. Because of lower reactivity of half ester, its mixture with aromatic amine can be stored for a long time in ambient temperature and at higher temperature (>100 °C) anhydride is reformed and reacts with the amine groups to produce a polyimide oligomer. PMR-30 and PMR-5 resins are widely used as commercial resins (marketed by SP Systems Imitec Inc., Hycomp. Inc., Eikos Inc.). The resin offers an upper use temperature greater than 300 °C and widely used for fabrication of carbon fibre-reinforced composites [54]. Second generation PMR resin (PMR-II) offers higher thermal stability compared to first-generation PMR resin (PMR-I) and can be synthesized in the similar way (as used for PMR-I) by replacing BTDA with fluorine-containing anhydrides such as 4,4'-(hexafluoroisopropylidene)-diphthalic acid, 4,4'-(2,2,2-trifluoro-1-phenylethylidene) diphthalic anhydride [55, 56].

As it is applicable for any thermoset resin, polyamide resins are required to be cured at high temperature to achieve high glass transition temperature. In fact the resin has to be cured at higher than the targeted glass transition temperature for a longer time. It is not possible to get a network with glass transition temperature higher than the cure temperature. Because when the glass transition temperature reaches the curing temperature, the vitrification takes place and the reaction becomes very slow due to diffusion controlled reaction mechanism. Therefore, it is necessary to cure at a sufficiently high temperature for a longer time to get a high crosslink density which results high glass transition temperature. The glass transition temperature of the network tends to increase with increasing cure temperature. For example, a polyimide resin (PMR-15) cured at 270 and 300 °C for a few hours exhibits T_g values 190 and 280 °C, respectively. Post-curing at higher temperature is necessary for further increase in glass transition temperature. The crosslinking reaction of polyimide involves various steps such as imidization, isomerization and double Diels-Alder adduct formation. In the case of nadic end-capped polyimides, retro Diels-Alder reactions of norborene end group take place leading to the formation of maleimide groups and cyclopentadiene [57].

3.1.2.4 Bismaleimide resin

Bismaleimide resins are monomers or prepolymers with two or more maleimide end groups. Bismaleimides offer advantages of both epoxy resins in terms of processability (autocave moulding) and thermal stability like polyimides. Bismaleimide resins display excellent physical property retention in hot/wet condition, flame retardancy and easy processability [58]. Bismaleimide resins are prepared by reacting maleic anhydride with a suitable amine and reaction proceeds through bismaleimic acid intermediate. The examples of commercial resins are Kerimid 601 (marketed by Rhone Poulenc), which is a product of bis(4-maleimidophenyl) methane (BDM) and MDA. Depending on the nature of amines, different types of bismaleimide network can be made. Hu and co-workers [59] investigated a BMI resin, comprising 2,2'-diallylbisphenol A (DBA) and 4,4'-bismaleimidodiphenylmethane (BDM) which shows improved thermal stability, mechanical and dielectric properties. The curing mechanism of the resin involves multiple reactions, for example, "Ene" and "Diels-Alder" reactions between maleimide group of BDM and allyl group of DBA and the self-polymerization of BDM via C=C bonds as shown in Fig. 3.6.

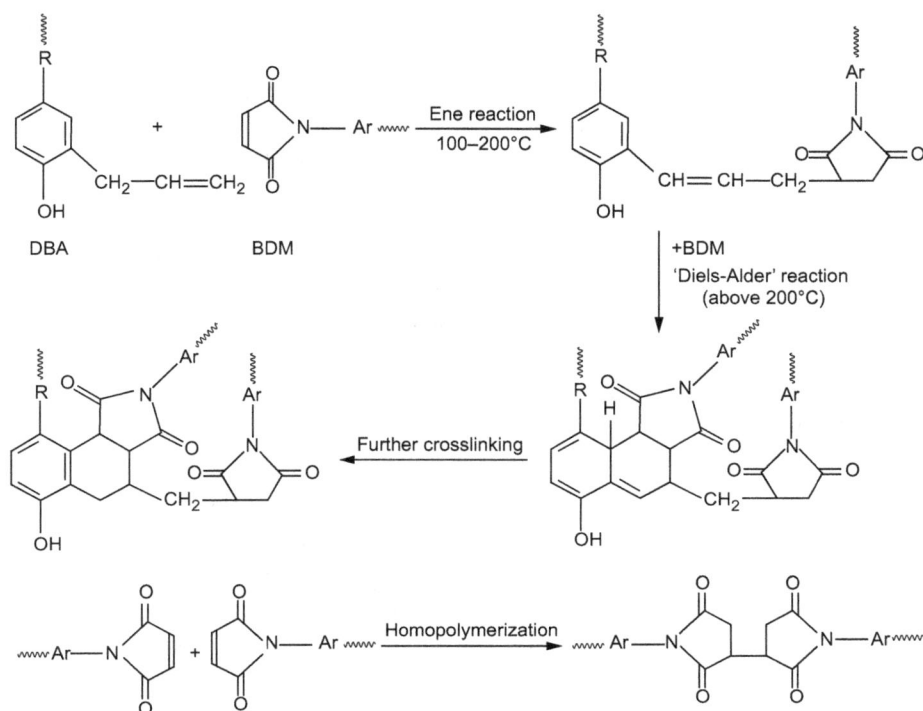

Fig. 3.6: The chemical reactions among BDM and DBA (reprinted with permission from J. Hu, A. Gu, G. Liang, D. Zhuo, L. Yuan *eXPRESS Polymer Letters* (2011) 555–568. © 2011 Online Publishing).

The common issue of low toughness for all the thermoset networks is also applicable to BMI resin. The research efforts have been made to resolve the issue. Use of BMI monomer containing ether group such as 2,2'-bis[4-(4-maleimidephen-oxy) phenyl] propane (BMPP) with conventional monomer used for making BMI resin is reported to result in significant enhancement in toughness [60]. Incorporation of optimal concentration of BMPP with BDM and DBA, the impact strength and critical stress intensity factor (GIC) can be enhanced up to 103% and 85%, respectively, without a significant sacrifice in thermal property (i.e. glass transition temperature of about 300 °C). A number of research papers have also been published addressing issue of processability of BMI resin using asymmetric monomer for the synthesis namely 2-[(4-maleimidophenoxy) methyl]-5-(4-maleimidophenyl)-1,3,4-oxadiazole (Mioxd) and 2-[(4-maleimidophenoxy) phenyl]-5-(4-maleimido-phenyl)-1,3,4-oxadiazole (*m*-Mioxd), 2,3,3'4'-oxydiphthalic dianhydride (a-ODPA). Various combinations of symmetric anhydride, asymmetric anhydride, symmetric amine and asymmetric amine can be used [61–63]. The asymmetric monomers hinder the packing of the molecular chains and thereby reduce the crystallinity, melting point and viscosity of the uncured BMI resins. As a result, solubility of the resin in low boiling solvents such as 1,4-dioxane and chloroform is significantly improved. However, such improvement in solubility is associated with a significant sacrifice in thermal stability and service of the related composites. The glass transition temperature of the asymmetric precursors-based bismaleimide is found to be close to 250 °C, which is much lower than polyimide resin. In addition to use of BMI resin for structural application, it is also used to modify the thermomechanical properties of other resins like phenolic and epoxy. When an allyl group containing addition phenolic resin is blended with bismaleimide resin, chemical bonding between phenolic and bismaleimide takes place through Alderene reaction leading to the formation of a crosslinked network. The properties of the blends in terms of thermal stability and impact strength can be tuned as a function of blend composition depending of the application as a wide variation in glass transition temperature (150–290 °C) and impact strength (175–230 kJ/m) can be generated. If thermal stability is the prime requirement we can select a blend with higher concentration BMI and if high impact strength is required for a particular application then a blend with higher concentration of phenolic has to be chosen.

3.1.2.5 Cyanate ester resin

Cyanate ester resin (CE) are characterized by the presence of two or more cyanate functional groups (–O–C≡N–) having an aromatic or cycloaliphatic backbone, which are known for good thermal stability. The monomer required for synthesizing CE resin was first reported by Grigat et al. [64] and commercial production of CE resin started sometime in the mid-1980s. CE resins are generally prepared by reacting cyanogen chloride with an alcohol or phenol in the presence of a tertiary amine [65, 66]. Diethylcyanamide is formed as a gaseous by-product. The formation of the by-product is more when cyanogen bromide is used instead of cyanogen chloride. Cyanogen bromide is preferable from a handling point of view as it is a solid substance unlike cyanogen chloride, which is a gas. In order to reduce the formation of the undesirable by-products, the reaction is carried out at low temperatures for a longer time to reduce the formation diethylcyanamide. The hydroxyl compounds used as a precursor for the reaction can also act as a catalyst for the trimerization of CE resin product. That is why for the preparation of CE resin with long shelf life, it is necessary to ensure almost complete consumption of phenol or alcohol in the reaction [67, 68]. The unreacted phenol present in the resin catalyses the curing of CE resin and reduces the shelf life. The curing of CE resin takes place through a thermally activated cyclotrimerizaton reaction at a temperature above 200 °C, which is basically an addition type of reaction and resulted in the formation of three-dimensional crosslinked network structure. It may be noted that cure temperature can be reduced by using a suitable catalyst. Like other thermoset the curing reaction is highly exothermic in nature, so adequate precaution should be adopted while processing the CE resin. The network structure comprises triazine rings, which are chemically bonded to the backbone structure through ether linkages. The presence of ether linkage is responsible for higher toughness of cured CE resin compared to bismaleimide or polyimides. Unlike in the phenolic resin where volatile gases are formed during curing, CE resin undergoes curing via an addition reaction and hence there is no issue of void formation observed during curing of CE resin. CE resin exhibits outstanding water absorption properties because of the symmetric cyclic structure and a very less number of polar groups it shows exceptionally low water absorption and dielectric loss. The cured CE resins when subjected to adequate post-curing offers a glass transition temperature of 350–400 °C and sustain a thermal ageing of more than 100 h at 300 °C with any significant loss in mechanical properties.

3.1.2.6 Phthalonitrile resins

Phthalonitriles (PN) are comparatively new class of high-temperature-resistant thermosetting resin family, which have drawn considerable research interest. Because of their high thermal stability and outstanding mechanical properties and superior fire retardancy many potential applications [69–76] in the area of high-temperature structural composites and adhesives have been proposed. Similar to cyanate ester resin, PN resin has a chemical structure in which aromatic and heterocyclic rings linked through. The high-temperature resistance of PN resins is the outcome of its chemical structure having aromatic and heterocyclic rings linked with flexible ether linkages form the backbone. PN resins are basically the product of reaction between 4-nitro-phthalonitrile and an aromatic dihydroxy compound in the presence of a basic catalyst, for example, K_2CO_3 in polar aprotic solvents like DMSO. The initial product which is formed is called "Stage A" resin. The same can be transformed into a stable prepolymer (green-coloured) known as Stage B resin. The Stage B resin can be stored for indefinite time. This property is quite different compared to other high-temperature thermoset resin which expires after a time of about a year at room temperature. Therefore, it is necessary to store them in subambient temperature. Another advantage is the ease of processing for fabrication of composites using various techniques. Phthalonitriles are potential high-temperature materials which work for thousands of hours at 200 °C, hundreds of hours at 300 °C and minutes at 540 °C without a considerable sacrifice in mechanical properties. In addition these materials retain useful properties under particular conditions like pressure or vacuum; mechanical loading, radiation, chemical or electrical exposure, all at temperatures ranging from subambient to above 500 °C. Unlike other thermosetting resins which are readily available in commercial market by a number of industries, PN resins are commercialized only by few companies, for example, M/s. Maverick Corporation, M/s. Cardolite Corporation, USA, M/s. JFC Technologies, USA.

Although PN resins offer many useful properties as mentioned above, there are some technical challenges which are related to very high curing temperature, slow and complex curing reaction [77–79]. Therefore, new tools and machineries are to be fabricated to process the composite parts using PN resin systems. In addition, more research and development initiative are required to tailor the chemistry in such a way so that to make these resins user-friendly encompassing variety of processing methods. The curing process and thermal stability can be improved [80] by using a novel curing process with mixed curing agents. With this objective, a new type of phthalonitrile-etherified resole resin from resole resin and 4-nitrophthalonitrile was reported [81]. The curing reaction is accelerated resulting a decreased cure time compared to those of phthalonitrile resin due to synergistic effect among the phenolic hydroxyl groups, hydroxymethyl and nitrile groups. Self-curing triphenol-A-based phthalonitrile resin precursor has also been reported [82], which can act as a flexibilizer and curing agent for phthalonitrile resin.

Chen and co-workers synthesized [83] a new phthalonitrile-based resin-containing benzoxazine (A-ph) using the aniline and 3-aminophenoxyphthalonitrile (3-APN) in a mole ratio of 3:1 as the ammonia source. The PN resin is ended with CE resin and the effect of bending on processability and fracture behaviour was investigated. The chemical structure of the PN resin and CE resin is shown in Fig. 3.7. As the resin blend is cured copolymerization between PN and CE resins takes place. The probable polymerization processes and mechanisms are presented in Fig. 3.8. It can be seen that the ring-opening polymerization of oxazine rings occurred preferentially in the presence of trace hydrogen (Fig. 3.8a). Then, more active hydrogen and amine groups are generated, which would significantly promote the ring-forming polymerization of cyanate and nitrile groups (Fig. 3.8b–d). In the end, the crosslinked network would be formed with various heteroaromatic rings (Fig. 3.8e) including isoindole, triazine. In the blend system, no clear boundary between ring opening of oxazine ring and self-crosslinking polymerization of CE was observed. However, copolymerization occurred between PN and CE resins due to ring-opening polymerization of oxazine and ring-forming polymerization of cyanate and nitrile groups. When the concentration of CE was more than 40 wt%, the self-polymerization of cyanate would dominate the ring-forming polymerization. First ring-opening polymerization of oxazine occurs and then the ring-forming polymerization of cyanate and nitrile groups would be encouraged by active hydrogen and amine structures generated from ring-opening of oxazine rings. It was concluded than addition 1 wt% CE resin with the PN resins resulted in optimum thermomechanical properties.

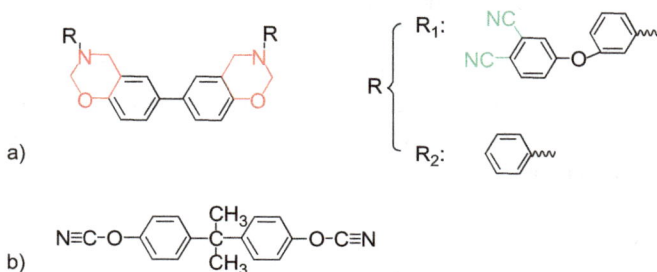

Fig. 3.7: Structures of (a) phthalonitriles-containing benzoxazine (A-Ph) and (b) cyanate ester (CE) (reprinted with permission from L. Chen, D. X. Ren, S. J. Chen, H. Pan, M. Z. Xu, X. B. Liu, *eXPRESS Polymer Letters* (2019) 456–468. © 2019 Online Publishing).

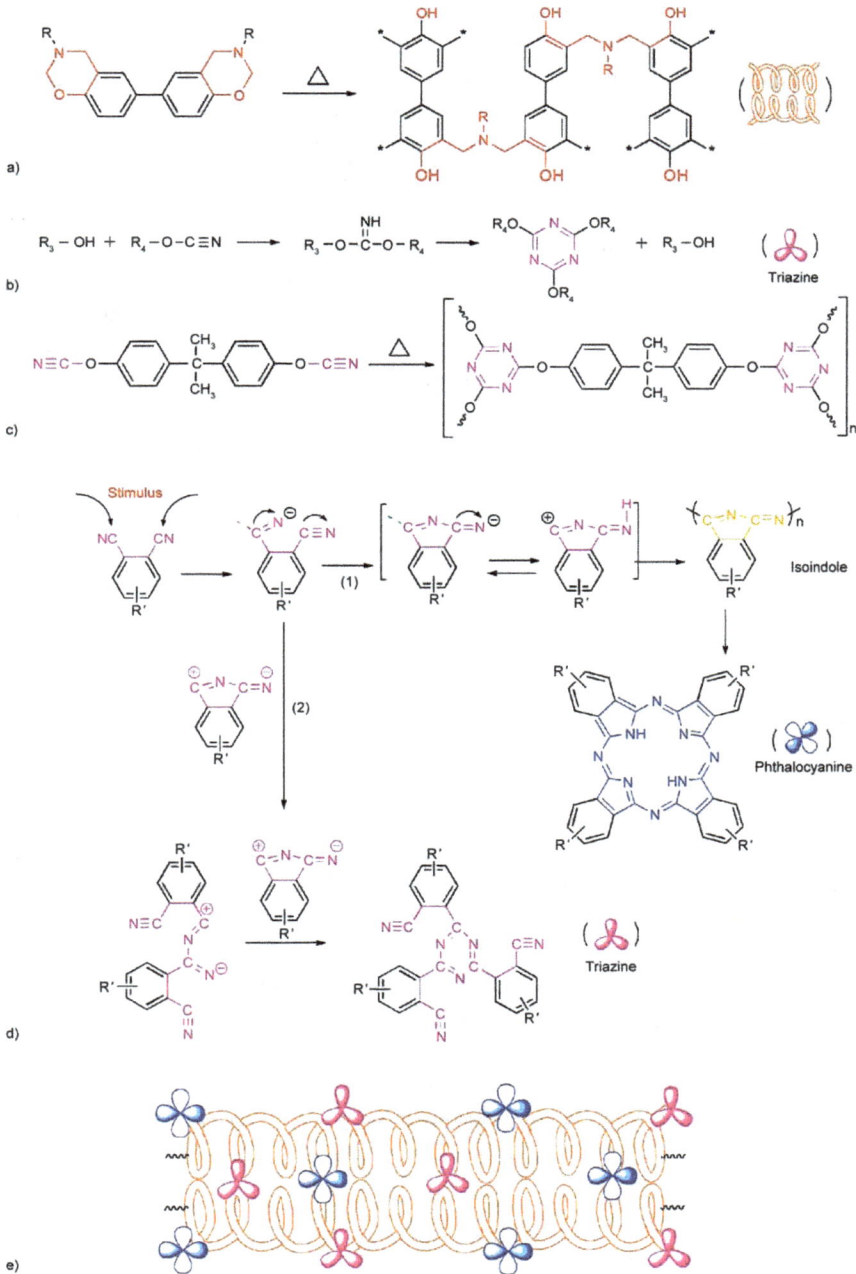

Fig. 3.8: (a) The Mannich bridge structure of ring-opened benzoxazine rings in A-ph; (b) and (c) cyanate esters form a triazine ring structure through phenolic hydroxyl groups and its thermal polymerization,

3.2 Kinetics of thermoset curing

Conventionally, the study of a chemical reaction by thermo-analytical tools needs a suitable reaction rate model which is either derived from the reaction mechanism or empirically described on the basis of experimental data, particularly for complex multi-step process. However, generally the mathematical models assume constant values of activation energy (E) and pre-exponential factor (A) throughout the reaction process with a fixed function of conversion $f(\alpha)$, for example, first order. Such approach may not actually represent a complex chemical reaction involving a solid core or where solidification (vitrification) takes place from a liquid reactant mass, for example, in a curing reaction of epoxy oligomer with a liquid diamine leading to an epoxy thermoset. This is because, with the progress of the reaction, several factors such as heat flow, autocatalysis and increasing crosslinking influence E and A for the most probable $f(\alpha)$. Hence, these parameters no longer remain independent of each other as assumed for simple reactions. In reaction involving vitrification, the rate of reaction is controlled more by diffusion (mass transfer) of the reacting species in the solid state rather than kinetic process. Near the gel point, the kinetic parameters before and after vitrification will be different. Such complexities may result in approximate fit for more than one model, leading to different model dependent values of E and A. Additionally, neither the activation energy nor the pre-exponential factor is constant during the progress of the reaction.

3.2.1 Isoconversion kinetics

Since E and A are not constant during the course of curing, these parameters are to be determined at every conversion to describe the curing process. Such conversion dependent kinetics treatment is called isoconversion kinetics. Using the same basic kinetic equation with Arrhenius temperature dependence of rate constant, the parameters E and A are determined from the DSC thermogram of a curing reaction. Since one particular model of conversion function $f(\alpha)$ must be valid for the reaction, it is advantageous to eliminate the function from the equation variable terms. The method then becomes a model-free kinetics. A and E are determined from identical conversion (α) from a set of experiments by a thermo-analytical tool. The study of the reaction can be in isothermal mode, several of them at different constant temperature or

Fig. 3.8: (continued)
respectively; (d) the nitrile group in A-ph forms an isoindole ring, a phthalocyanine ring and a triazine ring by the action of the stimulus; (e) the possible structures of the composites. A general formula for the representation of alkyl by "R", R3, R4' (reprinted with permission from L. Chen, D. X. Ren, S. J. Chen, H. Pan, M. Z. Xu, X. B. Liu, *eXPRESS Polymer Letters* (2019) 456–468. © 2019 Online Publishing).

non-isothermal mode with several runs with constant heating rate. The isothermal mode has a disadvantage that during the preheating from ambient to the isothermal set temperature, the sample reactants may undergo some change or reaction and hence the total enthalpy evolved is not possible to capture and the kinetic parameters may be inaccurate. This is particularly important for high-temperature curing of thermosets. Second, non-isothermal study always enables one to calculate the changes in activation energy at various temperatures as the reaction progresses, which in isothermal study would require large number of isothermal experiments.

Two independent studies by Ozawa [84, 85] and Flynn and Wall [86] first used this method for calculation of activation energy of a reaction by thermo-analytical methods at constant heating rates as a non-isothermal process. Subsequently, there were many reports in the literature for determination of activation energy, and a few are given here as references [87–94] and also the pre-exponential factor of Arrhenius expression of rate constant-temperature relationship [95].

The general kinetic equation can be stated as follows:

$$\frac{da}{dt} = k(T)f(a) \tag{3.1}$$

For a non-isothermal reaction, the expression will be

$$\frac{da}{dt} = Ae^{-E/RT}f(a) \tag{3.2}$$

where R is universal gas constant and A is the pre-exponential factor, also referred to as Arrhenius frequency factor.

In a DSC experiment, the heat flow rate (dH/dt) is recorded against time at a constant temperature (isothermal curing) or against temperature using a dynamic temperature run (non-isothermal curing) at a constant heating rate. In both the methods, the heat generated at any instant normalized by the total exotherm is taken as the conversion at that instant:

$$\alpha = \frac{\Delta H_t}{\Delta H_0} \tag{3.3}$$

and hence,

$$\frac{da}{dt} = \frac{1}{\Delta H_0}\frac{dH_t}{dt} \tag{3.4}$$

dH_t/dt being simply the Y-axis value in the DSC plot at any instant t. The enthalpies ΔH_t and ΔH_0 are the area under the DSC thermogram up to the instant t and total area of the thermogram, respectively.

3.2.1.1 Friedman's method

Equation (3.2) can be solved for E by differential method. This method is termed as Friedman's method [96]:

$$\ln \left[\frac{d\alpha}{dt}\right]_{\alpha,j} = \ln[A_\alpha f(\alpha)] - \frac{E_a}{RT_{\alpha,j}} \tag{3.5}$$

The subscript α implies that the data are taken for same extent of conversion from non-isothermal reaction experiment with different constant heating rates (β), where

$$\beta = \frac{dT}{dt}$$

so that

$$\frac{d\alpha}{dt} = \left(\frac{dT}{dt}\right)\left(\frac{d\alpha}{dT}\right) = \beta\left(\frac{d\alpha}{dT}\right) \tag{3.6}$$

and j in eq. (3.5) represent the particular thermal programme, for example, at a heating rate (β) of 5, 10, 15, 20 . . . K/min, etc.

In eq. (3.5), the first term in the right hand side is constant, since at the particular conversion, A and the function $f(\alpha)$ are constant. From a non-isothermal DSC thermogram, the sets of $(d\alpha/dt)$ –$1/T$ for identical α at heating rates (β) 5, 10, 15 and 20 °K/min are calculated and the plot of eq. (3.5) represents a straight line for each set of α. The slope of the plot is $-E_\alpha/R$, from which E_α is calculated. The result gives an E-dependency on conversion. Since $f(\alpha)$ is not evaluated for calculation of E_α, the result is independent of the model describing $f(\alpha)$. For isothermal DSC, the $d\alpha/dt$ –$1/T$ sets are taken at identical conversions in each thermogram of several isothermal temperatures (e.g. 100, 120, 130 and 140 °C).

3.2.1.2 Kissinger method

Considering eq. (3.2), Kissinger [97] used the condition of the conversion rate maxima by taking derivative of the rate and equating to zero:

$$\frac{d}{dt}\left(\frac{d\alpha}{dt}\right) = 0 = \frac{d}{dt}\left[Ae^{-E/RT}f(\alpha)\right] \tag{3.7}$$

or,

Hence,

$$\frac{E\beta}{RT_p^2} = A\left[\frac{d}{dt}f(\alpha)\right]_p \exp\left(-\frac{E}{RT_p}\right) \tag{3.8}$$

The Arrhenius pre-exponent A and the derivative of $f(\alpha)$ at peak are constant due to the fixed conversion at the peak. Therefore, taking logarithm of eq. (3.8), we get

$$\ln\left(\frac{\beta}{T_p^2}\right) = \ln\left(A\left[\frac{d}{dt}f(\alpha)\right]_p\right) - \ln\left(\frac{E_p}{R}\right) - \frac{E_p}{RT_p} \tag{3.9}$$

A plot of the left-hand side against $1/T_p$ would result in a straight line and the activation energy can be calculated from the slope of the line.

Since the activation energy is calculated without $f(\alpha)$, it is also a model-free method. Generally, if the curing mechanism is same for the envelop of the temperature considered, then it has been observed that the peak rate corresponds to a particular conversion irrespective of heating rate (β). However, the activation energy thus calculated is not applicable to conversions other than at peak rate. Otherwise, being a differential method, it should yield same value of activation energy as Friedman method at the conversion corresponding to the peak rate.

3.2.2 Integral form of model-free kinetics

The model-free kinetics of isoconversion can be in the integrated form of eq. (3.2). The integration of eq. (3.2) gives

$$g(\alpha) = \int_0^a \frac{d\alpha}{f(\alpha)} = \frac{A}{\beta}\int_{T_0}^T \exp\left[-\frac{E_a}{RT}\right]dT \tag{3.10}$$

or

$$g(\alpha) = \left(\frac{AE_a}{\beta R}\right)p\left(\frac{E_a}{RT}\right) = \left(\frac{AE_a}{\beta R}\right)p(x) \tag{3.11}$$

where $x = \dfrac{E}{RT}$ and $p(x)$ is the integral defined as

$$p(x) = \left[\frac{e^{-E/RT}}{E/RT} + \int_{-\infty}^{-E/RT}(e^{-x}/x)dx\right] \tag{3.12}$$

The integral in eq. (3.10) cannot be analytically solved. However, there can be either a numerical integration or an approximate value of the integral may be used. Doyle [98] Murray and White [99] gave approximate values. Coats and Redfern [100] used an approximation by assuming an nth-order kinetic expression. A rational approximation up to four degrees of x is more frequently used for better accuracy which is given by Senum and Yang [101, 102]. Various solutions to eq. (3.10) based on approximations give minor variations in the value of activation energy.

From eq. (3.11), taking logarithm,

$$\log[g(\alpha)] = \log(AE/R) - \log\beta + \log p[x] \tag{3.13}$$

A table of x (=E/RT) and corresponding –log $p(x)$ is given by Toop [103] for the first approximate value. With Doyle's approximation [97] for the value of log $p(x)$:

$$\log p\left(\frac{E}{RT_i}\right) \cong -2.315 - 0.457\frac{E}{RT_i} \tag{3.14}$$

and value of $g(a)$:

$$g(\alpha) = 7.03\frac{AE_a}{\beta R}(10^{-3})\exp(-1.052x) \tag{3.15}$$

3.2.2.1 Flynn-Wall-Ozawa method

Flynn-Wall-Ozawa (FWO) method [87, 88] is based on Doyle's approximation as shown in eqs. (3.14) and (3.15). Taking logarithm of eq. (3.15), we have

$$\ln(\beta) = \ln\left(\frac{AE_a}{R}\right) - \ln[g(\alpha)] - 4.9575 - 1.052\frac{E_a}{RT} \tag{3.16}$$

At a constant conversion a, the functions $g(a)$, A and E_a are constant. Therefore, eq. (3.16) can be written as

$$\ln(\beta) = \text{Cons.} - 1.052\frac{E_a}{RT} \tag{3.17}$$

Therefore, when a particular extent of decomposition is to be taken for life time assessment, the TGA runs for several heating rates β_1, β_2, . . . should be taken and corresponding temperatures T_1, T_2, . . . should be noted at that constant extent of conversion (weight loss). Subsequently, a plot of log(β) vs 1/T should yield activation energy E_a from the slope of the curve. If more than one conversion is taken, then we should ideally have parallel lines for each conversion; otherwise the data scattering will be too high to determine a reliable activation of energy at various conversions.

FWO method is not so accurate because of the Doyle's approximation. However, it gives comparatively acceptable result.

3.2.2.2 Kissinger-Akahira-Sunose (KAS) method

Integral form of KAS method [103] uses the approximate integral value of $p(x)$ from Murray and White [98] as

$$p(x) \cong \frac{e^{-x}}{x^2}$$

The KAS equation is supposed to give more accurate estimation of activation energy:

$$\ln\left(\frac{\beta}{T^2}\right) = \ln\left(\frac{AR}{E_a}\right) - \ln[g(\alpha)] - \frac{E_a}{RT} \tag{3.18}$$

$$\ln\left(\frac{\beta}{T^2}\right) = \text{Cons.} - \frac{E_a}{RT} \tag{3.19}$$

A plot of left side of the equation against $1/T$ will be a straight line with a slope E_a/R and the value of E_a is calculated therefrom.

3.2.2.3 Starink method

A comparison of all approximation of $p(x)$ was done by Starink [104] where it is seen that all the approximations are derived from the series expansion of $p(x)$ by different researchers. The following approximation given by Starink for $p(x)$ is supposed to be more accurate:

$$p(x) \cong \frac{\exp(-1.0008x - 0.312)}{x^{1.92}} \tag{3.20}$$

and corresponding expression for estimate of activation energy:

$$\ln\left(\frac{\beta}{T^{1.92}}\right) = \text{Cons.} - 1.0008\frac{E_a}{RT} \tag{3.21}$$

The plot of the left side against $(1/T)$ would give a straight line with the slope $(1.0008E_a/R)$. It is seen that the Starink method is almost similar to KAS method, with a small difference in the power of T on the left side expression and also a minute difference for the slope.

3.2.3 Advanced isoconversion models

The integral methods use integration of the function $p(x)$ at every conversion for 0 to t. Therefore, there is always adding up of errors in each infinitesimal step. This results in so-called flattening of data and gives rise to errors. Vyazovkin [105–107] suggested a modification of the determination of activation energy by integration of a small interval of conversion and a non-linear iterative process to determine activation energy more accurately. The small interval of conversion ($\Delta\alpha$) can be 0.01. He proposed minimization of the following function by iteration to arrive at a successively constant value of activation energy in each conversion:

$$\phi(E_a) = \sum_{i=1}^{n} \sum_{j\neq1}^{n} \frac{J[E_a, T_i(t_\alpha)]}{J[E_a, T_j(t_\alpha)]} \tag{3.22}$$

where J is the integral defined as

$$J[E_a, T_i(t_a)] = \int_{t_a - \Delta a}^{t_a} \exp\left[\frac{E_a}{RT_i(t)}\right] dt \tag{3.23}$$

Cai and Chen [108] subsequently used a linear iterative method, also using small conversion interval ($\Delta a = 0.01$).

$$g(a, a - \Delta a) = \frac{A_{a-\Delta a/2}}{\beta} \int_{T_{a-\Delta a}}^{T_a} \exp\left[-E_{a-\Delta a/2}/RT\right] dT$$

$$= \frac{A_{a-\Delta a/2}}{\beta} \left[\int_0^{T_a} \exp(-E_{a-\Delta a/2}/RT)dT - \int_0^{T_{a-\Delta a/\Delta a22}} \exp\left(-E_{a-\Delta a/2}/RT)dT\right] \tag{3.24}$$

Budrugeac [109] proposed the following iterative process with small conversion interval:

$$\ln\frac{\beta}{T_2 - T_1} = \ln\frac{A}{g(a_2) - g(a_1)} + \ln R_1 - \frac{E}{RT_2} \tag{3.25}$$

Details of all the above methods and the advantages, limitations, applicability and the methods to carry out a meaningful kinetics by thermal analysis suggested by various researchers and organizations are discussed in Chapters 6 and 7.

3.2.3.1 Pre-exponential factor

The pre-exponential factor is considered as the frequency of collisions of reacting species. This is assumed to be constant in case of unhindered collisions, particularly in gaseous or liquid state. However, in a progressively vitrifying reaction mass, the number of collisions may not be same as the reaction progresses. It is thus important to study this parameter when considering a curing reaction. In order to get a complete reaction rate from DSC thermogram, we need the pre-exponential factor (A_a) of Arrhenius equation of rate constant. From eq. (3.5), A_a can be calculated from the intercept as follows:

$$\ln(A_a) = I_a - \ln f(a) \tag{3.26}$$

where I_a is the intercept. The function $f(a)$ can be defined by many mathematical expressions mostly derived from basic theories as tabulated below in Table 3.2. Therefore, using different $f(a)$ selected from the list, one can get various values of A_a. Therefore, a set of $E_a - \ln(A_a)$ pair for each conversion can be obtained. It is generally observed that the relationship is linear

$$\ln(A_a) = a + bE_a \tag{3.27}$$

This is termed as compensation effect for variation of model specific values of E_a and A_a. However, the relationship could be a simple polynomial function as well.

3.2.3.2 Model fitting

It can be seen that the values of E_α and A_α are independent of heating rate. Hence, in order to investigate the most suitable model $f(\alpha)$ satisfying the pair of $E_\alpha - A_\alpha$ for various conversions, calculated values of conversions from eq. (3.9) can be compared with the conversions (α) taken from the DSC thermogram for any heating rate. Equation (3.9) is actually the integrated form of eq. (3.2):

$$\alpha = A_\alpha \int_{T_{i=1}}^{n} \exp\left[-\frac{E_\alpha}{RT_\alpha}\right] f(\alpha) dt \qquad (3.28)$$

The integral $\int \exp\left[-\dfrac{E}{RT}\right] dt$ can be evaluated by some approximation and such approximate integrals are listed in several literatures [97–102, 104, 105]. It should be noted that the function $f(\alpha)$ is constant for a constant conversion. However, for convenient analysis, numerical integration is done to obtain the conversion at various temperatures corresponding to chosen conversions of the DSC thermogram with one heating rate. An average $f(\alpha)$ vs α of all heating rates can be calculated and compared with the corresponding average calculated values using eq. (3.9). For that, the average values of T_α, A_α, $d\alpha/dt$ and $f(\alpha)$ for each conversion should be calculated. The calculated conversion function $f(\alpha)$ using eq. (3.9) can be tried to fit with a model in Table 3.2.

Tab. 3.2: Some reaction models $f(\alpha)$ used for study of reaction kinetics of solids.

No.	Name of the model	Symbol	$f(\alpha)$
1	Power law	P4	$4\alpha^{3/4}$
2	Power law	P3	$3\alpha^{2/3}$
3	Power law	P2	$2\alpha^{1/2}$
4	Power law	P2/3	$(2/3)\,\alpha^{-1/2}$
5	First order	F1	$1 - \alpha$
6	Second order	F2	$(1 - \alpha)^2$
7	Avrami–Erofeev	A2	$2(1 - \alpha)[-\ln(1 - \alpha)]^{1/2}$
8	Avrami–Erofeev	A3	$3(1 - \alpha)[-\ln(1 - \alpha)]^{2/3}$
9	Avrami–Erofeev	A4	$4(1 - \alpha)[-\ln(1 - \alpha)]^{3/4}$
10	Contracting Cylinder	R2	$2(1 - \alpha)^{1/2}$
11	Contracting Sphere	R3	$3(1 - \alpha)^{2/3}$
12	1-Dimensional diffusion	D1	$(1/2)\,\alpha^{-1}$
13	2-Dimensional diffusion	D2	$[-\ln((1 - \alpha)]^{-1}$
14	3-Dimensional diffusion	D3	$(3/2)(1 - \alpha)^{2/3}[1-(1 - \alpha)^{2/3}]^{-1}$
15	Modified Sestak-Berggren	SB	$\alpha^m (1 - \alpha)^n$

3.2.3.3 Case study

An epoxy resin of average epoxy equivalent of 180 is being cured by a polyether diamine of molecular weight 800 under a non-isothermal DSC programme at heating rates 5, 10, 15 and 20 °C/min with a nitrogen flow. The sample and the cell initially were cooled down to −30 °C and the runs were taken up to 250 °C in each experiment. Subsequently, the cell was cooled down to −30 °C and the run was repeated to ensure completion of the curing reaction. The data was recorded as W/g (*dH/dt* per g) against temperature. The base line correction was done solving a linear equation taking the coordinates of start to end of the DSC peak in the thermogram. Figure 3.9 shows the four thermograms of the non-isothermal DSC study. The noticeable beginning of the reaction can be seen at about 70–75 °C and completed at around 250 °C. The peak heat flow rate takes place at higher temperature as the heating rate is increased.

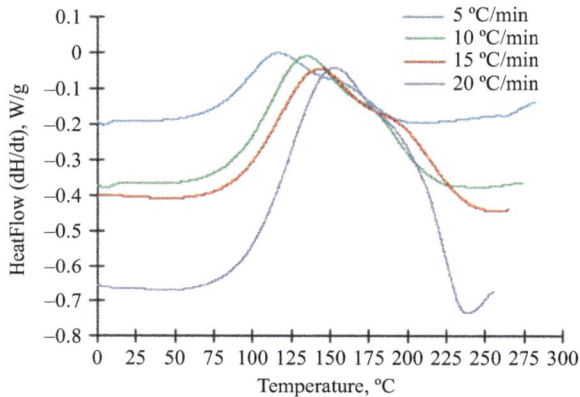

Fig. 3.9: DSC thermogram of epoxy curing at four heating rates (non-isothermal).

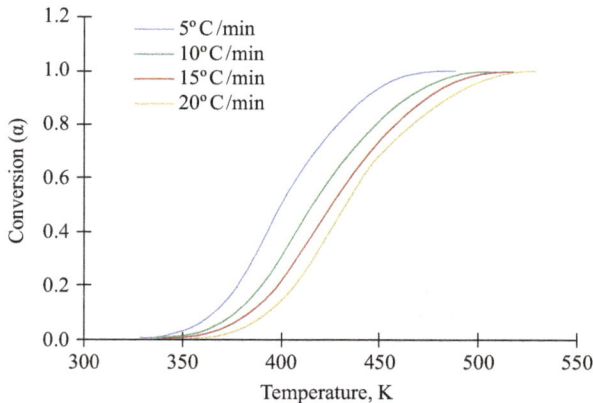

Fig. 3.10: Conversion (extent of reaction) of epoxy curing by non-isothermal DSC scan.

The enthalpies of reaction were calculated and are shown in Table 3.3. The conversion (α) corresponding to the temperatures for each heating rate plot is calculated using eqs. (3.3) and (3.4) and plotted as $T - \alpha$ in Fig. 3.10. The conversion lines are almost parallel, indicating identical reaction mechanism of curing for all heating rates. The conversion at a particular temperature is higher for slower heating rate.

Tab. 3.3: Enthalpy of curing of epoxy at four heating rates: non-isothermal DSC.

Heating rate (Δ), °C/min	5	10	15	20
Enthalpy of curing (ΔH_0), J/g	120.76	134.28	145.27	156.63

The kinetic parameters calculated from the thermograms according to one differential method of Friedman using eq. (3.5) and one integral method by Starink using eq. (3.21). The kinetic plots for these two methods are shown as Fig. 3.11 (Friedman) and Fig. 3.12 (Starink) at conversions from 0.05 to 0.95. Both the kinetic plots are almost perfectly linear for all conversions, indicating fairly good DSC thermogram data, baseline correction and calculated values of ΔH_0 and conversion-dependent ΔH from $\alpha = 0.05$–0.95.

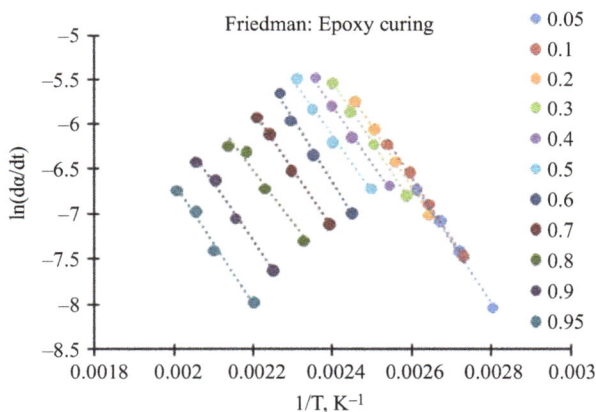

Fig. 3.11: Kinetic plot according to Friedman method: non-isothermal DSC study.

The conversion-dependent activation energy was calculated for these two methods in the same conversion range and plotted against conversion in Fig. 3.13. The activation energy Friedman method has overall range of 50–60 kJ/mol, but have some variations around 50–60% and 80% conversion, and with average value of 54.5 kJ/mol while Starink method has resulted in almost constant activation energy ranging from 52 to 56 kJ/mol with an average of 54 kJ/mol, same as Friedman method. The variation of the activation energy for Friedman method is obviously due to the differential method, where even

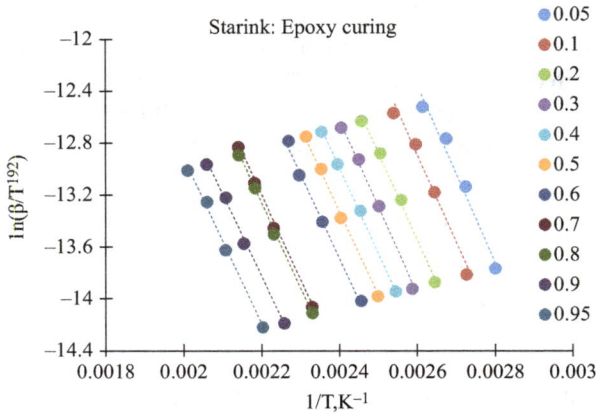

Fig. 3.12: Kinetic plot according to Starink method: non-isothermal DSC study.

minute error in da/dt data can cause such variation, while in Starink method of approximate integration, the cumulative errors are averaged out.

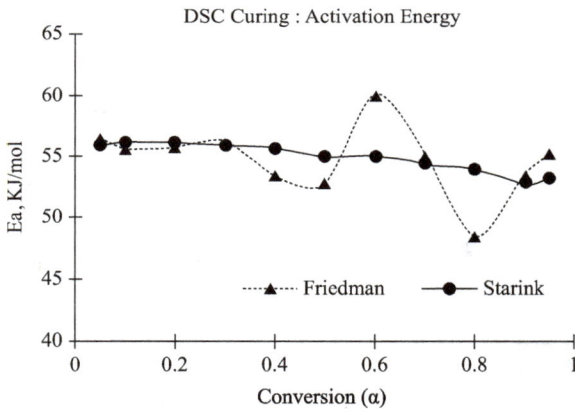

Fig. 3.13: Conversion-dependent activation energy of epoxy curing by Friedman and Starink methods.

The compensation parameters were calculated to obtain a nearly perfect linear relationship of E_a and A_a when the sets of these two parameters are calculated for all the function $f(a)$ in Table 1 (except Sestak-Berggren model). The calculation was done using the following rearranged equation:

$$\ln\left(\frac{1}{f_k(a)}\frac{da}{dt}\right) = \ln A_{k,a} - \frac{E_{k,a}}{RT_{k,a}}$$

(3.29)

In the above expression, k stands for the particular model $f(\alpha)$ taken from Table 1. The da/dt and T values correspond to a heating rate. Therefore, for the four heating rates, we have four sets of calculated values of compensation parameters. The plots of the compensation parameters are shown in Fig. 3.14.

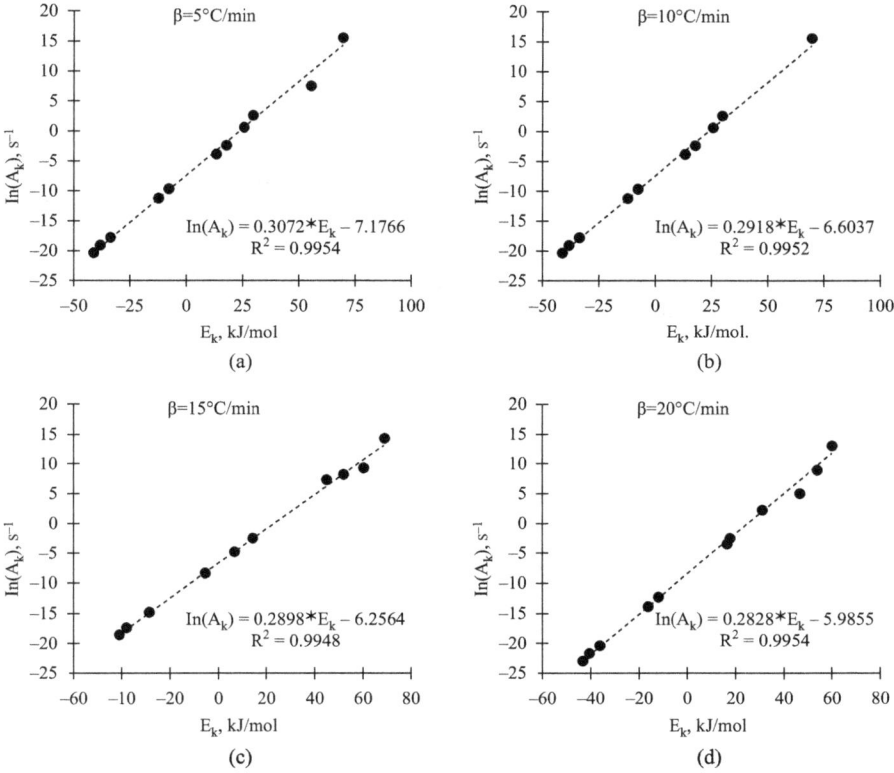

Fig. 3.14: Compensation parameters: (a) 5 °C/min, (b) 10 °C/min, (c) 15 °C/min and (d) 20 °C/min.

Subsequent to the four sets of $E_\alpha - A_\alpha$ of the four heating rates, $f(\alpha)$ for the heating rates were calculated using the rearranged equation:

$$f(\alpha) = \frac{(da/dt)_\alpha}{A_\alpha \exp[-E_\alpha/RT_\alpha]} \qquad (3.30)$$

Hence, we have four $f(\alpha)$ calculated values for four heating rates. We need to have one function only for the reaction because the reaction model should not depend on heating rate or any thermal programme. In order to do an averaging, we can find average A, T, α, da/dt and calculate $f(\alpha)$ for each heating rates and then calculate average $f(\alpha)$.

In this case, we resort to the first method. The average values, fitted to linear or polynomial equations, are given in Tab. 3.4.

Tab. 3.4: Average values: best-fit equations.

Parameters	Best-fit equation for average values of four heating rates	R^2
$\alpha - T/1{,}000$	$\alpha = -1023\,(T/1{,}000)^3 + 1{,}280\,(T/1{,}000)^2 - 552.68\,(T/1{,}000) + 69.965$	0.9986
$\alpha - E_a$	$E_a = 3.4275\,\alpha^3 - 9.0946\,\alpha^2 + 2.512\,\alpha + 55.848$	0.965
$E_a - \ln(A_a)$	$\ln(A_a) = 0.2943\,E_a - 6.5056$	1.00
$\alpha - da/dt$	$da/dt = 0.0103\,\alpha^3 - 0.0239\,\alpha^2 + 0.0142\,\alpha + 5E\text{-}05$	0.996
$\alpha - f(\alpha)$	$f(\alpha) = -86.957\,\alpha^6 + 269.66\,\alpha^5 - 333.44\,\alpha^4 + 215.35\,\alpha^3 - 76.909\,\alpha^2 + 10.028\,\alpha + 2.1427$	0.9998

Having the average $f(\alpha)$ from the above parameters for $\alpha = 0.05$ to $\alpha = 0.95$ range, we now observe that the shape of the $f(\alpha)$ against α is S-shaped, which indicates that it should be an autocatalytic reaction, and we find a best fit using Sestak-Berggren model (No. 15 in Table 1). The calculated values of $f(\alpha)$ and the fitted Sestak-Berggren model are shown in Fig. 3.15. The regression coefficient $R^2 = 0.9954$ suggests that the model is quite in agreement with the experimental data.

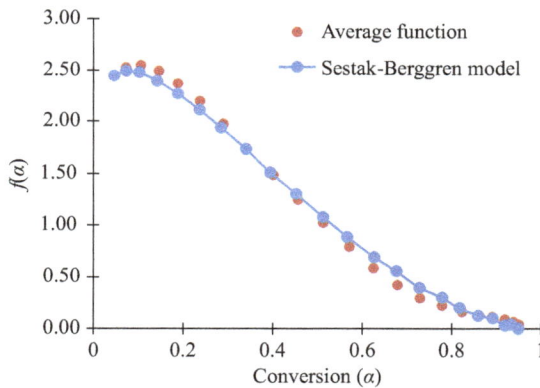

Fig. 3.15: Model fitting: Sestak-Berggren model for epoxy-curing study. Comparison with the calculated $f(\alpha)$ values as studied by non-isothermal kinetics.

3.2.4 Prediction of T_g

The glass transition temperature is the temperature below which it is glassy (hard) and above which it behaves like a soft rubbery material. Thermodynamically glass transition is a second-order transition at which the second derivative of free energy (volume, heat capacity) versus temperature plot is continuous. The second derivative of free energy (volume, heat capacity) versus temperature plot is discontinuous for melting, which is a thermodynamically first-order transition. For melting the first derivative of free energy (enthalpy) versus temperature plot is discontinuous. Polymers being long chain molecule cannot be more than 80% crystalline. Therefore polymers are either amorphous or semi-crystalline. Thermosets are mostly amorphous in nature and characterized by a glass transition [110, 111]. The glass transition is the consequence of segmental motion and called α-relaxation. According to reptation theory polymer chains can be considered to be made up with a large number of segments having number of carbon atom ranging from 20 to 40 nos. As the temperature increases the free volume of the polymer increases and at a certain temperature the free volume becomes sufficient to start the segmental motion. The segments undergo motion keeping the end-to-end distance unaltered like the snake does. The temperature is called glass transition temperature (T_g). Below T_g also some relaxations take place. These are called β- and γ-relaxations.

The modulus of a polymer drastically decreases when it attains the T_g. A composite cannot work for structural applications if the glass transition temperature of the matrix resin is lower than service temperature even if it is made with extremely high modulus fibre. Because the composite will lose its dimensional stability once the T_g is attained. As mentioned earlier the resin should be selected in such a way that the glass transition temperature of the resin should be at least 50 °C higher than the service temperature. For an application, it may so happen that a material has to work partly at a low temperature and partly at a high temperature. In that case, the higher service temperature has to be considered for selection of resin system for fabrication of composite for such applications. Therefore prediction and determination of T_g is very important from composite application point of view. The various methods of determination of T_g will be discussed in Chapter 6. The prediction of T_g of a thermoset network will be discussed here.

The state of a thermoset resin is determined by its chemical conversion as a result of curing or its extent of cure. From the process control point of view, it is important to know the relationship between the chemical conversion of a thermoset and its glass transition temperature. A unique relationship between the T_g and chemical conversion independent of cure temperature and thermal history has been reported for many thermosets although it does not hold well for all the thermoset resins. The one-to-one relationship between T_g and the conversion implies that either the network structure does not change with cure temperature or the associated change in network structure does not have any significant effect on T_g. Such relation is found to hold good for epoxy/amine systems. However, the temperature-independent relationship is reported to be not valid for epoxy/dicyanamide [112], phenolic/hexamethylene tetramine and UPE resins [113]. For the thermosetting resin systems for which the relationship exists, measurement of conversion is equivalent to the measurement of T_g. Both the data can be generated from DSC analysis and compared. T_g can be measured from a more sophisticated instruments like dynamic mechanical thermal analysis (refer to Chapter 6 for detailed information). The T_g measurement can be used as a practical means for the determination of the degree of cure. This is especially useful where the determination of heat of reaction is erroneous due to the loss of crosslinker or for analytical errors.

With the advancement of curing reaction, it is very easy to understand that the T_g of the resin will increase. However, the goal is to quantitatively predict the T_g of a resin as a function of cure conversion. Several models have been proposed to correlate the T_g with the conversion or extent of curing (α). With the increase in conversion, the concentration of reactive functionalities decreases, crosslinks or junction points is formed leading to the departure from Gaussian behaviour. The steric hindrance arises on the chain conformation at high crosslink densities. The models basically are based on the statistical description of network formation and calculation of the concentration of junction points of different functionalities as a function of conversion. However, one issue, which complicates the calculation and not fully resolved, is that whether to consider all the junction point or only those which are elastically effective.

Dibenedetto equation has been successfully utilized to correlate the experimental values of T_g as a function of conversion for many thermosetting resins like epoxy, phenolics. The relation is presented below [114]:

$$\frac{T_g - T_{g0}}{T_{g\infty} - T_{g0}} = \frac{\lambda\alpha}{1 - (1-\lambda)\alpha} \tag{3.31}$$

where T_{g0} is the T_g of the resin mixture before cure, $T_{g\infty}$ is the T_g obtainable after maximum possible curing and λ is an adjustable parameter. A similar equation can be derived considering the isobaric heat capacity change as a variable as given below [115, 116]:

$$T_g = \frac{\alpha\Delta c_{p\infty} T_{g\infty} + (1-\alpha)\Delta c_{p0} T_{g0}}{\alpha\Delta c_{p\infty} + (1-\alpha)\Delta c_{p0}} \tag{3.32}$$

where Δc_{p0} and $\Delta c_{p\infty}$ are the change in heat capacity corresponding to T_{g0} and $T_{g\infty}$.

Comparing the two equations we get

$$\lambda = \frac{\Delta c_{p\infty}}{\Delta c_{p0}} \tag{3.33}$$

As the curing reaction advances, the heat capacity change decreases. Montserrat [117] proposed an equation to correlate $\Delta c_p(T_g)$ with the T_g of the network as given below:

$$\Delta c_p(T_g) = x + \frac{b}{T_g} \tag{3.34}$$

Neglecting the constant (x), which may be applicable for a particular case we get

$$\lambda = \frac{\Delta c_{p\infty}}{\Delta c_{p0}} = \frac{T_{g0}}{T_{g\infty}} \tag{3.35}$$

Combining eqs. (2.31) and (2.34) and rearranging we get

$$\frac{1}{T_g} = \frac{(1-\alpha)}{T_{g0}} + \frac{\alpha}{T_{g\infty}} \tag{3.36}$$

This is an equation similar to the Fox equation, which is widely used to predict T_g of a copolymer as a function of composition. This simple rule-of-mixture equation can very precisely predict the glass transition temperature of a copolymer; however, it cannot precisely explain the experimental results of glass transition temperatures obtained by partially curing the thermoset. Therefore, a modified equation is proposed [118, 119]:

$$\frac{1}{T_g} = \frac{(1-\alpha)}{T_{g0}} + \frac{\alpha}{T_{g\infty}} + c\alpha(1-\alpha) \tag{3.37}$$

This equation predicts the increment of the glass transition temperature of cured thermoset at various stages of cure (from monomer stage to network polymer).

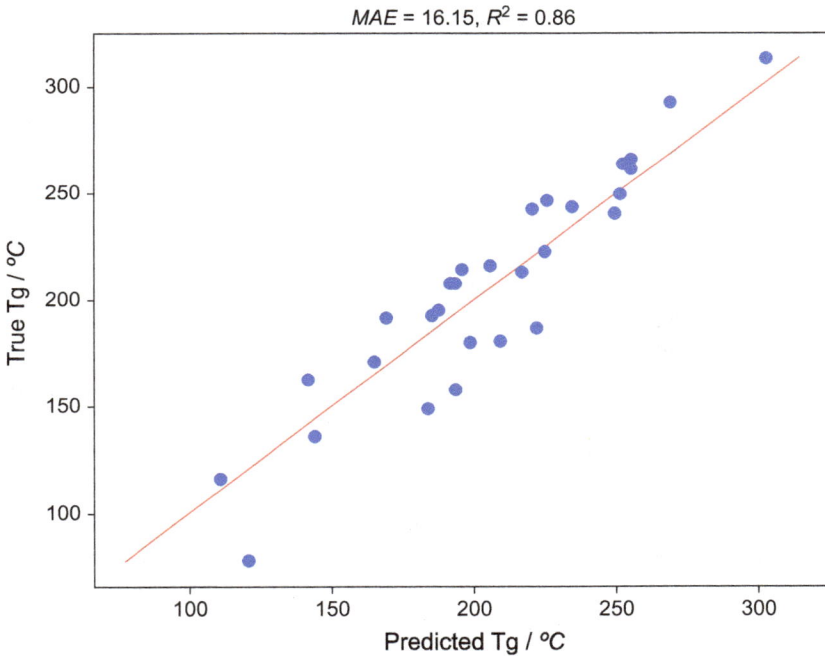

Fig. 3.16: Comparison between experimental (true) and predicted T_g using the optimized ensemble model with composition 53% GBR and 47% KRR. The predictions were performed for the test (29 samples) using the model trained with the training set (65 samples). MAPE = 9.38% (accuracy = 90.62%) [127].

A number of modelling and simulation works have been initiated to predict the T_g of a thermoset resins by using group additive property method [120, 121] and a more general quantitative structure–property relationships [122, 123]. A model for prediction of the dependence of T_g on the stoichiometric ratio of co-monomers in amine-cured epoxy systems was also developed [124]. Recently, machine learning (ML) approach has been utilized for the prediction of T_g of thermoset resins, which are based either on very small datasets and/or predominantly on datasets obtained using molecular dynamics [125, 126]. For example, artificial neural network has been used to optimize the method of prediction of T_g for multicomponent epoxy resin systems [125]. Meier et al. [127] used machine learning with the aim to predict the T_g for resin-hardener thermosets using an ML ensemble approach based solely on theoretical molecular descriptors that can be easily calculated for any molecular structure present in a thermoset. They have performed the predictions for the 29 samples using the model trained with the training sets. The performance of the trained ensemble model evaluated using 53% gradient boosting regression (GBR) AND 47% kernel ridge regression (KRR) is shown in Fig. 3.16, where mean absolute error (MAE), and R^2 are equal to 16.15 °C and 0.86, respectively, indicating about 90% accuracy.

3.3 Classification of thermoset composites

Thermoset composites can be broadly classified into three categories: (i) particulate composites; (ii) fibre-reinforced plastic (FRP) composites and (iii) nanocomposites [128, 129] depending on the reinforcement. In particulate composites, various particulate fillers like silica, carbon black, calcium carbonate, glass bead, glass balloons and silicon carbide are used and processed by mechanical mixing followed by casting, compression moulding, matched-die moulding, etc. FRP composites are of two types: chopped fibre-reinforced composite and long fibre-reinforced composites. Chopped fibre-reinforced composites are processed through dough moulding compound (DMC). DMCs contain about 25–40 wt% chopped fibre. The component or products which require very low permeability, for example, valve and pump, DMCs are used because long fibre-reinforced composites which typically contain more than 60 wt% and permeability particularly at high pressure is comparatively higher due to the presence of interfacial free volume. Recently author's research group has developed FRP composite ball valve using DMC comprising epoxy resin and a combination of chopped glass and carbon fibre [130]. It may be noted that fabric-based FRP composites failed to stop leakage at high pressure (20–30 bar). The DMC-based fabrication process is described below.

A typical recipe for making DMC comprises a thermoset resin (1.0 kg), required amount of curing agent, ammonium polyphosphate 25 g, melamine cyanurate 15 g and 11 g antimony trioxide. All the ingredients are mixed in a drum using a mechanical stirrer. Then 200 g wollastonite and halloysite clay 25 g were added and thoroughly mixed. Mixing takes about 30 min. Then the chopped glass or carbon fibre (360 g) are added and mixed with the help of a sigma kneader. The mixing should be done for about 10 min. The DMC can be used to prepare sheet for coupon test or component by using compression moulding. Generally about 10 kg/cm^2 pressure is applied for moulding and cure and post-curing are to be carried out depending on the cure schedule of the resin to be used for fabrication. The process has to be optimized taking resin manufacturer's data sheet as guidelines especially for thick and complex component. It may be noted that the tools used are to be extremely precise to achieve very low dimension tolerance of ±50 μm. The process is described in Fig. 3.17.

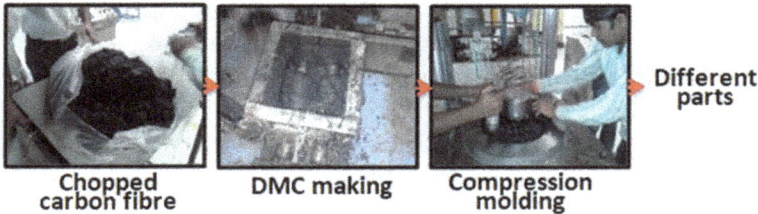

Chopped carbon fibre   **DMC making**   **Compression molding**   **Different parts**

Fig. 3.17: Description of processing of DMC for manufacturing of FRP component.

The long fibre-reinforced composites are made using uniaxial fibres or biaxial (2D) or 3D fabric and processed using various techniques like wet lay-up followed by compression moulding, prepreg lay-up with vacuum bagging and autoclave moulding, filament winding with oven-curing, pultrusion, resin transfer moulding (RTM), vacuum bag moulding, liquid composite moulding (LCM) and structural reaction engineering moulding, which will be discussed in detail in the subsequent sections. Such FRP composites offer very high strength and opportunity to tailor the material properties through the control of fibre and matrix combinations and processing conditions. The sizes of reinforcement for particulate and FRP composites are in the micron level. So they are called microcomposites. When reinforcement sizes are further reduced to nanolevel (1–100 μm) then the resultant composite is called nanocomposites. It is not necessary that for a nanocomposite enforcement sizes are nano in all dimensions; at least in one dimension,, it should be in the nanosize. For example, carbon nanotube is nano in diameter but micron size in length. Therefore a nanocomposite can be defined as the composite in which one of the constituents is in nanoscale in at least one dimension. Nanocomposites are prepared by using nanofillers like nanosilica, nanocalcium carbonate, nanotitania, nanoclay, carbon nanofibres and carbon nanotubes and processed by mechanical mixing and sonication followed by casting or compression moulding. Processing of long fibre-reinforced composites is elaborated as given below.

3.4 Processing of FRP composites

3.4.1 Wet lay-up moulding

This method involves combination of resin and reinforcment fibre by usual hand lay-up or spray-up techniques. The process is described schematically in Fig. 3.18. The reinforcement (uniaxial or 2D or 3D fabric or rovings) are cut into the desired size and shape using a scissor or knife. The reinforcement of desired shape is kept on a clean mould and impregnated with liquid resin and hardener mixture. A suitable mould release agent like silicone grease has to be applied to ensure easy removal of the product from the mould. Depending on the design of the products, the reinforcements are applied layer by layer and laid on the mould. The mould is then kept inside a compression mould press and subjected to cure at required temperature under pressure. The cure temperature has to be decided from the product data sheet or considering the chemistry of the resin. Similar process when carried out without pressure is called contact moulding, in which a gel coat of about 0.5-mm thick resin-rich layer for providing a smooth, hard surface prevents the fibre reinforcements coming to the surface. The gel coat can be provided with a particular colour on request depending on the application using a suitable colourant. Reinforcement of desired size and shape is then laid over the resin layer.

The process involves the application of alternate layers of resin and fibre reinforcements until the required thickness is built up. The fibres are continually rolled to ensure proper wetting of the fibres, expulsion of air and consolidation of the materials. The glass content of 25 to 30 wt% can be achieved. However, press moulding can produce a product containing up to 65 wt% or 45 vol% glass content. The major advantages of this process are low capital investment, fast cycle time and capability to produce product with good finish and high content of the reinforcing fillers and structural integrity. The technique is versatile and can be easily learned and suitable for a wide range of fibres and resins. However, this process is not an automated one and involves human error, and therefore the composite materials made using this method suffer from the lack of consistency and quality. The quality of composite made by this method is very much dependent on skill of operator. The composites made using this process cannot satisfy the criteria for high-end applications like aerospace. In order to increase the production rate, spray-up technique (using multiple headed spray gun) is used instead of a hand lay-up technique.

| Resin & Fabric | Hand lay-up by Brush | Hand lay-up by Roller | Close Mould |

Fig. 3.18: Description of wet-lay process for fabrication of composites.

4.3.2 Resin transfer moulding

The resin transfer moulding involves filling a mould cavity by injecting a resin through one or several, points. The number of points is decided based on the size of the component. The reinforcements are placed in the interior of the mould and the mould is closed and locked firmly. Because of closed mould process, it allows faster gelling as compared to contact moulding process. It is possible to achieve good toler-ances and control of emission of volatiles by using a closed mould. RTM differs from the compression moulding in the sense that in RTM the resin is inserted into the mould containing the reinforcement, whereas in compression moulding, both rein-forcement and resin remain in the mould, where heat and pressure are applied. In the RTM process, the positive injection pressure is applied to process the fibre-reinforced composite and the injection pressure is significantly higher than the envi-ronmental pressure.

RTM requires a long filling time which is a disadvantage for the production of large-scale composite parts with a high fibre volume fraction. As solution to this prob-lem reported an effective process incorporating the method of compression into RTM known as compression RTM (CRTM) has been developed [131, 132]. The idea is to en-hance the fibre volume fraction and to reduce the mould filling time. Unlike in RTM, in which the resin is injected into a closed mould,, in CRTM, the resin is injected into a partially closed mould cavity filled with the loose preform and thus the resin injection time is reduced due to the lower flow resistance offered by the loose preform. After the required volume of resin is injected, the rigid mould platens are brought together at a constant speed or clamping force, driving the resin through the remaining dry preform and compacting the laminate to the final thickness of the part.

Another adaptation of the resin transfer moulding (RTM) process is vacuum-assisted RTM (VARTM) which exploits vacuum pressure to draw off resin to the im-pregnate performs [133, 134]. The set-up comprises vacuum bag, flow distribution me-dium, peel ply, sealing tape and resin tubing. Peel ply is used to cover the preform. The flow distribution medium layer (which is connected to the resin injection port) is applied on the top of the peel ply to enhance the resin infusion speed. An omega-shaped or a helical open tube can also be used as a resin injection line source. The vent port is required to be placed on top of the peel ply above the preform. The vac-uum port should be opened after closing the injection port in order to apply the vac-uum inside the assembly. A debulking process is generally adopted by cyclically compressing and relaxing the preform to achieve better compactness of the fibre pre-form by removing entrapped air. After completion of curing process as per the pre-scribed cure schedule, the vacuum is turned off and the composite part is demoulded. In this process, the pressure difference between the vacuum pressure and the envi-ronmental pressure is utilized as driving force to infuse the resin into the preform. The same is also used to compress the preform and keep them in place against the mould. In this process making arrangement for larger mould is much easier com-pared to simple RTM. In other words VARTM offers greater flexibility to alter moulds as per the prescribed design compared to simple RTM. VARTM process has advantages of both RTM in terms of reduced emission of volatile organic compounds, clean han-dling, high quality product, repeatability as well as wet lay-up method in terms of scalability and flexibility of the process. VARTM is widely used for the fabrication of composite products in both the civil and defence sectors.

Chang et al. [135] reported a hybrid process of bag compression and RTM, called vacuum assisted CRTM (VACRTM). VACRTM uses an elastic bag placed between the upper mould and the preform. Various stages in VACRTM are identical to RTM, but not to the filling process as shown in a schematic diagram (Fig. 3.19). The filling process of the VACRTM is composed of two phases including resin injection and flexible bag compression. During resin injection the preform is preplaced in the mould cavity and the vacuum is drawn into the mould cavity. After enough volume of resin is injected, the inlet is closed. A compression pressure is applied on the bag that compacts the preform to the final part thickness and drives the resin through the remaining dry preform. The basic difference between VARTM and VACRTM is that in VARTM the bag materials replace the upper mould and resin is infused into the preform under vacuum, whereas in VACRTM, the resin is injected into a closed mould with vacuum assistance and then a flexible bag compacts the preform in constant pressure. It was reported that for an UPE/E glass composite, VARTM and VACRTM processes are associated with infusion time and flexural strength of 1066 s, 448 s and 1,150 kgf/cm^2, 1,315 kgf/cm^2 indicating 58% increase in infusion time and 10% enhancement in flexural strength.

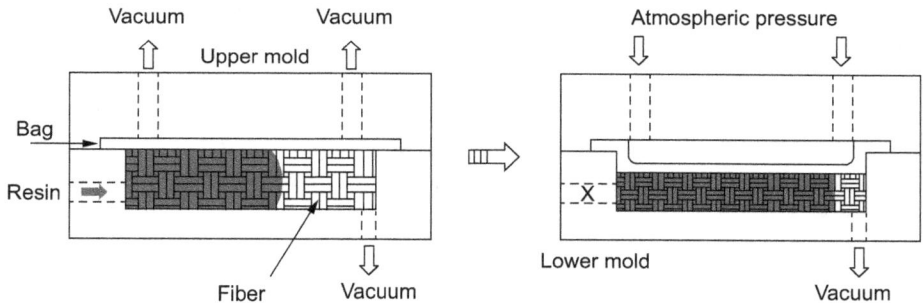

Fig. 3.19: Schematic diagrams for VACRTM (reprinted with permission from Chih-Yuan Chang and Hung–Jie Lin *Journal Polymer Engineering* 32 (2012) 539–546. © 2012 De Gruyter, Publishing Company).

3.4.3 Pultrusion

Pultrusion is a simple processing technique for the fabrication of thermosetting resin-based FRP composite products using continuous fibre or fabric. A pultrusion is a well-established industrial process mainly utilized for making composite products like ladders, stanchions, pipes, aerial booms, building parts using UPE or vinyl ester resins [136–138]. A typical pultrusion process [139] is depicted in Fig. 3.20, which comprises a creel for supplying fibre, a resin tank, forming dies, machined die with a temperature control facility, puller and saw for cutting the product from the continuous system. The continuous fibre is passed through a resin tank containing a mixture of resin and suitable curing agent. After passing through the resin bath the resin-impregnated fibres or fabrics were then pulled through a series of forming dies. The final die is heated to cure the resin system in order to produce a rigid composite structure. The cured profile is continuously pulled out of the die which provides the driving force for the impregnation of fibres to be forced through the die. Hence the raw materials are basically pulled through the combining, shaping and curing operations; therefore, the process is called pultrusion.

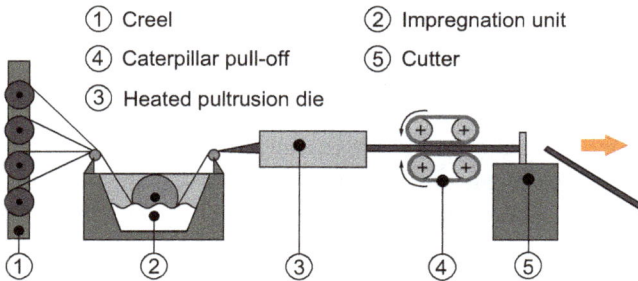

Fig. 3.20: Basic pultrusion process (reprinted with permission from Maik Gude, Florian Lenz, Andreas Gruhl, *Science Engineering Composite Materials* 22 (2015) 187–197. © 2015 De Gruyter, Publishing Company).

The process has a similarity with extrusion which is generally used for the processing of thermoplastics. The difference is that in extrusion the thermoplastic is forced through the die from inside by a rotating screw; on the other hand in pultrusion the thermoset material is pulled through the die orifice from the outer side. Both solid and hollow products with high stiffness and strength can be made using this technique. Typical products made by pultrusion are ladders, stanchions, pipes, aerial booms and building parts. Pultrusion offers the advantages like high specific strength of the product, easy handling and a variety of product profiles. Pultruded FRP composites have a relatively higher fibre volume ratio (60–70 vol %) compared to the laminates made by the wet lay-up technique dis-

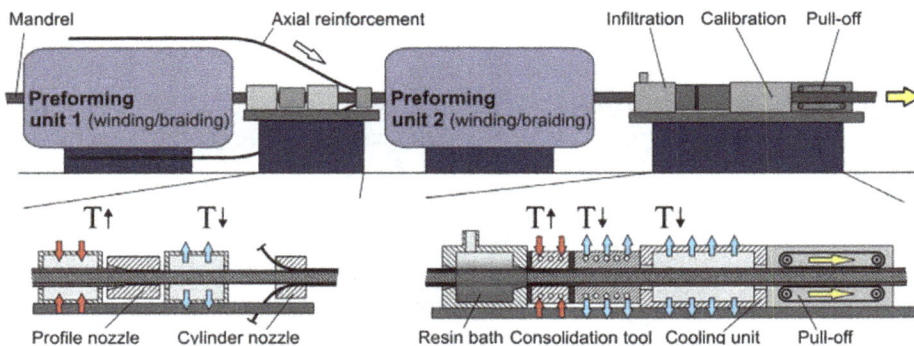

Fig. 3.21: Process principle for a continuous pultrusion of profiled driveshaft bodies (reprinted with permission from Maik Gude, Florian Lenz, Andreas Gruhl, *Science Engineering Composite Materials* 22 (2015) 187–197. © 2015 De Gruyter, Publishing Company).

cussed above (up to 50 vol%). The mechanical strength is known to increase with fibre volume ratio and hence high strength FRPs for civil infrastructure can be made by pultrusion.

For conventional pultrusion, no continuous industrial process is known. Gude et al. [139] published a concept in which established lay-down processes are combined with the state-of-the-art infusion processes to form a production arrangement. This arrangement eliminates the manual process steps and utilizes highly productive and automated preforming processes such as braiding or continuous winding. It also contains a novel device for the continuous preforming of the profiled geometry, which is validated experimentally on a laboratory scale. This way a significant reduction of part cost combined with an enhancement of the reproducibility of the manufacturing can be achieved. The general concept for serial manufacturing feasible for both preforming processes is depicted in Fig. 3.21, which comprises the steps such as (i) preforming of the profiled laminate on continuously fed mandrels; (ii) forming of the profile; (iii) supply of the axial reinforcement; (iv) preforming of the cylindrical outer laminate; (v) infiltration in a pultrusion nozzle; and (vi) consolidation, straightening and pull-off.

3.4.4 Filament winding

Filament winding is another continuous processing technique, which is widely utilized to fabricate structures like pressure vessels, drive shaft and radomes used in defence sectors.

The process can be automated to produce high-quality components and high-volume products. Filament winding involves winding of reinforcement in hoop direction unlike in pultrusion the same is in the longitudinal direction. In this process, resin-impregnated reinforcement is laid down in the form of rovings or tapes onto a mandrel. A continuous length of roving or tape of fibres is passed through a bath containing resin and curatives. The excess resin is squeezed out before winding over a mandrel. The components of different sizes and shapes can be produced by using a specified mandrel. The winding process is to be carried out till the desired thickness is achieved required for a particular application. After completion of initial curing (as per the recommended cure schedule), the component is pulled off the mandrel and subjected to the post-curing treatment. A number of patterns like helical, hoop or polar can be generated depending on the manner and direction of fibre lay down. Xu et al. [140] used filament winding process for making polyurethane-based gasbag. They used four-axis winding machine and the winding pattern is shown in Fig. 3.22. The measured data are as follows: the distance between adjacent gauges is 50.4 mm, and the width of yarn is 50 mm in this experiment; then, the precision of the winding path is 0.4 mm. Generally, the mandrel is rotated continuously in one direction and the fibre source reciprocates parallel to it. In order to achieve the total coverage of the mandrel surface, synchronization of rotation and reciprocation has to be established. A judicious selection of winding type, mandrel design and winding equipment is absolutely necessary for the successful development of a product using a filament winding process.

(a) The linear of one layer winding

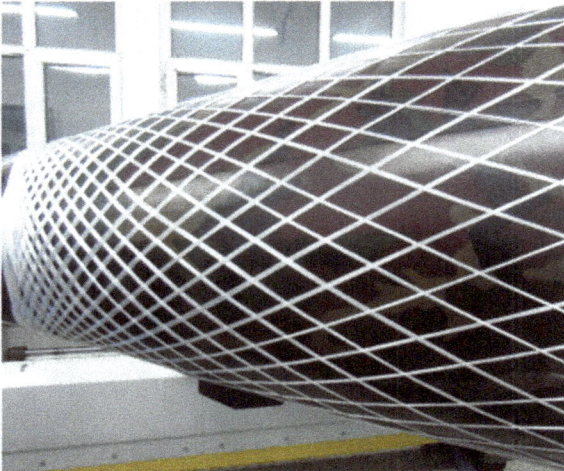

(b) The linear of cone

Fig. 3.22: Filament winding mandrel (reprinted with permission from Jiazhong Xu, Meijun Liu, Hai Yang, Tianyu Fu, and Jiande Tian *Science Engineering Composite Materials* 26 (2019) 540–549. © 2019 De Gruyter, Publishing Company).

A comprehension of the complex curing process of thermosetting resins can be made possible by using analytical models. All models may not be correct to simulate actual condition, but they are very useful to reduce the number of rejection and predict both, the process behaviour and its final characteristics. Marchetti and co-workers [141] reported the models, which represent the whole process in the form of mathematical equations and solved them numerically to explore the effects of a systematic variation of process parameters, without any scrap for experimental controls or for rejects. It also allows to compare different simulation trials and to point out advantages, defects or critical points for any specific production cycle. A numerical code structural flow chart is shown in Fig. 3.23.

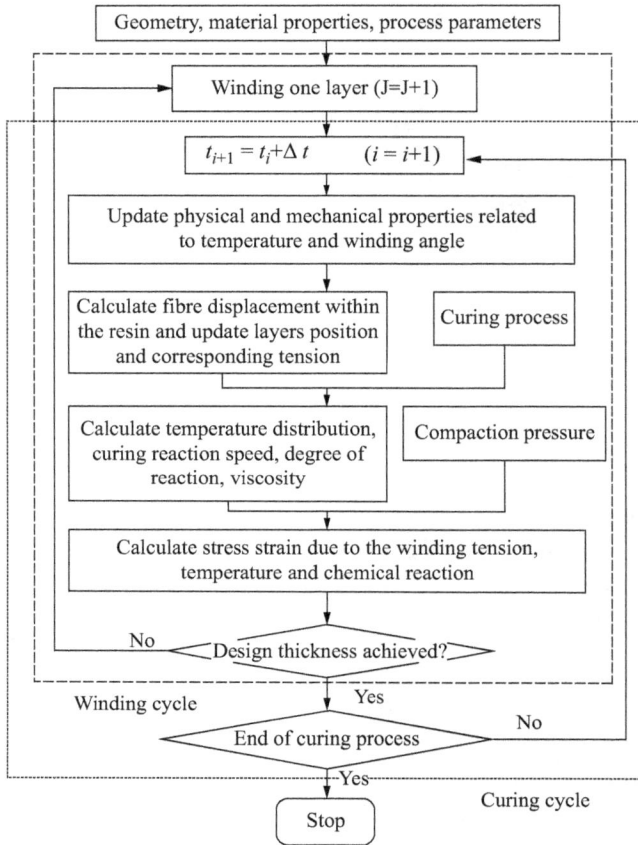

Fig. 3.23: Flow chart of filament winding numerical code of simulation (reprinted with permission from G. Ikonomopoulos and M. Marchetti, *Science Technology Composite Materials* 7 (1998) 215–222. © 1998 De Gruyter, Publishing Company).

3.4.5 Resin film infusion bolding

Resin film infusion (RFI), is one of the most promising methods for the fabrication of composites in aerospace, automotive and military applications. It is a cost-effective technique for the fabrication of complex-shaped parts resolving several critical concerns of conventional liquid composite moulding methods. RFI also ensures near-zero void fractions because of better compaction and local flow of the polymer resin. In this method, the thermoset resin is cast to film forms and the same is used as a matrix unlike in the wet lay-up process where liquid resin is used [142, 143]. Although in principle resin film moulding can be used for any thermoset resin, the process for making partially cured film with uniform thickness has only been standardized industrially for epoxy resin. A suitable mixture of solid and liquid epoxy and hardener can be cast into a partially cured tacky film which can be used to make composite.

Once the firms are ready they can be stored in similar way as followed for storage of prepregs. For the fabrication of composite, the resin firms are transferred to fabric layers (over a metallic mould plate), ensuring that the film just sticks to the fabric [144]. The process is depicted in Fig. 3.24. The process is repeated to yield sandwiches of resin films with two fabric layers on both sides. Such sandwiches are cut to the desired dimension and placed on one another to build desired thickness. The sandwich stack over the mould is then cured by vacuum bagging technique inside an oven. Heated moulds can replace the use of ovens for large structures. The limitation of RFI process is that uniform resin films are difficult to with filled epoxy system and therefore the process cannot be used for many functional composites which requires high loading of functional fillers. Second the epoxy system used for making resin film is generally a combination of solid and liquid epoxy. Solid epoxies due to their higher epoxy equivalent generate a network with lesser crosslink density and glass transition temperature. Therefore composites with very high glass transition temperature (>200 °C) are difficult to make using this process.

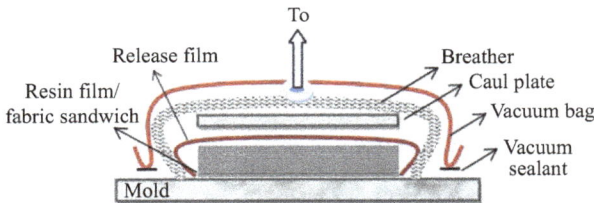

Fig. 3.24: Schematic of the resin film infusion process [144].

3.4.6 Autoclave moulding

In autoclave moulding, the pressures and temperature required for moulding is provided by an autoclave. As we have discussed for the other manufacturing processes the curing is carried out either by using oven or self-heating set up namely the placement of the heating fabric. An illustration [145] of the energy required to cure composite parts using various manufacturing techniques using oven, autoclave and self-heating set-up is shown in Fig. 3.25. Autoclave processing is a robust and well-understood process because of considerate research and wide spread industrial applications especially in the aerospace industries [146–148]. In this process the reinforcement is impregnated with matrix resin and partially cured in such a way that it forms a tacky semi-solid stage as B-stage. The reinforcement with impregnated partially cured resin is called pre-preg. The prepreg on heating the resin viscosity decreases and it undergoes curing to produce composite depending on the desired design. Using various types of reinforcement, a wide variety of prepreg can be made, for example, unidirectional prepreg, multidirectional tape-prepreg, woven fabric prepreg and tow prepregs.

The resin has to be selected considering the application of resulting composite. If the application is for general-purpose use then the prepreg is called general-purpose prepreg and if the same is for high performance use such as aerospace then the prepreg is called high performance prepreg. Whether it is general purpose or high performance, prepregs are generally prepared by solvent impregnation method. The reinforcement is passed through a bath containing the resin solution and allowed soak with the resin. The role of the solvent is to reduce the viscosity of the resin to ensure required impregnation and wetting of the reinforcement. The impregnated reinforcement (fabric) is then allowed to pass through a series of nip-rollers to sqeeze the excess resin. During this process majority of solvent is removed. The rest of the solvent present in the prepreg is removed by passing it through a heated drying chamber. The complete removal of solvent is to be ensured because even traces of solvent present in the prepreg may initiate bubble (defect) formation during processing. Therefore, selection of solvent and adjustment of nip gap and drying condition are to be addressed properly to develop successful prepreg technology. In order to

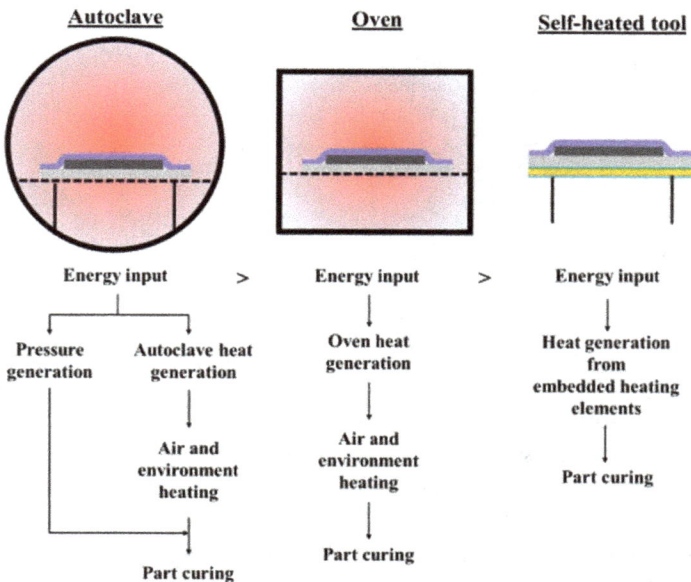

Fig. 3.25: An illustration of the energy required to cure composite parts using various manufacturing techniques (reprinted with permission from Nithin Jayasree, Sadik Omairey, and Mihalis Kazilas *Science Engineering Composite Materials* 27 (2020) 325–334. © 2020 De Gruyter, Publishing Company).

avoid contamination, prepregs are covered with polyester film from both sides before winding on a drum. It may be noted that prepregs are required to be stored in subambient temperature (<−10 °C). The maximum temperature for storing is decided depending on the resin system. If we look at the commercial market the cTech is the supplier of wide variety of prpreg-based various epoxy systems which includes toughened and flame-retardant epoxy, for example, MTM45-1, MTM47-1 and Cycom 5320. Gurit (Sprint ST94), Hexcel (Hexply M56) and Toray (2510) are the other suppliers of epoxy-based prepreg. Tencate deals with epoxy (BT250E, TC250y and TC275TC350-1) as well as cyanate ester (TC420, TC800 BMI+)-based prepregs.

As discussed above autoclave moulding offers the highest level of uniformity and remains a benchmark for manufacturing of composite as far as the quality and reliability of the composite part are concerned. However, autoclaves involve significant costs for acquisition, operation and tooling, particularly for large parts. Compared to other processing technique autoclaves moulding offers less flexibility terms of manufacturing of parts because the potential part designs are constrained by available vessel sizes and production rates are restricted by scheduling. Hence the process is not being practical for medium and small-scale industries although the same is extensively used in aerospace sector. Therefore recent trend in prepreg manufacturing technology is to move from autoclave moulding and adopt out-of-autoclave process to produce autoclave quality products.

As a result a new generation prepregs have been introduced [149] which can be processed by vacuum bag moulding and offer properties close to the same processed with autoclave. By introducing such prepregs the process can be cost-intensive and the production costs can be considerably reduced. In addition such processes are compatible with a varying range of lower-cost cure set-ups, including conventional ovens, heating blankets and heated tools. Hexcel and Henkel have commercialized the new generation epoxy-based (TC250) and benzoxazine-based (Loctite BZ) prepregs, respectively. However, some considerations related to these new generation prepregs are required to be discussed in this context. In autoclave moulding very high pressure is used which takes care of errors and imperfections in lay-up and material factors, such as absorbed moisture, air entrapment and out time. When we are replacing autoclave moulding with vacuum bag moulding it is necessary to take extra care for material storage and handling, along with a skilled workforce and attention to lay-up procedures. This is absolutely necessary for successful fabrication of autoclave moulding quality part using vacuum bag moulding. It may be noted that prepregs being partially cured a composite product undergoes significant compaction during cure. This large thickness change from the as-laid-up state can lead to wrinkling and fibre bridging especially in contoured parts [150].

3.5 Toughened thermoset composites

Thermoset composites have poor damage tolerance due to inherent brittle nature by virtue of the 3D network structure. When a crack is generated in such composite structure it propagates very fast leading to a catastrophic failure. Therefore a lot of research have been made to enhance the impact strength of thermoset composites [151–159]. If we consider the impact properties of individual constituents then we find a synergistic enhancement. For example the notched Izod impact an epoxy–glass fibre composite is 800–900 J/m which is several times higher than the value for cured epoxy or glass (<20 J/m). Thus the impact behaviour of a composite is quite different compared to other mechanical properties like tensile strength or modulus where we get a good agreement with the predicted value from rule-of-mixture. This behaviour can be explained by considering various energy-absorbing or energy dissipating mechanisms, which operate when an impact load is subjected to a composite structure. The mechanisms include rearrangement of the reinforcement, creation of damaged areas, matrix fibre delamination fibre breakage, fibre pull-out and fibre debonding.

There are four important ways to improve the impact strength or damage tolerance of a thermoset composites namely (i) use of high strain fibre; (ii) interleaving, that is, insertion of interlaminer "interleaf" layers); (iii) stitching, that is, introduction fibre in z-direction; (iv) resin toughening. The impact strength of a thermoset composite can be improved by using high strain fibre. For example, polymeric fibre like Kevlar fibre or spectra fibre (ultra high polyethylene fibre)-based composites exhibit higher impact strength compared to inorganic fibre (like glass or carbon) composites using same matrix. There are two issues with such fibres. First processing technology for such fibre is costly. Second, use of high strain fibre produce composite with lower modulus which restricts their applications in many high performance engineering applications. As will be discussed in the subsequent sections, the modulus of a composite mostly depends on modulus of fibre reinforcement. The high strain fibre-based composites find application for ballistic applications like bullet proof laminate.

The second approach is interleaving in which a tough resin layer, whisker or nanofibres are incorporated in the interface which introduce plastic deformation into the elastic interface. Abdelal et al. [160] investigated the effect of SiC whiskers-based interleaving on Mode I interlaminar fracture toughness of carbon fibre-epoxy composites manufactured using hand lay-up vacuum bagging process. It was observed that the composites containing 5 wt% whiskers exhibited 67% increase in the crack initiation interlaminar fracture toughness G_{IC}. Thermoplastic and liquid rubbers have also been successfully utilized as interleaving materials [161–163]. The use of 3-carboxy phenyl maleimide-styrene copolymer-based interphase in graphite epoxy-based composite resulted in two-fold increase in critical strain energy factor (K_{IC}) while maintaining the interlaminar shear strength at around the same value as for controlled composite. However, the use of interleaving layer may affect the impregnation behaviour of the reinforcement due to reduction of in-plane and out-of-plane permeability of the fabric reinforcement leading to a decrease in process efficiency and final properties of the composite.

Wimolkiat and Bell [164] electro-polymerized a high-temperature thermoplastic (3-carboxy phenyl maleimide-styrene copolymer) interphase onto graphite fibre and evaluated the fibre-reinforced epoxy composites. The improvement in critical strain energy factor of about 100% and notched impact strength of about 60% were achieved. Other thermoplastic and liquid rubbers have also been successfully utilized as interleaving materials [161–163]. However, the use of such fillers in high concentration (>5 wt%) required for achieving the desired toughness is associated with a significant deterioration in modulus and glass transition temperature. Another issue is that an excessive interlaminar toughener between plies may reduce the in-plane and out-of-plane permeability of textile reinforcements, which will affect resin impregnation characteristics and decrease the process efficiency of liquid composites moulding (LCM) processes. Polyamide-6 (Nylon 6) was used for interleaving of carbon-woven fabric by Zhao et al. [165]. The Nylon 6 powder is dissolved in ethanol and applied onto the surface of carbon fabric by spray technique. The ethanol was removed by drying at 70 °C. It was observed that incorporation of 5 wt% of Nylon-6 tackifier resulted in enhancement G_{IC} and G_{IIC} from 0.147 and 1.243 kJ/m^2 to 0.619 and 1.482 kJ/m^2, respectively.

The third approach is stitching in which fibre is introduced in Z-axis as well. In order to study the effect of stitching a series of vinyl ester/glass composites were made using 3D fabric, 2D fabric and a combination of two and they are subjected to a blast test. The photographs showing the condition of the FRP composites after the blast test is shown in Fig. 3.26. We can see that 2D fabric-based laminate had broken into pieces whereas the 3D-based laminate is able to retain its integrity even after experiencing the blast. It is observed that extensive delaminations have taken place which are actually the source of absorption and dissipation of mechanical energy. In case of hybrid composite we can see the intermediate situation. This clearly indicates that the impact strength or damage tolerance of a composite can be significantly enhanced by stitching approach.

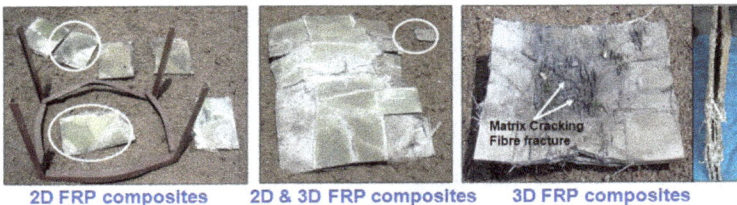

2D FRP composites   2D & 3D FRP composites   3D FRP composites

Fig. 3.26: Photographs taken after blast testing of FRP composites made with 2D, 3D fabric and the combination.

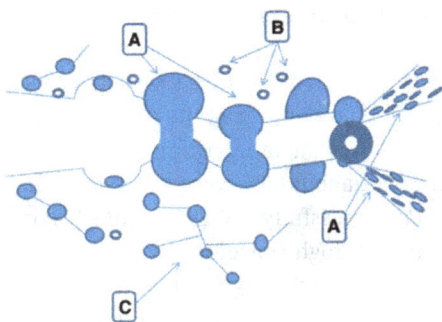

Fig. 3.27: Three mechanisms for CSRs to improve the impact proper ties of composites: (A) deformation, (B) cavitation and (C) crack termination (reprinted with permission from Hyeongcheol Park, Hana Jung, Jaesang Yu, Min Park and Seong Yun Kim *e-Polymers* 15 (2015) 369–375. © 2015 De Gruyter, Publishing Company).

The fourth approach is resin toughening which is much easier approach compared to the other three. The basic idea is to make the resin tough which can be translated into composite toughness. The resin can be toughened in two ways: (i) making the resin ductile by changing chemistry or crosslink density; (ii) dispersing micron-size rubber particles uniformly into resin matrix. The advantage of second approach is that the higher toughness can be achieved without compromising the thermal property, that is, T_g. Kim and co-workers reported [166] the effect of incorporation of core shell rubber (CSR) particles on the impact strength of carbon fibre-reinforced plastic composites fabricated using the vacuum-assisted resin transfer moulding process. The impact strength versus rubber concentration plot is shown in Fig. 3.27. As we can see from the plot, an 87% increase in impact strength was reported for incorporation about 20 wt% of CSR. They have attributed this increase in impact performance to three mechanisms as given in Fig. 3.28, namely (i) the energy consumption due to deformation of rubber phase; (ii) cavitation followed by shear deformation which absorbs energy; (iii) dispersed particles uniformly serve as crack terminators to stop crack growth. Ratna and co-workers [167] reported similar results using hyperbranched polymer-based toughening agent in epoxy/glass fibre system. For details of matrix toughening, the readers may refer recently published book [6] of one of the authors (DR).

3.6 Thermoset nanocomposites

Thermoset nanocomposites are thermoset matrix composites, in which one of the incorporated filler is in nanosize (1–100 nm) at least in one dimension. Nanosize fillers are called nanofillers. It we look at the literature of thermoset nanocomposites, we will find four types of systems namely (i) themoset resin + nanofiller; (ii) thermoset resin + nanofiller + fibre; (iii) thermoset resin + nanofiller + rubber; (iv) thermoset

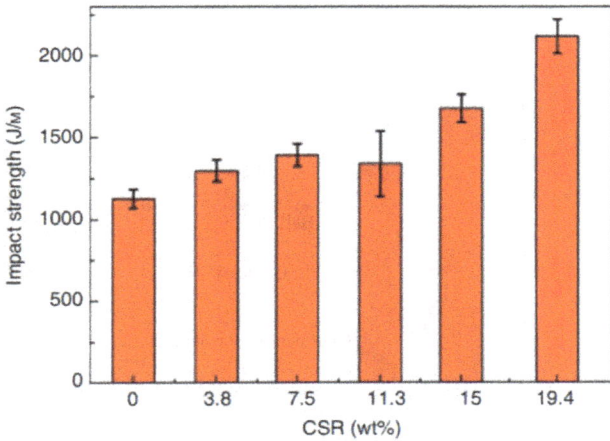

Fig. 3.28: Impact strength of CFRPs with CSR content (reprinted with permission from Hyeongcheol Park, Hana Jung, Jaesang Yu, Min Park and Seong Yun Kim *e-Polymers* 15 (2015) 369–375. © 2015 De Gruyter, Publishing Company).

resin + nanofiller + rubber + fibre. In this section, those systems with fibre will only be covered since the book is focused on fibre-reinforced composites. The other systems are already discussed elsewhere. The various nanofillers used are single-walled carbon nanotube (SWCNT), multiwalled carbon nanotube (MWCNT), graphene, graphene oxide (GO), layered silicates or nanoclay, polyhedral oligomeric silsesquioxane (POSS), carbon nanofibre and so on. The nanofillers can be incorporated in various ways either with the matrix or directly with the reinforcement. Recently, Zhang et al. [168] published a review on carbon nanotube (CNT)-modified FRP composites and elaborated different methods of incorporation of CNT as shown in Fig. 3.29. The same is applicable for other nanomaterials as well.

In the previous section it was discussed that introduction of ductile tackifier, for example, nylon 6 in carbon fibre-reinforced epoxy composites resulted in significant improvement in interlaminar fracture energy. When GO is incorporated in the tackifier, the enhancement in fracture energy is found to be more compared unfilled tackifier. Incorporation of 0.5% GO resulted G_{IC} and G_{IIC} from 0.619 and 1.482 kJ/m^2 to 0.679 and 1.557 kJ/m^2, respectively. It may be noted that GO is incorporated into the tackifier by in situ polymerization ring-opening method [169]. In this work graphene oxide is incorporated into the fibre along with a tough thermoplastic. Similar work is reported with graphene without any tough thermoplastic [170]. Graphene sheets were prepared by a simple thermal reduction method in air and used as interleave materials for epoxy/CF composite. It was observed that the G_{IC} is enhanced up to 145%.

Molnar et al. [171] investigated the effect of polyacrylonitrile nanofibrous interlaminar layers on the impact properties of unidirectional and woven carbon fabric (CF)-reinforced epoxy (EP) matrix composite. The electrospinning set-up used for the

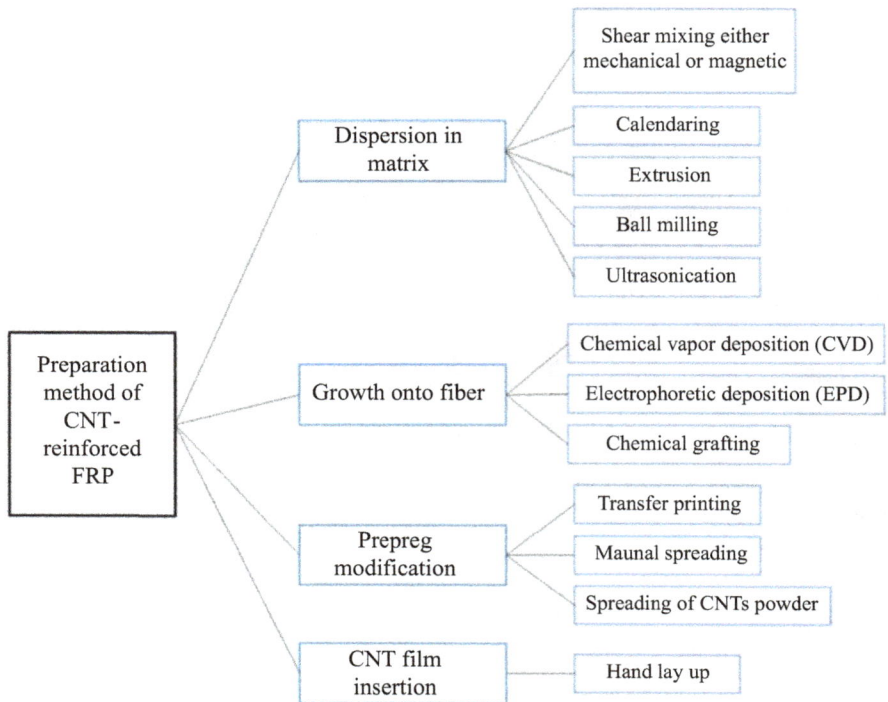

Fig. 3.29: Different preparation methods of CNT-reinforced FRP (reprinted with permission from Jianbin Li, Zhifang Zhang, Jiyang Fu, Zhihong Liang and Karthik Ram Ramakrishnan *Nanotechnology Reviews* 10 (2021) 1438–1468. © 2021 De Gruyter, Publishing Company).

preparation of carbon nanfibres is shown in Fig. 3.30. Electrospinning was carried out in an air-conditioned chamber with controlled humidity of 20%. The flexural stress versus strain plot is shown in Fig. 3.31. We can see that for both UD fibre and fabric the flexural strength is increased as a result of interleaving. However, the flexural strength increases only in case UD fibre but decreases for fabric. A summary of the nanomaterials used in various fibre-reinforced thermoset resin systems and their effect on fracture toughness of the resulting composites are presented in Table 3.5.

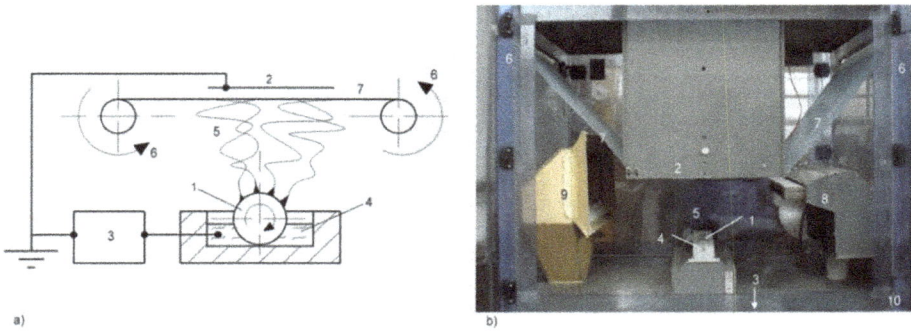

Fig. 3.30: Electrospinning set-up used to collect nanofibres on the surface of reinforcing materials: (1) rotary electrode; (2) grounded plate electrode; (3) high-voltage power supply; (4) polymer solution; (5) fibre formation space; (6) rollers; (7) moving belt collector; (8) air inlet; (9) air outlet; (10) chamber (reprinted with permission from K. Molnár, E. Kostáková, L. Mészáros, *eXPRESS Polym. Lett.* 8 (2014) 62–72. © 2014 Online Publishing).

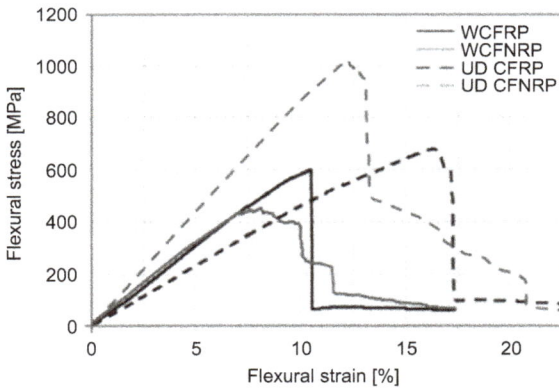

Fig. 3.31: Typical flexural curves of the prepared composites (reprinted with permission from K. Molnár, E. Kostáková, L. Mészáros, *eXPRESS Polymer Letters.* 8 (2014) 62–72. © 2014 Online Publishing).

Tab. 3.5: Improvements in interlaminar fracture energy of fibre-reinforced thermoset nanocomposites.

Nanofiller dispersion techniques in interleaves/interlayer	G_{IC} (kJ/m^2)	G_{IIC} (kJ/m^2)	Reference
Interlayer with CNF powder (20 g/m^2)	26%	–	[172]
Interlayer with CNT paste (20 g/m^2)	20%	–	[173]
Interlayer with CNT/epoxy composite (0.5 wt%)	−5%	–	[174]
Interlayer with CNT/epoxy composite (0.3 wt%)	48%	–	[175]

Tab. 3.5 (continued)

Nanofiller dispersion techniques in interleaves/interlayer	G_{IC} (kJ/m^2)	G_{IIC} (kJ/m^2)	Reference
Interlayer with sprayed CNT (1.32 g/m^2)	32%	–	[176]
Interlayer with sprayed CNT (0.047 WT%)	21%	–	[177]
Interlayer with CVD grown CNT arrays (for IM7 carbon fabrics)	43%	–	[178]
Interlayer with CVD grown CNT arrays (for AS4 carbon fabrics)	36%	–	[178]
Interlayer with short carbon fibre tissue (for fibre length of 0.8 mm)	99%	–	[179]
Interlayer with short carbon fibre tissue (for fibre length of 4.3 mm)	63%	–	[179]
Interlayer with carbon tissue or CNT-modified fibre tissue for carbon fibre tissue	−5%	–	[180]
Interlayer with carbon tissue or CNT-modified fibre tissue for CNT-modified tissue	35%	–	[180]
Growth of CNTs on CF by CVD	46%	–	[181]
Growth of CNTs on CF by flame	67%	–	[182]
Interlayer with GO/E composite (2 g/m^2)	108%	–	[183]
Interlayer with graphene/epoxy composite(1 wt%–1 g/m^2)	145%	–	[170]
Interleaving with carbon fabric Polyamide/GO 5.0/0.5 wt%	300%	25%	
Interleaving with carbon fabric Polyamide/GO 5.0/1.0 wt%	410%	26%	
SWCNTs/0.1 wt%	13%	28%	[184]
f-CNTs		23%	[185]
COOH-MWCNTs	24%	11%	[186]
CNTs 0.02 wt%	24%	–	[187]
CNTs 0.047 wt%	47%	–	[187]
CNT films (two methods to prepare CNT films/laminate curing)	21%	42%	[188]
CNT films (RTM process)	No change	120%	[189]
COOH-MWCNTs 0.5 wt%	25%	–	[190]
COOH-MWCNTs 1 wt%	20%	–	[190]
COOH-MWCNTs 1.5 wt%	17%	–	[190]
COOH-MWCNTs	69%	–	[191]

Tab. 3.5 (continued)

Nanofiller dispersion techniques in interleaves/interlayer	G_{IC} (kJ/m^2)	G_{IIC} (kJ/m^2)	Reference
CNT films	–	94%	[192]
CNTs	26%	–	[193]
CNTs	–	53%	[194]
Bucky paper	–	53%	[195

References

[1]   Bakker, A. A., Jones, R., Callinan, R. J. Compos. Struct. 1985, 15, 154.

[2]   Jang, B. Z. Sci. Eng. Composite Mater. 1991, 2, 29.

[3]   Ratna, D., Chongderand, T. K., Chakraborty, B. C. Polym. Int. 2000, 49, 815–819.

[4]   Ratna, D. Rapra. Rev. Rep. 2005, 16, 1.

[5]   Okamoto, M. Rapra. Rev. Rep. 2003, 14, 1–166.

[6]   Ratna, D. Recent Advances and Applications of Thermoset Resins. London, Elsevier, 2022.

[7]   Gum, W. F., Rinse, W., Ulrich, H. eds. Reaction Polymer, Munich, Germany, Hanser Publishers, 1992, p. 838.

[8]   Bhatnagar, R., Verma, I. K. J. Therm. Anal. Calorimetry. 2005, 35, 1241.

[9]   Baffa, F., Borrajo, J. J. Appl. Polym. Sci. 2006, 102, 6064.

[10] Miyagawa, H., Mohanty, A., Burgueno, R., Drzal, L. T., Misra, M. J Polym. Sci. Part B Polym. Phys. 2007, 45, 698–704.

[11] Varelidis, P. C., McCullough, R. L., Papaspyrides, C. D. Compos. Sci. Technol. 1999, 59, 1813–1823.

[12] Hexion-Chemicals. EPON™ resin 828. Trade Lit Broch. 2005, 3942, 1–8.

[13] Shibata, K. Recycling of carbon fiber and epoxy resin from carbon fiber reinforced plastics Kumamoto University, 2014.

[14] Morgan, R. Structure-property relations of epoxy used as composite matrices. In: Dusek, K. ed. Advances in Polymer Science, 51st ed., Berlin Heidelberg, Springer Verlag, 1986, pp. 1–40.

[15] Morgan, R. J., Kong, F. M., Walkup, C. M. Structure-property relations of polyethertriamine-cured bisphenol-A-diglycidyl ether epoxies. Polymer (United Kingdom), 1984, 25, 375–386.

[16] Hughes, J. D. H. The carbon fibre/epoxy interface – a review. Compos. Sci. Technol. 1991, 41, 13–45.

[17] Aspin, I., Brownhill, A., Bugg, D. Soc. Chem. Ind. 2003, 1–4.

[18] Gonçalez, V., Barcia, F. L., Soares, B. G. J. Braz. Chem. Soc. 2006, 17, 1117–1123.

[19] Kultzow, R., Foxhill, S. Cycloaliphatic Epoxy Resin. In: Thermoset Resin Formulator Association. Woodlands, Texas, TRFA Publisher, 2007, pp. 1–6.

[20] He, H., Li, K., Wang, J., Gu, J., Li, R. Polym. Compos. 2011, 16, 227–235.

[21] Sprenger, S., Kothmann, M. H., Altstaedt, V. Compos. Sci. Technol. 2014, 105, 86–95.

[22] Zhiyuan, Y. Mater. Sci. Adv. Compos. Mater. 2018, 1, 1–6.

[23] Sukanto, H., WisnuRaharjo, W., Ariawan, D., Triyono, J., Kaavesina, M. Open Eng. 2021, 11, 797–814.

[24] Son, D. R., Raghu, A. V., Reddy, K. R., Jeong, H. M. J. Macromol. Sci. Part B. 2016, 55, 1099.

[25] Cousinet, S., Ghadban, A., Fleury, E., Lortie, F., Pascault, J. P., Portinha, D. Eur. Polym. J. 2015, 67, 539.

[26] Sultania, M., Rai, J. S. P., Srivastava, D. Eur. Polym. J. 2010, 46, 2019–2032.

[27] Jahandideh, A., Muthukumarappan, K. Eur. Polym. J. 2017, 87, 360.

[28] Gardziella, A., Pilato, L. A., Knop, A. Phenolic resins. 2nd ed., Heidelberg, Springer-Verlag, 2000.

[29] Fisher, H. L. Chemistry of Natural and synthetic rubbers. New York, Reinhold Publishing Corporation, 1958.

[30] Hiltz, J. A., Kuzakand, S. G., Waitkus, P. A. J. Appl. Polym. Sci. 2001, 79, 385.

[31] Patel a, J. P., Zhao, C. X., Deshmukh a, S., Zou, G. X., Wamuo, O., Hsu, S. L., Schoch, A. B., Carleen, S. A., Matsumoto, D. Polymer. 2016, 107, 12–18.

[32] Reghunadhan Nair, C. P. Prog. Polym. Sci. 2004, 29, 401.

[33] Dekar, A., Stretz, H., Koo, J. Polym. Mater. Sci. Eng. 2000, 83, 105.

[34] Yan, Y., Shi, X., Liu, J., Zhao, T., Yu, Y. J. Appl. Polym. Sci. 1651, 83(2002).

[35] Liang, G. Z., Fan, J. J. Appl. Polym. Sci. 1999, 73, 1623.

[36] Dao, B., Hodgkin, J. H., Morton, J. H. High Perfor. Polym. 1997, 9, 413.

[37] Augustine, D., Mathew, D., Reghunadhan Nair, C. P. Eur. Polym. J. 2015, 71, 389–400.

[38] Augustine, D., Mathew, D., Reghunadhan Nair, C. P. Eur. RSC Adv. 2015, 5, 91254–91261.

[39] Holly, F. W., Cope, A. C., Am, J. Chem. Soc. 1944, 66, 1875.

[40] Schreiber, H. Ger Offen. 1973, 2, 225, 504.

[41] Brunovska, Z., Liu, J. Macromol. Chem. Phys. 2015, 5, 3709–3719.

[42] Rucigaj, A., Alic, B., Krajnc, M., ebenik, U. Express Polym. Lett. 2015, 9, 647–657.

[43] Andreau, R., Reina, J. A., Ronda, J. C. J. Polym. Sci. Part A: Polym. Chem. Ed. 2008, 46, 3353.

[44] Burke, W. J., Glennie, E. L., Weatherbee, C. J. Org. Chem. 1964, 29, 909.

[45] Yan, Z. Z., Wang, H., Cheng, J., Fang, Z. ACS Appl. Polym. Mater. 2021, 3, 3392–3401.

[46] Chernykh, A., Liu, J. P., Ishida, H. Polymer. 2006, 47, 766–7669.

[47] Takeichi, T., Kano, T., Agag, T. Polymer. 2005, 46, 12172.

[48] Agag, T., Takeichi, T. J Polym Sci Part A. 2007, 45, 1878–1888.

[49] Gong, W., Zeng, K., Wang, L., Zheng, S. X. Polymer. 2008, 49, 3318–3326.

[50] Mark, H. F., Biikles, N. M., Overberger, C. G., Menges, G. eds. Encyclopedia of Polymer Science and Engineering. 2nd ed., Vol. 10, New York, Wiley, 1989.

[51] Mittal, K. L. ed. Polyimides: Synthesis, Characterization and Applications. New York, Plenum, 1984.

[52] Cheng, S. Z. D., Lee, S. K., Barley, J. S., Hsu, S. L. C., Harris, F. W. Macro Mol. 1991, 24, 1883.

[53] Cheng, S. Z. D., Arnold, F. E., Zhang, A., Hsu, S. L. C., Herris, F. W. Macro Mol. 1991, 24.

[54] Han, Y., Fang, X. Z., Zuo, X. X. Express Polym. Lett. 2010, 4(11), 712–722.

[55] Brydson, J. A. Plastic Materials. Sixth ed., Butterworth- Heinemann, 1982, pp. 499–512.

[56] Allen, S. G., Bevington, J. C. eds. Compr. Polym. Sci. 1989, 5, 423.

[57] Hu, A. J., Hao, J. Y., He, T., Yang, S. Y. Macromolecules. 1999, 32, 8046.

[58] Lauver, R. W. J. Polym. Sci. 1979, 17, 2529.

[59] Margolis, J. M. ed. Advanced Thermoset Composites: Industrial and Commercial Applications. New York, Van Nostrand Reinhold Co, 1986.

[60] Hu, J., Gu, A., Liang, G., Zhuo, D., Yuan, L. Express Polym. Lett. 2011, 5(6), 555–568.

[61] Qu, C., Zhao, L., Wang, D., Li, H., Xiao, W., Yue, C. J. Appl. Polym. Sci. 2014, 40395, 1–8.

[62] Petrovicki, H., Stenzenberger, H. D., Harnier, A. V. In Proc. 32nd SPE Ann. Techn. Conf. Society of Plastics Engineers, San Francisco, 1974, 88.

[63] Stenzenberger, H. D. J. Appl. Polym. Sci.: Appl. Polym. Symp. 1977, 31, 91–104.

[64] Yu, P., Zhang, Y.-L., Yang, X., Pan, L.-J., Dai, Z.-Y., Xue, M.-Z., Liu, Y.-G., Wang, W. Polym. Eng. Sci. 2019, 59, 2265–2272.

[65] Grigat, E., Putter, R. Ger. Pat. 1963, 1, 746.

[66] Hamerton, I. ed. Chemistry and Technology of Cyanate Ester Resins. London, Blackie Academic & Professional (Chapman and Hall), 1994.

[67] Fang, T., Shimp, D. Prog. Polym. Sci. 1995, 20, 61.

[68] Shimp, D. Polym. Mater. Sci. Eng. 1994, 70, 561.

[69] Abed, J. C., Mercier, R., McGrath, J. E. J. Polym. Sci., Part A Polym. Chem. 1997, 35, 977.

[70] Derradji, M., Ramdani, N., Zhang, T., Wang, J., Gong, L. D., Xu, X. D., Lin, Z. W., Henniche, A., Rahoma, H., Liu, W. B. Polym. Compos. 2017, 38, 1549.
[71] Laskoski, M., Schear, M. B., Neal, A., Dominguez, D. D., Ricks-Laskoski, H. L., Hervey, J., Keller, T. M. Polymer. 2015, 67, 185.
[72] Ma, J.-Z., Cheng, K., Lv, J.-B., Chen, C., Hu, J.-H., Zeng, K., Yang, G. Chin. J. Polym. Sci. 2018, 36, 497.
[73] Shan, S., Chen, X., Xi, Z., Yu, X., Qu, X., Zhang, Q. High Perform. Polym. 2017, 29, 113.
[74] Xu, S., Han, Y., Guo, Y., Luo, Z., Ye, L., Li, Z., Zhou, H., Zhao, Y., Zhao, T. Eur. Polym. J. 2017, 95, 394.
[75] Augustine, D., Chandran, M. S., Mathew, D., Nair, C. R. Polyphthalonitrile Resins and their High-End Applications. Amsterdam, Elsevier, 2018, pp. 577.
[76] Bulgakov, B., Sulimov, A., Babkin, A., Kepman, A., Malakho, A., Avdeev, V. J. Appl. Polym. Sci. 2017, 134, 44786.
[77] Hu, Y., Weng, Z., Qi, Y., Wang, J., Zhang, S., Liu, C., Zong, L., Jian, X. RSC Adv. 2018, 8, 32899.
[78] Dominguez, D. D., Jones, H. N., Keller, T. M. Polym. Compos. 2004, 25, 554.
[79] Sastri, S. B., Keller, T. M., Polym, J. Sci. Part A: Polym. Chem. 1885, 36, 1998.
[80] Wu, D., Zhao, Y., Zeng, K., Yang, G. J. Polym. Sci. Part A: Polym. Chem. 2012, 50, 4977.
[81] Wu, Z., Han, J., Li, N., Weng, Z., Wang, J., Jian, X. Polym. Int. 2017, 66, 876–881.
[82] Zhang, D., Liu, X., Bai, X., Zhang, Y., Wang, G., Zhao, Y., Rong, L., Mi, C. E-Polymers. 2020, 20, 500–509.
[83] Hu, Y., Weng, Q. Y., Wang, J. Y., Zhang, S. H., Liu, C. RSC Adv. 2018, 8, 899–908.
[84] Chen, L., Ren, D. X., Chen, S. J., Pan, H., Xu, M. Z., Liu, X. B. Express Polym. Lett. 2019, 13, 456–468.
[85] Ozawa, T. A. New Method of Analyzing Thermogravimetric Data. Bull. Chem. Soc. Jpn. 1881–1886, 38, 1965.
[86] Ozawa, T. Kinetic analysis of derivative curves in thermal analysis. J. Therm. Anal. 1970, 2, 301–324.
[87] Flynn, J. H., Wall, L. A. J. Polym. Sci. B Polym. Lett. 1966, 4, 323–328.
[88] Vyazovkin, S. Thermochim. Acta. 1992, 211, 181–187.
[89] Flynn, J. H. J. Therm. Anal. 1983, 27, 95–102.
[90] Vyazovkin, S., Lesnikovich, A. I. Thermochim. Acta. 1992, 203, 177–185.
[91] Vyazovkin, S., Linert, W. J. Chem. Inf. Comput. Sci. 1994, 34, 1273–1278.
[92] Vyazovkin, S., Dollimore, D. J. Chem. Inf. Comput. Sci. 1996, 36, 42–45.
[93] Vyazovkin, S., Wight, C. A. Thermochim. Acta. 1999, 340-341, 53–68.
[94] Khawam, A., Flanagan, D. R. Thermochim. Acta. 2005, 436, 101–112.
[95] Stanko, M., Stommel, M. Polymers. 2018, 10, 698.
[96] Sbirrazzuoli, N. Thermochim. Acta. 2013, 564, 59–69.
[97] Friedman, H. L. J. Polym. Sci. C Polym. Symp. 1964, 6, 183–195.
[98] Kissinger, H. E. Analy Chem. 1957, 29, 1702–1706.
[99] Doyle, C. D. J. Appl. Polym. Sci. 1962, 6, 639–642.
[100] Murray, P., White, J. Trans. Brit. Ceram. Soc. 1955, 54, 204–238.
[101] Coats, A., Redfern, J. Nature. 1964, 201, 68–69.
[102] Senum, G. I., Yang, R. T. J. Thermal Anal. 1977, 11, 445–447.
[103] Perez-Maqueda, L. A., Criado, J. M. J. Therm. Anal. Calorim. 2000, 60, 909–915.
[104] Toop, D. J. IEEE Trans. Electr. Insul. 1971, 6, 2–14.
[105] Akahira, T., Sunuse, T. T. Joint Res. Rep. Chiba Inst. Technol Chiba, 1971, 16, 22–31.
[106] Starink, M. J. Thermochim. Acta. 2003, 404, 163–176.
[107] Vyazovkin, S. J. Comput. Chem. 1997, 18, 393–402.
[108] Vyazovkin, S. J. Therm. Anal. Calorim. 2006, 83, 45–51.
[109] Cai, J., Chen, S. J. Comput. Chem. 2009, 30, 1986–1991.
[110] Budrugeac, P. Thermochim. Acta. 2010, 511, 8–16.
[111] Billmeyer, F. W. Textbook of Polymer Science. 3rd ed., Wiley, New York, 1984.
[112] Hayaty, M., Honarkar, H., Beheshty, M. H. Iran Polym. J. 2013, 22, 591–598.

[113] Lin, Y. G., Gally, J., Sautereau, H., Pascault, J. P. Crosslinked Epoxies. In: Sedlecek, B., Kahobek, J. eds. Berlin, Walter de Gruyter, pp. 147–168.

[114] Adaboo, H. E., Williams, R. J. J. Appl. Polym. Sci. 1982, 29, 1327–1334.

[115] Nielsen, L. E., Macromol, J. Sci. Rev. Macromol. 1969, C3, 69.

[116] Pascault, J. P., Williams, R. J. J. J. Polym. Sci. Part B. 1990, 28, 85.

[117] Couchman, P. R. Macromolecules. 1987, 20, 1712.

[118] Montserrat, S. Polymer. 1995, 36, 435.

[119] Havlicek, I., Dusek, K. Crosslinked Epoxies. In: Sedlacek, B., Kahovec, J. eds. Berlin, de Gruyter, 1987, pp. 417–424.

[120] Stepto, R. F. T. ed. Polymer Networks, Blackie Academic and Professional. New York, 1998.

[121] Weyland, H. G., Hoftyzer, P. J., Van Krevelen, D. W. Prediction of the glass transition temperature of polymers. Polymer. 1970, 11, 79–87.

[122] Katritzky, A. R., Sild, S., Lobanov, V., Karelson, M. J. Chem. Inf. Comput. Sci. 1998, 38, 300–304.

[123] Katritzky, A. R., Rachwal, P., Law, K. W., Karelson, M., Lobanov, V. S., J. Chem. Inf. Comput. Sci. 1996, 36, 879–884.

[124] Camelio, P., Cypcar, C. C., Lazzeri, V., Waegell, B. J. Polym. Sci, Part A: Polym. Chem. 1997, 35, 2579–2590.

[125] Jin, K., Luo, H., Wang, Z., Wang, H., Tao, J. Mater. Des. 2020, 194.

[126] Higuchi, C., Horvath, D., Marcou, G., Yoshizawa, K., Varnek,. ACS Appl. Polym. Mater. 2019, 1, 1430–1442.

[127] Meier, S., Albuquerque, R. Q., Demleitner, M., Ruckdäschel, H. J. Mater. Sci. 2022, 57, 3991–14002.

[128] Ratna, D. Rapra. Rev. Rep. 2005, 16, 1.

[129] Okamoto, M. Rapra. Rev. Rep. 2003, 14, 1–166.

[130] Ratna, D., Mishra, V. S. "Materials and Fabrication Process for Manufacturing Fibre Reinforced Polymer (FRP) Composite Ball Valves for Sea Water Flow Control" Indian Patent Application No.: 202111019568 dated April 28, 2021.

[131] Chang, C. Y. Adv. Compos. Mater. 2011, 20, 197–211.

[132] Young, W. B., Chiu, C. W. J. Compos. Mater. 1995, 29, 2180–2191.

[133] Bhatt, A. T., PGohil, P. IOP Conf. Ser. Mater. Sci. Eng. 2020, 1004.

[134] ArySubagia, I. D. G., Kim, Y. Sci. Eng. Composite Mater. 2014, 21, 211–217.

[135] Chang, C.-Y., Lin, H.-J. J. Polym. Eng. 2012, 32, 539–546.

[136] Baran, I. Pultrusion: State-Of-The-Art Process Models, ISBN 191024242X, Smithers RapraTechnoogy, London, 2015

[137] Starr, T. F. Pultrusion for Engineers. Cambridge, Woodhead, 2000.

[138] Peters, S. T. Handbook of Composites. 2nd ed, Chapman & Hall, London, 1998.

[139] Gude, M., Lenz, F., Gruhl, A., Witschel, B., Ulbricht, A., Hufenbach, W. Sci. Eng. Compos. Mater. 2015, 22, 187–197.

[140] Xu, J., Liu, M. J., Yang, H., Fu, T., Tian, J. Sci. Eng. Compos. Mater. 2019, 26, 540–549.

[141] Ikonomopoulos, G., Marchetti, M. Sci. Techno. Compos. Mater. 1998, 7, 215–222.

[142] Williams, C., Summerscales, J., S, G. Appl. Sci. Manuf. 1996, 27(7), 517–524. 4. B) Antonucci, V., Giordano, M., Nicolais, L., Calabro, A.

[143] Cusano, A., Cutolo, A., Inserra, S. Resin Flow monitoring in resin film infusion process. J. Mater. Process. Technol. 2003, 143-144, 687–692.

[144] Anand, A., Joshi, M. J. Indian Inst. Sci. 2015, 9, 233–247.

[145] Jayasree, N., Omairey, S., Kazilas, M. Sci. Eng. Compos. Mater. 2020, 27, 325–334.

[146] Drakonakis, V. M., Seferis, J. C., Doumanidis, C. C. Curing Pressure Influence of Out-of-Autoclave Processing on Structural Composites for Commercial Aviation. Adv. Mater. Sci. Eng. 2013, 2013, 1–14.

[147] Mouritz, A. P. Manufacturing of fibre–polymer composite materials. In: Mouritz, A. P. Ed. Introduction to Aerospace Materials Woodhead Publishing, 2012, pp. 303–337.

[148] Campbell, F. C. Chapter 7 – polymer matrix composites. In: Campbell, F. C. ed. Manufacturing Technology for Aerospace Structural Materials Oxford. Elsevier Science, 2006, pp. 273–368.

[149] Centea, T., Grunenfelder, L. K., Nutt, S. R. Compos Part A Appl Sci Manuf. 2014, 70, 132–154.

[150] Centea, T., Hubert, P. J. Compos. Mater. 2014, 48, 2033–2045.

[151] Slow, J. P., Shim, V. P. W. J. Compos. Mater. 1998) 1178, 32.

[152] Kessler, A., Bledzki, A. K. Polym. Compos. 1999, 20, 269.

[153] Peijs, A. A. J. M., Venderbosch, S. W., Lemstra, P. J. Composites. 1990, 25.

[154] Cantwell, W. J., Morton, J. Composites. 1991, 22.

[155] Patridge, I. K., Cartie, D. D. R. Compos. A. 2005, 36, 55.

[156] Hosur, M. V., Vaidya, U. K., Ulven, C., Tuskegee, J. S. Compos. Struct. 2004, 64, 455.

[157] Chen, J. C., Lu, C. K., Chiu, C. H., Chin, H. Composites. 1994, 25, 251.

[158] Dutra, R. C. L., Soares, B. G., Campose, E. A., Silva, J. L. G. Polymer. 2000, 41, 3841.

[159] Sohn, M. S., Hu, X. Z. Compos. Sci. Technol. 1998, 58, 211.

[160] Abdelal, N. R., Aljarrah, M. T. J. Ceram. Int. 2018, 44, 2700–2708.

[161] Kunz, S., Sayre, J., Assink, R. Polymer. 1982, 23(13), 1897–1906.

[162] Ma, J., Mo, M. S., Du, X. S., Dai, S. R. I. Polym. Sci. 2008, 110, 304–312.

[163] Thompson, Z. J., Hillmyer, M. A., Liu, J., Sue, H.-J., Dettloff, M., Bates, F. S. Macromolecules. 2009, 42, 2333–2335.

[164] Wimolkiatisak, A. S., Bell, J. P. J. Appl. Polym. Sci. 1899, 46(1992).

[165] Zhao, X., Chen, W., Han, X., Zhao, Y., Du, S. Compos. Sci. Technol. 2020, 191, 108094.

[166] Park, H., Jung, H., Yu, J., Park, M., Kim, S. Y. E-Polymers. 2015, 15, 369–375.

[167] Ratna, D. Compos. A. 2008, 39, 462–469.

[168] Li, J., Zhang, Z., Fu, J., Liang, Z., Ramakrishnan, K. R. Nanotechnol. Rev. 2021, 10, 1438–1468.

[169] Zhao, X., Li, Y., Chen, W., Li, S., Zhao, Y., Du, S. Compos. Sci. Technol. 2019, 171, 180–189.

[170] Xusheng, D., Zhou, H., Sun, W., Liu, H.-Y., Zhou, G., Zhou, H., Mai, Y.-W. Compos. Sci. Technol. 2017, 140, 123–133.

[171] Molnár, K., Kostáková, E., Mészáros, L. Express Polym. Lett. 2014, 8, 62–72.

[172] Yang, Y., He, F., Wang, M., Zhang, B. J. Mater. Sci. 1998, 33.

[173] Ratna, D., Kushwaha, R., Dalvi, V., Manoj, N. R., Samui, A. B., Chakraborty, B. C. J. Adhes. Sci. Technol. 2008, 22(1), 93.

[174] Takahagi, T., Ishitani, A. Molecular Characterization of Composite Interfaces. In: Ishida, H., Kumar, G. eds. New York, Plenum Press, 1985.

[175] Drzal, L. T., Rich, M., Lloyd, P. J. Adhes. 1983, 16, 1.

[176] Plueddmann, E. A. Silicate Coupling Agents. New York, Plenum Press, 1982.

[177] Chua, P. S. Polym. Compos. 1987, 8, 308.

[178] Shen, C., Liu, L. S., Yeh, J. T. Polym. Polym. Compos. 1999, 21, 21.

[179] Theocaris, P. S., Papanicolaou, G. C. Coll. Poly. Sci. 1980, 258, 1044.

[180] Zhang, H., Zhang, Z., Claudia, B. Compos. Sci. Technol. 2021, 64(2004).

[181] Hook, K. J., Agrawa, R., K., Drzal, L. T. J. Adhes. 1990, 32.

[182] Walter, A. J. Phys. Educ. 1972, 7(8), 491.

[183] Okamoto, M. Rapra. Rev. Rep. 2003, 14, 1–166.

[184] Ashrafi, B., Guan, J., Mirjalili, V., Zhang, Y., Chun, L., Hubert, P., et al. Compos. Sci. Technol., 2011, 71, 1569–1578.

[185] Davis, D. C., Whelan, B. D. Compos. Part B Eng. 2011, 42, 105–116.

[186] Shan, F. L., Gu, Y. Z., Li, M., Liu, Y. N., Zhang, Z. G. Polym. Compos. 2013, 34, 41–50.

[187] Zhang, H., Liu, Y., Kuwata, M., Bilotti, E., T, P. Compos. Part A Appl. Sci. Manuf. 2015, 70, 102–110.

[188] Deng, H., Wang, L., Feng, Y., Chen, M., Jiang, W. Aerosp. Mater. Technol. 2015, 45, 31–35.

[189] Liu, G., Hu, X., Chen, M., Zhang, P., Yu, R., Li, Q., et al. Acta Polym Sin. 2013, 10, 1334–1340.

[190] Borowski, E., Soliman, E., Kandil, U., Taha, M. Polymers. 2015, 7, 1020–1045.

[191] Liu, L., Yang, Q. Chin. J. Solid Mech. 2015, 36, 80–84.

[192] Yu, Y., Zhang, Y., Gao, L., Qu, S., Lyu, W. Acta Aeronaut. Astronaut. Sin. 2019, 40, 307–314.

[193] Abidin, M. S. Z., Herceg, T., Greenhalgh, E. S., Shaffer, M., Bismarck, A. Compos. Sci. Technol. 2019, 170, 85–92.

[194] Khan, S., Bedi, H. S., Agnihotri, P. K. Eng. Fract. Mech. 2018, 204, 211–220.

[195] Shin, Y., Kim, S. Appl. Sci. 2021, 11, 6821.

Chapter 4
Elastomer-based composites

Description of abbreviations

Abbreviation	Description
CBS	*N*-Cyclohexyl-2-benzothiazolesulphenamide
IPN	Interpenetrating polymer network
MBTS	Mercaptobenzothiazyl disulphide
MPa	Mega-Pascal
PE	Poly(ethylene)
PP	Poly(propylene)
phr	Parts per hundred rubber (by weight)
SEM	Scanning electron microscope
TEM	Transmission electron microscope
TMTD	Tetramethylthiuram disulphide
NR	Natural rubber
PB	Polybutadiene rubber
IIR	Butyl rubber
BIIR	Bromobutyl rubber
CR	Neoprene rubber
NBR	Nitrile rubber
PUR	Polyurethane rubber
GPF	General-purpose furnace black
HAF	High-abrasion furnace black
ISAF	Intermediate SAF
SAF	Super-abrasion furnace black
EPC	Easy processing channel black
MT	Medium thermal black
FT	Fine thermal black
CPVC	Critical pigment volume concentration
MT-LA	Mori-Tanaka and laminated analogy
SBR	Styrene-butadiene rubber
TCAT	Tyre cord adhesion test
VFT	Vogel-Fulcher-Tammann

4.1 Elastomer matrices

The categories of polymers which are soft and cross-linked lightly are in general termed elastomers. The older term "rubber" was basically used due to the special property of the elastomers to rub off pencil marks on papers, as was first observed by a vulcanized natural rubber. Although the natural rubber latex from sap of *Hevea brasiliensis* tree was known from ancient time, vulcanization of the sap was done only in 1839 by Charles Goodyear, although Mesoamericans are believed to be using

https://doi.org/10.1515/9783110781571-004

rubber balls, etc. as early as 1600 BC. Consequent to the invention of natural rubber, which is a polymeric form of a hydrocarbon known as isoprene, the first synthetic isoprene was developed in 1909 by Fritz Hofmann. Polybutadiene was synthesized by Sergei Vasilyevich Lebedev in 1910.

Rapid development in polymer science enabled inventions of many synthetic rubbers as homopolymer and copolymers. Some of the rubbers (elastomers) are listed in Tab. 4.1. In addition, tailoring of properties of the elastomers can be done through controlling the molecular weight, cross-linking chemistry, which is known as vulcanization, blending and modifications by irradiation, grafting, etc. A large variety of nonpolar and polar elastomers are developed and being successfully used in industrial and domestic products according to their suitability in use. Very voluminous use of rubber is definitely the tyre industry. Among the non-tyre sector, most applications are shoes, rubber mats, hose pipes, bushes, gaskets, large boats, bellows, packaging, anti-vibration and shock mounts, etc., and many others.

Since elastomers are lightly cross-linked, they are soft at ambient environment, and the glass transition temperature of a typical elastomer is below 0 °C. As a consequence of low cross-linking, the free volume of an elastomer at ambient conditions is quite appreciable, and hence it can take up quite a good amount of filler loading. The purpose of filler addition is to enhance the mechanical strength, reduce cost, improve certain other properties like electrical conductivity, microwave attenuation, barrier property and vibration damping apart from imparting different colours.

Cross-linking (vulcanization) processes for most elastomers are almost similar, by a combination of sulphur and accelerator such as mercaptobenzothiazyl disulphide (MBTS) tetramethylthiuram disulphide (TMTD) and *N*-cyclohexyl-2-benzothiazolesulphenamide (CBS). Compounding of elastomers is done with processing aids such as stearic acid with zinc oxide, to reduce shear force requirement while adding other ingredients including the fillers. Some fillers can interact with the elastomers due to their small size and chemical structure, resulting in significant enhancement of properties. These fillers are termed reinforcing fillers, such as carbon black and functionalized carbon nanotubes. Even many nanoparticles can significantly enhance the properties of elastomers, both polar and non-polar. Typical examples are fumed/nanosilica, naturally occurring clays such as montmorillonite, cloisite, bentonite, kaolinite, hectorite, halloysite and their organically modified versions. Minerals, graphite, mica, etc. are non-reinforcing fillers and are not preferred for enhancement of mechanical strength, but for some other functional requirements and cost.

Polarity and stereo specific position of the groups/atoms of the rubber play very important roles in inherent strength, flexibility, compatibility with fillers and, hence, the properties of the composites. Inherent strength of a polar rubber is obviously higher than a relatively non-polar rubber, for example, comparing the glass transition of carboxylated nitrile rubber with nitrile rubber, or comparing neoprene with natural rubber. Moreover, a 4-carbon atom straight chain diene rubber such as polybutadiene or butyl rubber is more flexible than polyisoprene, which is a polymeric form of five mem-

Tab. 4.1: Typical elastomers.

S. no.	Name	Chemical structure	Glass transition temperature (°C)
1	Natural rubber (NR)	Poly(*cis*-isoprene)	−70
2	Polybutadiene (PB) rubber	Poly(butadiene)	−90
3	Butyl rubber (IIR)	Copoly(butadiene-*cis*-isoprene)	−55 to −49
4	Bromobutyl rubber (BIIR)	Copoly(bromobutadiene-*cis*- isoprene)	−55 to −45
5	Neoprene rubber (CR)	Poly(chloro-*cis*-isoprene)	−36
6	Nitrile rubber (NBR)	Copoly(butadiene-acrylonitrile), 33% acrylonitrile	−38
7	Hypalon	Chlorosulphonated poly(ethylene)	−55
8	Styrene–butadiene rubber (SBR) 15% styrene	Copoly(styrene-butadiene)	−50
9	Viton rubber	Copolymers of fluorinated ethylene-propylene	−10 to −30
10	Silicone rubber	Poly(di-alkyl/aryl) siloxanes	−125 to −50
11	Polyurethane rubber (PUR)	Polyether or polyester polyurethane	−52 to −30
12	EPDM rubber	Copoly(ethylene-propylene-diene monomer)	−50
13	XNBR	Carboxylated nitrile rubber	−23
14	HNBR	Hydrogenated nitrile rubber	−25

bered isoprene units but with a pendent methyl group at the carbon attached to the double bond, which hinders axial rotation of the bond. In fluoro-elastomer (Viton), the fluorine atoms impart not only strong covalent bond, but also very high inter-chain secondary valence bonds, that results in comparatively higher glass transition than the parent polyethylene-propylene copolymers. However, every elastomer needs cross-linking of entangled chains, commonly termed as vulcanization, through initiation by a peroxide or sulphur to form free radicals and subsequently covalent bond between the entangled chains.

Another process of tailoring the properties is the formation of compatible blends with either other one or more elastomers or with engineering plastics. Typical examples are nitrile-PVC blends, nitrile-phenolic blends and PE/PP-EPDM blends. For semi-compatible or incompatible polymers, the formation of interpenetrating polymer network is a very common option, such as urethane-acrylates and nitrile-alkyl methacrylate. The elastomeric nature of blends and interpenetrating polymer networks (IPN) is tailored by compositional adjustment and cross-link densities. Further, to improve physical and functional properties, appropriate fillers are compounded with the elastomers prior to vulcanization. With the advancement of technology, a large variety of fillers different shapes and sizes are used with all types of elastomeric materials.

4.2 Particulate rubber composite

A composite of rubber with small particulate fillers is the most common form of rubber items. Among the fillers, reinforcing types are those which impart synergistic effect in augmentation of all properties of a vulcanizate. Two main particulate matter for reinforcement of elastomers are carbon black and graphite, both being allotropes of elemental carbon. Graphite is the purest form of carbon, as only C=C bonds are present in the graphite structure.

4.2.1 Carbon black

Carbon black of various sub-micron sizes is the most widely used reinforcing filler because of many reasons such as its strong secondary valence bond with elastomer molecules, wettability by almost all elastomers, uniform distribution in the matrix, low cost, high hiding power and a very effective improvement in strength and atmospheric stability. Three most important features of carbon black are (1) "particle size": average diameter of one carbon particle; (2) "structure": the agglomerated particle formation; and (3) "chemistry": the chemical elements and groups chemically bonded to the carbon. The reinforcement mechanism of carbon black is a manifestation of these three characteristics of the black. The reinforcement is mostly by physical adsorption of the viscoelastic rubber on active sites of the carbon black surface and possible chemical linkages through some functional groups such as hydroxyl and carboxyl on the carbon black surface [1–3]. Carbon black can be formed by various methods from various sources. The reinforcing ability and ultimate properties of a black-filled vulcanized elastomer depends on the type of carbon black and its source. A neat vulcanizate of natural rubber might have tensile strength of 1.5–1.75 MPa on vulcanization by 1.5% sulphur, but on addition of 30% HAF carbon black, the strength could be 9–12 MPa. Table 4.2 gives various forms and sources of carbon black used for reinforcement [4, 5]. The nomenclature of a carbon black is done according to its source/origin. Apart from normal strength augmentation, carbon black is also used for making conducting plastic like electromagnetically absorbing and also reflecting surface, anti-static rubber bands, floors, UV stabilizer, for example for polypropylene water tanks and also IR reflector panels.

Tab. 4.2: Carbon black: particle size and application.

Designation	Particle size (nm)	Agglomerate size (nm)	Average N_2 surface area (m2/g)	Application in rubber industry
General-purpose furnace black (GPF) N660	50–70	280–400	30–40	General-purpose rubber items and tyres, low heat build-up.
High-abrasion furnace black (HAF) N330	25–40	150–250	70–80	Medium-high reinforcement, low modulus, high elongation, good flex, tear and fatigue resistance; vibration damping compositions
Intermediate SAF (ISAF) N220	20–25	112–115		High-strength rubber goods, tyres, good abrasion resistant rubbers
Super-abrasion furnace black (SAF) N110	15–25	50–100	120–140	High reinforcement; tyre products of high abrasion resistance
Easy processing channel black (EPC) N300	30–35	100–200	80–240	Medium to high modulus, high reinforcement, tyre, hose, gaskets, hose
Medium thermal black (MT) N990	250–350	500–600	6–15	Low reinforcement, low modulus, hardness, hysteresis and tensile strength, high elongation and loading capacity, used in mechanical goods, belts, hose, gaskets and tyre inner liners
Fine thermal black (FT) N880	180–200	350–450	5–10	

4.2.2 Graphite

Another allotropic form of carbon is graphite, which is pure carbon, and a plate-type filler. Graphite is a soft material with a density of about 2.26 g/c.c. and available in different natural and synthetic varieties:
a) Amorphous graphite
b) Vein or plumbago graphite
c) Flake graphite
d) Synthetic graphite

a) **Amorphous graphite**
   In general graphite is a highly crystalline material. But based on the availability, for instance, the one found in the metamorphic rock in the form of a slate is not orderly arranged. They are also termed as "microcrystalline graphite".

b) **Vein graphite**

Vein graphite grows naturally from carbon precursor under high temperature and pressure inside the earth crust from a fluid called "pegmatite". The graphite so formed is as pure as 99% carbon with high crystallinity which does not exist in any other type of graphite. Due to high crystallinity, they are used in thermal and electrical applications.

c) **Flake graphite**

They are evolved as flakes or in the form of lamella, hence they are called as flake graphite. The carbon content in this type of graphite could be about 99%. Most of the flake graphite is used in manufacturing of pencil leads.

d) **Synthetic graphite**

Apart from the natural source, synthetic graphite is also very commonly used in rubber industry. Synthetic graphite is made from the carbon rich precursor under high temperature which facilitates the conversion of mesophase carbon in to graphite. The temperature involved in graphitization process is above 2,800 °C and in an inert atmosphere. Many carbon precursor are reported as the starting materials for graphite synthesis. To name a few cokes, petroleum products, natural or synthetic organic chemicals.

A typical graphite morphology by scanning electron microscope (SEM) is shown in Fig. 4.1. The plate-type shape is quite evident from the SEM micrograph. The structure of graphite is ring of six carbon atoms arranged in layers of sheets with specific spacing, as shown in Fig. 4.2. The inter-layer bonding force is Van der Waal's, which is weak secondary valence force. In the axial plane, the rings are connected by C-C and C=C covalent bonds. Because of weak force among the successive sheets, the graphite sheets can move relative to each other in shear mode, and this makes the graphite a soft material. This property is best utilized in various applications such as lubrication, mechanical vibration damping and tapes for plugging leakages in pipe joints.

Reinforcing efficiency of carbon black is more than graphite, which is a purer form of carbon, devoid of any organic functional groups, and its specific surface area and surface energy is far less than carbon black. A typical small size graphite flake can be 4–25 micron in length [6, 7] and for bigger size, as high as 290 microns [8], compared to all submicron sizes of carbon black structures.

4.2.3 Mineral particulate fillers

Apart from commonly used carbon black and graphite, many mineral fillers are also used in elastomer composites such as aluminium flakes, mica, iron oxide, alumina and silica for tailoring mechanical properties, colour and cost reduction, tribological and dynamic mechanical properties such as vibration damping, resilience or heat build-up. Generally, particle size of such micronized fillers is in the range of 1–30 mi-

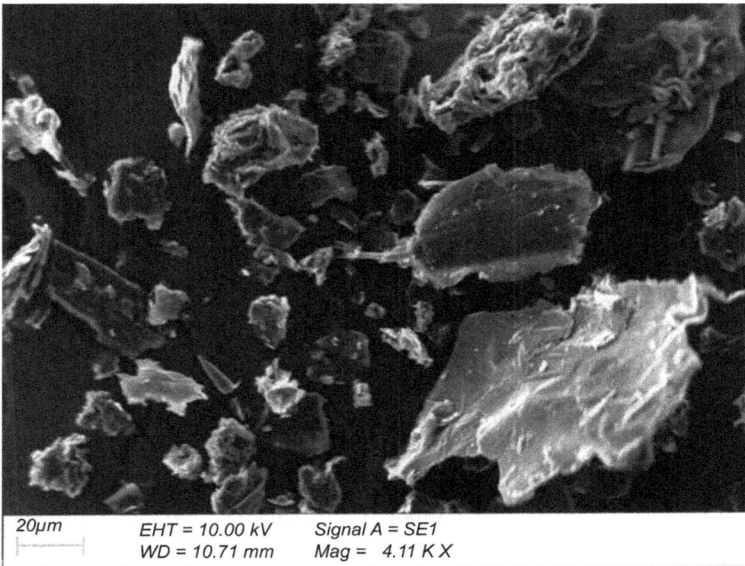

Fig. 4.1: SEM micrograph of a graphite sample.

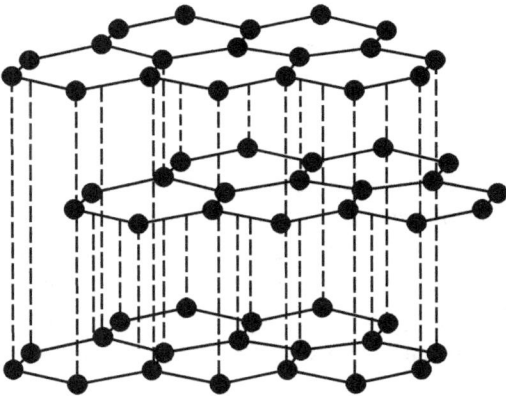

Fig. 4.2: Layered structure of graphite.

crons, beyond which these are not significantly effective reinforcement, since the physical forces of attraction with the elastomer molecule reduce with increasing size. Additionally, reinforcement by a neutral filler is reduced as the filler content increases, since the higher filler content reduces the continuity of the polymer matrix and hence failure occurs at the poor interfaces of large number of particle-polymer pairs [9].

In fact, the mechanical properties reduce with such higher particles compared to the neat elastomer, because the failure takes place at the poor interface of the polymer and filler particle. A study by Thongsang et al. [10] on effect of different silica particle reinforcement in natural rubber showed that precipitated silica has greater effect in improvement in mechanical and tribological properties than fly ash silica because of its higher particle size. At 40 parts per hundred gram of rubber (phr) loading of 40–74 mm sized fly ash silica, the NR showed a tensile strength of 8.1MPa, while same loading of a fly ash silica at a particle size of less than 25 mm showed tensile strength of 10 MPa and precipitated silica of lower size, approximately 15 mm [11], showed a tensile strength of 24 MPa, which clearly shows the effect of lower particle size on reinforcement.

Unlike carbon black and organically silanized silica, addition of micronized particulate minerals may not have any significant reinforcing effect in tensile or flexural strength, but compression modulus is much enhanced. In tension or bending mode, the polymer chain-filler interaction governs the reinforcement, and neutral minerals only contribute to van der Waals forces, while carbon black or silanized silica has some surface polar groups which contribute to greater interaction with the elastomer.

An interesting aspect of ceramic fillers in polymers is heat shielding effect due to interaction of the ceramics with the polymer at burning temperature, a series of reactions between the polymer matrix and the added ceramic powder produce a ceramic protective layer of dense ceramic matrix [12–17]. Ren et al. [12] studied the ablation and heat shielding property of EPDM rubber-based composites containing various proportions of ceramic fillers such as zinc borate, zirconium oxide, silicon carbide, fumed silica and kaolin. The mechanical strength was reduced with increasing filler content, as expected, but the thermal conductivity reduced drastically and the content of ceramic fillers improved the thermal stability and ablation properties.

Similarly, the thermo-oxidative stability of silicone rubber can be enhanced by adding ceramic particulate fillers such as ferric oxide, cerium oxide or their mixed oxides [18, 19], silica, zirconia, ceria [20], carbon nanotube modified iron oxide [21, 22] as reported in literature. The tensile strength or elongation at break does not change much on addition of the mineral and ceramic particulates of even very small size (crystalline size of 9–11 nm). However, on addition of only 4 phr of ceria-ferrite in a silicone rubber, the thermos-oxidative ageing showed that the tensile strength and % elongation was retained to at least 55–60% of original strength, after ageing at 300 °C for 24 h. The reason for improvement in thermal resistance on addition of metal oxides is trapping of free radicals generated on thermal excitation of the polymer, more so, when a ceramic like ceria-ferrite is used, due to synergistic effect of $Ce4+$ and $Fe3+$ cations, which was confirmed by XPS spectra [19]. A transmission electron microscope (TEM) image of the ceramic particulate of Ce-FeO is shown in Fig. 4.3. The image shows the crystallites of approximate 5–10 nm size in soft agglomerate form [19].

4.3 Properties of particulate composites

Particulate fillers of micron and lower sizes form an isotropic composite when compounded with the soft elastomer. Due to low modulus of the elastomers (of the order of 10^6 Pa), the mixing, dispersion and distribution of micronized particles in elastomers are easy, efficient and less energy consuming than in engineering thermoplastics. It is known that smaller the particle, more is the tendency of the particle to attain spherical shape, and subsequently higher surface area per unit volume and hence, better bonding force with the elastomer. Without any polar interaction, or presence of dipoles or hydrogen bonding or chemical reaction, the force of attraction is van der Waals force.

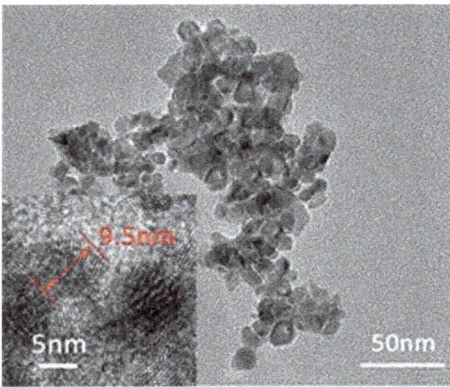

Fig. 4.3: TEM image of a Ce-FeO ceramic particulate filler for silicone rubber. (taken from fig. 3(a) in Ref. [19], with permission from M/s De Gruyter).

However, for perfect surface interface, sometimes tackifying bonding agents are used, such as oligomers of hydroxyl terminated resorcinol-formaldehyde and phenol-formaldehyde (resol). All necessary and desired properties of a particulate composite depend on this filler-elastomer interaction. Secondly, the properties also depend largely on the concentration of the filler in the elastomer. Up to a certain loading, known as critical pigment volume concentration (CPVC), the "wetting" of each filler particle by the elastomer will be possible, resulting in optimum property, and beyond that CPVC, there can be dry surface of the fillers, resulting in more filler-filler interaction, thereby reducing the reinforcing efficiency. Fillers are used mainly to alter important structural properties such as hardness, mechanical strength, flexibility, modulus of elasticity, tear strength, impact energy, creep compliance, stress relaxation, heat build-up, resilience and dynamic viscoelasticity. Other properties can be functional such as dielectric strength, thermal stability, electrical and thermal conductivity, microwave attenuation and acoustic properties. However, the effect of particulate filler reinforcement on mechanical properties of elastomer composites is most widely studied and discussed.

4.3.1 Hardness

The most common measuring method of hardness of elastomer composites is by the extent (cm) of indentation into the rubber surface by a rigid metallic of an indentor. The hardness is then expressed by arbitrary units of Durometer hardness Shore 'A' and 'D' for softer and harder elastomers respectively. Both scales are calibrated in a range of 0 to 100. A hardness of 40 in Shore D is approximately equivalent to 90 in Shore A. The indentation and durometer hardness (Shore A) are related as [23]

$$H = \frac{\Delta L}{0.025\,\text{mm}}, \quad \Delta L = L_0 - L \tag{4.1}$$

where DL is the indentation by the pressing tip into the rubber surface, measured in mm. L_0 and L are initial and final reading of the indentor position. Therefore, for an indentation of 1.25 mm, the Shore A hardness is 50 and at 2.5 mm indentation, the hardness reaches 100, which is the maximum in the Shore A scale.

The relation between the force applied and the indentation in measuring durometer hardness is given by [23, 24]:

$$F = 0.55 + 3\Delta L \text{ or } F = 0.55 + 0.075HA \tag{4.2}$$

for Shore A, and

$$F = 17.78\Delta L \text{ or } F = 0.4445HD \tag{4.3}$$

for Shore D scale. The unit of force is N (Newton).

Therefore, for a Shore A hardness of 50, the corresponding indentation is 1.25 mm and the force required is 4.3 N, and if the same force is required for a polymer in a Shore D scale indenter, then the Shore D hardness will be 9.7 and the indentation will be 0.242 mm.

The hardness is directly a reflection of the modulus of elastic deformation of the elastomer. As the elastomer is reinforced by the particulate filler, the hardness increases, and mostly monotonically till the critical particle loading is reached, beyond which the increase could be nonlinear.

A typical mechanical property evaluation was carried out with a blend of nitrile rubber (33% acrylonitrile content) and polyvinyl chloride in the approximate ratio of 70:30 by weight, reinforced with both HAF carbon black (N330) and graphite of 2 micron in equivalent diameter. The Durometer hardness (Shore A) of the two types of composites are shown in Fig. 4.4, both containing 10–45 phr fillers, cured under similar proportions of sulphur, accelerator and other ingredients, so that the difference of the composites between them and with the neat vulcanizate can be observed. The initial effect of graphite and HAF carbon black reinforcement at 10 phr is quite significant, and the rate of hardness increment subsequently is almost similar for both the reinforcements, but the rate of increase is somewhat less with graphite compared to carbon black at 45 phr. The reason for similarity is simply because the resistance to indentation is largely offered by the filler in compression, and is not a reflection of reinforcing efficiency.

4.3.2 Elastic modulus

The hardness is an indirect indication of Young's modulus and also related to compressive modulus of filled elastomeric composites. Qi et al. [23] used a non-linear FEM simulation to correlate Shore A hardness with Shore D and stress-strain behaviour of elastomers. The authors experimented with polyurethane rubber (PUR) and polybutadiene (BR) rubber formulations with varying hardness and arrived at a semi-empirical model on the relation of hardness and Young's modulus, as:

$$\log_{10} E = 0.0235 S_A - 0.6403 \tag{4.4}$$

for S_A = Shore A, 20 < S_A < 80, and
$\quad S_A$ = Shore D + 50, 80A < S_A < 85D

where E is the Young's modulus, can be taken as initial slope of a stress-strain curve of an elastomer composite, and S_A is durometer hardness (Shore A).

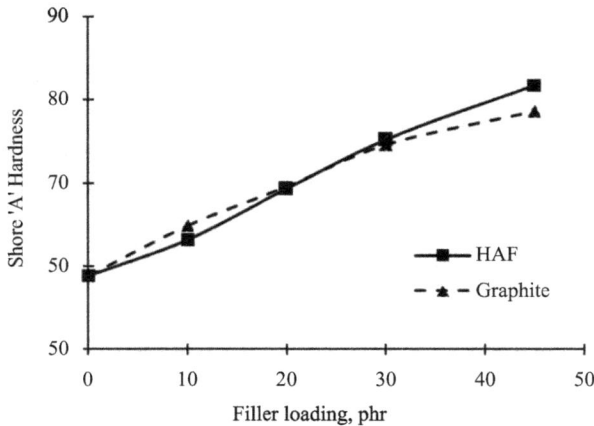

Fig. 4.4: Durometer hardness of carbon black and graphite composites of an NBR-PVC (70/30) blend vulcanizate.

Since the elastomers exhibit non-linear elastic behaviour, it is therefore best to consider the value of Young's modulus which does not change till an upper limit of strain, such as up to maximum 1% for nitrile rubber and NBR-PVC 50/50 blends as reported by Murali et al. [25]. The constants of the above empirical expression derived by the authors may not be same for all elastomers, since the constant would depend on some factors such as polarity of the filler and the matrix and filler-filler interaction due to different extent of agglomeration, etc. Murali et al. found the empirical expression for nitrile rubber-carbon black as

$$\log(E) = 0.0308 S_A - 1.1653 \tag{4.5}$$

Gent [26] has proposed a relationship as

$$E\left(\text{kg/cm}^2\right) = \frac{56 + 7.66 S_A}{2.67 r (254 - 2.54 S_A)} \tag{4.6}$$

where r is the radius of the round tip of the indentor used for measuring Shore A hardness, which is normally 0.0515 mm. Taking this value and by rearranging the units of E as MPa (N/mm2), we have a modified expression by Gent Expression as:

$$E(MPa) = \frac{5.4936 + 0.747852 S_A}{34.92627 - 0.349263 S_A} \tag{4.7}$$

The above expression is not in agreement with very soft rubbers, where indentation is high. BS 903 (British Standard 903) gives the following expression, valid for Shore A range from 0 to 100, correspondingly elastic modulus of any value [27]:

$$S_{IHRD} = 100 erf\left(3.186x10^{-4} E^{\frac{1}{2}}\right) \tag{4.8}$$

where *erf* stands for the error function and E is Young's modulus in Pa (N/m2). IHRD hardness and Shore hardness are not uniformly related, but differ according to the sample thickness [28].

In experiments with standard ASTM button sample, the thickness is generally 12 mm and the diameter is about 28–30 mm. For such thickness, the relation is approximately given as:

$$S_{IHRD} = S_A + 4 \tag{4.9}$$

Briscoe and Sebastian [29] derived a relation among indenting force, indentation, durometer hardness and Young's modulus of rubber and arrived at the following equation:

$$F = 0.55 + 0.075 S_A \tag{4.10}$$

$$F = \frac{2E^*}{\tan\theta}\left\{ha\tan\theta - \frac{a^2}{2}\left[\frac{\pi}{2} - arc\sin\left(\frac{b}{a}\right)\right] + \frac{b}{2}\sqrt{a^2 - b^2}\right\} \tag{4.11}$$

$$E^* = \frac{E}{1 - v^2} \tag{4.12}$$

The relation of Young's modulus and Shore A hardness was determined with experimental engineering values of a NBR-PVC blend vulcanizate and its HAF carbon black and graphite composites, with 0, 10, 20, 30 and 45 phr loading. The Young's modulus was determined for each sample at 1% strain, since up to 1%, the modulus was almost constant for each composition. The experimental values of Shore A hardness vs. Young's modulus was plotted along with the theoretical predictions by Qi, Gent, BS903 and Briscoe and Sebastian for comparison and extent of agreement at various hardness levels. The IHRD hardness obtained in BS903 method was converted to Shore A hardness by using eq. (4.6), since the experimental button for hardness measurement was 12 mm thick, as suggested by Morgans et al. [28]. The results are shown in Fig. 4.5.

The relation of Shore A hardness and tensile modulus in this case is best described by the following empirical expressions, similar to Qi et al., but with different coefficients:

$$\text{HAF reinforced composite: } \log_{10}(E) = 0.0348S_A - 1.778 \tag{4.13}$$

$$\text{Graphite reinforced composite: } \log_{10}(E) = 0.3334S_A - 1.6356 \tag{4.14}$$

The increase in modulus with hardness is almost similar for carbon black and graphite, but the values for the carbon black composites are higher at the same hardness.

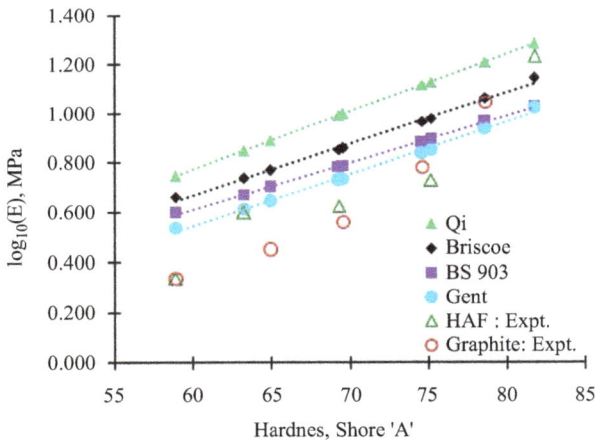

Fig. 4.5: Hardness–elastic modulus relationship of HAF and graphite composites of NBR-PVC blend.

The theoretical models are not exactly in agreement with this particular elastomeric composite, possibly because the blend is not a perfect elastomer, as PVC is added substantially. Among the theoretical models, prediction of modulus by Qi et al. is highest and that by Gent is lowest, rest are in between these two extremes for same hardness values. For these two particulate elastomeric composites, at low hardness, near about 60 Shore A, prediction by Qi is 57% higher compared to Gent and at high hardness, near 75–80 Shore A, it is 90% higher. However, at about 75 Shore A, the elastic modulus of HAF and graphite composites are similar to those predicted by Gent, Briscoe and BS 903.

4.3.3 Filler effect on reinforcement

The reinforcement effect in a softer elastomer, such as nitrile rubber is less because the softer matrix is poorer in transfer of load to the rigid filler compared to a harder matrix. Figure 4.6 shows the dependence of tensile modulus at varied filler content for NBR-HAF and NBR-graphite composites, and Fig. 4.7 for NBR-PVC blend vulcanizate. While NBR neat has a modulus of 1.17 MPa at 1% strain, NBR-PVC neat has 2.18 MPa at the same strain. The slope of the NBR-PVC-HAF plot is 0.28 while that of NBR-HAF is 0.168, and for NBR-PVC-Graphite, the slope is 0.209 compared to 0.095 for NBR-Graphite. These clearly shows that reinforcement by particulate fillers for softer elastomers are less effective, and that carbon black has higher reinforcing efficiency than graphite.

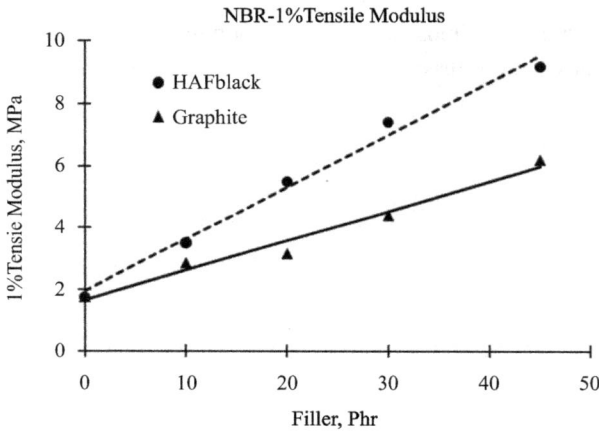

Fig. 4.6: Filler reinforcing effect in NBR: HAF black and graphite.

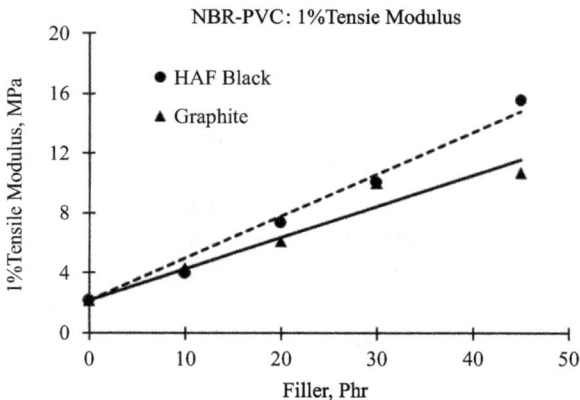

Fig. 4.7: Filler reinforcing effect in NBR-PVC: HAF black and graphite.

In order to determine the quantitative contribution of the particulate filler on change in elastic modulus of the elastomer particulate composite, many theories are developed, in which the simple mixing rule is the most fundamental:

$$E_C = V_m E_m + V_p E_p \tag{4.15}$$

However, this does not hold good for particles like carbon, graphite or minerals because the Young's modulus of the filler is much higher, generally 1,000 times or more than all elastomers and hence the transfer of force between the filler to the matrix is not much effective, thus drastically reducing the reinforcing effect of the filler. There is an attempt to incorporate the so-called efficiency factor of the filler in the simple mixing rule to obtain a more appropriate predictive equation:

$$E_C = E_{p(\text{eff})} V_p + E_m V_m \tag{4.16}$$

where $E_{p(\text{eff})}$ is the effective filler modulus, which can be obtained from experimental data by plotting E_C against V_p, provided $V_m = (1-V_p)$, which means that total volume of other ingredients in the compounding of the rubber are not significant. The effective fibre modulus is higher for higher modulus of the polymer, due to improvement in the stress transfer.

In 1945, Guth [30] proposed a non-linear relationship among the composite modulus, matrix modulus and concentration of the filler in the composite, considering the spherical particles of carbon black and no filler-filler interaction or agglomeration. The relationship is just analogous to the expression of viscosity of a liquid medium with suspended spherical particles. In case of elasticity, the suspended particles actually perturb the elastic stress and strain and hence the modulus increases. However, beyond a certain concentration of the filler, there will be linear agglomeration and subsequently, on still higher concentration, formation of cubical blocks of carbon black was observed, thereby reducing the reinforcing effect.

$$E_C = E_m \left[1 + 2.5c + 14.1c^2 \right] \tag{4.17}$$

However, when there is agglomeration of carbon black particles in the elastomer, the modified expression for the composite modulus is [30]:

$$E_C = E_m \left[1 + 0.67f.c + 1.62f^2c^2 \right] \tag{4.18}$$

where f is the aspect ratio (length/breadth) of the particle agglomerate. Here, c can be taken as a volume fraction instead of mass concentration.

A typical plot of the NBR/PVC (50/50 blend) – HAF carbon black composite modulus at 1% strain for various volume fractions of the HAF is shown in Fig. 4.8 and compared with eq. (4.17) which is without considering any agglomeration of the filler, and eq. (4.18) taking the aspect ratio of the carbon black clusters as 6.0 as suggested by Guth [30] and a fitting aspect ratio of 7.25 to obtain a better agreement with the experimental values (dark circle). It can be seen that the difference in modulus with or without considering the aspect ratio as a variable, the modulus values are not very different at low volume fractions, as eq. (4.17) is ideally applicable for no cluster formation, which is partly true for low volume fraction of filler. The lower values obtained using eq. (4.17) and also with eq. (4.18) with $f = 6$ could be due to the fact that the clustering do occur, at higher filler concentration and the cluster to matrix interaction is more for more clusters or bigger clusters. It is known that carbon black reinforces a polymer by van der Waals force as well as though polarity due to presence of some polar organic groups on the carbon black particles. However, the aspect ratio of 7.25 does not fit perfectly after 100% strain as seen for only 10 phr HAF containing NBR/PVC vulcanizate, as shown in Fig. 4.9, although the trend in change of modulus followed the same pattern. At higher strain, the elastic energy required to stretch is high due to covalent bond stretching of the polymer chain. Therefore, consideration of the physical elasticity is valid only at low to moderate stretching.

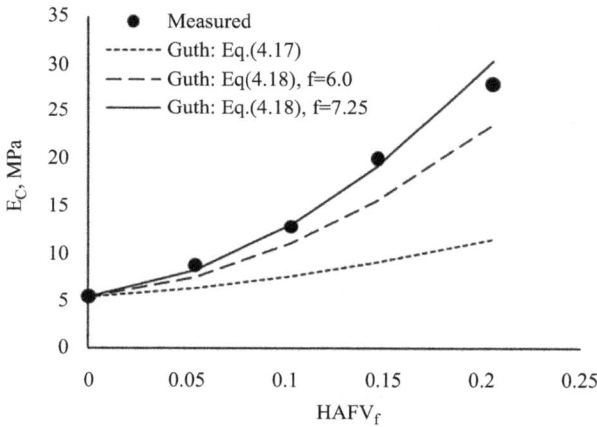

Fig. 4.8: Tensile modulus of NBR-PVC (1% strain) with HAF filler: experiment and theory: Guth [30].

In the above models, the elastic modulus of the reinforcing particles is not considered, possibly because the modulus of particulate fillers is very high compared to that of the elastomers. However, some theoretical analyses of the composite modulus were developed where particle modulus was included. Some such models are described here.

Isostress theory [31] gives the following expression for the particulate composite modulus:

$$\frac{E_C}{E_m} = \frac{1}{V_p\left(E_m/E_p - 1\right) + 1} \tag{4.19}$$

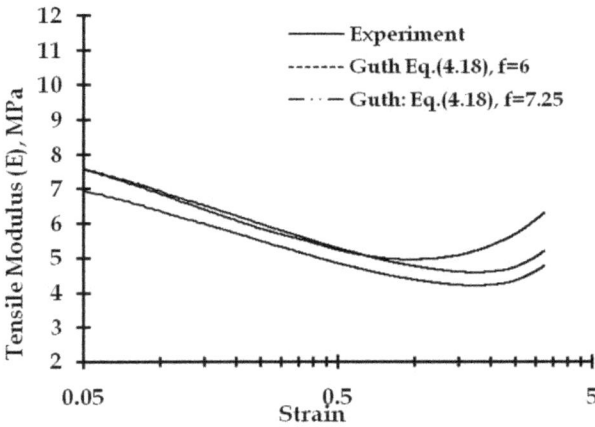

Fig. 4.9: Composite modulus vs. strain of NBR/PVC (50/50)-10 phr HAF: experiment and theory by Guth [30} with f = 6 and 7.25.

and the Poissons ratio of the composite, $v_C = \dfrac{v_p V_p + v_m (1 - V_p) E_p / E_m}{V_p + (1 - V_p) E_p / E_m}$ (4.20)

Kerner theory [31, 32] gives the following expression for the particulate composite modulus:

$$\frac{E_C}{E_m} = \frac{\dfrac{V_p E_p}{(7 - 5v_m) E_m + (8 - 10v_m) E_p} + \dfrac{V_m}{15(1 - v_m)}}{\dfrac{V_p E_m}{(7 - 5v_m) E_m + (8 - 10v_m) E_p} + \dfrac{V_m}{15(1 - v_m)}}$$ (4.21)

and shear modulus:

$$\frac{G_C}{G_m} = \frac{(7 - 5v_m)(G_m + V_p G_p - V_p G_m) + (8 - 10v_m) G_p}{(7 - 5v_m) G_m + (8 - 10v_m)(G_p + V_p G_m - V_p G_p)}$$ (4.22)

where V, E, G and n represent volume fraction, elastic (tensile) modulus, shear modulus and Poisson's ratio respectively and the suffixes C, p and m represent the composite, particulate reinforcement and the elastomeric matrix respectively. For all practical purpose, Poisson's ratio of an elastomer can be taken as 0.5 without much error.

A different, simple model was proposed by Ramsteiner and Theysohn [33] as

$$\frac{E_C}{E_m} = \left[1 + \frac{V_f}{1 - V_f} \left(\frac{15(1 - v_m)}{8 - 10v_m} \right) \right]$$ (4.23)

and a similar expression was given by Kerner-Nielsen [33, 34] as

$$\frac{E_C}{E_m} = \left[1 - V_f^{2/3} + \frac{V_f}{\left(\frac{E_m}{E_f} - 1\right)V_f^{2/3} + V_f^{1/3}}\right] \qquad (4.24)$$

However, these equations from 4.19 to 4.24 are applicable to the neutral fillers which do not reinforce by any attraction other than van der Waals. Therefore, reinforcement by carbons is not predicted by the above expressions. Rather a model similar to Guth [30] is better to describe the reinforcement of elastomers with carbons. Figure 4.10 clearly shows the large difference in the composite modulus (normalized by matrix modulus) experimentally found and the theories. The Kerner equation (4.21) and eq. (4.23) by Ramsteiner and Theysohn gave the same result, hence, only the latter is taken here.

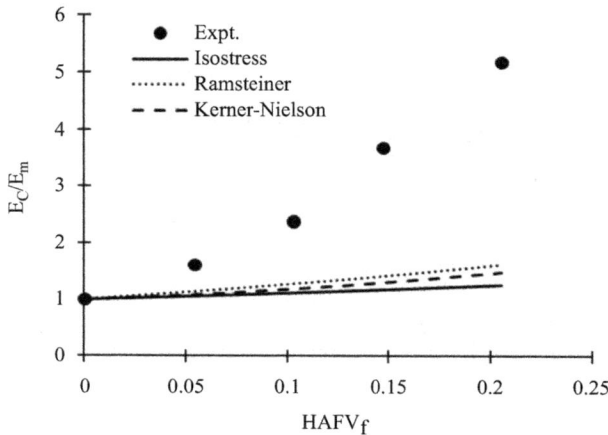

Fig. 4.10: Experimental and theoretical tensile modulus considering filler effect: NBR-PVC 50/50 vulcanizate filled with HAF black.

4.4 Short fibre rubber composites

A very useful reinforcement of elastomers is by short fibres of many materials, such as glass, aramid (Kevlar), nylon, carbon, polyester, asbestos, cellulose, jute, coir and wood fibre. The efficacy of the reinforcement depends on several factors, such as modulus ratio of filler and the elastomer matrix, uniformity in dispersion, filler-rubber interaction (compatibility), agglomeration of the short fibres, length to diameter ratio (aspect ratio) of the fibre, orientation of the fibre in the elastomer matrix and retention of the fibre length in the final product. The dispersion is counteracted by formation of bundles due to filler-filler interaction and poor fibre-matrix interface.

The poor interface results in fibre pull-out during tensile force as the fibre simply slips past the matrix due to both low force of attraction and large difference in strain. In general, hydroxyl terminated low molecular weight phenol-formaldehyde and resorcinol-formaldehyde resins are used during rubber compounding for improvement of bonding the filler to the rubber. However, special fibres of cellulose are pre-treated with resorcinol-formaldehyde resin such as in a commercial cellulosic fibre "Santofloc". Carbon fibres are better compatible with rubbers than others because of presence of some organic groups in the fibre, which can provide better bonding and reinforcement to the elastomeric matrix, which is also basically a carbon-carbon chain. Non-polar rubbers are difficult to be reinforced by the fibres unless the phenolic bonding agents are used. The ratio of the length to diameter (aspect ratio) of the fibre is important in reinforcement efficiency. Below an aspect ratio of 40, the filler behaves as a particulate filler, and beyond 250, it is difficult to incorporate in the rubber matrix. In most cases, the aspect ratio of the fibres is between 40 and 150 before mixing. Brittle fibres such as glass and carbon are likely to break during rubber compounding and the aspect ratio is drastically reduces thereafter. Murty and De [35] showed that the original length of 9 mm (diameter 0.33 mm) of a glass fibre has reduced below 0.6 mm with diameter reduced to 0.01 to 0.08 mm after mixing with SBR, whereas for the jute fibre, the aspect ratio was reduced from 150 to 40 (length = 1.5 mm), but the diameter remained unchanged. The study showed best dispersion for jute fibre length of 1.5–2 mm and 0.2 mm for glass fibre. The tensile strength of glass fibre-SBR composite was much less, almost half of that for jute fibre-SBR composite at identical fibre content.

The use of short fibre in enhancing the properties of rubber was known from early days, when environmentally friendly, fibrous, powdered cellulose, known as Solka-Floc, derived from purified wood pulp was used in rubber composition [36, 37]. However, the use of short fibre was not explored till seventies. Boustany and coworkers studied various short fibres for reinforcement of different rubbers with regard to dispersion, fibre length, orientation and properties in early and mid-1970s [38–42]. In a comparative study, Boustany and Arnold [43] reported NR, SBR, NBR, NR/Chlorobutyl, NR/BR,Neoprene and NR/SBR rubber composites of a commercially available short fibre of treated unregenerated cellulose, known as "Santoweb" with other short fibres such as chopped impregnated glass fibre, Solka-Floc (cellulose), nylon floc, asbestos and chopped polyester fibre. The study focused on comparison of vulcanization characteristics and properties such as resistance to swelling in solvents, stiffness, tensile strength, elongation at break, tear strength, creep, flex fatigue life using cyclic load at various fractions of failure load, etc. Cellulosic natural fibres can have a wide spectrum of properties and reinforcing efficiency when used in elastomers. The reinforcement depends on the original strength and stiffness of the cellulosic source. For example, the wood fibres from hard wood are having lower diameter and higher stiffness compared to soft wood fibres [44], brown paper pulp fibres are stiffer than filter paper tissue fibres and cotton fibre [37].

Geethamma and coworkers developed natural rubber-coir fibre composites and studied the effects of fibre length, orientation and alkali treatment on mechanical and dynamic mechanical properties [45, 46]. Varghese et al. [47] studied the effect of bonding on the mechanical properties of natural rubber-sisal fibre (lignocellulosic fibre) composite, where the fibre was chemically modified by acetylation and a resorcinol-hexamethylene tetramine (Hexa) combination was used for better adhesion. The treatment resulted in at least 20–30% increase in tensile strength and modulus compared to the untreated fibre when resorcinol-Hexa was used, and in the absence of this adhesion promoter, the treated fibre composite had 100% higher strength than the untreated fibre composite. Treated cellulosic short fibre from hard wood source can reinforce a rubber much better than nylon, acrylic, rayon, polypropylene and polyester (PET) short fibres as shown by Coran et al. [40]. Ismail et al. [48] studied bamboo fibre filled natural rubber composite with and without phenolic resin as bonding agent and reported 20–30% increase in properties with the bonding agent compared to the composition without the bonding agent. However, the mechanical properties of the composite were only 50% of the pristine natural rubber, possibly because the fibre had a very broad size distribution and irregular in shape, which might have reduced the load transfer from the fibres to the matrix. However, the hardness increased from 35 to 70 and the modulus improved from 1.25 MPa to about 4.5 MPa for 50 phr bamboo fibre loading with phenolic resin as bonding agent. Similar observation is made by Cracium et al. [49] for fibrous fir wood saw dust filled EPDM composites cured by electron beam and also be peroxide, where the tensile strength of the composites irrespective of curing system, was less than unfilled EPDM. However, the report showed that curing by electron beam at low radiation dose (100–200kGy) yielded better mechanical property than peroxide cured composite.

One important elastomer short fibre composite is in tyre tread application, where short fibres of aramid [50], carbon [51] and nylon [52] are used in rubber compounds. The tyre tread rubber composition should have a balance of low heat build-up and rolling resistance, which are contradicting features since in the first case, the loss energy should be low, and the second feature needs a higher loss factor. The balancing is possible by selection of short fibre as a filler to reinforce the dynamic mechanical properties, apart from Young's modulus. Mohammed et al. [53] studied aramid fibre reinforced nitrile rubber for gasket application. The inclusion of the aramid in the rubber in addition to nanosilica (10 phr fixed) and carbon black (30 phr fixed), greatly improved with fibre content till 20 phr with respect to several properties desired for gasket application, such as strength, compression set, load-deflection in compression and tear strength. However, resorcinol-hexa combination was also used as adhesion promoter for the fibre with the rubber.

4.4.1 Mathematical predictions

Several mathematical models are developed to predict the fibrous composites of polymers, considering simple mixing rules, modified mixing rules and models by Halpin-Tsi, Tsai-Pagano and others, taking into account the fibre orientation and aspect ratio (L/d) of the fibre and empirical efficiency factors of reinforcement. The efficiency factor comes from consideration of two aspects, one that of randomness of fibre orientation in a three dimensional composite body and the other for the effectiveness of load transfer from the fibre to the matrix. The load transfer is more as the elastic modulus of the polymer matrix is higher. For elastomer-short fibre composites, the efficiency of load transfer is far less compared to engineering plastics and thermosets.

Coran et al. [40] suggested a simple equation of the rubber-cellulose short fibre composite relating various parameters of the fibre and matrix rubber as:

$$E_C = E_m \left[1 + K f V_f \{ 26 + 0.85 (l/d) \} \right] \qquad (4.25)$$

where E_C is the Young's modulus of the elastomer composite, E_m is that of the elastomer, V_f is the volume fraction of the short fibre, l/d is the aspect ratio of the fibre, f is a function of average fibre orientation and K is a constant.

The equation is based on the assumption that the fibres being much stiffer compared to the rubber matrix, the extension of the fibre under tensile force is neglected. The equation shows that the volume fraction, orientation and aspect ratio of the fibre play very important roles. In a load bearing application, the fibre orientation must be as far as possible in the direction of force. This may be possible in tension, but not in flexure, particularly for thick sections, since the mid plane of a thick specimen would undergo shear deformation, apart from the compression at the upper (concave) surface and tension of the lower (convex) surface.

It can be seen from eq. (4.25) that the modulus linearly increases with the volume fraction at a fixed aspect ratio and also linear with aspect ratio at the fixed volume fraction of the fibre. However, neither case is applicable in practice, since at high aspect ratio and also at high volume fraction, the uniformity in mixing and ability to mix the long fibres in a polymer matrix becomes very difficult and impractical, thereby reducing the efficiency of reinforcement.

The most used expression is given by Halpin-Tsai equations for the longitudinal (E_L) and transverse (E_T) moduli of aligned short-fibre composites:

$$\frac{E_L}{E_m} = \frac{1 + 2(l/d)\eta_L V_f}{1 - \eta_L V_f}, \quad \frac{E_T}{E_m} = \frac{1 + 2\eta_T V_f}{1 - \eta_T V_f} \qquad (4.26)$$

where

$$\eta_L = \frac{(E_f/E_m) - 1}{(E_f/E_m) + 2(l/d)} \tag{4.27}$$

and

$$\eta_T = \frac{(E_f/E_m) - 1}{(E_f/E_m) + 2} \tag{4.28}$$

Tsai and Pagano (1969, 1994) showed that the modulus for randomly oriented short-fibre composites can be predicted approximately as:

$$E_C = \left(\frac{3}{8}\right) E_L + \left(\frac{5}{8}\right) E_T \tag{4.29}$$

where E denotes the modulus, V is the volume fraction and the subscripts C, m and f denote composite, matrix and fibre, respectively, l is length, subscripts L and T denote longitudinal and transverse directions., d is diameter of fibre, h_L and h_T are the efficiencies of the fibre reinforcement in longitudinal and transverse directions.

Following observations can be made from eqs. (4.26) to (4.29):
a) The longitudinal efficiency factor (h_L) reduces on increasing the aspect ratio (l/d) of the short fibre.
b) The transverse efficiency factor (h_T) depends only on elastic moduli of fibre and polymer matrix.
c) The differences in type of forces of attraction between the fibre and the polymer such as van der Waals force, polar force, hydrogen bonding, etc. are not reflected in efficiency factors.
d) The reinforcement asymptotically increases with increase in fibre aspect ratio.
e) After a particular aspect ratio, the reinforcement ceases to improve, irrespective of the fibre content.
f) With higher fibre to polymer modulus ratio (E_f/E_m), the aspect ratio should be higher for optimum reinforcement.
g) The longitudinal modulus depends both on the volume fraction of the fibre and the aspect ratio.
h) The transverse modulus depends only on the volume fraction of the fibre.
i) At a very low aspect ratio (~1), when the particle tends to be spherical, the longitudinal and transverse moduli are equal, signifying isotropic composite just like carbon black loaded rubber.

4.4.2 Examples

Two fibres of different Young's moduli are chosen: Polyester short fibre and Aramid short fibre. Aramid fibre has about 30 times higher Young's modulus than that of polyester fibre. Their characteristics are given below.

(1) Polyester: Fibre diameter = 15 mm, l/d = 70, 100, 150, 200, 250, 300, 350, 400 and 500.
Young's modulus E_f = 920 MPa
Loading: 5, 10, 20 and 30 phr
(2) Aramid: Fibre diameter = 12 mm, l/d = 150, 200, 250, 350, 450 and 550.
Young's modulus E_f = 2700 MPa
Loading: 1, 1.5, 2 and 3 phr

The polyester short fibre was used in a nitrile rubber (NBR) matrix at 5, 10, 20 and 30 phr. The NBR was vulcanized with sulphur-accelerator system and loaded with 30 phr HAF black. The tensile modulus of the NBR vulcanizate without fibre is 7.4 MPa at 1% strain.

The aramid short fibre was incorporated in natural rubber at 1, 1.5, 2 and 3 phr. The rubber was vulcanized with sulphur-accelerator system and was loaded with 43 phr HAF carbon black. The tensile modulus of the NR vulcanizate without fibre is 12.5 MPa at 100% strain.

The theoretical prediction of the composite modulus of NBR-Polyester short fibre composite is shown in Fig. 4.11 against the fibre aspect ratio for various fibre content. It is seen that the increase in predicted modulus ratio (composite to rubber) is not linear with aspect ratio and the modulus increase attained almost a constant value beyond l/d = 300. The modulus is enhanced by at least 2 times of original value (without fibre) at even 5 phr of the short fibre. However, when the modulus ratio is plotted against volume fraction of the fibre, it showed a linear behaviour, as seen in Fig. 4.12. The minimum fibre volume fraction of 0.034 corresponds to 5phr loading and maximum fibre volume fraction of 0.174 corresponds to 30 phr loading. The slope of the linear plot in Fig. 4.12 is seen to be higher for higher aspect ratio, signifying progressively higher reinforcement for higher fibre length. For the lowest loading at 0.034 volume fraction (5phr), the reinforcement efficiency was not much different for all the aspect ratios, as the ratio of composite to rubber modulus varied from 1.92 to 2.47 at the aspect ratio from 70 to 500. The predicted modulus of NR- aramid short fibre composite vs. aspect ratio (l/d) at various loading of fibre is shown in Fig. 4.13. The fibre has a Young's modulus of 27 GPa, which is typical of aramid, and NR has a very low tensile modulus, about 12.5 MPa at 100% strain. Because of the very high modulus of the fibre, we have taken very small quantity of fibre for analysis, 1, 1.5, 2 and 3 phr only. It is seen from Fig. 4.13 that the tensile modulus of the composite at 3 phr is quite high, at least 3–4 times the pristine rubber vulcanizate. The composite modulus is continuously increasing in this case with increase in aspect ratio, possibly because of very low volume fractions (0.0047 to 0.0141). At such low loading, the longitudinal and composite moduli were compared at the aspect ratio of 250 and also with the transverse modulus, as shown in Fig. 4.14. The transverse modulus, which depends on ratio of fibre to matrix modulus, is quite nearly equal to 1, signifying no enhancement due to the fibre. As it is also discussed, the transverse modulus is not dependent on aspect ratio.

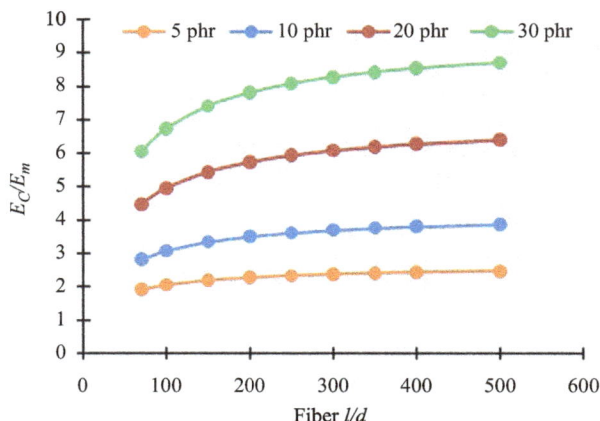

Fig. 4.11: Variation of elastic modulus of NBR-polyester short fibre composites with fibre aspect ratio: theoretical prediction.

However, in practical experimentation with short fibre, different results are obtained as reported in literature. Ryu and Lee [54] showed that the theoretical prediction of nylon short fibre reinforced chloroprene rubber (CR) according to the above equations are not in agreement with experimental results since the modulus ratio of reinforcing fibre to the rubber is more than 100. Also, it was seen that the maximum modulus enhancement was for the aspect ratio (l/d) of 300, beyond which the modulus value decreased, which is not observed in Fig. 4.11, where the modulus increased asymptotically with aspect ratio and linearly with fibre volume fraction (Fig. 4.12). In addition, the maximum composite modulus obtained experimentally was much higher than the theoretical prediction for identical fibre characteristics and loading, possibly because of higher matrix-fibre interaction due to polarity and presence of bonding agent.

Similarly, Kashani [50] studied NR-aramid short fibre composite and showed that with a fibre of aspect ratio of 250 and at a 1 phr loading of fibre, the ratio of longitudinal tensile modulus to that of the rubber vulcanizate is 1.063 only, whereas from Fig. 4.14, the same aspect ratio and fibre loading showed a modulus ratio of 2.93. For the transverse modulus, the calculation showed a ratio of 1.014 at 1 phr loading, but Kashani obtained a ratio of 0.982, which is quite low, in fact this was a deterioration of the original tensile modulus. In short fibre composite, there is a reason for this variation of composite modulus with fibre length. According to Shear Lag theory by Rosen [55, 56], the stress from the fibre end is transferred gradually towards the length and attains maximum at a particu-

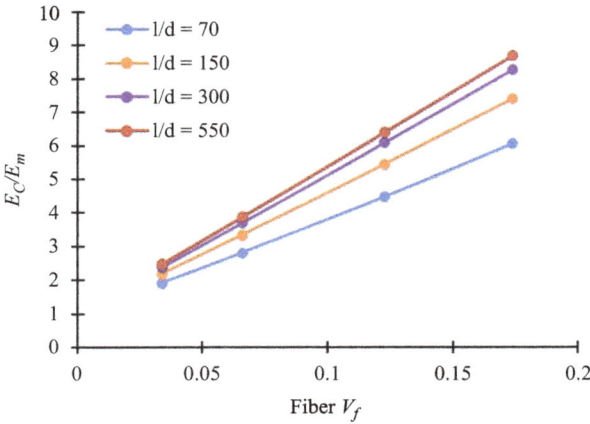

Fig. 4.12: Variation of elastic modulus of NBR-polyester short fibre composites with fibre volume fraction: theoretical prediction.

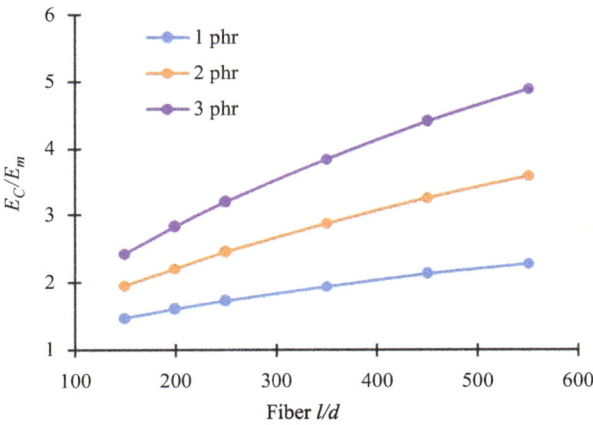

Fig. 4.13: Tensile modulus vs. aspect ratio for NR-aramid short fibre composite at various loading of the fibre.

lar length, which is the minimum length of the fibre required to obtain maximum composite modulus. This minimum length of fibre is given by.

$$\frac{l}{d} = \left[\frac{1}{2} \frac{E_f}{G_m} \frac{1 - V_f^{1/2}}{V_f^{1/2}} \right]^{1/2} \tag{4.30}$$

where G_m is the shear modulus of the matrix. Accordingly, taking Fig. 4.11 corresponding to example (1), and assuming the shear modulus of the NBR to be approximately one-third of the tensile modulus (Poisson's ratio ≈ 0.5) the minimum l/d required for

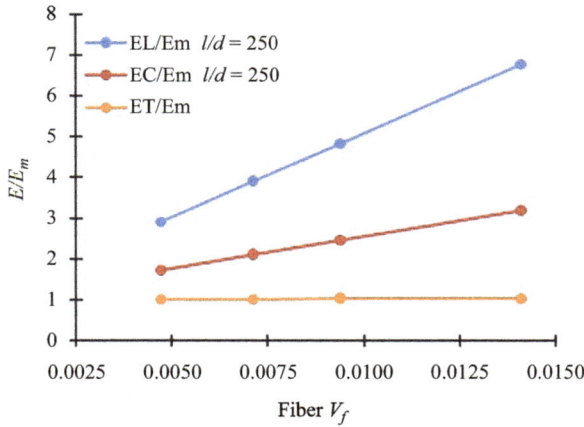

Fig. 4.14: Variation of composite modulus as ratio with rubber matrix vs. fibre volume fraction.

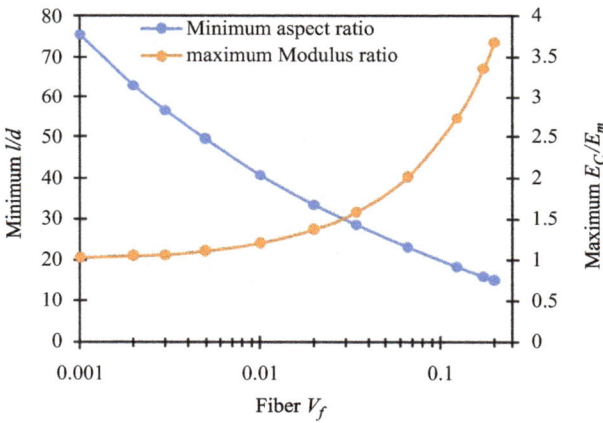

Fig. 4.15: Minimum fibre aspect ratio for optimum composite modulus at various fibre volume fraction.

maximum improvement in modulus is plotted against the fibre volume fraction V_f and corresponding maximum improvement in the composite elastic modulus (normalized by matrix modulus) in Fig. 4.15.

From eq. (4.30), it is seen that the aspect ratio required for maximum load transfer is higher for a higher modulus ratio of fibre to matrix (E_f/G_m). In the example, the ratio of E_f/G_m was 373 and corresponding fibre length was only 0.6 mm for 1% polyester short fibre loading, because the optimum aspect ratio was calculated as 40 and the average fibre diameter was 15 mm. If we consider example(2) with aramid short fibre loading at 1% V_f in NR, then the E_f/G_m is about 6,378 and corresponding optimum fibre aspect ratio would be 170. The aramid fibre average diameter being 12 mm, the optimum length becomes 2.04 mm, which is also quite small. Lower the modulus ratio, lower will be the optimum aspect ratio, while higher the fibre loading, lower will be the optimum aspect ratio. In most cases, it is seen that a fibre aspect ratio of 100 to 200 is optimum. This is, of course the final average size after the rubber compounding and fabrication is complete, because stiff fibres are likely to suffer breakage during rubber compounding and vulcanization under high shear and compression. Also, too long fibres are difficult to mix in practice, and hence for a higher average diameter, the optimum length might not be possible to incorporate, or might reduce in length after elastomer composite processing.

With a generalized expression of the coefficients in eq. (4.29), the composite modulus can be stated as a general Halpin-Tsai model:

$$E_C = aE_L + (1 - a)E_T \tag{4.31}$$

where a is 3/8 for random 2-D orientation of the short fibre and 0.184 for random 3-D orientation. In eq. (4.19), $a = 3/8$ and hence $1 - a = 5/8$. However, Shokrieh and Sani [57] used Mori-Tanaka and laminated analogy (MT-LA) approach coupled together to calculate the short fibre composite with random orientation of the fibre. Mori-Tanaka method was used to calculate the composite stiffness matrix of aligned fibres and laminated analogy approach was then used to calculate the stiffness of the composite with randomly oriented fibre. Consequently, they proposed the following empirical equation for "a" in case of 3-D randomly oriented fibre reinforcement:

$$a = 0.13 + 0.0815V_f - 1.669\left(\frac{E_m}{E_f}\right) \tag{4.32}$$

Figure 4.16 shows the comparison of the theoretically predicted elastic modulus by Halpin-Tsai eq. (4.29) with eq. (4.31) for the short fibre composite of Example (1), taking the l/d ratio of the polyester fibre as 250. The coefficient "a" in eq. (4.31) was calculated taking the data of V_f, E_f and E_m from Example (1) and "a" calculated using eq. (4.32). Obviously, the modified value of "a" gave lower composite modulus than $a = 3/8$.

The authors [57] used several experimental data on short fibre composites of epoxy-carbon nanotube and epoxy-graphene and compared the experimentally determined composite modulus with those calculated using Halpin-Tsai and modified equations. The authors showed that using the modified a value as in eq. (4.32), the predicted composite modulus was in much better agreement with experimental values, with average error of only 2%, and maximum 6.7%, while on using $a = 3/8$, as in eq. (4.29), the error was quite high, varying from 2% to even 52%. However, in all the analysis, the necessary assumptions are:

(1) The fibres do not interact with each other to form agglomerates
(2) The fibre is uniformly dispersed and distributed in the matrix
(3) There is no residual stress between the fibre and the matrix
(4) There is perfect bond between the matrix and the fibre
(5) The matrix and the fibre are linearly elastic
(6) There is no frictional loss of strain energy (Rayleigh or Coulomb damping) at the fibre-matrix interface

4.5 Steel chord reinforced rubber composite

Steel cords are extensively used in rubbers for making high-performance radial tyres and conveyor belts. In both the applications, the axial strength required to be higher than bending or shear strength. In a tyre, the circumferential force results in hoop stress which is many times higher than the radial stress of the tyre. In case of conveyor belts, the tension of the belt on tight side can be very high compared to other stresses. Both the items are subjected to dynamic stress while in use. A simple sketch of a single ply steel cord-rubber composite is shown in Fig. 4.17. The cord is made of a number of steel filaments twisted together as shown in Fig. 4.18. The end of the cord has an equivalent radius calculated from the total area of the circular cross section of the filaments.

The cords are unidirectionally embedded in the rubber matrix by *in situ* vulcanization process. The rubber item such as the tyre or the belt experiences tensile force in the direction of the cord. Needless to say, the contribution of the steel cords in transverse or other perpendicular direction is not significant.

The strength of the composite in the direction of the cord lay-up is dependent on the cord parameters, number of plies and adhesion of the steel cord with the rubber. The cross-sectional diameter of the cord is such that the cord can withstand the tensile force it has to experience in actual application. The number of plies can be two or more depending on the thickness and force requirements. The adhesion is a more complicated issue due to the chemistry of both the rubber and the steel cord. It has been observed that the adhesion of steel to the rubber is not adequate to avoid catastrophic failure/tearing of the tyre/belt. The failure can also take place as a result of fatigue due to constant dynamic loading of the tyre and the belt while in use.

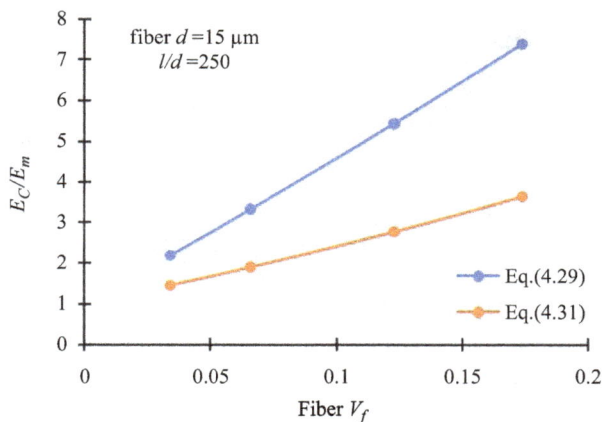

Fig. 4.16: Comparison of the NBR-polyester short fibre composite. Elastic modulus calculated using eqs. (4.29) and (4.31).

Fig. 4.17: A unidirectional steel cord-rubber composite.

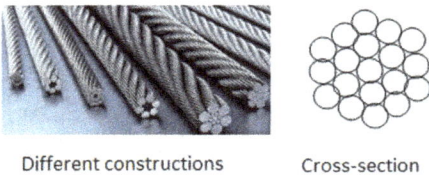

Different constructions   Cross-section

Fig. 4.18: Typical steel cords used in rubber-based unidirectional composites in tyres and conveyor belts.

4.5.1 Adhesion of steel cord to rubber

There are many reports on methods of improvement of adhesion between the steel cord and rubber. Normally, tyres are mostly made of styrene-butadiene rubber (SBR) and natural rubber (NR), and both are nonpolar. The adhesion with the steel is not very good for such high-performance application. Most modification of steel cord for adhesion improvement is brass plating. Adhesion of brass with rubber is being studied since seventies and Van Ooij [58] published a good review of the subject. In a comparatively recent review, the mechanism of adhesion of brass with rubber was elaborately discussed by Vandou et al. [59]. The technology of brass-coated steel cord-rubber composite for radial tyre is widely used because the failure of a tyre is mostly due to rubber degradation rather than loss of adhesion. In fact, the method of making such composite is quite perfected in the tyre industry and there have been few reports of the mechanism of adhesion of brass with rubber. Brass consists mainly of copper (about 66%) and Zinc (about 34%), and both react with the sulphur (curative) used in the rubber compound for vulcanization to form corresponding sulphides, Cu_2S and ZnS, thus bonding the brass with the sulphur linked rubber through the Cu_2S/ZnS-Sx-Rubber bridge [59, 60]. However, copper sulphide is more effective as adhesion promoter. Amorphous precipitated Silica, tackifier such as phenol/resorcinol formaldehyde and organocobalt compounds together is reported to enhance the adhesion between brass-coated steel cord and NR. For example, Shi et al. [60] used up to 3 phr cobalt decanoate and observed lower hysteresis and improved static and dynamic (200,000 cycles) adhesion force by about 30% presumably because it suppresses the formation of ZnS. Phenol formaldehyde was reported to be more effective compared to resorcinol, but the improvement was about 20% in static pull-out force and about 12% in dynamic pull-out force (200,000 cycles).

4.5.2 Mathematical models

The analysis and prediction of a cord-rubber composite is quite challenging. Even for a unidirectional cord reinforced rubber, the very construction of the cord poses the difficulty in estimation of stresses and strains in the three directions, because the cords are not single filament, but a combination of twisted filaments and hence, even in a unidirectional load situation, there will be bending and twisting of the rubber-cord composite and analytical methods are only approximation of actual stresses and fatigue. The cord rubber composite is a highly anisotropic body requiring special consideration of elastic modulus determination/calculation in two axes for plain stresses (lateral dimensions much higher than thickness). A simple material model of elastic tensile and shear analysis (plane stress) is represented as

$$E_1 = E_c \phi_c + E_r(1 - \phi_c) \tag{4.33}$$

$$v_{12} = v_c \phi_c + v_r(1 - \phi_c) \tag{4.34}$$

$$E_2 = E_r[E_c(1 + \xi_1 \phi_c) + \xi_1 E_r(1 - \phi_c)] / [\xi_1 E_r(1 + \phi_c/\xi_1) + E_c(1 - \phi_c)] \tag{4.35}$$

$$G_{12} = G_r[G_c(1 + \xi_2 \phi_c) + \xi_2 G_r(1 - \phi_c)] / [G_c(1 - \phi_c) + \xi_2 G_r(1 + \phi_2/\xi_2)] \tag{4.36}$$

$$v_{21} = v_{12} E_2 / E_1 \tag{4.37}$$

where v represents the Poisson's Ratio, ξ_1 and ξ_2 are factors depending on the geometry of the cord and the spacing, while ϕ_c is the volume fraction of the cord in the rubber ply and is given by Walter [61]:

$$\phi_c = \pi n R^2 / t \tag{4.38}$$

In a planar ply with unidirectional cord alignment, the volume fraction of the cord is same as the ratio of the cross-sectional area of the cords to the cross-sectional area of the ply.

For circular cross section of the cords and as the elastic modulus of rubber is much lesser than that of steel cord, the above equations for E_2 and G_{12} can be approximated as:

$$E_2 = E_r(1 + 2\phi_c)/(1 - \phi_c) \tag{4.39}$$

$$G_{12} = G_r[G_c + G_r + (G_c - G_r)\phi_c] / [[G_c + G_r - (G_c - G_r)\phi_c]] \tag{4.40}$$

4.5.3 Examples

Example 1

The tensile data of an experimental elastomer composition of nitrile rubber vulcanized with sulphur accelerator and having 45 phr carbon black is listed below.

UTS: 13.7 N/mm^2, elongation at break: 355% and tensile modulus (Er) at 1% strain: 9.15 N/mm^2. The Poisson's ratio is assumed as 0.48 a ambient condition. Hence Gr: 3.09 N/mm^2.

Material property of steel cord (neglecting the contribution by a thin layer of brass, as the case may be) is: shear modulus G_c: 82 × 10^3 N/mm^2.

The transverse elastic modulus and Shear modulus of the ply material is calculated for various volume fractions of the cords and shown in Fig. 4.19.

It is observed from Fig. 4.19 that the rate of increase in both moduli is enhanced to great extent beyond 0.5 volume fraction of cord. However, ultimate tensile strength of the rubber matrix or the bond strength between the rubber and the cord has to be taken into account while considering the validity of such theoretical estimation, and should be verified by an experiment.

Pidaparti [62] studied bending of cord-rubber 2-ply laminate composite using three-dimensional finite element method for four point bending and compared the results with experimental data and also with classical mathematical models of simple and cylindrical bending theory. The material model used was originally by Shield and Costello [63], derived by energy consideration and the elastic moduli are thus given by

$$E_1 = E_m + c_1 E_c \phi_c \left[1 - \frac{6c_2 c_3 \phi_c (1 + v_m)}{c_1 \left[\overline{Ah}^2 \overline{E} + 6c_4 \phi_c (1 + v_m) \right]} \right] \tag{4.39}$$

$$E_2 = \frac{4E_m [c_1 E_c \phi_c + E_m] + \left\{ [24(1 + v_m) \phi_c E_c] / \overline{Ah}^2 \right\} (c_4 E_m + C E_c \phi_c)}{4E_m + 4(1 - v_m^2) \phi_c c_1 E_c + \left\{ [24(1 + v_m) \phi_c] / \overline{Ah}^2 \overline{E} \right\} \left\{ c_4 E_m + C E_c \phi_c (1 - v_m^2) \right\}} \tag{4.40}$$

$$G_{12} = G_m (1 - \phi_c)$$
$$v_{12} = v_m \tag{4.41}$$

where E_m and G_m are the tensile and shear moduli of the rubber, v_m is the Poisson's ratio of the rubber, and $\overline{A} = A_c / R^2$, $\overline{h} = h/R$ and $\overline{E} = E_m / E_c . A_c$ is the area of the cord, h is the thickness of the ply, R is the outer radius of the cord and f_c is the volume fraction of the cord in the unidirectional ply. The constants c_1, c_2, c_3, c_4 and $C = c_1 c_4 - c_2 c_3$ are used to describe extension-twisting coupling and are determined analytically [63].

To study the mechanical properties of cord reinforced conveyor belt specimen of thickness of 30 mm, Golovanevskiy and Kondratiev [64] developed a modified tensometer and experimentally found the tensile strength and in-plane shear strength. The authors used simple rule of mixing to predict the elastic tensile modulus of a two-ply steel cord-rubber unidirectional composite:

$$E = \frac{E_1 S_1 + E_2 S_2}{S} \tag{4.42}$$

where E_1 and E_2 are elastic moduli of the cord and the rubber, S_1, S_2 and S are the cross-sectional areas of the cord, rubber and the ply sample respectively. Here, $S_2 = S - S_1$.

The accuracy of the prediction by the above equation was 98% compared to the experimental results, as reported.

The torsion experiment on bar and a square plate sample was done by fixing one end and twisting the other end and. The deflection d at the point of force F is determined by the experiment and the in-plane shear modulus was related to these parameters as:

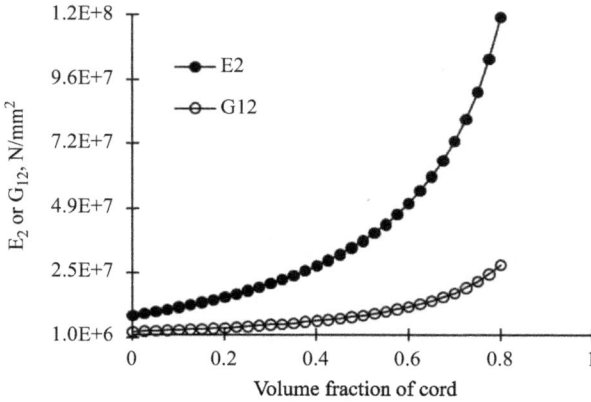

Fig. 4.19: Transverse and shear modulus of unidirectional cord-rubber single ply composite vs. volume fraction of cord in Example 2.

$$\delta h^3 = \frac{3Fl^2}{G_{xy}} \tag{4.43}$$

where h and l are the thickness and the side of the sample, respectively. The authors used the following expression as reported by Vasiliev and Morozov [65] for prediction of the in-plane shear modulus for a unit structural element with dimensions as $2p$ and $2l$ and a normalized radius r_b:

$$G_{xy} = \left[1 - \frac{r_b}{p} + \frac{r_b}{p}k\left(\frac{2k}{\sqrt{k^2-1}}\arctan\sqrt{\frac{k+1}{k}} - \frac{\pi}{2}\right)\right]G_p \tag{4.44}$$

where G_p is the in-plane shear modulus of the rubber. Here, k is defined as

$$k = \left[\frac{r_b}{l}\left(1 - \frac{G_p}{G_b}\right)\right]^{-1} \tag{4.45}$$

The above expression was fairly accurate with only 13% difference with experimental result as reported [64].

Example 2

Taking the material property data from Example 1, and assuming $p = 20$ mm and $l = 15$ mm and various normalized radii r_b, the elastic tensile modulus and in-plane shear modulus of a single ply unidirectional steel cord-rubber composite is computed and the result shown in Figs. 4.20 and 4.21, respectively.

Figure 4.20 was the plot according to eq. (4.42), which is a simple mixing rule and a straight line relating volume fraction of cord with the composite tensile modulus. The assumption is that the bond strength between the cord and the rubber is not a limiting factor, which is not true for a real composite ply, while Fig. 4.21 is based on eqs. (4.44) and (4.45) and shows a non-linear relation between the cord volume fraction and the in-plane shear modulus of the composite ply.

The failure of such cord-rubber composites is reported in literature using both static and dynamic tests. Basically, the failure is either for the failure of the rubber matrix as such or failure of adhesion bond between the cord and the rubber matrix. The strength and elastic tensile and sear modulus of rubber and improvement of adhesion of steel cord with the rubber matrix is already discussed. The static tests of adhesion can be using a specific test tyre cord adhesion test (TCAT), which is a modification of original ASTM D 2229 and further improved by Nicholson et al. [66, 67]. The test specimen is a rubber bar in which two cords are inserted from two ends as shown in Fig. 4.22. A finite element analysis is done to calculate stresses at various length of the sample. For low strain, the stress-strain of rubber is modelled as a Hookean behaviour, while for higher strain, Mooney-Rivlin model was used in the FEM by Ridhi et al. [66]. Obviously, at the rubber-cord interface, the stress is maximum and at the other end of the cord inside the rubber, the stress is minimum.

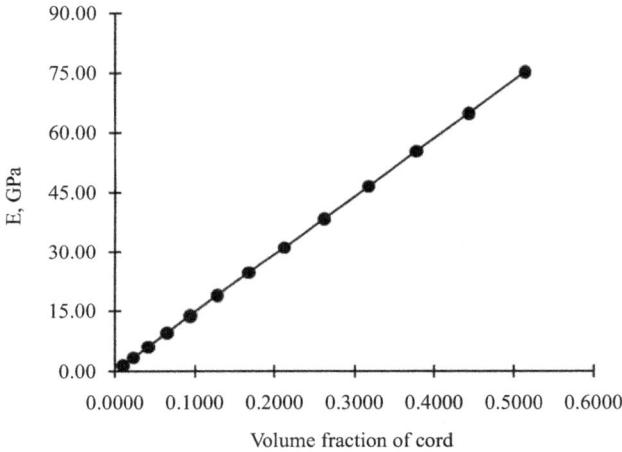

Fig. 4.20: Elastic modulus of steel cord-rubber unidirectional single ply composite as a function of volume fraction of cord.

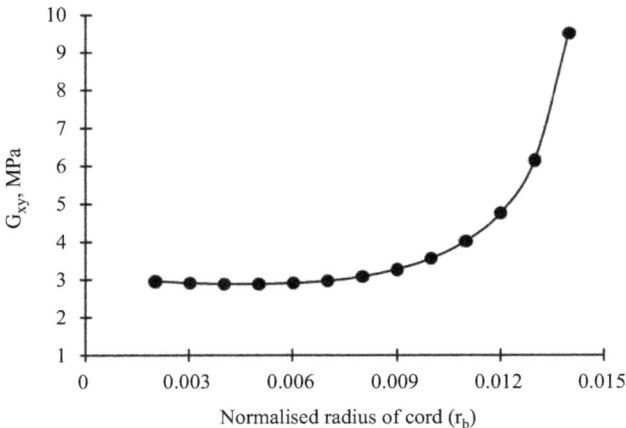

Fig. 4.21: In-plane shear modulus of steel cord-rubber single ply composite as a function of cross-sectional radius of cord.

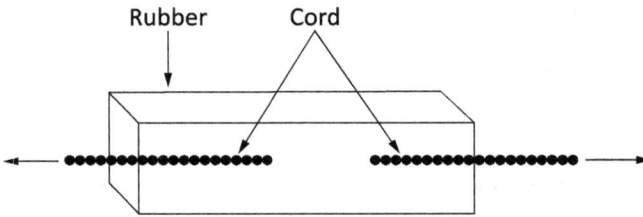

Fig. 4.22: A sketch of sample for modified TCAT test.

Frank et al. [68] also used a multiscale FEM modelling to assess the stresses, strains and failure of a conveyor belt of steel cord rubber unidirectional composite, with a set-up of a conveyor belt loop around two identical rotating drums and with an applied force. The rubber was modelled as a Mooney-Rivlin solid and the rule of mixture was used to determine the elastic modulus of the belt. The rotation of the drums was controlled (velocity nearly 6 m/s) using a special arrangement to avoid vibration of the belt. The tensile and bending stiffness were calculated by 2 scale FEM models of a splice of the belt. The authors reported that the highest stress is observed at the bending point of the belt, in the contact of the drum, and about 15% higher than flat region. The experimental set up used in this method can be applied to study dynamic fatigue of the belt on running several cycles.

During the use of conveyor belts and tyres, the rubber cover undergoes wear and tears while the core steel cord-rubber interface suffers damages and crack development. A conveyor belt cold is as long as 100 km, and also in a mining ore area and might be a part of the underground operation. The damage identification of the belt is thus a continuous process with time to identify the defects being developed with time and the extent of damage caused in the belt core. Recently, Blazej and coworkers [69] have reported a method of damage detection in conveyor belts using digitized magnetic signatures of the belt sections in 2D view with the help of a software, such as DiagBelt. Damage causes change in magnetic field around the steel cord and this change is reproduced in a digital data system. It is an improved version of the older analogue system. The diagnostic bar contains many magnetic sensors to identify the locations and estimate the extent of damage in the belt sections minutely.

4.6 Elastomer nanocomposite

Nanometric size of a reinforcement material is more effective in enhancing strength and other properties of elastomers. In fact, carbon black, which is the most common reinforcing filler in rubber industry, is basically a nanoparticle, but remains as agglomerates as can be seen in Tab. 4.1. Carbon black particles are essentially spherical in shape. There are nanometric carbons in the shape of tubes, fibres, platelets, etc., which ae very effective in small quantities and have gained importance as reinforcing fillers apart from conventional carbon black. In addition, there are more types of nanoparticles such as minerals, ceramics, fumed silica, natural clays, synthetic metal nanoparticles, metal oxides, etc. There has been enormous interest among the scientific community and industry about the nanocomposites of elastomers for several reasons, such as quantum jump in some properties, new avenues and opportunities to enhance functional performance of modern systems. A very important aspect is reduction in weight penalty, which is an obvious advantage allowing light weight construction of all types of vehicles, thus saving on fuels and pay-loads. Nanosized materials owing to their high surface energy per unit volume, can reinforce a rubber to much greater extent, not only in respect of strength, but also in electromagnetic properties, conductivity, barrier property, viscoelastic damping etc. However, smaller the size, more difficult it is to avoid or break agglomeration, which reduces the nano-effect to large extent. Secondly, health hazard of handling such small particles is well known. Nevertheless, elastomer industry has adopted a large-scale application of nano-reinforcing fillers other than carbon black. Most common nanocomposite of elastomers are organically modified clay nanocomposite, functionalized or pure carbon nanotube composite and graphene and graphene oxide nanocomposites.

4.6.1 Elastomer /nanoclay composites

Clay is a naturally occurring mineral, consists mainly silicates of aluminium, magnesium, potassium, sodium, iron, etc. Nanoclays are nanosized particles of natural clays. There are classifications of natural clays based on crystal structures and chemical composition. For example, the following common clays are widely available and used in synthesis of nanocomposites. Some examples of naturally occurring clays and their approximate empirical formulas are given below:

Phyllosilicates:
Kaolinite:$Al_4Si_4O_{10}(OH)_8$
Bentonite (mixture of minerals including smectite),
Chlorite: $Al_2Mg_3Si_4O_{10}(OH)_8$
Talc: $Mg_3[Si_4O_{10}(OH)_2]$,
Mica:$KAl_2[AlSi_3O_{10}(OH)_2]$,
Sepiolite: $Mg_4[Si_6O_{15}](OH)_2\ 4H_2O$
   Smectite: Component of bentonite, known as a 2:1 layered clay,2-dimensional arrays of silicon-oxygen tetrahedra and 2-dimensional arrays of aluminium- or magnesium-oxygen-hydroxyl octahedra.
Montmorillonite (di-octahedral): $Mg_{0.33}Al_{1.67}[Si_4O_{10}(OH)_2](Ca, Na)\ x\ (H_2O)_n$
Hectorite (tri-octahedral)$Na_{0.3}(Mg,Li)_3Si_4O_{10}(OH)_2$
Saponite: $Ca_{0.25}(Mg,Fe)_3((Si,Al)_4O_{10})(OH)_2\ .n(H_2O)$,
kaolin & Halloysite: $(Al_2[Si_2O_5(OH)_4])$

Montmorillonite (MMT) is the most abundantly available clay in the earth crust and hence, very cost-effective. They are preferred also due to simple processability and can cause significant improvement in performance when a polymeric composite is formed with even minute (<10%) quantity of the nanoclay. They are layered silicates containing two coordinated tetrahedral silicon atoms, with a sandwiched octahedral sheet of mostly aluminium or magnesium oxide [70]. Natural MMTs is hydrophilic due to the presence of inorganic cations on their surface and thereby not compatible with hydrophobic polymers. Generally, the length of a nano MMT is less than 2 μm, while width and thickness are approximately 50 nm and 0.96 nm. The layer spacing of natural MMT is about 1.23–1.54 nm as observed by X-ray diffraction study [71].

In order to compatibilize MMT with elastomers, organic modification is done to MMT to increase the interlayer spacing, and to create hydrophobic surface, for better interactions with organic polymers. Cloisite series of organically modified MMT are widely used, with varying hydrophobicity as: Cloisite 15A > 20A > 25A > 10A > 93A > 30B > Cloisite Na+ (hydrophilic).Therefore, a selection can be made to use various Cloisites in polar and non-polar elastomers.

Typical chemical structures of the organic modifiers are shown in Fig. 4.23 [72].

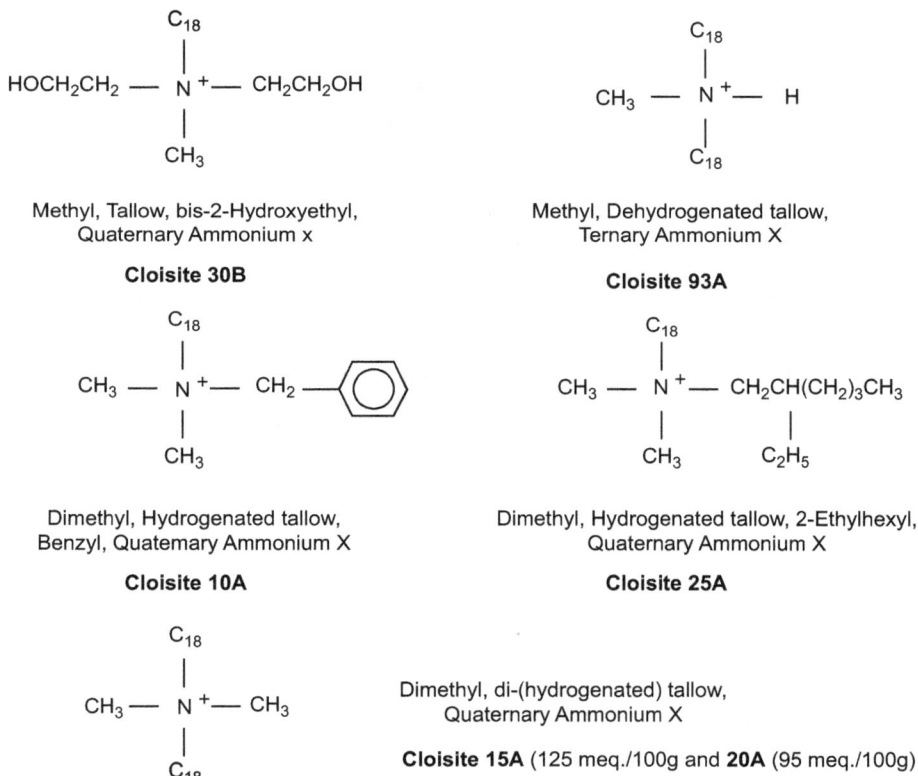

Methyl, Tallow, bis-2-Hydroxyethyl, Quaternary Ammonium x

Cloisite 30B

Methyl, Dehydrogenated tallow, Ternary Ammonium X

Cloisite 93A

Dimethyl, Hydrogenated tallow, Benzyl, Quaternary Ammonium X

Cloisite 10A

Dimethyl, Hydrogenated tallow, 2-Ethylhexyl, Quaternary Ammonium X

Cloisite 25A

Dimethyl, di-(hydrogenated) tallow, Quaternary Ammonium X

Cloisite 15A (125 meq./100g and **20A** (95 meq./100g)

Fig. 4.23: Chemical structures of organic modifiers for various Cloisite types.

The cation of the natural clay (MMT) is exchanged with the quaternary ammonium ion which is attached to derivatives of the tallow (C_{18} chain) and thus this long chain moiety causes an increase in the d-spacing of the silicate layers. The d-spacing can be increased to 2–3 nm by such organic modification of the clay [73]. There can be different extent of layer separation, for example intercallation, where the layers are not completely sepa-rated and still maintain some order, while exfoliation means the layers are completely separated by the long chain organic compound. The intercallated and exfoliated clays are compounded with elastomer in the unvulcanized state, when the elastomer is linear with low molecular weight and further reduced by severe shearing of the elastomer in rubber processing units such as two roll mill and internal mixer. This process causes further exfoliation of the clay as the oligomer enters the clay gallery. The similar effect can also be observed when the elastomer is dissolved in a solvent. However, a good solvent may not be preferred due to high hydrodynamic volume of the elastomer chain causing higher viscosity. The second problem of solution process is difficulty in removal of trace solvent from the clay-rubber mixture. Considering the normal compounding process of rubber industries, the mixing should be done in an internal mixer or a two-roll mill at a somewhat elevated temperature, avoiding degradation of the elastomer. A pictorial representation of the stages of the original clay, intercalated clay and exfoli-ated clay is shown in Fig. 4.24.

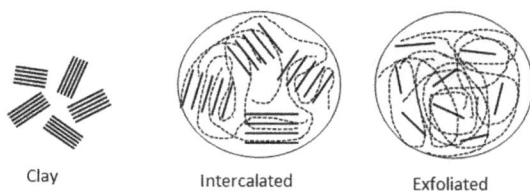

Clay   Intercalated   Exfoliated

Fig. 4.24: Clay modification by large organic molecules and oligomers.

Since the organic modifier and the oligomer (low molecular weight elastomer) can enter the gallery of the clay, the resulting composite (termed "clay nanocomposite") is better reinforced than that by simple mixture of a particulate filler. However, due to such intercalation/exfoliation, there is a limit of the clay concentration for maximum effective reinforcement, beyond which the clay behaves as a simple particulate filler. There are many studies to establish the optimum clay content in rubber, and mostly it is 2–6% of the rubber. A few are cited in references [73–77].

Nevertheless, the platelet-type structure of the clay can still enhance some properties beyond optimum level, such as barrier to fluid diffusion, vibration damping and fire retardancy, to mention a few. The tortuous path offered by the platelets and the hydrophobic nature of the modified clay offer greater resistance to diffusion and decrease solubility of small molecules into the composite [78], while in vibration, the platelets with adjacent rubber chain behave as micro-constraining layer damper due to mode conversion from longitudinal to shear strain of the soft rubber in between the of bundles of the stiff clay [79]. In addition, the clay acts as synergistic additive with fire retardants such as polyphosphate, aluminium tri-hydroxide and magnesium hydroxide, etc. [80, 81] and also without any other fire-retardant additives [82].

4.6.2 Elastomer /carbon nanotube composites

One of the most effective reinforcements of elastomers are done by carbon particles. The atomic structure of carbon allows it to form very different types of clustering and chemical association. Apart from spherical shape of carbon black, it can form the shape of a football with five and six membered rings, known as "Fullerene", tube shaped particles called "Carbon nanotube", both single tube (SWNT) and many concentric tubes together (MWNT), fibres, and purer forms of clustered plates such as graphite and single platelet called graphene. A typical pictorial representation of carbon nanotubes is shown in Fig. 4.25. The single wall nanotube is actually a cylindrically folded graphene sheet and multiwall tubes are concentric cylindrical arrangement formed by several graphene sheets [82].

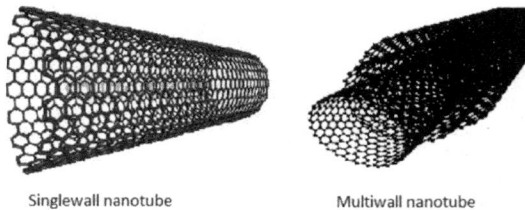

Singlewall nanotube   Multiwall nanotube

Fig. 4.25: Carbon nanotubes.

Single-walled carbon nanotubes have on an average diameter of 0.5–1.5 nm, while multiwall nanotubes might have as high as 100 nm diameter, length of a nanotube can be anything from 0.5 nm to several mm. Density of isolated SWNT is 2.1 g/c.c. and that for MWNT are found to be 2.1 g/c.c. [83]. However, depending on the structural arrangement and dimensions of the nanotubes, the density, mechanical properties vary widely. The carbon nanotubes are anisotropic and therefore, the Young's modulus and Poisson's ratio are varying as 1–6 GPa and 0.28–0.36 in three directions [84]. Qian et al. reported very high Young's modulus of 1.2 TPa and tensile strength of 50–200 GPa [85].

Carbon nanotubes are often functionalized in order to enhance compatibility with the elastomers so that the reinforcing effect is significant at very minute quantity of the nanotubes. For example, a mixture of nitric and sulphuric acid is used to oxidize the surface, introducing -COOH group, thus making it compatible with polyurethanes and polyimides, while grafting of the polybutylene onto the nanotube is done to compatibilize with polybutylene, acid treatment, while organo-silane treatment on CNT is done to compatibilize with vulcanized rubber and amine modification is done for compatibilize with polyamide and epoxy polymers. Dispersion of carbon nanotubes in polymers is very critical, since the size of the nanotubes is very small, agglomeration is quite high, particularly polar modified CNTs. However, in maximum cases, it is observed that only 0.1–1% functionalized CNT is effective to significantly improve the Young's modulus and electrical conductivity. Even 1% silane-modified CNT can enhance the Young's modulus of rubber by 28% and for most polymers, 1% functionalized CNT is enough for achieving the percolation threshold in electrical conductivity (10^{-10} to 10^{-6} S/cm) [86]. Malekie and Ziaie [87] calculated the electrical percolation threshold as 0.13–0.25% SWCNT compared to experimental results of 0.13–0.7% depending on the type of polymer, taking the electrical conductivity of the SWCNT as 10^4–10^7 S/m and that of the selected polymers in the range of 10^{-16} S/m for polystyrene to 10^{-12} S/m for Polyethylene terephthalate. Li et al. grafted polyurethane chain on MWCNT surface and when used in NR vulcanizate at 5 phr, the thermal conductivity increased by 43% [88]. Li and coworkers [89] also modified MWCNT by grafting butyl acrylate-a-methyl methacrylate glycidyl methacrylate (BA-MMA-GMA) terpolymer onto the MWCNT surface and the modified MWNCT (OMWCNT) was used up to 2.5 phr in NR/SBR blend vulcanizate as reinforcement. The mixing of the OMWCNT and other vulcanizing ingredients with the blend elastomer blend was done in a conventional two-roll mill. Properties of the reinforced vulcanizate were maximum at 1.5 phr OMWCNT. The tensile strength improved from 23.7 MPa to 26.6 MPa, elongation at break improved from 375% to 450% and degradation temperature increased in thermogravimetric analysis up to 70 °C after the onset. The effect of the OMWCNT reinforcement was very obvious in dynamic mechanical analysis, as the reinforcement caused a decrease in loss factor maximum value from 0.575 to about 0.525, without the shift of the peak temperature (broad peak covering −55 °C to −47 °C). This is the result of reduction in segmental mobility of the reinforced chain.

4.6.3 Elastomer /graphene nanocomposites

Graphene can be defined as a single sheet of graphite structure, and hence much smaller in layer thickness, and in addition, have quite smaller in lateral dimensions. The ideal structure of graphene is a single atom thick sp^2-hybridized carbon-carbon planar sheet. The most common crystalline size of a nanographene particle is 20–200 nm. The thickness of the single layer is about 0.335 nm as seen by X-ray diffraction [73]. However, due to nanometric size, graphenepractically consists of few sheets and the thickness of such bundle can be few nanometres. Characteristics of a typical commercial graphene nanoparticle produced by a combination of chemical vapour deposition and chemical method are:

Purity >99%, thickness of a 3–6-layer graphene nanoparticle is about 3–8 nm, density is approximately 2.2 g/cm^3, surface area is about 180 m^2/g, tensile strength is about 130 GPa, thermal conductivity is approximately 5,000 W/m.K and electrical conductivity is about 10^{8}Siemens/m [90].

Graphene is easy to synthesize from microcrystalline graphite by several methods such as by ultracentrifuging the graphite particles in an organic solvent to obtain the exfoliated single graphene platelets, or by chemical vapour deposition [90], or by the common "Hummer-Offeman" method [91], using oxidation of graphite by mixture of potassium permanganate, sodium nitrate and sulphuric acid, to obtain graphene oxide, which is polar. A modified Hummer method uses a mixture of sulphuric and phosphoric acid with the permanganate for oxidation reaction [92]. To reduce the polarity, the graphene oxide is reduced by thermal process to remove some oxygen atom as CO/CO$_2$ in controlled manner to obtain a preferred ratio of C/O in the reduced graphene oxide (rGO). The properties of the rGO are more like a pure graphene, with some polarity due to presence of oxygen.

Pristine graphene is a honeycomb of carbon rings, without any defect and perfectly non-polar, which makes it unsuitable for reinforcement of elastomers, due to low interactive bond strength with the rubber. Graphene oxide, on the other hand, is somewhat polar and can reinforce rubbers, especially polar rubbers such as nitrile rubber, carboxylated nitrile rubber, neoprene, acrylic rubber and hypalon. However, reduced graphene oxide and surface modified GO are suitable for reinforcing non polar rubbers such as natural, butyl and SBR. Figure 4.26 shows a typical grapheme nanosheet image in TEM and a schematic representation of three grapheme nanosheet layers.

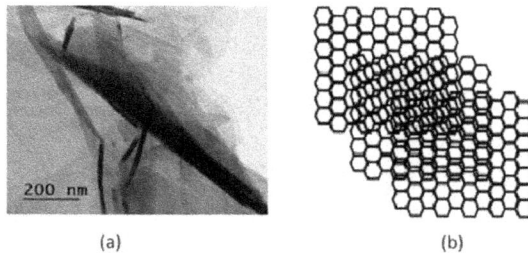

(a)   (b)

Fig. 4.26: Grapheme (a) TEM image of a commercial grapheme and (b) schematic of a three layer graphene nanosheets.

Surface modification of GO can be done by chemical reaction with suitable vinyl rubber. For example, Tang et al. synthesized SBR/graphene composite by introducing GO into SBR latex, thereafter reducing the GO by hydrazine hydrate [93]. Esmiazadeh et al. synthesized silicone rubber-graphene nanosheet composite using a coupling agent 3-aminopropyl triethoxysilane, mixed in a two-roll mill commonly used for rubber compounding, and vulcanized by dicumyl peroxide [94]. Berki et al. synthesized GO by modified Hummer method as aqueous suspension and added to NR latex, and compounded, vulcanized in a conventional process and compared this nanocomposite with block NR-GO suspension mixture, compounded and vulcanized conventionally and observed that the latex process resulted in higher reinforcing effect than block NR processing. The authors observed that the dynamic modulus at various strains and the ultimate tensile property is better for latex NR-GO nanocomposite. In fact, the tensile and tear strength reduces on addition of GO in NR block -GO nanocomposites [95]. Wang et al. used N,N, bis (3-aminopropyl)methylamine for simultaneously grafting onto the GO and reducing it, and thereafter used NR latex/rGO nanocomposite. The vulcanization of the rubber/rGO was done by conventional sulphur-accelerator curing system. The developed composite was compared with a latex rubber/ GO system reduced by hydrazine hydrate and cured similarly [96]. It was observed that with similar rGO content, the developed process resulted in a more rubbery flexible high-strength elastomeric composite and hydrazine hydrate rGO reinforcement resulted in lower-strength, higher modulus elastomeric composite.

4.7 Rheology of elastomer composites

Elastomers are formed as randomly oriented long, entangled and flexible chains, mainly with low or no order in the chain arrangement, which make them nearly pure amorphous material. The free volume in an elastomer is quite high compared to engineering plastics and thermosets. Glass transition temperature of elastomers is below ambient, and therefore, the chain segments are at random motion like repetition, but without a neat displacement of the bulk. The macro-Brownian motion of the chain segments is a result of thermal excitation. Due to the thermodynamic state at ambient conditions, the elastomer shows two deformation phenomenon simultaneously, one which is like an elastic solid, where the internal stress is instantly developed in the elastomer when a strain is induced, and secondly, the chains can move like a viscous fluid under the external force, and the movement is opposed by the entangled chains, which is defined as viscosity. The strain thus develops with time under the constant stress and the rate of strain is proportional to the stress in the elastomer. Therefore, the elastomer is termed as a "viscoelastic" body. Hence, a viscoelastic body is represented by a combination of an elastic member, such as a spring and a viscous fluid member, such as a dashpot. The mathematical expression of the viscoelastic phenomenon can be either by a series arrangement of these members, which is Maxwell's Model or the arrangement can be parallel, which is Kelvin-Voigt (KV) model. However, a single model cannot describe a real elastomeric material, and less for its composite, where anisotropic behaviour is observed.

4.7.1 Viscoelastic properties

Figure 4.27(a) shows a Maxwell model, and Fig. 4.27(b) shows a Kelvin-Voigt model. In Maxwell's model, the strain is additive and the spring and dashpot experience the same force of deformation, while in KV model, the stresses are additive and the same strain is induced on both the arms. Since the elastomer chains can have viscous behaviour, the strain develops with time under constant or variable stress, and stress reduces due to rearrangement of the chains, if held at constant strain.

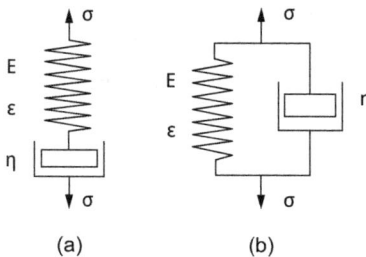

Fig. 4.27: Viscoelastic models: (a) Maxwell and (b) Kelvin-Voigt.

The constitutive equations relating the stress, strain and the strain rate in a viscoelastic body is represented as

$$\text{Maxwell: } \frac{d\varepsilon}{dt} = \frac{1}{E}\frac{d\sigma_s}{dt} + \frac{\sigma}{\eta} \tag{4.46}$$

$$\text{Kelvin-Voigt: } \sigma = \sigma_s + \sigma_d = E.\varepsilon + \eta\frac{d\varepsilon}{dt} \tag{4.47}$$

In treating the above equations, it is assumed that the modulus (E) and the viscosity (η) are independent of strain and strain rate, respectively, which in other words is an assumption that the body is a linear viscoelastic body. However, this assumption is valid for a low strain and strain rate for the elastomers and beyond that limit, the material shows non linearity. As an example, let us take nitrile rubber vulcanizate without any filler, the stress-strain relationship and secant elastic modulus is shown in Fig. 4.28. The relationship is neither a Hookean one, nor a pure Newtonian fluid. The non-linearity results from the viscous behaviour, i.e., increase in strain rate with increase in stress. From the figure, it is seen that the stress does not increase at the same rate after about 10% elongation and the modulus continuously drops on increasing the strain. The rubber chains resist deformation to maximum extent in the onset of stretching (force application), due to highly coiled and entangled configuration of the chains, and subsequent drop in modulus is due to relatively low force required by the uncoiled chains to move past one another showing large viscous strain at much slower rate of stress, and finally till break due to any defect or ideally due to covalent bond breaking. It can be better explained by plotting stress rate ($d\sigma/dt$) and strain rate ($d\varepsilon/dt$) as shown in Fig. 4.29. The rate of strain is perfectly linear with time, although the cross-head was made to move at a constant speed during the experiment. However, the nonlinearity in the stress rate is seen, which is due to the viscous behaviour of the rubber along with elastic response. The stress, strain and their rates (time derivative) were continuously varying in this experiment.

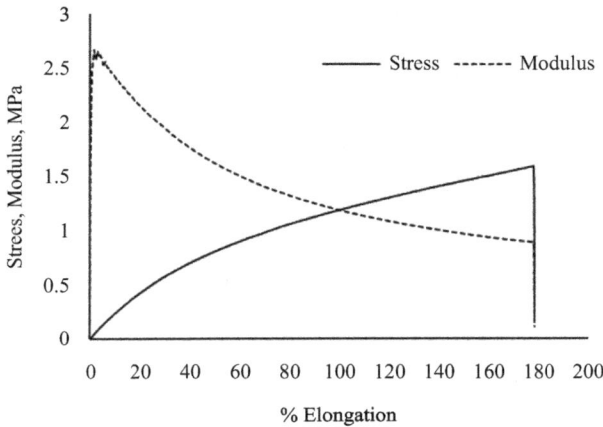

Fig. 4.28: Tensile property of nitrile rubber unfilled vulcanizate.

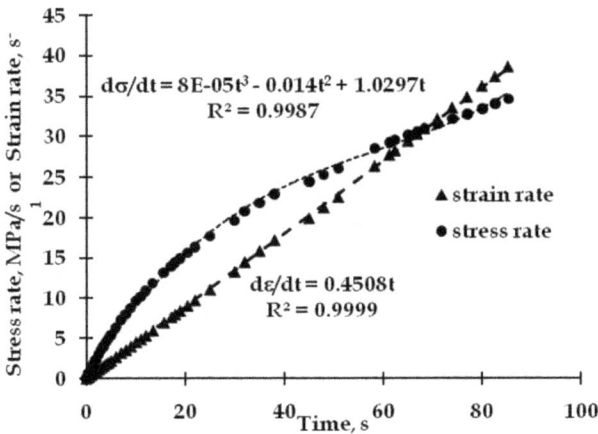

Fig. 4.29: Stress rate and strain rate of the nitrile rubber during tensile experiment.

The linear fit of the strain rate vs. time is quite accurate with $R2 = 0.9999$, and that of the stress rate is expressed as a third-order polynomial with excellent fit with $R2 = 0.9987$. The expressions are:

$$\text{Stress rate (MPa/s):} \frac{d\sigma}{dt} = 8 \times 10^{-5}t^3 - 0.014t^2 + 1.029t \qquad (4.48)$$

$$\text{Strain rate (s}^{-1}\text{):} \frac{d\varepsilon}{dt} = 0.4508t \qquad (4.49)$$

Such a real rubber shows a complex viscoelastic behaviour which cannot be explained by either Maxwell's or KV models, but may be a combination in series or parallel. A possible series combination similar to one form of four-element Burger model was attempted by Alkelani et al. [97]. The authors used stiffness and deformation instead of elastic modulus and strain to describe viscoelastic behaviour of a soft gasket. Figure 4.30 shows a series arrangement of Maxwell's and KV and a simple form of the final expression of such a combination is shown below:

$$\text{The total strain: } \varepsilon = \varepsilon_1 + \varepsilon_2 + \varepsilon_3 \qquad (4.50)$$

Here, ε_1 is the elastic strain of the Maxwell arm, ε_2 is time pendent strain due to viscous flow of the Maxwell arm, and ε_3 is a restricted strain of the KV arm, equal for the elastic and viscous parts. The stress σ is same for the Maxwell elements E_1 and η_1 and for the KV arm as well. The total strain ε can be expressed as

$$\varepsilon = \sigma \left[\frac{1}{E_1} + \frac{t}{\eta_1} + \frac{1}{E_2} - \frac{1}{E_2} \exp(-E_2/\eta_2)t \right] \qquad (4.51)$$

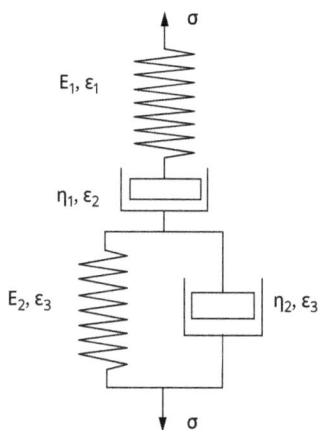

Fig. 4.30: A series combination of a Maxwell and KV models: four elements.

The term η/E is a characteristic of the rubber, termed as relaxation time τ, which is the time required by a rubber to internally reduce the stress to $1/e$ ($e = 2.718$) of the initial stress when the rubber is held at a fixed deformation, causing a stress relaxation of the elastomer chain segments.

The above expression may be fairly suitable for creep or relaxation behaviour of an elastomer, when the strain is not sufficiently large, near to the breaking point. This is because, for large deformation of the chains, the viscosity may change, and hence the values of η_1 and η_2 cannot be taken as constant in simulating large deformation.

There are some semi-empirical expressions defining real rubbers, such as Kohlrusch function for creep and relaxation [98]:

$$\text{Relaxation function: } \phi(t) = (G_0 - G_\infty)e^{-(t/\tau)^n} \tag{4.52}$$

$$\text{and creep function: } \psi(t) = (I_\infty - I_0)(1 - e^{-(t/\lambda)^m}) \tag{4.53}$$

Another semi-empirical set of equations are [98]:

$$\text{Relaxation function: } \phi(t) = \frac{G_0 - G_\infty}{1 + (t/\tau)^a} \tag{4.54}$$

$$\text{Creep function: } \psi(t) = (I_\infty - I_0)\frac{(t/\lambda)^b}{1 + (t/\lambda)^b} \tag{4.55}$$

In the above equations, G denotes shear modulus and I denote compliance, suffix 0 denotes initial value and ∞ denotes value after sufficiently long time. τ, λ, n, m, a and b are constants. Here, τ is the relaxation time in stress relaxation, and λ is retardation time, in creep.

The nature of the relaxation and creep are shown in Figs. 4.31 and 4.32, respectively. Equations (4.52) and (4.53) are used to construct the plots. Shear modulus and compliance values are taken from tests on a nitrile rubber vulcanizate without filler. The plots are time-dependent values of modulus and compliance. The time-dependent shear modulus was calculated as:

$$G(t) = G_\infty + \phi(t) \tag{4.56}$$

and time-dependent compliance was calculated as

$$I(t) = I_0 + \psi(t) \tag{4.57}$$

It is seen from the figures that the change in modulus and compliance initially takes place at a faster rate, expressed by eqs. (4.52)–(4.55), and tends to attain a constant value after a considerable time. The final value attains faster if the relaxation/retardation time is smaller and also depends on the exponents n and m. The finite values of modulus or compliance indicate that the vulcanized rubber behaves like a viscoelastic solid body, having a finite elastic energy, whatever small, and the elasticity of the rubber is not purely kinetic. A simple analysis of the Maxwell model reveals that his viscoelastic body is a fluid, since the stress becomes zero at infinite time, and hence the elasticity is more kinetic in nature, whereas the KV model does not predict a relaxation phenomenon at all, representing an elastic solid, but can predict creep under a constant force, and recovery of strain on withdrawal of external force, and has a limiting strain depending on the inherent elastic part as shown in Fig. 4.27(b).

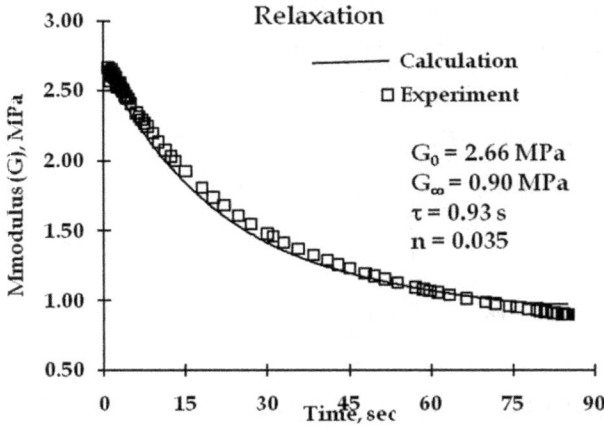

Fig. 4.31: Time-dependent shear modulus of nitrile rubber on relaxation.

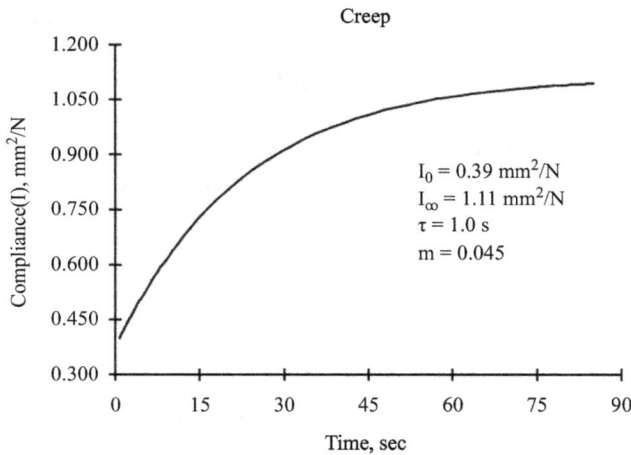

Fig. 4.32: Time-dependent shear compliance of nitrile rubber due to creep.

4.7.2 Dynamic viscoelasticity

Under time-dependent stress, for example a sinusoidal stress, the elastomers show strong viscoelastic loss due to internal friction of the chain segments, especially near the glass transition. According to the phenomenological relaxation theory, the segmental mobility increases with thermal agitation and freezes below the glass transition temperature (T_g). Correspondingly, at low frequency of stressing, the smaller segments relax faster than the time period of oscillation; hence, only bigger segments account for modulus and loss, both. On increasing the frequency, more and more segments can participate the relaxation process with the same time period of oscillation. Thus, both modulus and loss grow, and the loss reaches a maximum at a frequency, which is termed as α-relaxation frequency. In a temperature scale, the transition temperature (peak loss modulus or peak loss factor) shifts to higher frequency. This differentiates glass transition temperature to α-relaxation temperature, which is also known as DMA T_g. However, at very low frequency, the α-relaxation temperature would almost match with DSC measured T_g. A mathematical representation of the dynamic properties can be described reasonably by a three-parameter model of viscoelastic solid, such as Zenner model:

$$E^* = E_1 + \frac{\omega^2\tau^2 E_2}{1+\omega^2\tau^2} + \frac{i\omega\tau E_2}{1+\omega^2\tau^2} \tag{4.58}$$

$$\therefore E' = E_1 + \frac{\omega^2\tau^2 E_2}{1+\omega^2\tau^2} \tag{4.59}$$

$$E'' = \frac{\omega\tau E_2}{1+\omega^2\tau^2} \tag{4.60}$$

$$\tan\delta = \frac{\omega\tau}{(E_1/E_2)(1+\omega^2\tau^2)+\omega^2\tau^2} \tag{4.61}$$

It can be seen from the above equations, the storage modulus attains a maximum value at infinitely high frequency, while the loss modulus has a peak value at

$$\omega\tau = 1 \tag{4.62}$$

The energy stored in the viscoelastic body in dynamic strain per cycle is given by:

$$E_{\text{stored}} = \varepsilon_0^2 E' \tag{4.63}$$

The loss of mechanical energy per cycle due to the viscoelastic deformation is given by

$$E_L = \pi E'' \varepsilon_0^2 \tag{4.64}$$

The unit of the energy thus expressed here is energy per unit volume per cycle (J/m^3/cycle), when the modulus is expressed in Pa (N/m^2) and frequency (ω) in rad/s.

The stored (recoverable) energy and loss energy per unit time, are expressed as:

$$P_{\text{stored}} = \omega\varepsilon^2 E', \quad P_{\text{loss}} = \omega\pi\varepsilon^2 E'' \tag{4.65}$$

The dynamic resilience can be calculated as the ratio of the stored (and hence recovered) energy to the total input energy. The equation thus can be derived from eqs. (4.63) and (4.64):

$$R_D = \frac{E'}{(E' + \pi E'')} = \frac{1}{(1 + \pi \tan \delta)} \qquad (4.66)$$

Consequently, it is seen that the loss of energy (damping) and resilience are dependent on the temperature of the viscoelastic body and the frequency of the external dynamic force.

It should be noted that for all the above expressions, the elastic modulus is taken in tension mode in the linear viscoelastic region. The expressions are valid in other mode of deformations, such as shear mode. Although the shear modulus of an elastomer is about one-third of tensile modulus, the loss factor ($\tan \delta$) is same and the relaxation temperature (and frequency) is also same in all modes.

The relaxation time (τ) is a characteristic time of a polymer, and changes with temperature. The temperature dependence of relaxation time can be expressed by the Arrhenius equation:

$$\tau = \tau_0 \exp\left(\frac{E_a}{RT}\right) \qquad (4.67)$$

where E_a is the activation energy of relaxation, τ_0 is the relaxation time at infinite temperature and R is universal gas constant. The equation shows that the relaxation time decreases with increase in temperature. The assumption of this theory is that the segmental motion is not hindered by neighbouring segments or chains, which essentially means that the void volume should be more and hence it is not applicable to the temperatures in the glass transition region. A modification is made in Vogel-Fulcher-Tammann (VFT) law:

$$\tau = \tau_0 \exp\left[\frac{B}{T - T_0}\right] \qquad (4.68)$$

which is actually useful again, at a $T > T_g$, when the T_g is measured arbitrarily, though, at 10^{-2} Hz and the polymer properties measured such as modulus, creep compliance, viscosity etc., in the frequency range of 10^{-2} to 10^{14} Hz satisfy the VFT theory. The index B is not a fixed value and varies with temperature near the glass transition. The values of T_0 and B are to be determined by experiments.

A theory based on temperature dependence of free volume in a polymer was used by William, Landel and Ferry to obtain the WLF equation. The theory assumes that the free volume in a polymer does not change below glass transition temperature, and increases linearly with a definite coefficient of volume expansion and is valid only for segmental motion. The WLF equation is

$$\log\frac{\tau}{\tau_0} = \frac{C_1(T-T_g)}{C_2+T-T_g} \qquad (4.69)$$

where C_1 and C_2 are -17.44 and 51.6, respectively, when $T_g < T < (T_g + 50)$. The T_g is determined by DMA method at 10^{-2} Hz. The values are 8.86 and 101.6 when the reference temperature is $T_g + 50$ and $(T_g + 50) < T < (T_g + 100)$. Therefore, the WLF is valid up to $T_g + 100$.

The constants C_1, C_2 for above equations are not absolutely fixed and only can be determined by experiment for specific polymer, under a study from 10^{-2} Hz onwards for wide range, both in frequency and temperature.

However, in amorphous polymers, there are β and γ-transitions corresponding to lower order, non-coordinated motion and corresponding relaxation. For instance, the side group mobility, or a 4-carbon crank-shaft-type movement for β-transition and defects and end group mobility for γ-transition. Obviously, these transitions take place below the T_g. The frequency or the relaxation time for these transitions may satisfy Arrhenius temperature dependence:

$$\tau_\beta = \tau_0 \exp\left[\frac{E_\beta}{RT}\right], \quad \tau_\gamma = \tau_0 \exp\left[\frac{E_\gamma}{RT}\right] \qquad (4.70)$$

τ_0 is same for expressions of all transitions, α, β and γ. The γ-transition takes place at lower temperature than β-transition and activation energy of β-transition (E_β) is greater than that of γ-transition (E_γ). Consequently, at a fixed temperature, the relaxation times of these transitions are in the order $\tau_\gamma < \tau_\beta < \tau_\alpha$. However, the relaxation time for β and γ are very small in elastomers compared to α-transition, and only a theoretical estimation can be done, especially for γ- transition, rather than direct experimental measurement. In case of elastomeric materials, for all application purpose, most relevant is obviously the α-transition temperature and the corresponding frequency.

4.7.3 Effect of filler inclusion

The effect of fillers on viscoelastic behaviour is more complicated due to several factors such as polarity of the filler and rubber, filler-rubber interaction, agglomeration of filler, effective surface area of the filler particle, aspect ratio of the filler, uniformity of dispersion of the filler, orientation of the filler in the rubber (in case of non-spherical filler) and filler concentration. From the force transfer point of view, the important aspect is the difference in stiffnesses (elastic modulus) of the elastomer and the filler. The fillers are much stiffer than the elastomer matrix; hence, the load cannot be fully transferred from the matrix to the adhering filler particle. This reduces the efficiency of the reinforcement by the filler. The higher the modulus of the elastomer, higher will be the efficiency of reinforcement.

The effect of nanosized filler particle is widely studied now; hence, it is known that the elastomer nanocomposites have great promise as potential light weight, strong viscoelastic materials. Nanocarbons, nanoclay, nanoceramics, nanoparticles of metals, titanium dioxide, ferric oxides and ferrites are some examples.

The most common effect of filler inclusion in elastomer viscoelastic property is change of viscous part and enhancement of modulus. These changes might influence the dynamic viscoelastic loss by the elastomer composite. For example, as series of particulate composites of a nitrile-PVC blend was investigated for dynamic mechanical properties under a wide temperature and frequency range. Frequency scale master curves were generated by graphical method of time-temperature superposition at a reference temperature of 30 °C. High-abrasion furnace carbon black (HAF), graphene nanoplates and organically modified montmorillonite (OMMT) nanoclay were used to observe the effect on dynamic properties due to differences in reinforcement by these three fillers. All fillers are used at 10 phr for comparison.

Figures 4.33 and 4.34 show storage modulus and loss factor of the composites in frequency scale. The reinforcement by HAF black is that of a typical of common rubbers with almost ideal trend in dynamic properties. The storage modulus of OMMT filled composite is lowest, the loss factor peak value and also the frequency of peak is highest among all. The broadness of the loss factor peak is also highest. The behaviour indicates that the OMMT inclusion results in a more so-called rubbery character of the blend, whereas graphene inclusion improves the modulus at higher frequency, but reduces the loss factor peak value and width. The typical behaviour is like a stiffened composite. The rate of increase of storage modulus for graphene composite is quite significantly high at the transition (near about 50–100 Hz). The transition frequency for HAF composite is 83 Hz, while it reduced to 50 Hz for Graphene and increased to 100 Hz for OMMT composite. This indicates that the OMMT composite is most flexible and graphene composite is the stiffest.

Under a dynamic stress, despite the increase in the storage modulus, or to be precise, more elastic in nature, there is a possibility of enhancement of loss of viscoelastic energy, known as hysteretic damping, due to inclusion of reinforcing fillers, such as HAF, graphene or OMMT. For example, a dynamic mechanical analysis of a nitrile-PVC blend vulcanizate with and without carbon black filler is shown in Fig. 4.35. The loss modulus, which is a representation of hysteretic damping (refer eq. (4.64)), is progressively higher with higher HAF contents.

Correspondingly under the dynamic strain, the dynamic viscosity ($\eta' = G''/\omega$) is also higher for filled elastomers as can be seen from Fig. 4.36. The filler effect is visible in increase in dynamic viscosity on increasing filler content signifying more restriction on segmental mobility and increase in elastic energy. The viscosity of each composition reduces with frequency due to high strain rate as is observed in all oscillatory viscosity measurements.

The filler effect can be seen in non-ideal behaviour in relaxation under dynamic strain. Theoretically, the relaxation time is related to the frequency at the peak loss modulus (E''_{max}) as $\omega\tau = 1$ (eq.4.62).

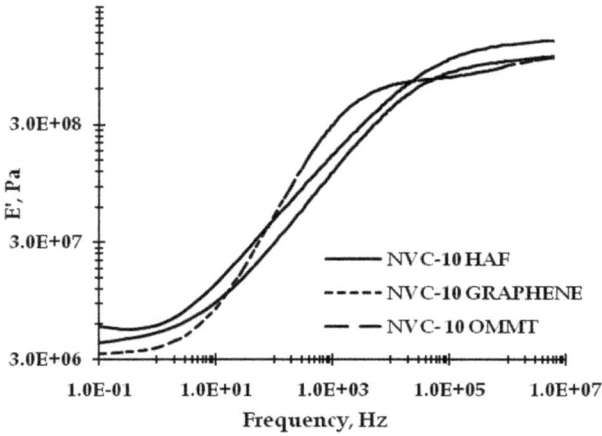

Fig. 4.33: Storage modulus of various particulate composites of NBR/PVC.

The frequency is termed as α-relaxation frequency and the corresponding temperature at which the frequency is measured, is termed as *α-relaxation temperature*, erroneously also referred as *glass transition temperature, T_g*. However, the loss factor (tan δ) peak is sometimes taken to define modified glass transition, and the two relaxation times are related as [99]

$$\tau_1 = \tau \left(\frac{G_0 G_\infty}{G_\infty - G_0} \right)^{\frac{1}{2}}$$

(4.71)

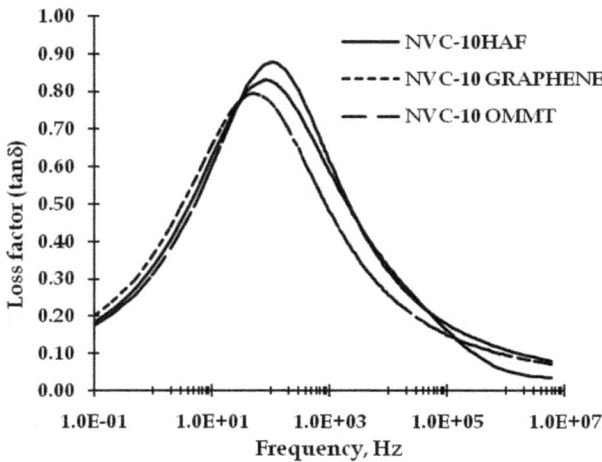

Fig. 4.34: Loss factor of various particulate composites of NBR/PVC.

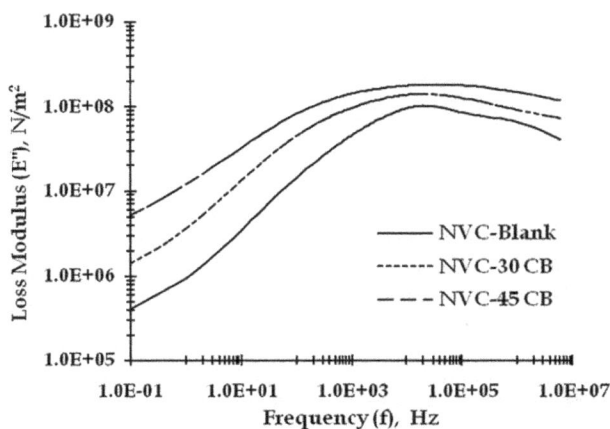

Fig. 4.35: Loss modulus of NBR-PVC blend vulcanizate unfilled and with carbon black filler.

where τ is the relaxation time related to G''_{max} and τ_1 is corresponding to $\tan\delta_{max}$. G_0 is the rubbery shear modulus (low frequency) and G_∞ is the glassy shear modulus (high frequency).

Ideally, a soft elastomer can be fairly described by the above relationship of the two relaxation times, and also for spherical, small fillers which influences the modulus isotropically, but in case of any heterogeneity and anisotropy arising from different shaped fillers, or their interaction, deviations from experimental findings is observed. To explain the deviation, a number of NBR-PVC unfilled and filled vulcanizates were subjected to dynamic mechanical tests with frequency multiplexing. Graphite, graphene nanoplates and OMMT nanoclay were used for this study. The loss modulus (E'') and $\tan\delta$ spectra in frequency scale were obtained at a reference temperature of 30 °C. Frequencies corresponding to E''_{max} and $\tan\delta_{max}$ were used to obtain corresponding τ and τ_1 respectively. Taking the rubbery and glassy modulus in frequency scale data, τ_1 was calculated with the index 0.5 as in eq. (4.59), and a fitting index was used to approximately match that of experimental value. The results are shown in Tab. 4.3. As expected, the relaxation time by E''_{max} is about 2–3 orders lower than that by $\tan\delta_{max}$, since the E'' peak appears at higher frequency (20,000–40,000 Hz) than $\tan\delta$ peak (75–150 Hz).

The revised matching values of τ_1 and the corresponding fitting index are listed in the last two columns. It is clear that the blend and the filled vulcanizates do not exactly follow eq. (4.59), as considerable deviations from experimentally obtained values of τ_1 are observed. The blank sample does not follow the ideal behaviour possibly because of a stiff PVC component in the blend which has a large difference in T_g compared to nitrile rubber. There is no regularity in the trend of indices, signifying non homogeneity in particle size, their distribution in the matrix and agglomeration.

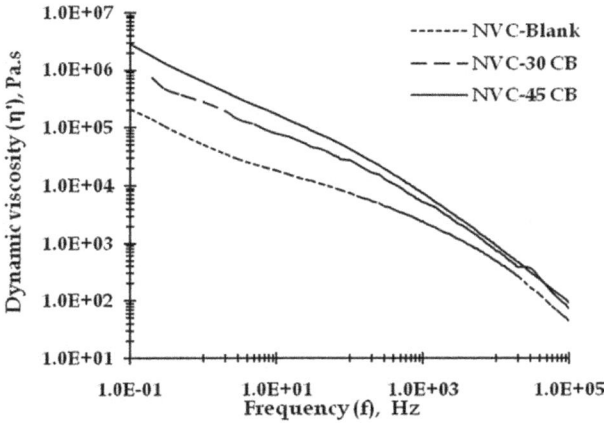

Fig. 4.36: Dynamic viscosity (η') of NBR-PVC blend vulcanizate unfilled and with carbon black filler.

All filled composites show higher modified relaxation time τ_1 compared to the blank sample, as expected, due to greater restriction of segmental mobility in case of filled compositions. The τ values, corresponding to E''_{max} on the other hand, are not consistent, as the experimental result shows. It is more useful to consider tanδ peaks for all relaxation-related calculations. Interestingly, the activation energy of relaxation is identical, whether one takes loss modulus peak or loss factor peak frequency.

The indices for organically modified montmorillonite nanoclay (OMMT)) filled composite are 0.45–0.49, which can be closest to eq. (4.59), possibly because OMMT reinforcement makes the blend more flexible and low modulus composite compared to carbon black or graphene nanoplates, as already observed (Figs. 4.33 and 4.34).

Considering the filler-matrix interaction in dynamic straining, Ziegel and Romanov [100] suggested a relationship of ratio of dynamic modulus of filled to unfilled polymer and volume fraction of the filler, where the term *interaction parameter* is introduced to differentiate the filler contribution to dynamic properties:

$$\frac{E'_c}{E'_0} = \frac{1+1.5V_f B}{1-V_f B} \tag{4.72}$$

where E' is dynamic (storage) modulus, subscripts c and 0 represent filled composite and unfilled rubber, V_f is the volume fraction of the filler and B is the interaction parameter, defined by:

$$B = \left(1+\frac{\Delta R}{R}\right)^3 \tag{4.73}$$

where $\Delta R/R$ is the relative change in the diameter of the filler due to interaction with the matrix rubber [99].

Tab. 4.3: Relaxation time of elastomeric composites from dynamic mechanical data.

Composition	By E"$_{max}$	By tanδ_{max}	τ_1 (s) by calculation		
	τ (s)	τ_1 (s)			
	Expmt.	Expmt.	Eq. (4.59)	Revised	Index
NPC-Blank	7.955E-06	1.090E-03	6.790E-03	1.170E-03	0.37
NPC-5OMMT	5.303E-06	2.121E-03	5.915E-03	2.930E-03	0.45
NPC-10 OMMT	3.182E-06	1.515E-03	3.774E-03	1.860E-03	0.45
NPC-20 OMMT	1.591E-06	1.591E-03	1.765E-03	1.530E-03	0.49
NPC-2 graphene	1.224E-05	1.273E-03	1.208E-02	1.330E-03	0.34
NVC73-5 graphene	7.955E-06	2.121E-03	8.795E-03	2.160E-03	0.4
NVC73-10 graphene	3.977E-05	3.182E-03	5.313E-02	2.980E-03	0.3
NVC73-20 graphene	1.326E-05	3.977E-03	1.540E-02	3.750E-03	0.4
NVC73 10 graphite	1.446E-05	3.002E-03	1.932E-02	2.970E-03	0.37
NVC73-45 graphite	1.3258E-05	3.977E-03	2.753E-02	3.780E-03	0.37

In case of spherical fillers with no agglomeration and with only van der Waals force of attraction, the interaction parameter linearly varies with a regular trend with volume loading of the filler. For a representative series of NBR-PVC blend vulcanizate with two types of filler, spherical carbon black and plate-type nanosized graphene filler was analysed for interaction parameter using eq. (4.60). Figures 4.37 and 4.38 show the interaction parameters for carbon black filled and graphene nanoparticle filled NBR-PVC 70/30 blend vulcanizate.

There is uniformity in the trend of the interaction parameter of carbon black at all volume loading in the elastomer vulcanizate (Fig. 4.37). The loading of HAF black was up to 45 phr (V_f = 0.19). The parameter reduces with volume loading, and it is expected because the storage modulus the reinforcing efficiency decreases since agglomeration increases with loading and the agglomerates cause more filler-filler interaction rather than filler-matrix interaction. The interaction is not so much different at the wide frequency variation (10 Hz to 1,000 Hz) as seen in Fig. 4.37, for carbon black reinforcement. On the other hand, the effect of the nanographene plate-type reinforcement is quite different as seen in Fig. 4.38. The inclusion of the graphene nanoplates was only up to 20 phr (V_f = 0.10). There is no uniformity in the trend of interaction parameter with volume fraction and also with respect to lower frequencies.

However, at low-frequency zone, the trend shows a maximum interaction at about 0.028 volume fraction. It was observed by Dharmaraj et al. [73] that the dispersion and mechanical property at this volume fraction of graphene nanoplate is most pronounced in NBR/PVC blend vulcanizate. Thereafter the interaction parameter reduces with the increase in volume fraction. One reason of lower value of the interaction parameter can be the formation of agglomeration, thereby increasing filler to filler interaction rather than filler to matrix. At higher frequencies (500 Hz and 1,000 Hz), the effect is not seen. However, higher the frequency, higher is the interaction, which is quite unique for the

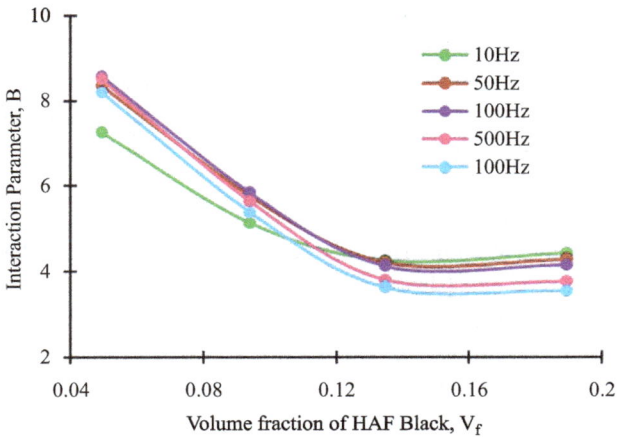

Fig. 4.37: Interaction parameters for carbon black (HAF) filled NBR/PVC blend vulcanizates.

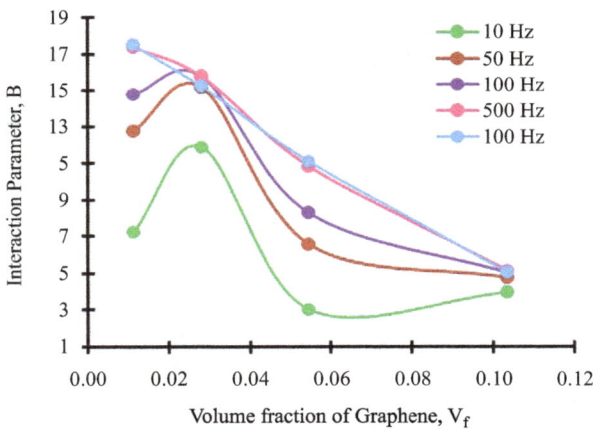

Fig. 4.38: Interaction parameters for graphene nanoparticle filled NBR/PVC blend vulcanizates.

graphene nanoplatelets. This explains the sharp rise in storage modulus of graphene composite as the frequency increases from 50 Hz to 2,000 Hz, as seen in Fig. 4.33.

4.8 Processing of elastomer composites

Elastomer processing is quite well established in the rubber industries for common raw rubbers, except the ones where a liquid-liquid reaction leads to controlled cross-linking and vitrification to form an elastomeric object similar to a vulcanized rubber. Typical examples are polyurethane and flexible epoxy networks.

The important aspect of the processing is dispersion and distribution of tiny filler particles in the raw rubber. Dispersion stands for separation of the particles including breaking of soft agglomerates, while distribution stands for uniformity in concentration of the filler in any portion of the bulk mixture. In the case of nanoclay and graphene nanosheets, which have layers of plate-type elements, the extent of intercalation and exfoliation decide the efficacy of dispersion.

Elastomer composites of carbon black, mineral fillers, metallic powders, graphite, etc., can be processed in a two-roll mill or an internal mixer. The sequence of adding the fillers is after the addition of processing aids (stearic acid and zinc oxide), but before the vulcanization system, such as peroxide curative or sulphur-MBTS-CBS. The mixing process of a polymer with any ingredient increases the temperature, hence there has to be a temperature controlling arrangement. A most useful temperature limit for mixing can be 70 °C to enable good mixing while restricting the scorch as well. The mixing of short fibre such as chopped glass, carbon, Kevlar, cellulose, jute or nylon is similar to carbon black, but there could be incompatibility with the rubber, especially for non-polar ones. It is generally overcome by addition of a phenolic tackifier such as resol (hydroxymethyl terminated phenol-formaldehyde oligomer) to the extent of only 3–5 parts by weight per 100 part rubber. The curing of such tackifier takes place by self-condensation simultaneously during vulcanization at about 140–170 °C. Gen et al. [101] treated short nylon fibres by glycidyl 3-pentadecenyl phenyl ether which is a renewable derivative product from cashew nut industry, and triethylamine in an aqueous medium before processing in a two-roll mill with sulphur-accelerator system to develop an SBR-C-Black-nylon hybrid composite for application in tyre tread. The different processes of treating short fibres for reinforcement in rubbers is discussed in Section 4.3.

Mixing of nanoparticles is more difficult because of the tiny size and hence, more clustering. Additionally, nanofillers cannot be uniformly distributed in large quantity as commonly used micron size fillers. In fact, a small quantity of the nanoparticles would have same reinforcing effect as large quantity of common fillers. There are several processing methods to mix the nanoparticles in an elastomer:

(1) Solution method: The rubber is dissolved in a suitable solvent, but not in a good solvent, so that the viscosity remains lower even at high solid content. Subsequently, the nanofiller is added slowly under high-speed stirring, generally 3,000–6,000 rpm, for 10–15 min depending on the filler type. The mixture is sonicated thereafter at ultrasonic bath using approximately 40 kHz sound, for 10–30 min, sometime even much longer period for homogenization and breaking soft agglomerates. Stirring can be employed simultaneously for better effect. The process of solution mixing is for small batches only since it is difficult to handle large solution volume, because the concentration of the rubber is to be kept low enough. Secondly, after the process, the rubber solution has to be dried off the solvent completely, which is again difficult in case of polar solvents.

(2) A common processing is by high temperature internal mixing. In this process, a temperature of 70–90 °C is enough to have sufficient shear deformation of the rubber and consequently better filler distribution during the shearing process. The disadvantage is possible breaking of fillers such as plate type (nanoclay, graphene, etc.), carbon nanotube and nanofibre.

(3) Two roll mills mixing are also employed to some extent nanofillers also [102]. The advantage is that in terms of mixing sequence, the rolls can be heated and the speed of the rolls can be controlled as required. It is the most flexible process, but in terms of shear deformation, less effective than internal mixers. The wastage is also more than internal mixer process.

(4) Rubber extrusion: rubber extrusion process is used where a definite cross-section and continuous length has to be maintained. The extruder has two parts, one the heated screw-barrel system and the dye at the output end which has the desired cross-sectional shape. The raw stock with all ingredients is fed through a hopper at the rear end of the screw-barrel and the rotation of the screw pushes the stock helically, shearing the mix and simultaneously moving towards the dye. A very small annular gap between the screw and barrel results in greater shear mixing. This process is well suitable for mixing nanoparticles in rubber. The extruder length of a rubber extruder is much less than plastic extruders. Therefore, a good mixing requires multiple pass operation. The main advantage is that the wastage of material is minimum in extrusion process.

(5) For liquid-liquid reaction leading to elastomeric product, the nanoparticles can be mixed in the base reactant, such as polyol (for polyurethane) or epoxy oligomer, dissolved in a small quantity of a solvent. The use of small amount of solvent is possible since the base reactants are liquid. The solvent reduces the viscosity of the base epoxy/polyol and the dispersion and distribution of the nanoparticles are facilitated. In such process, ultrasonication can be gainfully used. However, removal of the solvent is somewhat difficult because of the polar nature of the base epoxy/polyol, and may require vacuum application along with heat. The wastage is minimum in this process. However, solvent usage makes it relatively expensive in large scale operation.

(6) A shear mixing by triple roll mill for viscous base reactants is another option, where the nanoparticles can be effectively dispersed and distributed uniformly. This is generally used for processing of viscous liquids such as putty. The gap between the rolls can be quite narrow to enable good shear deformation and flow on the roll surface to disperse the nanofillers. The rolls can be continuously run to shear the adhered liquid layer on the three rolls and can be discharged afterword. The process can be repeated to observe a satisfactory smoothness checked by a Hegman Gauge (0–10 μm). The process is suitable for bulk production of elastomer nanocomposites in industrial scale. However, there may be some wastage of materials.

4.9 Application of elastomer composite

Rubber-carbon black composites are known from as early as nineteenth century (Charles Goodyear, 1844). Since then, innumerable varieties of rubber composites are developed, studied scientifically and engineered to very useful products for industry, defence and consumer market. Nanoscience and technology has even widened and enhanced the scope for more efficient use of rubber composites.

The applications of the rubber composites can be so wide that any general categorization is not appropriate. However, some broad areas can be defined as: (a) transportation industry, (b) heavy engineering, (c) chemical/petrochemical plants, (d) medical sector, (e) packaging and (f) civil construction. The unique properties of rubber are that it is highly flexible in stretching, resilient, can absorb mechanical energy and can provide a sealing effect. The other advantage is that it has almost similar density of water, which makes it a relatively light material. Rubber compositions are generally no conductor of heat and electricity; thus, they can provide good insulation at ambient and sub-ambient conditions. However, special composites of rubbers can be electrically semi-conductive, and these conducting elastomeric items have quite good applications.

The disadvantage of rubber composites is limitation of high temperature application. Almost all rubber composites can be used continuously up to about 120 °C, with reasonable detention of properties. Silicone and Viton rubber (fluoropolymer) can be used continuously up to about 200 °C. Due to the limitation, the study of degradation and life prediction of rubber composites is so important in product design and application.

Resilience, vibration/shock absorption, flexibility are the manifestation of thermal motion of the chain segments, causing viscoelastic deformation and absorption of mechanical energy. In most application, this becomes the main focus of functional characteristics of the elastomeric composite items. The second most important property is electrical property, based on which electromagnetic shielding and microwave propagations are effectively used in electronics field. However, both viscoelasticity and electromagnetic properties are important for defence application too.

4.9.1 Vibration damping

The most common and simple method of damping of structural vibration and making resilient mount for machines is by a rubber composite. It provides a passive damping material, by absorbing viscoelastic energy under a dynamic force, such as domestic kitchen mixer/grinder, hydraulic system, turbine, engine, vehicle, aircraft etc. Under kinematic motion, each of these generates vibration, and rubbers are known for reducing this vibration. Among the rubbers, those having specific groups which contribute more for damping are nitrile, butyl and acrylic rubbers. However, not all are useful as such without special reinforcement and modification. The useful reinforcements are plate-type fillers and modifications are blending, grafting, forming inter-penetrating network, etc. For example, nitrile-rubber-poly alkyl methacrylate IPNs, urethane-epoxy IPNs and nitrile-PVC blends are reported to be highly damping elasto-meric materials.

Among conventional fillers, plate-type particles such as graphite, aluminium flakes, mica, etc. provide better damping due to conversion of longitudinal strain to shear strain. Since shear modulus is one third of Young's modulus of rubbers, shear strain is more compared to longitudinal strain, and thus strain energy loss is also more. Additionally, the plate or filament-type material causes a friction damping effect, similar to Coulomb Dampingor Rayleigh Damping. The theories are based on loss of energy of friction between two adjoined surfaces by variable clamping force. The damping depends on the clamping force. In rubber composites with plate or filament-type fillers, the adhering force between the rubber and the filler is equivalent to the clamping force.

It is now quite obvious that use of nanoparticles of non-spherical shapes, even in small quantity, enhances damping capability to considerable extent compared to conventional reinforcing fillers like carbon, black and minerals. The higher efficiency of the nanoparticles is obviously due to the tiny size, corresponding to large surface area, which enhances the interaction with the elastomer many times compared to micronized particles. For example, consider a spherical particle of carbon, 10 μm in diameter, the specific surface area (surface area per unit volume) is 6×10^3 cm^{-1}, whereas, for a 50 nm sphere, the same parameter is 1.2×10^6 cm^{-1}, which is 3 order higher. This large difference is reflected in reinforcement properties of the nanoparticles. A disadvantage is, of course, the high association of nanoparticles because of high surface area, which is practically not possible to break totally and hence the efficiency can never be in proportion to the increase in surface area, be it damping or elastic modulus.

Few examples of vibration damping and acoustic application of elastomer-nanocomposites are discussed here. Sasikumar et al. used XNBR-MWCNT nanocomposite for study of high and low compression hysteresis for a structural damping application [103] and also for underwater acoustic sensor application [104]. The nanocomposite exhibited 300% increase in hysteresis loop area at 5% MWCNT compared to unfilled XNBR vulcanizate, corresponding to more than 200% increase in damping factor. However, on increasing the CNT content up to 10%, the frequency of loss peak shifted beyond 10,000 Hz and the loss factor value also reduces by about 50%. With variation of the composition, it is thus possible to either use for high structural damping or tuning of the operational frequency range of underwater sensors. Ratna et al. developed an elastomeric epoxy-clay nanocomposite mastic with plate-type fillers such as aluminium and mica, for structural vibration damping [105]. The system loss factor (a measure of damping) was seen to be about 0.1–0.25 in the frequency range of 50–3,000 Hz, when applied on a steel substrate at a thickness ratio of substrate: mastic: 1:2. Gu et al. developed a dual purpose polyurethane elastomeric nanocomposite coating with MWCNT which is mainly used for underwater acoustic absorption and anticorrosive protection too [106]. Addition of 0.4% CNT improved the corrosion resistance significantly in a 3.5% NaCl aq. solution, measured by electrochemical impedance spectroscopy while 1% CNT containing nanocomposite showed enhanced underwater acoustic absorption coefficient from 0.28 to 0.55 for pristine polyurethane, to 0.48–0.60, measured by a water filled impedance tube method in the frequency range of 2,000–6,000 Hz. Praveen et al. [107] observed a synergistic effect of carbon black and anisotropic reinforcing fillers such as short aramid fibre, organically modified montmorillonite nanoclay and graphite for vibration damping by a segmented polyurethane elastomer, prepared as hybrid nanocomposite from millable polyurethane, vulcanized conventionally by sulphur-accelerator curing system. While carbon black was used at 20 phr in all compositions, graphite was added at 10–20 phr and the other fillers were added at 5–10 phr. The system loss factor in a constrained layer damping arrangement was measured and it was observed that the combination of carbon black and any one anisotropic filler gave better damping than carbon black or other fillers alone. The system loss factor (damping) by CB-nanoclay combination at 800–2,500 Hz was 0.18–0.10, when 5 phr nanoclay was used.

4.9.2 EMI shielding and absorbing elastomer

Electromagnetic interference shielding is an essential protection of electronic equipment and consoles in a high-power electromagnetic field. The easiest way to shield an object is to cover it with a material which can simply reflect the EM wave back to the medium of origin, and not allow to be transmitted through the shield layer. A more effective way is to attenuate the wave inside the shield layer along with a small reflected part. Generally, metals are very good reflectors of EM waves irrespective of frequency, because of their high electrical conductivity, and copper is the best choice. In some cases, the reflection causes an increase in EM signature of the body which is vulnerable to detection by EM sensors. Secondly, metals are heavy and get corroded in atmosphere. Many defence structures use FRP composites in which the EM waves are transmitted even through considerable thickness and may affect the interior electronics of the object. EMI shielding is also required to safeguard the health of personnel and environment from the bad effects of the EM radiation. A solution to all these can be an EM absorbing-type shield, which can reduce the transmission and reflection simultaneously. The emergence of flexible polymeric EMI shielding has found a wide application in military and civil industry for electronic consoles. These can be light weight, durable and quite maintenance-free and cost effective compared to metallic shields. Carbonyl iron powder, ferrites, especially barium hexagonal ferrites with various modifications are so long being used as EM absorbing materials as pigment in elastomer matrix to make a flexible shielding material and also microwave absorbing coating on metallic substrates. Ferrites strongly show complex permeability and permittivity in wide band of microwave frequency. However, the density of ferrites is quite high, resulting in weight penalty. In order to make light weight shields, conducting polymers are preferred. Subsequently, nanoparticles of carbon in various forms such as nanotube, nanofibre, graphene nanosheets are more and more being used in elastomeric flexible matrices which are effective in small quantities and are light weight. Such nanocarbons have sufficient electrical conductivity, and hence the nanocomposite can be tailored to match the electrical impedance of free space (vacuum) and also good dielectric loss property for absorbing the EM energy in a wide frequency spectrum. Although the loss is less compared to ferrites, requiring higher thickness, the overall weight penalty is far less, since the ferrite/iron loading required for flexible shield/absorber is quite high. In addition to reflection and absorption, the third mechanism for reduction of EM wave intensity is multiple reflection of the wave inside the medium (shielding) because, at the interfaces, there is always some impedance difference to cause some reflection back to the medium in either direction. A simple diagram as Fig. 4.39 shows the three types of mechanism.

The shielding effect is the fraction of incident energy transmitted to the other side of the shield. The effectiveness can be measured in terms of voltage in a vector network analyser (VNA). However, it is customary to convert the voltage intensity to power and express the power in dB scale, keeping in mind that power is proportional to square of intensity and dB is 10 times the logarithm (10 base) of power ratio, or 20 times the intensity ratio. The general expression for estimation of EMI shielding effect (SE) is (referring Fig. 4.39):

$$SE = 20 \log \left(\frac{V_0}{V_T} \right) dB \tag{4.74}$$

EM wave absorption is expressed as

$$Absorbed\,power\ =\ incident\,power\ -\ (reflected\,power\ +\ transmitted\,power) \tag{4.75}$$

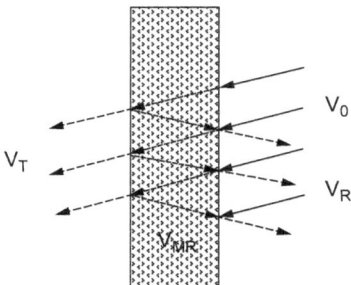

Fig. 4.39: Mechanism of EM wave shielding.

The absorption takes place in a medium due to dielectric loss for polymer nanocomposites, which are having desired conductivity. The free space (vacuum) impedance is about 377 Ohms, and if a composite has a volume conductance of about 1 mho/cm, then it can be used as a shielding material. The wave, on travelling inside the medium, suffers loss exponentially with distance travelled. This is expressed as a property called *skin depth* of the medium (e.g. a nanocomposite), which is defined as the thickness for which the incident intensity reduces to ($1/e$) of original value ($e = 2.718$) and is related to the frequency of the EM wave and material parameters by:

$$\text{Skin depth: } \delta = \frac{1}{\sqrt{\pi f \sigma \mu}} \tag{4.76}$$

where f is the EM wave frequency, σ is the electrical conductivity (in $\text{Ohm}^{-1}\text{m}^{-1}$) of the shield, μ is the magnetic permeability, of the shield ($\mu = \mu_0 \mu_r$, where μ_r is the relative magnetic permeability of the shield and μ_0 is permeability of free space ($=4\pi \times 10^{-7}$ H/m). From eq. (4.77), it is seen that the skin depth reduces with frequency, permeability and conductivity. The exponential decay along the shield thickness suggests that smaller conducting particles in the elastomer should be better for shielding effect. The unit dimension of the filler should be equal or less than the skin depth at the highest frequency in the desired band. However, the thickness of the composite shield must be more than the skin depth to ensure high shielding and absorbing effect. In general, EM frequency band may range from 300 MHz to 110 GHz for most applications from microwave ovens to high-frequency satellite communication and military radars. However, definite frequency bands are applicable for different purpose, hence such a wide range shielding is not practically required. In defence application, for example, mostly 1–40 GHz range is used for surveillance, detection, naval warfare, air defence and fire control. The effect of thickness of the shielding composite on absorption is given by

$$SE_A = 20\log_{10} e^{t/\delta} \tag{4.77}$$

The reflection loss is given by

$$SE_R = -10\log_{10}\left(\frac{\sigma_T}{16 f \varepsilon_r \mu_r}\right) \tag{4.78}$$

The multiple reflection loss is given by

$$SE_{MR} = 20\log_{10}\left(1 - e^{-2t/\delta}\right) \tag{4.79}$$

In order to design a composite for low-frequency shielding, the skin thickness can be made lower if the loss parameters μ'', ε'' and the total conductivity σ_T are higher.

The magnetic permeability and dielectric permittivity are frequency-dependent complex parameters:

$$\mu = \mu' - j\mu'' \tag{4.80}$$

$$\varepsilon = \varepsilon' - j\varepsilon'' \tag{4.81}$$

where μ' is real part of the complex magnetic permeability, μ'' is the imaginary or loss part and similarly, ε' is real part of permittivity or dielectric constant and ε'' is the imaginary part of the permittivity or the dielectric loss. The operator j is the imaginary number $= \sqrt{-1}$, used to indicate that mathematically, the imaginary or loss part is 90 degree out of phase with the real part in both eqs. (4.80) and (4.81).

Like the permeability, the dielectric constant is also defined as: $\varepsilon = \varepsilon_0 \varepsilon_r$, where ε_0 is the dielectric constant of free space $= 8.85 \times 10^{-12}$ F/m.

The velocity of EM wave in a medium is related to the permeability and permittivity of the medium as

$$c = \frac{1}{\sqrt{\mu_0 \varepsilon_0}} = 2.998 \times 10^8 \text{ m/s} \tag{4.82}$$

The impedance of the shield is the ratio of the amplitudes of the magnetic to the electric field and is given by

$$Z_{\text{SH}} = \sqrt{\frac{j\omega\mu}{\sigma + j\omega\varepsilon}} \tag{4.83}$$

While the free space impedance is given by (noting that $\sigma=0$ for free space):

$$Z_0 = \sqrt{\frac{\mu}{\varepsilon}} = 377\Omega \tag{4.84}$$

The EM shielding and absorbing capability can be theoretically calculated as a coefficient of absorption (α), mathematically expressed as:

$$\alpha = \frac{\sqrt{2}\pi f}{c} \sqrt{(\mu''\varepsilon'' - \mu'\varepsilon') + \sqrt{(\mu''\varepsilon'' - \mu'\varepsilon')^2 + (\mu''\varepsilon'' + \mu'\varepsilon')^2}} \tag{4.85}$$

The reflection loss can be estimates by the impedances of free space and the shield:

$$SE_R = 20\log_{10} \frac{Z_{\text{SH}} - Z_0}{Z_{\text{SH}} + Z_0} \text{ dB} \tag{4.86}$$

The loss can be enhanced if the reflected wave from the air-shield interface and that reflected back from the conducting backing plate have a phase difference of 180°, thereby creating a destructive interference. This happens when the thickness of the shielding layer is quarter wavelength ($\lambda/4$). For a low-frequency EM wave, the wavelength is quite large, for example 30 cm for 1 GHz EM wave, hence the thickness required for quarter wavelength criteria is 7.5 cm, which is quite thick in many applications. However, the ideal absorber thickness (tm) for excellent absorption at a matching frequency due to phase cancellation is calculated by the following relation [108]:

$$t_m = \frac{nc}{4f_m\sqrt{|\varepsilon_r||\mu_r|}} \tag{4.87}$$

where n is an odd number, c is the velocity of light in free space and f_m is the matching frequency. Since all the parameters for EM shielding and absorption are directly related to the basic properties of the composite, viz., total electrical conductivity, complex magnetic permeability and complex dielectric permittivity, and of course, frequency of the EM wave and thickness of the composite shield. Since all parameters are frequency dependent, it is necessary to measure basic properties in a VNA using a wave guide with coaxial cables as shown in a schematic diagram as Fig. 4.40.

The VNA measures intensity of EM wave generated by an electrical energy as EM pulses. The pulse is fed to a wave guide by a coaxial cable and the receiving signal is collected by another coaxial cable on the opposite end of the guide and returned to the VNA for analysis. Four signals are measured which are termed as *S-parameters* and designated as S_{11} for forward reflection coefficient, S_{22} for backward reflection coefficient, S_{12} for forward transmission coefficient and S_{21} for backward transmission coefficient. These parameters are also expressed in terms of dB [109]:

$$SE_T = 10\log\left(\frac{1}{|S_{12}|^2}\right) = 10\log\left(\frac{1}{|S_{21}|^2}\right) = 10\log\left(\frac{1}{T}\right) \tag{4.88}$$

$$SE_R = 10\log\left(\frac{1}{1-|S_{11}|^2}\right) = 10\log\left(\frac{1}{1-|S_{22}|^2}\right) = 10\log\left(\frac{1}{1-R}\right) \tag{4.89}$$

$$SE_A = 10\log\left(\frac{1-|S_{11}|^2}{|S_{12}|^2}\right) = 10\log\left(\frac{1-|S_{22}|^2}{|S_{21}|^2}\right) = 10\log\left(\frac{1-R}{T}\right) \tag{4.90}$$

The effective absorbance by an EM shield should be calculated from the basic understanding that the summation of the energies of transmission, reflection and absorption is the total energy input. Therefore, the coefficients R, T and A are summed up to unit, and the fractional absorbance is the ratio of A to $1-R$. Hence, we can write [109]:

$$A_{\text{eff}} = \left(\frac{1-R-T}{1-R}\right) \times 100(\%) \tag{4.91}$$

4.9.3 Examples

Flexible EMI shield and absorbing composites are known to be used extensively for civil and military installations and equipment. Barium hexagonal ferrites and spinnel ferrites as such and with other conducting materials such as metallic powders, thin

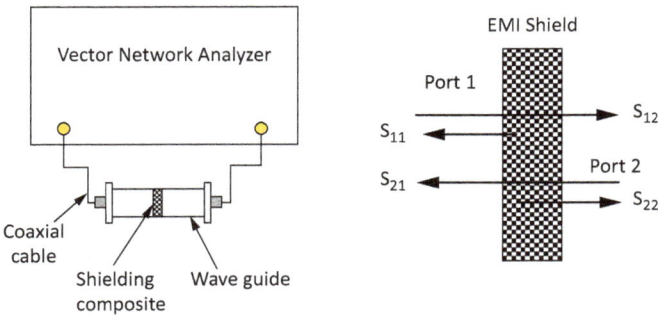

Fig. 4.40: Measurement of shielding parameters by VNA.

metal-coated fabric, reduced graphene oxide, carbon nanotubes, conducting carbon black and graphite the most used particles in rubber matrix for making such shield and absorber. There are a large number of literatures on this type of design and a few are described here.

Peymanfar and Rahmanisaghieh [108] used a barium ferrite $BaFe_2O_4$ pure and capped (using glucose as organic capping agent), treated at 650 and 850 °C in silicone rubber and measured basic properties such as complex permeability and permittivity and calculated the conductivity, impedance and the attenuation constant for 1.5 mm thick silicone rubber-ferrite composites. The authors reported attenuation constant (refer eq. (4.85)) about 10–25 in the frequency range of 8–18 GHz.

Kruzelak et al. [109] made a detailed review of various ferrite and conducting nanoparticle filled polymer composites for EMI shielding and MW absorbing application.

Morari et al. [110] used a ferrite and graphite in silicone rubber with a 300 µm thick polyester fabric having a Fe-Ni thin coating on one side and the 3 mm thick flexible composite was evaluated for EMI shielding effect. The measurement was done by horn antenna in an anechoic chamber for large size samples. The authors reported a negligible shielding effect in 1–18 GHz for ferrite-graphite fillers, compared to the metal-coated fabric containing rubber sheet, where average of 33 dB shielding effect was observed.

Joseph et al. [111] studied EMI shielding effectiveness of natural rubber and chlorobutyl rubber nanocomposite. The authors used carbon nanotube, nanofibre, nanoclay (Cloisite 10A) and carbon black to tailor the electromagnetic properties of the nanocomposites and reported about 25 dB shielding for chlorobutyl-CNT nanocomposite at a frequency range of 2.7–3.9 GHz.

Sykora et al. [112] used specially prepared Lithium ferrite and used in Nitrile rubber matrix at as high as 600 phr loading and compared the EM shielding property with NBR at similar loading of a standard manganese-zinc ferrite (H40). The elastic modulus (at 300% strain) and tensile strength decreased considerably beyond 200 phr loading, more so for manganese-zinc ferrite, but the flexibility increased considerably (monotonically up to 600 phr). The MW return loss was better at lower frequency for manganese-zinc-NBR composite (0.1 to 1.1 GHz) with sharp peaks of 30–35 dB. The peaks were shifted to lower frequency as the loading increased. Similarly, the return loss for Li-ferrite-NBR composite showed better at 1 to 1.2 GHz with sharp peaks of 30–34 dB. The peaks shifted to lower frequency as the loading increased.

Ray et al. [113] investigated ethylene-vinyl acetate elastomer–based nanocomposites with various carbon fillers such as carbon black, expanded graphite, carbon nanofibre and nanotube for EMI shielding effect in 8–12 GHz range. The authors reported that even 4% carbon nanofibre in EVA-CNF composite showed about 15 dB shielding effect and 16% composite showed 30–33 dB shielding effectiveness, while the return loss was only about 4–6 dB in his frequency range. Carbon nanotube-EVA composite (16% CNT) showed about 26 dB shielding effectiveness. Expanded graphite or carbon black was far less effective.

Joseph et al. [114] studied butyl rubber (IIR)-single-walled carbon nanotube (SWCNT) composite for EMI shielding performance in the frequency range of 8–18 GHz. The authors measured the conductivity and skin depth of the IIR nanocomposite samples with 0 to 8 phr SWNT. The conductivity of 8 phr composite was about 0.4 to 1.4 S/m, increasing with frequency, and the skin depth was reported to be 0.01–0.004 m diminishing with frequency. About 9–12 dB total shielding effect for IIR-8 phr SWCNT composite was observed.

Nguyen et al. [115] developed a flexible shielding composite of PDMS with free standing graphene foam structure decorated with F_3O_4 nanoparticle-intercalated the $Ti_3C_2T_x$ (T stands for groups like -O, -OH, -F) (MXene). The composite exhibited very high EMI shielding effect of 80 dB in 8.2–12.4 GHz (X-band) and about 77 dB in 26.5–40 GHz (Ka-band) due to high absorption of EM wave. With only graphene, the shielding effect was about 19 dB in these two frequency bands. However, the synthesis process of the nanofiller with all the ingredients and composite forming is quite complicated and intricate.

4.10 Conclusion

Elastomer-based composites are very widely accepted in all fields of application because of the flexibility, which is quite an advantage since any contour can be covered without a template and complex moulding. Secondly, elastomers processing is easy, low in energy consumption and overall cost. Since elastomers contain large free volume at ambient conditions, they can accommodate large proportion of fillers and additives of diverse shapes and this can result in better tailoring of any structural or functional property such as antistatic, piezoelectric, sound and vibration damping or EMI shielding. In addition, elastomers are quite stable in various atmospheric conditions, with high humidity and in cold or tropical countries. Most elastomer composites can be used at sub-zero temperatures, few of them up to −40 °C without developing brittleness. Recent research on smart materials-based elastomer composites is encouraging in terms of application in large scale even at large volume percent of the active ingredient. A typical example is 0–3 PZT-Neoprene composite for acoustic transducer with good flexibility and piezoelectric effect, and smart vibration damper. Acceptable piezoeffect is realized at a very high concentration of the PZT particle in the matrix and it is only possible in an elastomer of high flexibility. The piezoelectric property of the composite is utilized to advantages of sensing and dissipation of the strain energy to a load. The advantage of such smart damper is that on ageing, the efficiency does not decrease to the same extent with the deterioration of the viscoelastic damping effect. A different smart elastomer composite is a self-healing system since the microspherical inclusions in the elastomer is more conveniently done than in structural thermosets, maintaining the integrity of the item. A modification in the rubber processing ensures no breakage of the microcapsules containing the healing material, more convenient in polyurethane or silicone elastomer casting at ambient temperature. There can be more application of smart elastomer composite items as shape memory material such as nanocomposites of polyurethane elastomer. The shape memory effect with appropriate mechanism such as IR or UV irradiation can be very effectively achieved for cross-linked elastomer nanocomposites containing laser absorbing dyes, or by heat within the temperature limits of the elastomers.

Commercial acceptance of new elastomer composites will depend on energy saving, conservation, ecological and personnel safety, which are most priority of the present-day global needs.

References

[1] Parkinson, D. The reinforcement of rubber by carbon black. Br. J. Appl. Phys. 1951, 2, 273.
[2] Robertson, C. G., Hardman, N. J. Nature of carbon black reinforcement of rubber: perspective on the original polymer nanocomposite. Polymers. 2021, 13(4).
[3] Three main Properties of carbon Black. Mitsubishi Carbon Black: Mitsubishi Chemical corporation. https://www.m-chemical.co.jp/en/products/departments/mcc/carbonblack/product/1201075_7946.html. Accessed on 03 June 2022.
[4] Carbon Black. IARC Monograph on the Evaluation of Carcinogenic Risks to Humans. Lyon, France, 2010, Vol. 93, 43–49.
[5] Reinforcing Carbon Blacks. https://en.wikipedia.org/wiki/Carbon_black. Accessed on 03 June, 2022.
[6] Panda, J. N., Bijwe, J., Pandey, R. K. Variation in size of graphite particles and its cascading effect on the performance properties of PAEK composites. Compos. Part B Eng. 2020, 182, Article I.D. 107641. https://doi.org/10.1016/j.compositesb.2019.107641.
[7] Shojaeenezhad, S. S., Farbod, M., Kazeminezhad, I. Effects of initial graphite particle size and shape on oxidation time in graphene oxide prepared by Hummers' method. J. Sci-Adv. Mater. Dev. 2017, 2(4), 470–475.
[8] Liu, B., Zhang, D., Li, X., He, Z., Guo, X. H., Liu, Z., Guo, Q. Effect of graphite flakes particle sizes on the microstructure and properties of graphite flakes/copper composites. J. Alloys Compd. 2018, 766, 382–390.

[9]   Sombatsompop, N., Thongsang, S., Markpin, T., Wimolmala, E. Fly ash particles and precipitated silica as fillers in rubbers. I. Untreated fillers in natural rubber and stirene-butadiene rubber compounds. J. Appl. Polym. Sci. 2004, 93, 2119–2130.

[10]   Thongsang, S., Vorakhan, W., Wimolmala, E., Sombatsompop, N. Dynamic mechanical analysis and tribological properties of NR vulcanizates with fly ash/precipitated silica hybrid filler. Tribol. Int. 2012, 53, 134–141.

[11]   Brochure, "Hi-Sil Reinforcing Fillers", M/s PPG Silica Products, Monroeville, PA 15146 USA. https://www.ppgsilica.com/getmedia/3d8b6fdf-df2f-4379-b2da-94c384ea3433/HiSil233Dand233GD Brochure.pdf.aspx. Accessed on 01 July, 2022.

[12]   Ren, J., Qin, Y., Peng, Z., Li, Z. Influence of composite structure design on the ablation performance of ethylene propylene diene monomer composites. e-Polymers. 2021, 21(1), 151–159.

[13]   Song, J. Q., Huang, Z. X., Qin, Y., Wang, H. H., Shi, M. X. Effects of zirconium silicide on the vulcanization, mechanical and ablation resistance properties of ceramifiable silicone rubber composites. Polymers. 2020, 12(2), 496. https://doi.org/10.3390/polym12020496.

[14]   Ren, J. W., Song, J. Q., Fu, H. D., Peng, Z. W., Qin, Y. Preparation and properties of boron nitride/ silicone rubber ceramizable composites. China Rubber Ind. 2020, 67(3), 163–169.

[15]   Yang, D., Zhang, W., Jiang, B. Silicone rubber ablative composites improved with zirconium carbide or zirconia. Compos. Part A Appl. 2013, 44, 70–77, 16/j.compositesa. 2012.09.002.

[16]   Rallini, M., Puri, I., Torre, L., Natali, M. Thermal and ablation properties of EPDM based heat shielding materials modified with density reducer fillers. Compos.-A: Appl. Sci. Manuf. 2018, 112, 71–80.

[17]   Anyszka, R., Bieliński, D. M., Pędzich, Z., Zarzecka-Napierala, M., Imiela, M., Rybinski, P. Processing and properties of fireresistant EPDM rubber-based ceramifiable composites. High. Temp. Mater. Proc. 2017, 36(10), 963–969.

[18]   Fei, H. F., Han, X. J., Liu, B. Z., Gao, X. Y., Wang, Q., Zhang, Z., Xie, Z. Mechanism of the antioxidation effect of α-Fe_2O_3 on silicone rubbers at high temperature. RSC Adv. 2016, 6(10), 7717–7722.

[19]   Wu, L., Zhang, Y. Enhanced thermal oxidative stability of silicone rubber by using cerium-ferric complex oxide as thermal oxidative stabilizer. e-Polymers. 2019, 19, 257–267.

[20]   Hayashida, O., Kazuhiro, O., Atsuhito, K. Silicone rubber composition having excellent heat resistance. 2015, European Patent No. EP2554585 B1, dt. 15-04-2015, assigned to Shin-Etsu Chemical Co. Ltd. Tokyo, Japan.

[21]   Li, H. Y., Tao, S., Huang, Y. H., Su, Z. T., Zheng, J. P. The improved thermal oxidative stability of silicone rubber by using iron oxide and carbon nanotubes as thermal resistant additives. Compos. Sci. Technol. 2013, 76, 52–60.

[22]   Zhang, X., Zhang, Q., Zheng, J. P. Effect and mechanism of iron oxide modified carbon nanotubes on thermal oxidative stability of silicone rubber. Compos. Sci. Technol. 2014, 99, 1–7.

[23]   Qi, H. J., Joyce, K., Boyce, M. C. Durometer hardness and the stress-strain behaviour of elastomeric materials. Rubber Chem. Technol. 2002, 7, 419–435.

[24]   Standard test method for rubber property – Durometer hardness. ASTM D 2240 – 15.

[25]   Murali Manohar, D., Chakraborty, B. C., Begum, S. S. Hardness – Elastic Modulus Relationship for NitrileRubber and Nitrile Rubber – Polyvinyl Chloride Blends. In: Ganippa, L., Kartikeyan, R., Muralidharan, V., eds. Advances in Design and Thermal Systems. Springer, 2021, pp. 301–314.

[26]   Gent, A. N. Trans. Inst. Rub. Ind. 1958, 34, 46–57.

[27]   BS 903 Methods of testing vulcanised rubber Part 19 (1950) and Part A7, 1957.

[28]   Morgans, R., Lackovic, S., Cobbold, P. Understanding the IRHD and Shore Methods used in Rubber Hardness Testing. Presented at a meeting of the Rubber Division, American Chemical SocietyOrlando, Florida September 21–24, 1999.

[29]   Briscoe, B. J., Sebastian, S. K. Rubber. Chem. Technol. 1993, 66, 827–836.

[30]   Guth, E. Theory of filler reinforcement. J. Appl. Phys. 1945, 16, 20–25.

[31] Richard, T. G. The mechanical behaviour of a solid microsphere filled composite. J. Compos. Mater. 1975, 9, 108–113.

[32] Kerner, E. H. The elastic and thermoelectric properties of composite media. Proc. Royal Soc.-London, 69B, 808 (1956).

[33] Ramsteiner, F., Theysohn, R. On the tensile behaviour of filled composites. Composites. 1984, 15(2), 121–128.

[34] Nielsen, L. E. Mechanical Properties of Polymers. Reinhold Publishing Corporation, Chapman & Hall Ltd, London, 1962.

[35] Murty, V. M., De, S. K. Short-fiber-reinforced stirene-butadiene rubber composites. J. Appl. Polym. Sci. 1984, 29, 1355–1368.

[36] Goodloe, P. M., Reiling, T. L., McMutic, D. H. Rubber Age. 1947, 61, 697.

[37] Boustany, K., Coran, A. Y., Discontinuous cellulose reinforced elastomer. U.S. Patent No. 3,697,364, October 19, 1972.

[38] Boustany, K., Coran, A. Y., U.S. Patent No. 3,709,845, January 9, 1973.

[39] Coran, A. Y., Boustany, K., Hamed, P. J. Appl. Polym. Sci. 1971, 15, 2471.

[40] Coran, A. Y., Boustany, K., Hamed, P. Short-Fiber-Rubber Composites: The properties of oriented cellulose-fiber-elastomer composites. Rubber Chem. Technol. 1974, 47, 396–410.

[41] Boustany, K., Hamed, P. Rubber World. 1974, 171(2), 39.

[42] Boustany, K., Yaucher Coran, A., Preparation of discontinuousfiberreenforced elastomer. U.S. patent No. 3836412, September 17, 1974.

[43] Boustany, K., Arnold, R. L. Short fibers rubber composites: the comparative properties of treatedand discontinuous cellulose fibers. J. Elastomers Plast. April 1976, 8, I60–176.

[44] Hamed, P., Paul, C. Li, reinforcement of EPDM elastomers through discontinuous unregenerated wood cellulose fibers. J. Elastomers Plast. 1977, 9(4), 395–415.

[45] Geethamma, V. G., Joseph, R., Thomas, S. Short coir fiber reinforced natural rubber composites: effects of fiber length, orientation and alkali treatment. J Appl Polym Sci. 1995, 55, 583–594.

[46] Geethamma, V. G., Kalaprasad, G., Groeninckx, G., Thomas, S. Dynamic mechanical behavior of short coir fiber reinforced natural rubber composites. Compos Part A. 2005, 36, 1499–1506.

[47] Varghese, S., Kuriakose, B., Thomas, S., Koshy, A. T. Mechanical and viscoelastic properties of short fiber reinforced natural rubber composites: effects of interfacial adhesion, fiber loading, and orientation. J. Adhesion Sci. Technol. 1994, 8(3), 235–248.

[48] Ismail, H., Edyham, M. R., Wirjosentono, B. Bamboo fibre filled natural rubber composites: the effect of filler loading and bonding agent. Polym. Testing. 2002, 21, 139–144.

[49] Cracium, G., Manalia, E., Ighigeanu, D., Stelescu, M. D. A method to improve the characteristics of EPDM rubber based eco-composites with electron beam. Polymers. 2020, 12, 215.

[50] Kashani, M. R. Aramid-short-fiber reinforced rubber as a tire tread composite. J. Appl. Polym. Sci. 2009, 113, 1355–1363.

[51] Sandstrom, P. H., Westgate, W. K., Botts, B. P., Barnette, R. R., Tire with rubber tread composed of a primary and at least one lateral tread portion containing a dispersion of short carbon fibers. US patent No. US20070221303A1 (2007), assigned to The Goodyear Tire and Rubber Co. Akron USA.

[52] Cen, L., Lv, G., Tan, X., Gong, Z. Short nylon fibers waste modified with glycidyl 3-pentadecenyl phenyl ether to reinforce stirene butadiene rubber tread compounds. Adv. Polym. Technol. 2019, Article No. 5847292. https://doi.org/10.1155/2019/5847292.

[53] Mohammed, H. S., Elangovan, K., Subrahmanian, V. Studies on aramid short fibers reinforced acrylonitrile butadiene rubber composites. Ind. J. Adv. Chem. Sci. 2016, 4(4), 458–463.

[54] Ryu, S. R., Lee, D. J. Effects of fiber aspect ratio, fiber content, and bonding agent on tensile and tear properties of short-fiber reinforced rubber. KSME Int. J. 2001, 15(1), 35–43.

[55] Rosen, B. W., Dow, N. P. Fracture, Vol. 7. In: Leibowitz, H., ed. Academic press, New York, 1972, p. 612.

[56] Abrate, S. The mechanics of short-fiber-reinforced composites: A review. Rubber Chem. Technol. 1986, 59, 384–404.

[57] Shokrieh, M. M., Sani, H. M. On the constant parameters of Halpin-Tsai equation. Polymer. 2016, 106, 14–20.

[58] Van Ooij, W. J. Fundamental aspects of rubber adhesion to brass-plated steel tire cords. Rubber Chem. Technol. 1979, 52, 605.

[59] Vandou, W. J., Harakuni, P. B., Buytaert, G. Adhesion of steel tire cord to rubber. Rubber Chem. Technol. 2009, 82, 315–339.

[60] Shi, X., Ma, M., Lian, C., Zhu, D. Investigation of the effects of adhesion promoters on the adhesion properties of rubber/steel cord by a new testing technique. J. Appl. Polym. Sci. 2013. 10.1002/APP.39460.

[61] Walter, J. D. Cord-rubber tire composites: Theory and applications. Rubber Chem. Technol. 1978, 51, 524–576.

[62] Pidaparti, R. M. V. Analysis of cord-rubber composite laminates under combined tension and torsion loading. Compos. Part B. 1997, 28B, 433–438.

[63] Shield, C. K., Costello, G. A. The effect of wire rope mechanics on the mechanical response of cord composite laminates: An energy approach. J. Appl. Mech. 1994, 61(3), 9–15.

[64] Golovanevskiy, V., Kondratiev, A. Elastic properties of steel-rubber conveyor belt. Exp. Tech. 2021, 45, 217–226.

[65] Vasiliev, V. V., Morozov, E. V. Advanced Mechanics of Composite Materials and Structural Elements, 3rd Edn. Ch.3. Elsevier Science, 2013, 51–132. ISBN:978-0-08-045372-9.

[66] Siddha, R. A., Roach, J. F., Erickson, D. E., Reed, T. F. Stress analysis of cord adhesion tests- a route to improved tests. Rubber Chem. Technol. 1981, 54(4), 835–856.

[67] Nicholson, D. W., Livingston, D. I., Fielding-Russell, G. S. A new tire cord adhesion test. Tire Sci. Technol. 1978, 6(2), 114–124.

[68] Martin Frank, S., Pletz, M., Wondracek, A., Schuecker, C. Assessing failure in steel cable-reinforced rubber belts using multi-scale FEM modelling. J. Compos. Sci. 2022, 6(34).

[69] Blazej, R., Jurdziak, L., Blazej, A. K., Kozlowski, T. Identification of damage development in the core of steel cord belts with the diagnostic system. Sci. Rep. 2022, 11, 12349. https://doi.org/10.1038/s41598-021-9153-z. Source: www.nature.com/scientificreports/.

[70] Montmorillonite -Wikipedia. https://en.wikipedia.org/wiki/Montmorillonite. Accessed on 03 Sept, 2022.

[71] Borralleras, P., Segura, I., Aranda, M. A. G., Aguado, A. Influence of experimental procedure on d-spacing measurement by XRD of montmorillonite clay pastes containing PCE-based superplasticizer. Cem. Concr. Res. 2019, 116, 266–272.

[72] Kamena, K., Product commercialization: One Nanostep at a time. PPT Presentation in FF Conference. 1998. Southern Clay Products. Source:https://www.powershow.com/view/3bb1ff-MDVhM/Product_Commercialization_One_Nanostep_at_a_Time_powerpoint_ppt_presentation. Accessed on 04 September, 2022.

[73] Dharmaraj, M. M., Chakraborty, B. C., Begum, S. The effect of graphene and nanoclay on properties of nitrile rubber/polyvinyl chloride blend with a potential approach in shock and vibration damping applications. Iran. Polym. J. 2022. https://doi.org/10.1007/s13726-022-01064-6.

[74] Barghamadi, M., Ghoreishy, M. H. R., Karrabi, M., Mohammadian-Gezaz, S. Modeling of non linear hyper viscoelastic and stress softening behaviors of acrylonitrile butadiene elastomer/polyvinyl chloride nanocomposites reinforced by nanoclay and graphene. Polym. Compos. 2021, 42, 583–596.

[75] Hanhua, L., Li, W., Guojun, S. Study of NBR/PVC/OMMT nanocomposites Prepared by Mechanical Blending. Iran. Polym. J. 2010, 19(1), 39–46.

[76] Rajasekar, R., Pal, K., Heinrich, G., Das, A., Das, C. K. Development of nitrile butadiene rubber-nanoclay composites with epoxidized natural rubber as compatibilizer. Mater. Des. 2009, 30, 3839–3845.

[77] Hussain, F., Hojjati, M., Okamoto, M., Gorga, R. Review article: Polymer-matrix Nanocomposites, Processing, Manufacturing, and Application: An Overview. J. Compos. Mater. 2006, 40, 1511–1575.

[78] Jia, C., Zhang, L. Q., Zhang, H., Lu, Y. L. Preparation, microstructure, and property of silicon rubber/organically modified montmorillonite nanocomposites and silicon rubber/OMMT/fumed silica ternary nanocomposites. Polym. Compos. 2011, 32(8), 1245–1253.

[79] Manohar, D. M., Chakraborty, B. C., Begum, S. S., Natarajan, R., Chandramohan, S. Study on the effect of nanoclay as reinforcing filler for nitrile rubber – polyvinyl chloride blend: frequency response of dynamic viscoelasticity and vibration damping. Iran. Polym. J. 2022. https://doi.org/10.1007/s13726-022-01074-4.

[80] Zhang, J., Hereid, J., Hagen, M., Bakirtzis, D., Delichatsois, M. A., Fin, A., Castrovinci, A., Camino, G., Samyn, F., Bourbigot, S. Effects of nanoclay and fire retardants on fire retardancy of a polymer blend of EVA and LDPE. Fire Saf. J. 2009, 44(4), 504–513.

[81] Liu, Y., Fang, Z. Combination of montmorillonite and a Schiff-base polyphosphate ester to improve the flame retardancy of ethylene-vinyl acetate copolymer. J. Polym. Eng. 2015, 35(5), 443–449.

[82] Ghazinezami, A., Jabbarnia, A., Asmatulu, R. Fire retardancy of polymeric materials incorporated with nanoscale inclusions. Proceedings of the ASME 2013 International Mechanical Engineering Congress and Exposition IMECE2013 November 15–21, 2013, San Diego, California, USA.

[83] Lu, Q., Keskar, G., Ciocan, R., Rao, R., Mathur, R. B., Rao, A. M., Larcom, L. L. Determination of carbon nanotube density by gradient sedimentation. J. Phys. Chem. B. 2006, 110(48), 24371–24376.

[84] Al Hasan, N. H. J. mechanical and structural properties of carbon nanotubes: a molecular dynamic study. Int. J. Tech. Phys. Probl. Eng. 2022, 14(2), 380–385.

[85] Qian, D., Wagner, G. J., Liu, W. K., Yu, M. F., Ruoll, R. S. Mechanics of carbon nanotubes. Appl. Mech. Rev. 2002, 55, 495–533.

[86] Ma, P. C., Siddiqui, N. A., Marom, G., Kim, J. K. Dispersion and functionalization of carbon nanotubes for polymer-based nanocomposites: A review. Compos. Part A. 2010, 41, 1345–1367.

[87] Malekie, S., Ziaie, F. A two-dimensional simulation to predict the electrical behavior of carbon nanotube/polymer composites. J. Polym. Eng. 2017, 37(2), 205–210.

[88] Li, M., Jiang, J., Lu, X., Gao, J., Jiang, D., Gao, L. Natural rubber reinforced with super-hydrophobic multiwall carbon nanotubes: obvious improved abrasive resistance and enhanced thermal conductivity. J. Polym. Eng. 2022, 42(8), 688–694.

[89] Li, M., Tu, W., Chen, X., Wang, H., Chen, J. NR/SBR composites reinforced with organically functionalized MWCNTs: simultaneous improvement of tensile strength and elongation and enhanced thermal stability. J. Polym. Eng. 2016, 36(8), 813–818.

[90] ADL- Graphene Technical Data Sheet. Ad-Nano Technologies Private Limited. Shimoga- 77222, Karnataka, India. www.ad-nanotech.com. Accessed on 16 Sept 2022.

[91] Hummers, W. S., Offeman, R. E. Preparation of graphitic oxide. J. Am. Chem. Soc. 1958, 80(6), 1339–1339.

[92] Marcano, D. C., Kosynkin, D. V., Berlin, J. M., Sinitskii, A., Sun, Z., Slesarev, A., Alemany, L. B., Lu, W., Tour, J. M. Improved synthesis of graphene oxide. ACS Nano. 2010, 4(8), 4806–4814.

[93] Tang, Z., Zhang, L., Feng, W., Guo, B., Liu, F., Jia, D. Rational design of graphene surface chemistry for highperformance rubber/graphene composites. Macromolecules. 2014, 47, 8663–8673.

[94] Esmizadeh, E., Arjmandpour, M., Vahidifar, A., Naderi, G., Dubois, C. Preparation and characterization of silicone rubber/graphene nanosheets nanocomposites by in-situ loading of the coupling agent. 2019, 53(24), 3459–3468.

[95] Berki, P., László, K., Tung, N. T., Kocsis, J. K. Natural rubber/graphene oxide nanocomposites via melt and latex compounding: Comparison at very low graphene oxide content. J. Reif. Plast. Compos. 2017, 36(11), 808–817.

[96] wang, J., Zhang, K., Hao, S., Xia, H., Lavorgna, M. Simultaneous reduction and surface functionalization of graphene oxide and the application for rubber composites. J. Appl. Polym. Sci. 2018, 136, 47375.

[97] Alkelani, A. A., Housari, B. A., Nassar, S. A. A proposed model for creep relaxation of soft gaskets in bolted joints at room temperature. J. Press. Vessel Technol. February, 2008, 130, 011211–1.

[98] Vinogradov, G. V., Malkin, A. Y. Rheology of Polymers –Viscoelasticityand Flow of Polymers. Mir Publishers, Moscow, 1980.

[99] Perepechko, I. Acoustical Methods of Investigating Polymers. Mir Publishers, Moscow, 1975.

[100] Ziegel, K. D., Romanov, A. Modulus reinforcement in elastomer composites I. Inorganic fillers. J. Appl. Polym. Sci. 1973, 17, 1119–1131.

[101] Gen, L., Lv, G-z, Tan, X.-W., Gong, Z.-L. Short nylon fibers wate modified with glycidyl 3-pentadecenyl phenyl ether to reinforce styrene butadiene rubber tread compounds. Adv. Polym. Technol. 2019, Article ID 5847292. https://doi.org/10.1155/2019/5847292.

[102] Hanhua, L., Li, W., Guojun, S. Study of NBR/PVC/OMMT nanocomposites Prepared by Mechanical Blending. Iran. Polym. J. 2010, 19(1), 39–46.

[103] Sasikumar, K., Manoj, N. R., Mukundan, T., Khastgir, D. Hysteretic damping in XNBR-MWNT nanocomposites at low and high compressive strains. Compos. Part B. 2016, 92(1), 74–83.

[104] Sasikumar, K., Manoj, N. R., Mukundan, T., Khastgir, D. Design of XNBR nanocomposites for underwater acoustic sensorapplications: Effect of MWNT on dynamic mechanical properties and morphology. J. Appl. Polym. Sci. 2014, 131(18), Article ID. 40752. https://doi.org/10.1002/app.40752.

[105] Ratna, D., Barman, S., Kushwaha, R. K., Chakraborty, B. C. Viscoelastic Mastic for Free Layer Damping (FLD) of structural vibrations. Patent No. 300740, dt 22/11/2010.

[106] Gu, B.-E., Huang, C.-Y., Shen, T.-H., Lee, Y.-L. Effects of multiwall carbon nanotube addition on the corrosion resistance and underwater acoustic absorption properties of polyurethane coatings. Prog. Org. Coatings. 2018, 121, 226–235.

[107] Praveen, S., Bahadur, L., Yadav, R., Billa, S., Patro, T. U., Rath, S. K., Ratna, D., Patri, M. Tunable viscoelastic and vibration damping properties of a segmented polyurethane synergistically reinforced with carbon black and anisotropic additives. Appl. Acoust. 2020, 170, Article ID: 107535. https://doi.org/10.1016/j.apacoust.2020.107535.

[108] Peymanfar, R., Rahmanisaghieh, M. Preparation of neat and capped $BaFe_2O_4$ nanoparticles and investigation of morphology, magnetic, and polarization effects on its microwave and optical performance. Mater. Res. Exp. 2018, 5 (10). https://doi.org/10.1088/2053-1591/aadaac.

[109] Kruzelak, J., Kvasnicakova, A., Hlozekova, K., Hudec, I. Progress in polymers and polymer composites usedas efficient materials for EMI shielding. Nanoscale Adv. 2021, 3, 123–172. https://doi.org/10.1039/d0na00760a.

[110] Morari, C., Balan, I., Pintea, J., Chitanu, E., Iordache, I. Electrical conductivity and electromagnetic Shielding effectiveness of silicone Rubber filled with ferrite and graphite powders. Prog. Electromag. Res. 2011, 21, 93–104.

[111] Joseph, T. M., Mariya, H. J., Haponiuk, J. T., Thomas, S., Esmaeili, A., Sajadi, S. M. Electromagnetic interference shielding effectiveness of natural and chlorobutyl rubber blend nanocomposite. J. Compos. Sci. 2022, 6, 240. https://doi.org/10.3390/jcs6080240.

[112] Sýkora, R., Babayan, V., Ušáková, M., Kruželák, J., Hudec, I. Rubber composite materials with the effects of electromagnetic shielding. Polym. Compos. 2016, 37(10), 2933–2939.

[113] Ray, M., George, J. J., Chakraborty, A., Bhowmick, A. K. An investigation of the electromagnetic shielding effectiveness of ethylene vinyl acetate elastomer reinforced with carbon nanofillers. Polym. Compos. 2010, 18(2), 59–65.

[114] Joseph, N., Janardhanan, C., Sebastian, M. T. Electromagnetic interference shielding properties of butyl rubber-single walled carbon nanotube composites. Compos. Sci. Technol. 2014, 101, 139–144.

[115] Nguyen, V. T., Min, B. K., Yi, Y., Kim, S. J., Choi, C. G. MXene(Ti3C2TX)/graphene/ PDMS composites for multifunctional broadband electromagnetic interference shielding skins. Chem. Eng. J. 2020, 393, Article ID: 124608. https://doi.org/10.1016/j.cej.2020.124608.

Chapter 5
Smart composites

Description of abbreviations

Abbreviation	Description
DGEBA	Diglycidyl ether of bisphenol-A
CFRP	Carbon fibre-reinforced plastic
GFRP	Glass fibre-reinforced plastic
FRP	Fibre-reinforced plastic
PU	Polyurethane
PTMG	Poly(tetramethylene) glycol
PCLD	Passive constrained layer damping
VEM	Visco-elastic material
CLD	Constrained-layer damping
APCLD	Active-passive constrained layer damping
PVDF	Poly(vinylidene fluoride)
PZT	Lead zirconate titanate
PLZT	Lead lanthanum titanate zirconate
BIIR	Bromobutyl rubber
NBR	Nitrile rubber
SHM	Structural health monitoring
CSA	Camphor sulphonic acid
PLA	Poly(lactic acid)
ESD	Electrostatic discharge
PPS	Polyphenylene sulphide
PP	Polypropylene
ABS	Acrylonitrile-butadiene-styrene
SWCBNT	Single-walled carbon nanotubes
MWCNT	Multiwall carbon nanotube
FIT	Functionalized interleaf technology
GNP	Graphene nanoplates
PLA	Polylactic acid
EM	Electromagnetic
SE	Shielding effectiveness
RRCS	Radar reflecting composite structure
RACS	Radar-absorbing composite structure
RSC	Radar cross section
VNA	Vector network analyser
SWCNT	Single-walled carbon nanotubes
CB	Carbon black
WIPO	*World Intellectual Property Organization*
SHE$_s$	Self-healing elastomers
SHDC	Self-healing dental composite
CBNCs	Conducting polymer bio-nanocomposites
PTFE	Polytetrafluoroethylene
EMC	Electromagnetic compatibility

https://doi.org/10.1515/9783110781571-005

5.1 Self-healing Composites

Damage of composites during service and the consequent failure is very common, but undesirable. In a catastrophic failure, the event might turn fatal to human users and also to the machine/system as such. It is well established that failure by structural disintegration occurs due to the presence of microdefects or cracks that subsequently grow very fast due to extremely high stress intensity at the crack tip, resulting in a catastrophic failure. In many applications, it becomes almost impossible to handle cracks that appeared on a composite's surface due to difficulty in access or conditions that cannot be attended to, such as a flying or underwater object or a large ship structure.

A possible solution to such problems could be the use of so called 'Smart Composite', which can heal automatically by some mechanism as soon as a crack forms. The idea of self-healing has originated from healing of a wound on a human skin. However, it takes several days to heal even a small cut on the skin. A more interesting example is the regeneration/re-growing of the tail of a lizard. In order to accomplish healing in a reasonable time and to maintain structural integrity after a healing process, self-healing smart composite must have a few important characteristics such as fast curing, identical or nearly same properties in the cured state as in the parent matrix and should have sufficient storage life in the uncured condition, embedded in the host composite. For an underwater object, an additional requirement could be the capability to cure in the presence of water.

The most common self-healing process is to incorporate microcapsules as hollow spherical beads, containing the healing oligomer that would cure on exposure to atmosphere as the capsule breaks on cracking of the composite. Therefore, the liquid oligomer must be capable of flow and should fill up the crack. Subsequently, a chemical reaction of the oligomer with a second reactant or a curing agent is to be fast and might be catalysed by atmospheric moisture or oxygen or in combination. In a different mechanism, the oligomer might contain a catalyst, and on exposure to atmosphere, the oligomer could be cured in the presence of oxygen, moisture and the catalyst. There can be a number of reaction systems such as isocyanate-terminated polyurethane pre-polymer, which is cured by moisture, alkyd resin, cured by catalysts such as cobalt octoate, moisture-curable epoxy-urethane pre-polymer, etc. A different self-healing composite may contain vascular tubes running the healing fluids inside and perform the same healing process as the capsule, on exposure due to cracking. A second mechanism is the intrinsic healing, where a reversible chemical or a physical bond formation of a polymer takes place, either by a sharp increase in viscosity of the pH-sensitive micro-gels, or by swelling of the shape memory polymers or by the melting and solidification of thermoplastic materials [1]. Intrinsic healing has the advantage of repeated cycles of healing at the same place of occurrence, which is not possible by capsule or vascular healing methods [2].

There were attempts to develop self-healing fibre-reinforced composites using hollow glass fibres [3] but with very low efficiency (5%) due to the extensive damage of fibres when the healed composite was again tested for impact strength. Some researchers developed combined solid and hollow glass fibre-reinforced composite with epoxy and hardener separately in hollow fibres, with a higher healing efficiency of 73–87% [4, 5]. A different approach was to incorporate a healing ply in the FRP laminate [6] where the healing ply contained the epoxy-/mercaptan-filled hollow polypropylene tubes. The FRP was made by hand layup. The healing efficiency was determined by comparing the flexural strength of the original and the healed composite. A 62% healing efficiency was achieved by heating the laminate at 70 °C during the healing process. The gas evolved during the reaction of epoxy and mercaptan on healing exerted pressure to fill up the cracked FRP by the healing fluid quite effectively. The incorporation of epoxy (DGEBA) and amine hardeners and catalysts separately in different microcapsules and using them for healing in FRP laminates was studied by some researchers with very high efficiency (>100%) [7, 8].

Jhanji and co-workers [9] developed self-healing FRP composites using microcapsules containing an epoxy and a hardener in separate capsules. Two types of self-healing composites were made, one with carbon fibre (CFRP) and the other with glass fibre (GFRP), both containing hollow glass fibre with healing agents. The fabrication of the composites was by simple hand lay-up, without any compression force. Only 2.5% (by weight) hollow fibres were used in the matrix resin. The flexural strength therefore was not compromised to any significant extent. The authors described the steps in making the microcapsules and the self-healing FRP composites. The time taken for healing (complete curing of the healing agents (epoxy and amine hardener) was 24 h. The results showed very high healing efficiency (>100%) after a 35% loss of flexural strength that occurred due to the damage. The synthesis of microcapsules is generally done by forming cross-linked rigid thermoset hollow capsules in situ in the presence of the healing resin in a liquid form, separately for the epoxy solution and the hardener. A typical microcapsule material is melamine-formaldehyde/ urea-formaldehyde or resorcinol-formaldehyde thermoset resins.

The process of healing by the epoxy-hardener healing reactants, either in tubes or in microcapsules, often require heating after filling up the cracked FRP so that the curing is proper, and the strength and impact energy are comparable to the undamaged pristine FRP (without the healing system). This process is external healing process. The determination of strength after the healing is mostly done by flexural tests through three-point or four-point bending. The creation of crack or damage on the FRP laminate is generally by low impact in the range of 1–5 Joules, or by an indentation force of 1–2 kN. However, healing efficiency is always low (below 60%) when the impact force/energy for damage is high [8].

The efficiency of the healing process is measured by the ratio (in %) of the flexural strength of the healed composite and the original composite without the healing compounds. This is because in most application of FRP panels, the composites experience flexural force and, consequently, the possibility of damage of the capsules in use. Secondly, the flexure mode evaluation can predict the difference in interlaminar shear strength due to the incorporation of the healing capsules/tubes. As such, after the incorporation of the healing capsule or tubes, the strength and/or toughness is compromised to some extent, unless the proportion of the healing capsules/tubes is small [9].

The design of structural elements with FRPs decides the minimum strength and modulus of elasticity in tension, bending and shear required, according to the application. Addition of such microcapsules in the composite poses two difficulties. The first, during the processing, there is a possibility of breaking the rigid but brittle thin capsules/tubes, particularly for mixing with the resin matrix of the composite for a uniform distribution in the matrix, and the second is the fabrication by pressure- or shear-intensive moulding method such as compression moulding. Therefore, the tubes or the capsules are required to be strong enough to withstand the forces during the above processing. A typical example is shown in a report on a self-healing coating development [10] where the authors determined the mechanical stability in terms of percentage of the capsules (by weight) ruptured during mixing of two different solutions, with variation in the rpm of the stirrer, viscosity of the mixing solution and the duration of mixing. Heating during the healing process is sometimes undesirable, especially when the structure is airborne or under water or is an unattended large surface like the superstructure of a ship. Special epoxy system of low viscosity and faster reaction rate should be used in such cases so that the process need not be monitored or assisted. The most disadvantage of external healing is that the process is not repeatable at the same site of damage because once the healing reaction is complete; there would not be any healing ingredient as capsule or tubes at that site.

Another consideration of healing by two different reactant-filled capsules is the adequate mixing on release of these reactants when a crack or damage in the composite occurs. For example, the epoxy and the amine hardener are encapsulated separately so that, on cracking, these are released and supposed to be mixed automatically so that the curing reaction can take place. Heating the composite after damage can help in reducing the viscosity of the epoxy resin and hardener for better mixing and a faster reaction, but thorough mixing of these two components is not ensured in all cases. This might explain the large differences in the healing efficiency reported in various literatures where such dual capsules are used for external healing process in self-healing FRPs.

Internal healing process offers repeatable healing at the same site in the case of repeated damage. Reversible bond formation is such an example. A typical reversible covalent bond formation was reported by Lin and co-workers [11]. The method was a thermo-reversible Diels-Alder reaction for a recyclable polyurethane-Halloysite nanocomposite. A nanocomposite with furan pendent polyurethane and a maleimide-terminated cross-linker was prepared with 0–2% halloysite content. The healing conditions are heating at 90 °C for 5 min, followed by 48 h at 65 °C, resulting in 93% healing (recovery of tensile strength compared to the original composite). Three cycles of healing were carried out, following the Diels-Alder reaction. The nanocomposite with 1% halloysite was best in mechanical properties as well as in self-healing efficiency. Jia and Gu [12] developed a carbon nanotube composite with flexible polyurethane (PU), with a random copolymer (polyol) of ε-caprolactone and D,L-lactide, hexamethylene diisocyanate, along with the chain extenders butane diol and poly(tetramethylene) glycol (PTMG). The PU was an iso-cyanate-terminated product. The self-healing polyurethane was made by reacting this PU with 2,2-dithiodiethanol. The resulting PU undergoes disulphide bond exchange revers-ibly on heating, and reverted back to the original structure on cooling. The advantage is that it can heal repeatedly during use.

5.2 Smart structural composites

One of the most important requirements of a structural composite is smart vibration damping. The term smart used in this special application indicates a process of en-hancement of vibration attenuation, irrespective of the frequency of vibration. In a passive vibration damping, the viscoelastic loss of the damping material (a polymeric composition, in general) is highly dependent on the frequency range of vibration and temperature of the application.

In order to have a smart response, irrespective of such imposing conditions, the damping can be made active using a reverse bias of vibration – attenuated by con-verting the vibration acceleration to electric current, or by a shunt without electrical biasing voltage, or using a magnetic panel and electromagnet, or by a damper with an electrorheological or magnetorheological fluid, or a shear-thinning fluid. Principles of some active and active-passive damping are discussed here in brief.

5.2.1 Passive constrained layer damping (PCLD)

Passive constrained layer damping (PCLD) is quite common in industrial, automotive, defence and aerospace applications. In this method, a viscoelastic core material (VEM) is sandwiched between the stiff vibrating substrate and the constraining layer that is equally stiff. The VEM undergoes shear deformation under vibrating force be-cause its elastic modulus is far less than the substrate or the constraining layer. A typ-ical example is a vulcanized rubber strip (VEM) sandwiched between two metallic strips. The damping is dependent on many parameters such as the geometry of the CLD construction, temperature and frequency range, dynamic viscoelastic properties of the VEM, fraction of the area covered, etc. [13–19].

5.2.1.1 Active control system
An active control uses an electrical system that senses the vibration intensity and phase, and by either feed-forward or feed-back method supplies the vibration (acceleration or dynamic displacement) with an identical intensity with 180°reverse phase for a complete destructive interference. The resulting intensity would be nil in ideal condition. Figures 5.1 and 5.2 show typical feed-back and feed-forward control systems.

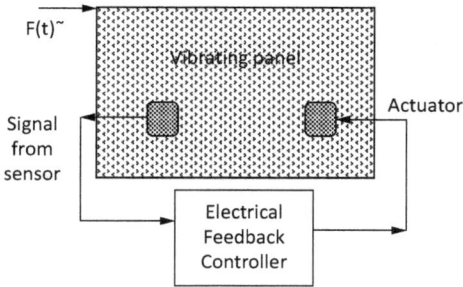

Fig. 5.1: Feed-back control system for vibration damping.

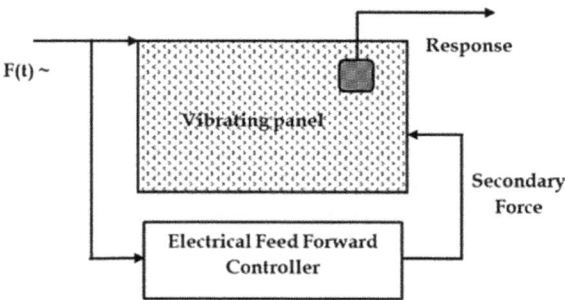

Fig. 5.2: Feed forward control system.

5.2.1.2 Active-passive constrained layer damping (APCLD)

In the construction of the active-passive CLD arrangement, a viscoelastic material (VEM) that is normally a composite of an elastomer is used as the passive damping element, and a stiff constraining panel is used on top of it, just as a normal CLD construction. The signal from the vibrating substrate is picked up by a piezoelectric transducer (accelerometer) and an electrical controller feeds a reverse signal to the substrate through a piezoactuator. The feed-forward system feeds a secondary force to the substrate on analysing the primary force signal in a feed-forward electrical controller. The response is measured by an accelerometer attached to the constraining layer of the damping CLD construction. The signals are amplified by an amplifier, which is part of the controller system. The gain of the amplifier is fixed according to the signal received, which is dependent on the vibration intensity at different frequencies.

There are many designs of this hybrid damping CLD, with variations in the placement of the sensor and actuator in the substrate-VEM-constrained layer construction. Figure 5.3 shows one such arrangement of a combined active-passive CLD. The constraining layer is fitted with the piezoelectric actuator for damping the low-frequency vibration that is transmitted through the VEM with more or less undamped intensity.

The piezo-sensor layer is just above the vibrating panel to pick up the vibration signal constantly for analysis of magnitudes and phases corresponding to the dominant frequencies.

The advantage of the active control is that damping at very low frequency is very effectively accomplished and, in fact, lower the frequency, better the efficiency. Also, low bandwidth of vibration frequency can be better controlled. Secondly, active control practically eliminates the vibration at a narrow band of frequency, even at resonance. The active mode has a greater advantage for very minute vibration intensities, even at the nanometric scale at low frequency, which is required for some optical applications such as holographic recording. A more practical advantage is that in the case of an electrical system failure, vibration damping is still taken care of by the VEM, although sacrificing the lower frequency damping.

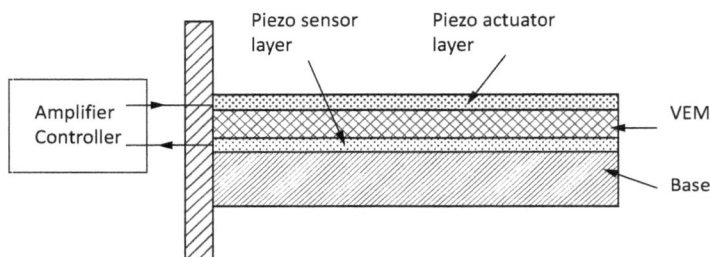

Fig. 5.3: A typical active-passive CLD arrangement with a piezo-layer as the constraining layer.

The disadvantages are that the reverse dynamic force is very critical in terms of phase angle because in some discrete frequencies, it might amplify the vibration intensity due to constructive interference rather than attenuate it. Secondly, active control is not effective in higher frequencies due to the very minute time lag and complexity of the large number of frequencies to be controlled. The use of passive VEM is therefore required to accomplish a wide-band vibration damping. The third disadvantage is that the active control cannot be employed in large vibrating systems like the engine room on the deck of a ship or a large turbine base in hydroelectric plants because the energy required for feeding back to the substrate is very high, requiring a large controller and actuator. The most important disadvantage in commercial applications is the complexity of construction and the associated cost. Passive CLD is very cost effective and is still preferred over this combination of active and passive damping. However, where low-frequency vibration is very critical and for relatively small structural elements, active-passive CLD is very useful. A typical comparison of the transmissibility plots of a passive (viscoelastic) damping and active damping at low frequency is shown in Fig. 5.4 below (values taken arbitrarily for the purpose of illustration only).

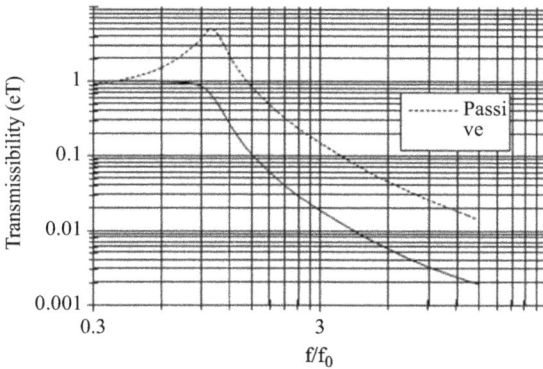

Fig. 5.4: Vibration transmissibility of a typical passive viscoelastic damper and an active damper.

In active control, the amplification of vibration acceleration intensity is not observed because of the fully destructive interference of waves, but in the passive damping, the amplification at resonance (natural frequency) occurs, irrespective of the material damping factor; only the extent of transmissibility will reduce more at resonance as the damping factor increases. The other modes, particularly beyond a certain frequency (generally below 1,000 Hz), are partly damped by passive CLD, since the active control is difficult to achieve at higher frequencies.

Plump and Hubbard Jr. [20] introduced this active constrained layer in a ACLD system and, at present, large number of research articles are published on this subject due to the growing interest of industrial applications. For example, Lam et al. [21] and Shen [22] designed simultaneous damping system by a passive viscoelastic core and an active control system. To make a relatively robust system, brittleness of ceramic piezoelectric materials can be replaced by polymeric, flexible sensors. Baz [23] used a thin piezoelectric polymer sensor (PVDF) between the VEM and the substrate.

The subject of polymeric composite for such passive-active combined damping has many interdisciplinary aspects such as quantification of electromechanical parameters of piezoelectric sensors and actuators, dynamic viscoelasticity of polymer composites (both frequency- and temperature-dependent loss properties) and an electrical unit with a control algorithm. Finite element methods are used by many researchers, for example by Trindade et al. [24], van Nostrand et al. [25] and Varadan and co-workers [26]. Valey and Rao [27] made a comparison among active, passive and combined active-passive damping systems, while Trindade and Benjeddou [28] made a comprehensive review of the active-passive CLD system. The authors highlighted that a passive damping system with a damping factor of 2–3% can be coupled with an active control system to enhance it even up to 15–20%, but at a maximum 400 Hz.

A modification of the above passive-active CLD can be used as a viscoelastic material that can accommodate a high concentration of piezoelectric ceramic powders, for instance, lead zirconate titanate (PZT), lead lanthanum titanate zirconate (PLZT) or lead magnesium niobate – lead titanate (PMN-PT) to make a 0–3 composite as VEM damping layer, and a piezoceramic constraining layer for enhanced active control. In a 0–3 composite, however, the solid ceramic filler loading has to be quite high for achieving adequate piezoelectric properties and this can possibly be done with bromobutyl rubber (BIIR), which has inherently high structural damping capability.

5.2.2 Magnetic CLD

Another modification of the smart CLD could be a magnetic VEM for a steel base and constraining layer. The VEM would be an elastomeric composite filled with fine particles of a permanent magnet (consisting of neodymium, iron and boron, for example) and magnetized to make a magnetic VEM. In an alternate construction, the VEM could be a metamaterial having short, thin strips of the polled permanent magnet embedded in it. The advantage of such a VEM is that there is no need to have any adhesive for bonding. Secondly, if there is an active control by an electromagnetic system, then the magnetic attraction can increase the shear deformation of the VEM, thus enhancing the shear–strain energy (loss) and reduce longitudinal strain (reduction of vibration amplitude). The electromagnet with an electrical control can be placed laterally in a non-contact arrangement so that the polarity of the electromagnet changes with the dynamic strain as conceptualized, calculated and experimented by Zheng et al. [29]. The authors could achieve very good agreement on their theoretical estimation of mode shapes (17–85 Hz) and system loss factors (0.04–0.12), with experimental findings. The achieved system loss factors for a CLD at such a low frequency are a definite indication of the effect of frictional damping due to electromagnetic force control. A typical passive magnetic CLD and active-passive magnetic CLD construction is shown as Fig. 5.5.

This arrangement is also applicable up to a frequency level, beyond which the passive damping is dominant. This magnetic CLD has two major disadvantages, one, the substrate must be a magnetic material like steel (baring austenitic stainless steel), and the arrangement of a non-contact electromagnet and control system is a little complicated. Loading of ferrite in the VEM is possible if the VEM is quite flexible, such as a soft rubber. The magnetic powder can be a ferrite-type material, including barium hexagonal ferrite or strontium-substituted barium ferrite, or a similar permanent magnet material. The rare-earth-type magnetic powder could be samarium cobalt material or powder of neodymium-iron-boron-type material.

The VEM can be selected from those that are known to damp vibration in the practical temperature and frequency range. For example, nitrile rubber (NBR), ethylene vinyl acetate copolymer (EVA), butyl and halo butyl rubbers, polyurethane rubbers, nitrile-PVC blends, carboxylated nitrile rubber, etc., which are useful for mid- to high-frequency, or from thermoplastic elastomers such as thermoplastic polyurethane elastomer, etc. – those that can damp at low-frequency vibrations.

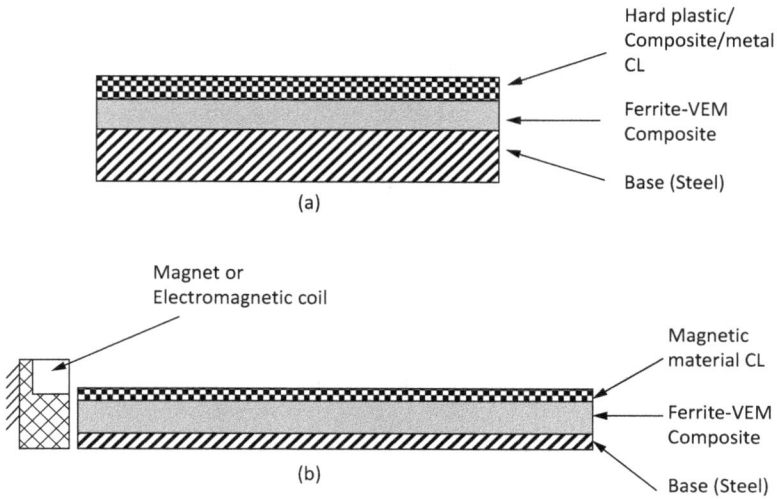

Fig. 5.5: (a) Passive magnetic CLD and (b) passive/active magnetic CLD.

Hansaka and co-workers [30, 31] developed a series of magnetic VEM composites using a permanent magnetic powder, magnetized to have a residual magnetic flux density of about 25 gausses to about 15,000 gausses. The constraining layer is preferably an engineering plastic, including nylon (polyamide resin), polycarbonate, polyacetal, polyethylenepolypropylene, ABS resin, etc. or a thermoset composite such as epoxy-FRP with a Young's modulus of 3GPa, which can be bent as per the rail surface contour. The substrate is a steel rail. Because of the magnetic attraction, the damping layer does not require an adhesive and the damping is also enhanced by the lateral sliding friction damping along with the viscoelastic damping. The passive magnetic CLD construction has some major advantages as the magnetic flux density does not change significantly with temperature (except if heated to Curie temperature, which is generally very high), and is independent of the frequency of the vibration. There is no active electromagnetic control system; therefore, the construction is simple and can be applied to actual large systems such as rail lines.

5.2.3 Structural health monitoring (SHM)

An active vibration control by piezoelectric sensors and actuators has an offshoot – monitoring vibration (dynamic strain or sudden change in strain) of a composite panel in applications where the panels are subjected to continuous vibration, such as the wings of an aircraft made of FRP composites. Under constant vibration, the amplitude vs. frequency has a particular signature (finger print) when the panel has no defects. A series of piezo sensors attached to the different parts of the panel transmit the vibration strain and they are recorded in a multichannel analyser. Each plot of frequency-dependent amplitude is monitored continuously to observe any change in the pattern of vibration amplitude. Any defect, in terms of a crack or added mass or even a sudden impact on the panel would get reflected in the vibration signature at the neighbourhood of the defect, which can be captured by the nearby sensor. The common defects of interlaminar shear failure, internal crack in a thick panel section, etc. are very precisely detected by such health monitoring system. A typical such vibration signature with and without defect is shown in Fig. 5.6. The data are arbitrarily chosen as an example only. The distortion of the wave is near the first natural frequency of the aluminium FRP sandwich beam as shown here. The defect size decides the frequency range at which the deviation would be prominent. In this case, a sharp cut was deliberately made to cause a prominent change in the 10–50 Hz frequency zone.

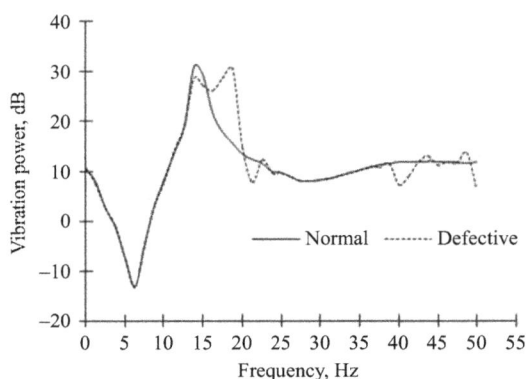

Fig. 5.6: Vibration signature of a panel with and without a defect.

The SHM technology of composites is widely used with sonic and ultrasonic waves for primary aircraft structures [32], large bridges [33], etc. Montalvão et al. [34] and Singh and Sehegal [35] reviewed the SHM of composites for various special applications.

5.3 Electrically conducting composites

Polymeric composites can be tailored to have sufficient, or even high electrical conductivity. Polymers are, in general, insulating materials, both electrically and thermally. However, some polymers having conjugated double bonds in the polymer chain show electronic conductivity, which is a result of inter-chain hopping of electrons of the pi bonds. The hopping is further facilitated by a dopant, mostly anionic, for example, simple HCl doping in polyaniline (PANI).

For a polymeric composite to conduct electricity, there has to be a continuous connected path of electron flow that requires a minimum content of the conducting component, termed as the percolation threshold concentration. It is determined by the abrupt increase in conductivity with a small change in concentration. As an example, aneoprene rubber (CR)-PANI composite with various loadings of PANI doped with camphor sulphonic acid (CSA) developed for microwave-absorbing flexible sheets was studied for in-plane conductivity and plotted as in Fig. 5.7. The percolation threshold concentration is about 15% PANI, doped with CSA, since the conductivity is changed from about 10–11 S/cm to about 10–7 S/cm at that concentration.

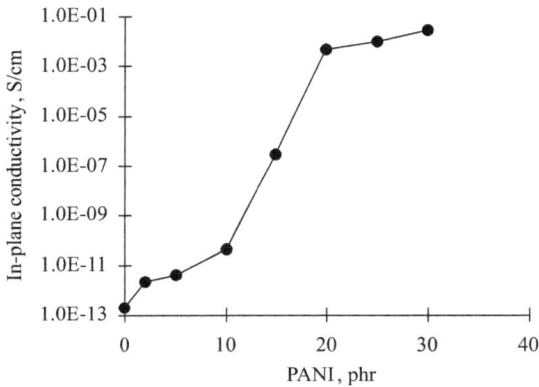

Fig. 5.7: Percolation threshold in electrical conductivity of NBR-PANI composites.

Percolation theory predicts the relationship between the resistivities of the composite and the conductive filler volume as [36]:

$$\rho = \rho_0 (v - v_c)^t \text{ for } v > v_c \qquad (5.1)$$

where ρ is the resistivity of the composite, ρ_0 is the resistivity of conductive filler, v and v_c are filler volume and the percolation threshold, respectively, and t is a critical exponent. The critical exponent, t, based on theoretical prediction, varies from 1.6 to 2.0, while for CNT-filled nanocomposites, the value is between 0.7 and 3.1 [36].

The advantages of electrically conducting composites are many, for example, a wide range of conductivities, flexibility to accommodate any contour, light weight, biocompatibility for medical applications, ease of manufacturing, cost effectiveness, etc. The conducting composites can be made with a choice of matrix polymer, which can be conducting like ionically doped polyaniline, poly (alkyl thiophenes), polypyrrole, etc. or using a common polymer such as rubber or plastic. Electrically, the doped conducting polymers resemble a semiconductor and the incorporation of these organic materials in an insulating polymer matrix can be of use in many critical applications such as in lightning arresters, sensors, detectors, etc. For example, 30% PANI (by volume) in epoxy resin was used in an epoxy-carbon fibre FRP to make a lightweight conducting composite for lightning arresters. The conducting polymer was doped by camphor sulfonic acid to enhance the conductivity to few S/cm levels [37]. However, most conducting polymers are not film-forming and are used as conducting fillers in common rubbers and plastics. Chavan et al. [38] developed an electrically conducting composite polyaniline and an epoxy resin cured with maleic anhydride, and Jain et al. used FEP-polyaniline conducting composite for chlorine gas sensing [39] and humidity sensing [40]. Shukla et al. [41] reported a nanocomposite of polyaniline matrix loaded with crystalline tin oxide nano particles for humidity sensing. Sivaraman et al. [42] developed an all-solid supercapacitor using a sulfonated poly(ether ether ketone)-poly(3-methyl thiophene) conducting composite. While sulphonated PEEK is electrolytically conducing, poly(thiophene) with anionic doping is an electronically conducting material.

Dubey and co-workers [43] made a comprehensive review of electrically conducting polymer bio-nanocomposites for biomedical and other applications where various biopolymers and biodegradable polymers like PLA, polylactides, polycaprolactone, poly(lactide-co-glycoside), chitosan, cellulosic materials, gelatin, collagen, heparin, etc. are used with conducting polymers, fibres, carbon nanotubes, graphene and metal fillers for a wide range of applications such as humidity sensing, vapour sensing, water purification, chemical analysis, enzyme-sensing, agriculture, drug delivery, tissue engineering, etc. Other types of electrically conducting composites are made of common elastomers, engineering plastics and also thermosets that are normally insulators with conducting fillers such as carbonyl iron, copper, gold, silver, nickel and many alloys in micron size as well as with nanoparticles, graphene nanosheets, graphite particles, fibres and fabrics, various forms of nano carbons, carbon fibres, fabrics, etc.

The most common electrically conducting composites are rubber-filled with conducting particles such as conducing carbon to tailor the resistance from 104 to 109 Ω for electrostatic discharge (ESD) and antistatic mats, wrist bands, floorings, etc. Conducting carbon black generally has electrical volume resistivity in the range of 10^{-1}–10^{2} Ω cm. These composites are made with any commercial rubber, for example, neoprene, nitrile, SBR, PU NR, EPDM, etc. Polar rubbers have a definite advantage of better compatibility with metallic particles. Antistatic mats help to protect static-sensitive components from electrostatic discharge. They are engineered to drain static discharge from items placed on their surface. A secondary benefit of mats is that they serve to protect the surface of the ESD-sensitive devices from wear and tear. In applications on board aircrafts and vehicles, these elastomeric sheets also help in damping undue surface vibrations. ESD flooring or ESD table top mats have a low upper limit for electrical resistance, between 105 and 108 Ω. The composite mat prevents the build-up of electrostatic charge in the body through a safe and very controlled discharge of static electric charge. They also suppress sudden electrical discharge between electrically charged objects, on contact. Anti-static mats have a low electrical resistance, between 105 and 109 Ω. The use of anti-static matting prevents the build-up of electrostatic charge in the body by dissipating this charge and thus prevents a sudden discharge between electrically charged objects, on contact.

Highly conducting thermoplastic composites are made of engineering plastics with very high loading of conducting metal particles. These composites are very special because of the high degree of filling by conducting metal powders, even up to 50 volume %, still retaining the flexibility of a typical engineering plastic. A wide range of plastics, such as Nylon 6, Nylon 66, polypropylene (PP), polyphenylene sulphide (PPS), acrylonitrile-butadiene-styrene (ABS), etc., and thermoplastic elastomers can also be used to make such a high-conducting composite. Conducting fillers can be metallic powders or molten metals processed into the molten plastic by extrusion, followed by injection moulding or thermoforming. Depending on the plastic, the composites can also be machined. The laminate form of the composites can be used in calendaring process as an intermediate layer in a functionally graded laminate, where electrical and thermal conductivities are graded layer-wise. The surface conductivity may be as high as 103 S/cm. Depending on the plastic and the conducting filler loading, the conducting plastic composites can be co-extruded as cable sheathing for EMI-EMC qualification.

The highly-filled conducting composites (except carbon fibre composites) are inferior to common structural composites in terms of flexibility and weight penalty, and also suffer from low fracture toughness. Graphite flakes offer very good composite conductivity, but suffer from weak interface with most polymers. Recent developments on single-wall carbon nanotubes (SWCNT), multi-wall carbon nanotube (MWCNT) and graphene platelets have opened opportunities to solve these problems. Carbon nanotube is a graphite sheet rolled into a cylindrical shape, either as single roll or multiple concentric rolls, and has high electrical conductivity due to the large number of free electrons on the surface due to the many Π bonds. The electrical conductivity of single-walled CNTs (SWCNT) has been reported to be on the order of 102 to 106 S/cm and that of multi-walled CNTs (MWCNT) as 103 to 105 S/cm [44]; CNT fibres have conductivities that could vary in orders of magnitude, ranging from 10 to 67,000 S/cm [45].

Earp and others [46] reported enhancement of the electrical conductivity of CNT using copper and observed that 10% addition to CNT sheets had a good dispersion and enhanced the conductivity by four times, while 90% copper by weight (by chemical vapour deposition method) enhanced the conductivity by 560 times compared to pristine CNT. Graphene is made of either a single layer or few layers of graphite sheets, and has electrical conductivity of about 1.22×10^5 S/cm for nine layers (3 nm thickness) and 7.14×10^5 S/cm for single layer (0.33 nm thickness), while graphene nanosheets, consisting of about 300 layers (100 nm thickness), have conductivity of about 7×10^3 S/cm, as calculated by Fang et al. [47] and the values are in good agreement with literature data. These nano carbons have other advantages such as high mechanical reinforcing effect on the polymer matrix due to their large surface area to efficiently increase the dispersive force as well as due to the functional groups present on their surface by chemical modification or by a manufacturing process to increase the polar attraction.

Hu et al. [48] used CFRP laminates with functionalized interleaf technology (FIT). The electro-less copper-nickel-plated polyester veils were used as the interleaves in high conducting interlayers. The in-plane and through-thickness conductivities improved from 74 S/cm to about 1,080 S/cm and 1.5×10^{-3} S/cm to 5.3 S/cm, respectively. In addition, the authors claimed significant improvement in fracture toughness as a result of this interleaf arrangement in the laminate. Ma and co-workers [49] developed hybrid carbon fibre/carbon nanotube (2%)–epoxy composite for enhanced lightning strike protection and achieved a surface conductivity of about 234 S/cm.

Kim and co-workers [50] demonstrated 3D printing by a conductive composite of polylactic acid (PLA) and graphene nanoplates (GNP). The PLA solution was mixed with the exfoliated GNP suspension in an alcohol and extruded as filament, which was utilised for 3D printing. The authors reported that the conductivity of a composite with threshold concentration of 1.49% GNP (by volume) was about 1 mS/cm. Park et al. [51] modelled the electrical resistivity of polymer composites with segregated structures. The authors used silica particles in micron size and also nanoparticles in a carbon nanotube-filled polymer composite and carried out numerical characterizations by Monte Carlo simulation for the prediction of percolation behaviour.

Microwave propagation characteristics of conducting composites are very special due to the diverse requirements in the ever expanding commercial markets of electronic gadgets and also in defence. In common electronic gadgets and power systems, EMI-EMC qualification is very important to avoid accidents and malfunctioning of equipment. Microwave propagation through materials is very important for the development of radar stealth technology of defence targets. Brief discussions are included here on the type of composites used in these areas.

5.3.1 EMI shielding

Electromagnetic (EM) waves of wide frequency ranges are generated due to the current flow (electron flow) through a conductor and electrical circuits, which may be in multiple phases and intensities. These EM waves are conducted and radiated from electrical gadgets, power circuits, high power cables, communication devices, etc. and interfere with the other electronic or conducting objects. In addition, the interference of EM wave adversely affects human health. Therefore, to protect the electronic items from such EM waves, the item is covered with an EM shield, which could simply be a reflecting surface. It is well known that metallic objects perfectly shield these waves by almost total reflection. Some shields are also made to absorb the EM waves by dielectric, magnetic or combined loss, but these are for selected frequency bands only and are not frequency-independent.

Despite good shielding effect, there are few disadvantages such as high weight of the shield, manufacturing difficulties, non-flexibility, etc. The development of electrically conducting polymers and EM-absorbing dielectrically and magnetically lossy ma-

terials like conducting polymers, CNT, graphene, carbon fibres, barium hexagonal ferrites, etc., it is possible to use polymer matrix composites for EMI shielding to obtain
an almost similar efficiency. The advantages are that these composites can be designed with a lower weight penalty, can be easily manufactured, can have flexible
sheet form and can offer maximum shielding by EM wave absorption, so that the EM
waves are reflected back less to the environment. A wide choice of electrically conducting and EM-absorbing materials as well as the type of polymer makes it a very
attractive subject for lightweight EMI shields.

For a metallic reflecting shield, the reflection takes place due to the interaction
with innumerable free electrons on the surface of the metal. For an absorbing shield,
the EM wave needs to enter the shield material and it can then be attenuated by dielectric or magnetic or combined-loss mechanism. Therefore, the EM impedance of
the shield material should be as close as possible to the free-space EM impedance of
377 ohms. All polymers have complex dielectric properties (complex permittivity) in
EM frequency ranges and hence the dielectric loss occurs in a polymer upon interaction with an incoming EM wave. The loss is inversely proportional to the frequency.
Some materials such as ferrites show complex permittivity as well as complex permeability (magnetic property), both depending on the EM frequency.

The EMI shielding effect is primarily quantified by the loss in EM wave transmission through a medium, and is termed as Insertion Loss (IL) or Transmission Loss
(TL). The propagation of EM wave in a medium can have three types of consequences,
transmission through the medium, reflection from the surface of the medium and absorption in the medium. The transmittance, reflectance and absorbance in terms of
power are given by

$$T = (V_T/V_I)2 \tag{5.2}$$

$$R = (V_R/V_I)2 \tag{5.3}$$

$$A = 1 - T - R \tag{5.4}$$

where V represents voltage and I, T, R and A represent incident, transmission, reflection and absorption.

EM shielding effectiveness is the transmission loss (the energy not transmitted
through the shield) due to the insertion of the shield. If the transmitted signal is V_T and
the incident signal is V_0, then the signal loss is $V_0 - V_T$. Therefore, the SE in EM propagation is given by

$$SE = 20 \log [V_T/ (V_0 - V_T)] \tag{5.5}$$

In a situation of measured reflection and transmission, the SE can have three components, the reflection loss (SE_R), the absorption loss (SE_A) and any loss due to multiple
reflections (SE_M) between the back surface and the front. Hence, SE can be given by

$$SE = SE_R + SE_A + SE_M \tag{5.6}$$

As is obvious, while the transmitted and the reflected signals can be experimentally measured, the absorption term (second in the right-hand of eq. (5.4)) can only be calculated, and not measured. The equation is valid for a semi-infinite medium having a large lateral dimension that is enough to neglect the diffracted/scattered waves (not measurable), and an infinite thickness.

To calculate the shielding effectiveness (SE) by measuring the transmission and reflection signals, we need to use the eqs. (5.2)–(5.4) and take the power ratio in terms of decibels (dB) as follows [52]:

$$\text{Transmitted power (dB)} = 10 \log(T) \tag{5.7}$$

$$\text{Reflected power (dB)} = 10 \log(R) \tag{5.8}$$

$$SE = 10 \log[(T/ \ (1-R)] \tag{5.9}$$

As example, a carbon fibre-epoxy CFRP is made as a flat panel of 1 m × 1 m × 6 mm thickness for a radar-reflecting composite structure (RRCS). The panel was evaluated for EM shielding efficiency in the frequency range of 2–20 GHz, covering the S-band, the X-band, the Ku-band and partly the K-band. The frequency-dependent transmission loss is shown in Fig. 5.8.

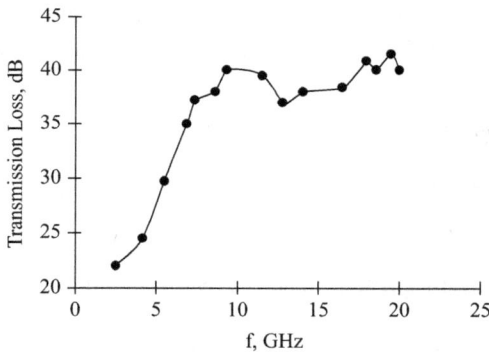

Fig. 5.8: EMI shielding by a radar-reflecting CFRP panel in the wide-frequency band.

The transmission loss presented in the figure is the EMI shielding efficiency, expressed as the ratio of the incident to the transmitted power in dB. Since the panel is a reflecting type, the EM shielding is the same as the loss in transmission since the absorbed power is negligible.

A second example is a panel made of a hexagonal barium ferrite and fine nickel powder in a glass fibre-reinforced FRP backed by a thin CFRP layer (reflecting backing) as a radar-absorbing composite structure (RACS) and was evaluated for radar transmission, reflection and absorption to account for shielding efficacy in the S-band and X-band (2–12 GHz). In this example, all the transmitted and reflected voltage ratios are arbitrarily selected, and do not represent actual data. Figure 5.9 shows the transmittance and reflectance as voltage ratios for the composite in the L-band and the X-band. Figure 5.10 shows the transmitted power and shielding effectiveness in the L- and X-bands by the same composite, calculated using eqs. (5.7) and (5.9). The negative sign indicates reduction in power (dB). The reflectance is also small and the reflected power (not shown in Fig. 5.10) would have similar reduction, as can be easily calculated using eq. (5.8).

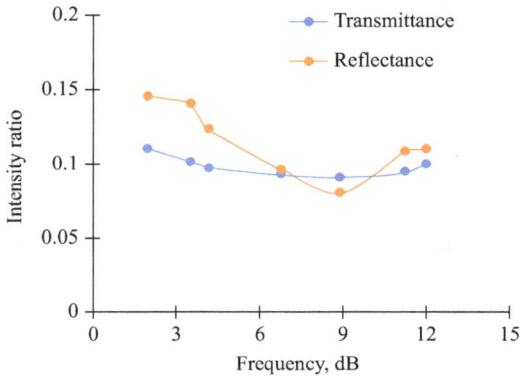

Fig. 5.9: Transmittance and reflectance of an absorbing composite.

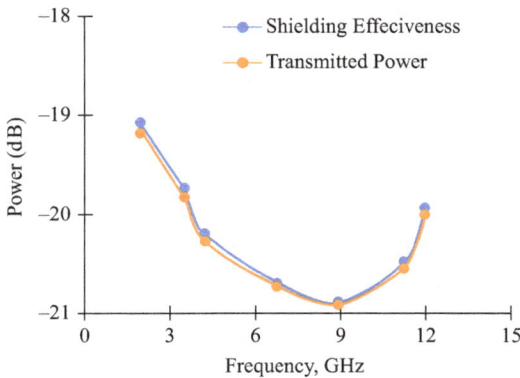

Fig. 5.10: Transmission and shielding effectiveness of a composite-based EMI shielding material.

In any conductor, the EM wave attenuates progressively as it travels in the medium, and the thickness at which the wave intensity is reduced by a factor of $1/e$ (=0.368) is called the "skin-depth" of that medium. Consequently, at the skin depth, the attenuation of power will be −20 [log (0.368)] = 8.68 dB. Skin depth is an absorption component and is given by [53]

$$\delta = \frac{1}{\omega} \left[\frac{\mu \varepsilon}{2} \left(\sqrt{1 + \left(\frac{\sigma}{\varepsilon \omega} \right)^2} - 1 \right) \right]^{-1/2} \tag{5.10}$$

where σ is the conductivity of the material, μ and ε are the permeability and permittivity, respectively, and ω is the angular frequency (rad/s) of the EM wave. Using these properties, it can be shown that metals have very low skin depth, for example for copper, $\delta = 0.634$ μm (microns) [53]. Therefore, a 300 μm thin copper sheet or leaf or even a fine mesh is enough to provide good EMI shielding.

The skin depth varies with frequency as given in the equation and the loss properties also vary with frequency. Therefore, to predict the absorption of EM wave in a composite EMI shield, the basic complex properties are to be investigated in terms of dependence in the desired range of frequency to find out the minimum thickness required for the desired attenuation.

The permeability and the permittivity is complex for lossy materials such as ferrites, and conducting polymers and alike, and hence for polymer matrix composites. These complex permeability and permittivity are represented by the real part and the loss part, which contains the i-operators:

$$\mu^* = \mu' + i\mu'' \tag{5.11}$$

$$\varepsilon^* = \varepsilon' + i\varepsilon'' \tag{5.12}$$

where the μ' and ε' are the real part, while μ'' and ε'' are the imaginary part, accounting for the energy loss. The ratio of the imaginary part to the real part is the characteristic loss factor in the complex magnetic and dielectric constants. In purely dielectric materials like polymers, magnetic loss is not applicable, therefore, the shielding effect with only dielectric absorbers are not as effective as with ferrites, which provide both types of loss in the EM field.

There are innumerable possibilities of using EMI shields with a wide range of electromagnetic shielding effectiveness (SE) at various frequency bands, from MHz to GHz levels, as reported in various literatures [54, 55]. The flexible rubbery composites for EMI shielding [56, 57] is very attractive because of ease of manufacturing, application by adhesive bonding in difficult contours and maintenance/replacement, which are not so easy for hard engineering plastics, thermoset composites and FRPs. As the industry of electronic gadgets is fast growing, flexible sheets for EMI shielding are more in demand. The advantage of rubber is the large void volume to accommodate the high filler loading as against graphite and other conducting fillers, which are in micron size. However, recent developments on nano-metallic fillers, carbon nano particles and graphene nanosheets resulted in more efficient conducting composites and EMI shields at very low loading in polymers. In this respect, graphene nanosheets, which are basically few layers of reduced graphene oxide (rGO), have proven to be the most promising conducting fillers for light-weight, flexible composites. They require quite a small proportion of the rubber matrix but the conductivity and the shielding effectiveness are very encouraging. Pure form of graphene (without any oxide) shows very good results in this respect. For example, Lu et al. made an EMI shield with flexible EPDM-graphene nanoplate composite using a cost effective method of mixing, ultrasonication and compression, wherein 8% GNP by weight was used and the EMI SE was 33–35 dB, covering a wide frequency range of 8–18 GHz. The shield was a 0.3 mm thick sheet and the percolation threshold for electrical conductivity was seen to be near 7% GNP loading and relatively constant beyond that. The mechanical strength also enhanced by five times on the addition of 8% GNP.

Conducting polymers are also potential materials for the development of EMI shields, but have to be augmented for the required conductivity and microwave reflectivity/absorption. Various conducting materials have been used with conducting polymers, especially PANI and PPy, for example, rGO, MWCNT, Fe_3O_4, barium ferrite, nanosilver, etc. [58]. The nanocomposites of conducting polymers with these particles are mostly synthesized or mixed in situ so that different specific morphologies are obtained, like the core-shell morphology with ferrite-SiO_2-PPy nanocomposite [59], MWCNT/PANI/graphene yields composite nanotubes by ball-milling [60] and epoxy/Fe_3O_4/PPy as river-like morphology by chemical polymerization [61].

All the above EMI shield materials mostly showed good shielding effect at X (8–12 GHz) and Ku bands (12–18 GHz) and very few are useful for low frequencies such as L-band (1–2 GHz) and S band (2–4 GHz); the corresponding wavelength ranges are 15–30 cm and 8–15 cm, respectively. Even for a quarter-wavelength cancellation method (destructive interference), the thickness of the panel required is quite high, minimum 2 cm (20 mm), which is considered as a very high thickness for a shield. In addition, even for an electromagnetically active composite, it can be appreciated following the eq. (5.10) on skin depth; the skin depth is inversely proportional to the frequency. However, among various developments, Sun et al. [62] have reported a PANI/CuS composite synthesized by the in situ polymerization method, which resulted in a flower-like morphology. The CuS particles were observed to be coated with PANI. The composite showed excellent EMI shielding properties below −18 dB in the range of 300 kHz to 3 GHz and at a thickness of only 3 mm. The authors claimed to obtain an EMI shielding of −45 dB at 2.78 GHz. This type of composites can be practically applied as a wide-band shield, especially the high shielding effectiveness at the low thickness of only 3 mm for such low frequency is remarkable.

In many cases of stationary objects, multilayer panels of conducting/absorbing/dielectric sheets of polymeric composites are used where the multiple reflections within the different layers due to large variation in electrical conductivity aids in high EM SE. This is particularly important where a structural element has to be made to replace the metallic structure, for example, an FRP structural unit with considerable thickness. Typical example is the large radar installations. A typical layered FRP composite consists of alternate CFRP and GRP, arranged in 3 or more layers, and a sketch of 4-layer shield is shown as Fig. 5.11, for which the estimated EM SE is about 17–24 dB in X-Band. Microwave propagation through the layers is adjusted by the complex permeability and permittivity of the basic constituent materials, layer thicknesses, type of chopped carbon fibres, and their weight fraction in the vinyl ester resin.

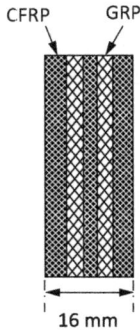

Fig. 5.11: A typical multilayer EMI shielding made up of vinyl ester thermoset composite.

In a similar principle, but different design, Liang et al. [63] developed an anisotropic FRP shielding panel, fabricated as flat, corrugated and corrugated sandwich chopped short carbon fibre (SCF) felt in epoxy resin. The shielding effectiveness changes significantly (38 dB to17 dB) with the polarization of the field as well as by the angle of the corrugated felt. Corrugated sandwich SCF felt composite consisting of two flat and one corrugated felt has excellent EMI SE (>70 dB) in X band, which results from multiple reflection in corrugated core.

Maiti and others [64] developed a novel technique that involves in situ polymerization of styrene/multiwalled carbon nanotubes (MWCNTs) in the presence of suspension polymerized polystyrene (PS)/graphite nanoplate (GNP) microbeads. The composite with only 1.5–2% MWCNT and GNP was capable of shielding effectivity of –20 dB in the X-Band (8.2–12.4 GHz). The nano size of the MWCNT/GNP may provide a larger interfacial area; therefore, the number of conductive interconnected network structures of CNT-GNP increases. Also, the high aspect ratio (L/D) of CNTs and GNP helps to create extensively continuous network structures that facilitate electron transport in the nanocomposites with very low fillers loading, maximum 2%. Investigation of the microwave properties of composites for EMI shielding is very important as the same principle, with the exception of reflection type shield, is used for radar stealth material development. All the fundamental properties such as complex permeability, permittivity, skin depth, etc. are the basis for the development of radar stealth materials.

5.3.2 Microwave-absorbing composites

Absorption of microwaves in the defence radar frequency bands is very important for providing stealth feature for defence targets, stationary or moving. Radar pulses are used to detect even a very small object, depending on the frequency of the radar wave. Pulses are used to isolate the incident and the reflected signals (volts, in terms of electric field), else, with continuous waves as the incident and the reflected, there will be a standing wave and will result in a complicated method of isolating the individual signal strengths. The radar reflections are captured either by monostatic or bi-static radars. Radar detection is thus always an active method, and this requires certain measures to reduce the reflection by add-on absorbers, shaping the surface, scattering of the wave by special design or by creating an envelope of a plasma to change the frequency band of the incoming wave. Out of these, the shaping of the surface is the most effective, but cannot be practically used in many situations. Add-on coatings are the next best solution. During the period immediately after the World War II, a large number of research articles were published on magnetic resonance of the different materials such as magnetic oxides, spinel (soft) and hard ferrites, alloys of metals such as iron, zinc, nickel, cobalt, strontium, manganese, etc. and the barium hexagonal ferrites with substitution of iron by these metals. Polymer-based composites came into real application in the early 1980s when special characterisation of small particles, cluster science, new polymer processing methods involving nano technology started gaining interest. The replacement of heavier fillers, particularly iron and ferrites, were effectively done by nano particles of carbon and single nanosheet of graphite, known as graphene. The result of using such tiny particles were dramatic improvement in electromagnetic, mechanical, thermal and fracture toughness properties, together with substantial weight and cost reduction. The quantity requirement of such nano fillers are so low that it is now possible to develop a small, stealth unmanned aircraft with acceptable weight penalty and almost similar strength as unfilled FRP. However, the stealth feature is specific to one or maximum two frequency bands, and cannot be for a wide band of an entire radar range (2–40 GHz). The problem of lower frequency absorption is the high thickness requirement, which cannot be accommodated in the case of criticality in dimension and mass.

Electromagnetic energy attenuates in a medium due to loss by both magnetic and dielectric losses, and numerically determined from complex permeability and complex permittivity. Moreover, the wave intensity reduces in the path of propagation in a medium due to multiple reflection and scattering. The attenuation of a wave in a medium while propagating inside it is defined by the attenuation constant (a), which determines the dissipation properties of materials, and can be expressed by

$$a = \frac{\sqrt{2}\pi f}{c} \sqrt{(\mu''\varepsilon'' - \mu'\varepsilon') + \sqrt{(\mu''\varepsilon'' - \mu'\varepsilon')^2 + (\varepsilon'\mu'' - \varepsilon''\mu')^2}} \tag{5.13}$$

The dielectric and magnetic complex properties are frequency-dependent terms. The MW energy loss is the highest at the corresponding resonance frequency. The magnetic resonance frequency of a material depends on the saturation magnetization (numerical value). By suitable tailoring of the magnetic molecule, for example by substitution of iron with some transition metals in a barium hexagonal ferrite, and also by changing the particle size, the resonance frequency can be tailored to the selected electromagnetic wave frequency. Consequently, the material would show maximum magnetic loss in that region of frequency. In a dielectric medium, the dipole relaxation causes the loss, and here too, the complex dielectric permittivity depends on the applied electric filed. The dielectric loss attains a maximum when the relaxation time coincides with the time period of the applied EM field. Therefore, a dielectric or magnetic material would show lossy property during the propagation of an electromagnetic wave through the material. The free-space impedance is 377 ohms, and the material must have similar impedance for the wave to propagate through the medium. Higher the difference, more will be the reflection and lesser will be transmission. This is the reason why metals and highly conducting composites provide EMI shielding only by reflection, as their impedance is much lower than that of the space.

The detectability of a target is measured in terms of the radar cross section (RSC). The RCS is a property of the target's size, shape and the material, and is a ratio of the incident and the reflected power. The equation for the reflection power ratio (with incident power) at the receiver (which is also the MW transmitter) is given by [65]

$$\frac{P_R}{P_T} = \frac{G_T G_R \lambda^2 \sigma}{(4\pi)^3 R_R^2 R_T^2 L} \tag{5.14}$$

where P_R is the received power in watts, P_T is the peak transmit power in watts, G_T is the transmitter gain in decibels, G_R is the receiver gain in decibels, λ is the radar-operating frequency wavelength in meters, σ is target's non-fluctuating radar cross section in square meters, L is the general loss factor in decibels that accounts for both system and propagation loss, R_T is the range from the transmitter to the target and R_R is the range from the receiver to the target; if the radar is monostatic, the transmitter and receiver ranges are identical.

$$\text{when } R_T = R_R, \text{ then } \frac{P_R}{P_T} = C\left(\frac{\sigma}{R}\right)^4 \tag{5.15}$$

The radar cross section (σ) of a flat panel with perfect reflection in perpendicular incidence is given by

$$\sigma = 4\Pi A2/\lambda2 \tag{5.16}$$

where λ is the wavelength of the MW and A is the area of the panel. For 1 m^2 panel area, the RCS of a reflector is 139.62 m^2 at 10 GHz.

A plot of the fractional range vs the reflection power loss in dB is shown as Fig. 5.12. The plot is useful to predict the possible range reduction of the radar-absorbing composites from a lab evaluation of reflection data.

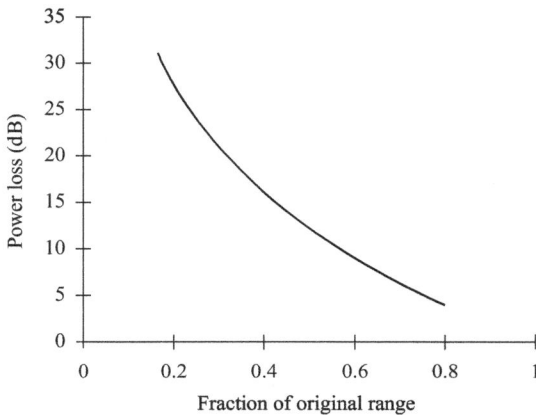

Fig. 5.12: Fractional detection range and the corresponding reflection power loss (dB) of the radar-absorber type target (refer eq. (5.14)).

From the figure, it can be seen that the detection range is reduced to 56% of the original range when the reflection loss is 10 dB. A simple calculation reveals that a 10 dB return loss corresponds to a 70% reduction in the reflection coefficient (intensity ratio) and a reduction of power by 90%. Intensity and power are taken in volt and watt, respectively. Therefore, the absorber reduces the detection range of the detection radar according to the reflected intensity.

In the case of polymeric and carbon-loaded dielectric lossy composites, the thickness of the composite can be determined by the following equation:

$$t_{comp} = \frac{\lambda_0}{4\sqrt{\varepsilon'_{filler}}} \tag{5.17}$$

The composite RAS should be backed by a perfect reflector (metal or CFRP) having a thickness more than its skin depth. For example, the skin depth of aluminium is 0.85 microns and the backing sheet made out of aluminium should be several multiples of the skin depth, say 100 microns.

The reflection and transmission parameters are determined by the S-parameters calculated from the voltage signals of the vector network analyser (VNA) in MW frequencies:

$$S_{ij} = \frac{V_i^-}{V_j^+}\bigg|_{V_k^+=0 \text{ for } k \neq j} \qquad (5.18)$$

Here, j is the input port of the wave and i is the port of the return wave. Therefore, there are four S-parameters, which are fractions (dimensionless) calculated from the VNA signals of port 1 and port 2 as S_{11}, S_{12}, S_{21} and S_{22} so that:

S_{11}: Reflection loss (fraction) out of port 1

S_{12}: Insertion loss (transmission loss) from port 2 to port 1

S_{21} Insertion loss from port 1 to port 2; for uniform transmission line in VNA, $S_{21} = S_{12}$

$S_{22} = S_{11}$ for uniform transmission line in VNA

Consequently, the reflection loss as a power ratio, expressed in dB (numerical value), is calculated as

$$RL = 20 \log(S_{11}) dB \qquad (5.19)$$

The absorbance is calculated as

$$A_a = \left(1 - 10^{RL/10}\right) \times 100\% \qquad (5.20)$$

To design an efficient radar-absorbing composite, both the magnetic and dielectric losses should be provided by a magnetic and a dielectric material. A very common example of magnetic lossy material is barium hexagonal ferrite and its many substituted versions, where strontium, cobalt, nickel, etc. can be used to tailor the saturation magnetization value in such a way as to result in a magnetic resonance in the desired range of radar frequency. The dielectric loss is offered by carbon particles, nano carbons, graphite, graphene, conducting metal powders, conducting polymers (PANI, PPy), etc.

The process of making the composite RASH sheet is usual rubber processing as shown in Fig. 5.13 by a block diagram. The mastication of rubber and mixing of the ingredients are done either by two-roll mill or an internal mixer, in a controlled temperature of maximum70 °C, followed by sheeting out as a thin compounded sheet and vulcanization in a mould using high pressure and about 150 –170 °C temperature.

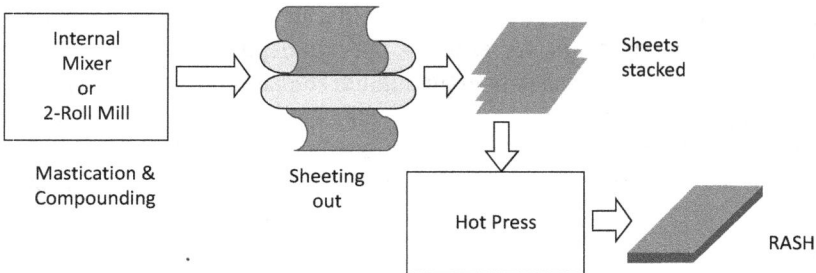

Fig. 5.13: Flexible RASH preparation using the rubber processing method.

The flow diagram for the fabrication of RACS is shown as Fig. 5.14 (a). The liquid resin (oligomer such as epoxy resin, unsaturated polyester, vinyl ester, polyol of polyurethane, etc.) are mixed with the microwave active particle fillers in a high-speed mixer as a pre-mix, followed by high shearing in a triple-roll mill, with a very narrow gap between the feeding roll and the adjacent roll; the mixing is continued till a homogeneous mass is obtained with the desired fineness (mark 7 in Hegman Gauge). For nano particles, the concentrations are very low, but the particles form agglomerates, which are difficult to break. Ultrasonication is one way, but requires a solvent because the resins are quite viscous. In a different method, the solvent-suspended nano particles are sprayed on the pre-forms (reinforcing fibre) as shown in Fig. 5.14 (b) and subsequently dried, followed by vacuum-assisted liquid resin infusion and high-pressure autoclave moulding.

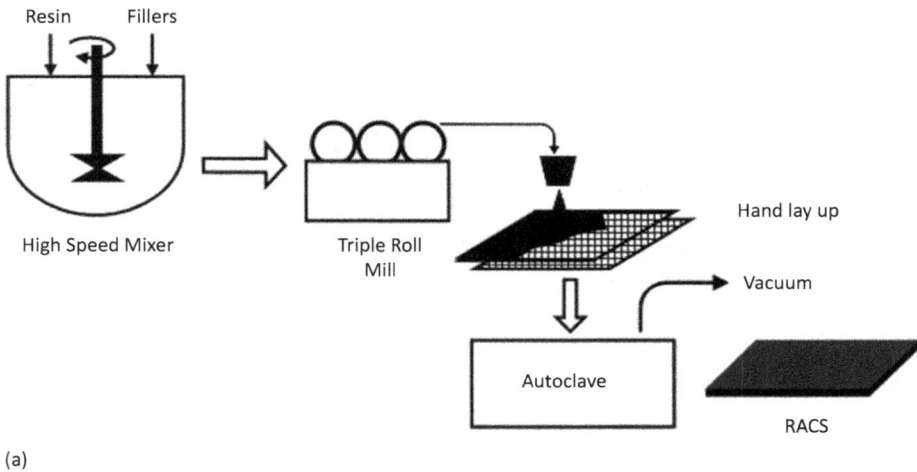

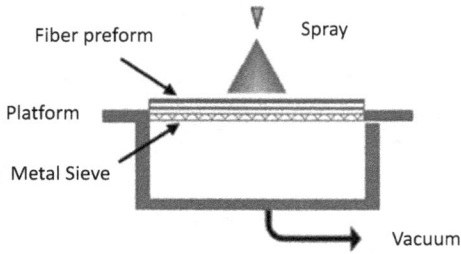

Fig. 5.14: Flow diagram of the fabrication of RACS: (a) pre-mixing method and (b) spray of nanoparticle suspension on fibre pre-forms.

Triple-roll milling is the next alternative, especially for large operations, since removal of the solvent is expensive and never perfect.

The processes described here can be scaled up from the lab process to a typical medium-to-large-scale process, where the manufacturing in bulk has to be efficient, without significant material loss and should also be cost effective with the desired homogeneity, optimum structural strength and functional properties. In the case of nano particles, the pre-mixing is done in a special machine, employing ultrasonic vibration with high-speed stirring. The operation has to be in a closed vessel with a closed batch feed and has to be isolated from the subsequent line of equipment. Protective gears for personnel are also recommended for handling nano fillers.

A typical barium hexaferrite powder of 2–8 µm size with a ferromagnetic resonance in the radar-frequency rang was used along with a very fine metal powder (1–5 µm size) to make radar-absorbing FRP structures and flexible sheets for different frequency bands, only to demonstrate that the same lossy material can be used for various bands by changing the thickness. Lower frequencies require higher thicknesses as also seen from eq. (5.10) of skin depth. The RASH sheets are of homogeneous composition without reflecting backing since these are applied on steel and aluminium structures. The FRP RACS composites are made of a layered structure of the composite with a special sequence of magnetic and dielectric layers. The composites are structural elements and are therefore backed by a CFRP layer of 1 mm, which is a perfect MW reflector. The thickness of the RASH for a minimum 10 dB return loss was determined from series of experiments, and it was observed that 6.5 mm thickness was required for 2–4 GHz, while 1.5 mm thickness was required for 12–18 GHz. For the RACS composite, for 2–4 GHz, the thickness required was 7 mm, for 8–12 GHz, 3 mm and for 12–18 GHz, 2.4 mm. The thicknesses do not suggest destructive interference due to quarter-wavelength ($\mu/4$) since the quarter-wavelength at say, 4 GHz is 18.75 mm, which is quite different. The velocity of the MW reduces as a result of the loss components of the magnetic and dielectric complex parameters and hence, the wavelength inside the lossy composite reduces, since the frequency remains unaltered.

The radar wave attenuation due to the absorption of microwaves by lossy materials is highly frequency dependent, with narrow bandwidth. In order to design a wide-band absorber, graded layers of dielectric and lossy materials are used, but with higher thickness if the frequency band has to extend to lower frequencies (<4 GHz). The various parameters that are required to be optimized for the development of broad-band absorber are the number of layers, the thickness of each layer and the loading of active material in each layer. The conducting backing of such flexible radar-absorbing sheet is not required since these are to be used as add-on absorbers on metallic structures. However, such RASHs because of their high density are most suitable for ground installation or ship superstructure but not suitable for most parts of an aircraft.

Atay [66] developed a microwave-absorbing composite with polyurethane resin filled with different contents of barium hexaferrite ($BaFe_{12}O_{19}$) and copper powder to observe the effect of filler on microwave properties. The ferrite was made by the sol–gel method. The particle sizes of both the fillers were few microns. The absorber sheet was 3 mm thick. The MW absorbance was determined in X-band (8–12 GHz) using a vector network analyser. The author reported % absorbance varying from 11% to about 17%, depending on the filler composition and the loading in the PU matrix.

Zhang et al. [67] designed a hybrid system of a frequency-selective surface (FSS) made of aluminium square sheets and hole perforation embedding in magnetic absorbing sheets to obtain a broad-band magnetic radar-absorbing polymer composite. The magnetic particles are Fe Co alloy filled in the polymer matrix. The magnetic composite layer was about 2.4 mm thick. The authors evaluated the microwave properties and the reflection loss in relation to the size of the hole, the position of the FSS layer and the width of the FSS. The result showed that all these parameters substantially change the reflection coefficient (in dB) in the frequency range of 2–18 GHz and an optimized design could offer a reflection loss of 10 dB or more in a wide range in this frequency scale.

Kim and Kim [68] designed a two-layer absorber composite using carbon fibre in epoxy resin for the CFRP and carbon black-filled silicone rubber as RASH. The carbon black-filled silicone rubber sheet was the front absorber layer, backed by the conducting reflector layer of CFRP composite. The complex permeability and permittivity of the layers were determined using a vector network analyser in the 4–12 GHz range. The authors observed that the performance of the absorber layer was influenced by the backing conducting layer. With increase in carbon black loading, the loss part of the permittivity of the rubber absorber increased, as expected. The MW reflection loss was distinctly varied with the thickness of the rubber-absorbing layer and as usual, the peak loss shifts to higher frequency for lower thicknesses. Considering all variables, a 2 mm thick absorber layer was almost the optimum one, with > 5 dB attenuation in the entire frequency range studied.

FRP composites with hard ferrites and metal powders have a distinct structural disadvantage of reduced fracture toughness and impact energy due to heavy loading of the filler to provide desired MW absorption. Table 5.1 shows typical flexural and impact properties of an unfilled FRP and a RACS using epoxy with E-glass or Kevlar and carbon fibre reinforcements, which were made as a multilayer panel with alternate layers of a dielectric and magnetic compositions. A 300-micron CFRP layer was used only as conducting backing. The flexural modulus increased since the hard particles impart rigidness, while the impact energy reduced by almost 40%. Kevlar fibre-based FRPs showed about 1.6 times higher impact strength compared to glass fibre FRP, and Kartal and Demirer [69] reported the same to be 1.75 times.

The effect of lower impact strength due to high loading of hard fillers is to reduce the damage tolerance in case of impact or shock, which is considered a very important aspect of transportation, airborne and military objects. The externally impacted force can change the MW absorption properties of an RACS.

Tab. 5.1: Mechanical properties of an FRP and its absorber version.

Composite	Thickness, mm	Density, kg/m3	Flexural strength, MPa	Flexural modulus, GPa	Izod impact energy, J/m
GFRP front with CFRP backing	3	1,720–1,760	320–375	22–24	1,270–1,310
RACS with GFRPCFRP backing	3.8	2,000–2,040	430–470	29–35	680–730
RACS with Kevlar and CFRP backing	3.8	1,650–1,680	178–185	17.5–22	1,100–1,150

Go et al. [70] had studied the damage tolerance of an RACS made of epoxy glass FRP with periodic pattern surface (PPS) using the low velocity impact test instrument and correlated the extent of damage with change in MW absorption. The PPS is electronically printed pattern on a polyimide film. The material of printing had specific inductance and capacitance to ultimately provide MW absorption depending on the thickness and complex permittivity of the spacer material. The reduction in MW reflection loss was approximately 18% on an impact of 45 J and the dent created on the RACS was about 0.15 mm.

Lighter absorbers are made with composites of conducting polymers with rubbers for flexible RASH or other thermosets with fibre reinforcements as RACS. MW performance of a flexible composite of polyurethane-PANI doped with phosphoric acid was determined at various loading. The best was a 2-mm-thick sheet with 65% (by weight) loading of the doped PANI as a pigment in PU. Figure 5.15 shows the reflection loss in X-band when the PU composite is backed by a 3-mm-thick mild steel panel.

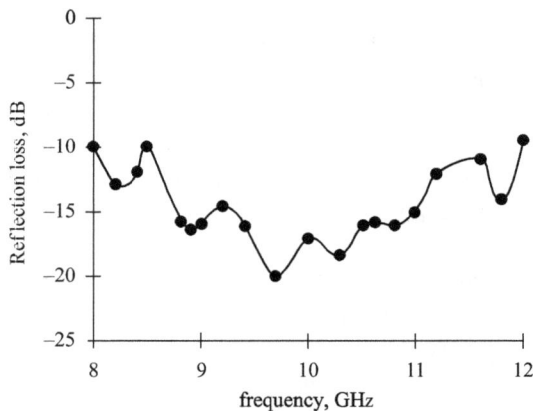

Fig. 5.15: MW reflection loss of a PU-conducing polymer composite flexible sheet.

Peymanfar and co-workers [71] developed a lightweight polyurethane foam-doped PANI composite for radar absorption in X- and Ku-Band. The authors prepared nano-structured doped PANI by sonochemical method. The PANI nanoparticles were mixed with the polyol and the PUF was made. The authors claimed as high as 85 dB attenuation at 10.2 GHz for a 4 mm-thick foam absorber and 56 dB at 13.8 GHz for a 2.8-mm thickness. Overall, the absorber showed >10 dB at 8–18 GHz for 3–5 mm thickness. The high attenuation by this absorber was attributed to the method of synthesizing the PUF-PANI nanocomposite.

Oyharcabal et al. [72] studied the effect of morphology of PANI on microwave absorption efficiency of epoxy-PANI composites. Three types of PANI were synthesized: globular (300-nm diameter), nanofibers and flakes type, having diameter/thickness in 150–300 nanometres and length of a few microns. The epoxy composite was prepared using an anhydride hardener for epoxy curing. Electrical conductivity was maximum (few S/cm) for the globular-shaped PANI. However, the percolation threshold was lowest for PANI flakes, slightly higher for the nanofibres, but the globular PANI showed a threshold at double the vol% compared to the nanofibres. Maximum reflection loss was shown by the epoxy-PANI flakes (by weight) at 8.8 GHz as 37 dB, whereas, the peak values for globular and fibre-based composites with identical loading of PANI were much lower, 7 and 8 dB, respectively, at the same MW frequency. The thickness requirement was also lowest for the flake-type PANI (<3 mm compared to about 3.2–3.3 mm for the other two types.

Jang et al. [73] evaluated a semi-cylindrical shaped epoxy-glass fibre FRP absorber, where a conducting polymer was used as an absorbing material. In order to avoid difficulties in handling and processing nanoparticles in RAS, the authors used circuit analogue (CA) sheet using conducting polymer materials to develop an easier method of applicability and also to establish absorbing efficiency in X-band at a thickness of about 3 mm total. The MW absorption was measured in terms of radar cross section in dBsm in comparison to a 500-micron-thick reflecting panel of carbon fibre-epoxy composite. It was shown that the radar cross section reduced significantly due to the special design of the CA in the FRP composite by a thin sheet of the conducting polymer (conductivity: 1,300 S/m). The corresponding reflection loss was minimum 65% (9 dB) and maximum 98.8% (20 dB).

Thinner absorbers made of polymer-based nanocomposites as flexible radar-absorbing sheets (RASHs) and radar-absorbing composite structures (RACSs) with nanofillers in the matrix are being studied extensively. The nanoparticles of different types of conducting carbons are multiwall and single wall carbon nanotubes (SWCNT and MWCNT), nanofibres and nanorods of carbon, exfoliated graphite, and reduced graphene oxide (rGO), which is as good as pristine graphene nanosheet. Apart from these, nanoparticles of some highly conducting metals like nanosilver are also useful since these are required in small quantities. A hybrid dielectric lossy material could be made with a conducting polymer synthesized in presence of a nanosilver or carbon nanotube. The other advantage of the nanoparticles is retention of mechanical property to almost unfilled material, and in case of reinforcing nanoparticles, the mechanical strength and gas barrier properties are also substantially enhanced without significant weight penalty. In printing of the conducting / dielectric lossy layer in circuit analogue and FSS technology, these nanoparticles are very appropriate as filled polymer ink, and can be tailored to very high MW absorption. There are, however, some disadvantages such as health hazard on handling in large quantity, difficulty in accuracy of measuring the quantity, loss of materials, particularly carbon-based nano particles and relatively high cost compared to micron size particles. Shivaraman and co-workers used exfoliated MWCNT and polymerized aniline [74] in its presence to obtain an MWCNT-PANI conducting nanocomposite material, and similarly, polymerised 3-methyl thiophene on MWCNT [75].

Hu et al. [76] prepared a simple hot pressed polycaprolactam (Nylon 6) with carbon black and glass fibre. The microwave absorption and mechanical properties of the composites were investigated. 8 wt% CB/GF/PA6 composite had a minimum reflection loss (RL) of 40.05 dB and the bandwidth of RL about 10 dB is 4.6 GHz. The microstructure of the composites was tailored by disc granulation method to obtain reticular microstructure composed of different particle sizes to improve the microwave absorption and mechanical properties of the composites. RL of 7 wt% granulated CB/GF/PA6 composite was increased to about −56 dB.

Choi et al. [77] studied E-glass/epoxy nanocomposites dispersed with carbonaceous conductive nanoparticles such as 3% carbon black (CB) and 0.3%, 2.5% and 3% carbon nanotubes (CNT) to develop an optimum absorbing composition. The authors measured the dielectric properties in the X-band(8.2 to 12.4 GHz) and the optimum design method was applied to select the nanocomposite candidates with high EM wave absorption, taking maxima at the centre frequency (10 GHz) of the X-band. The process of manufacturing the FRP RAS was by high-speed mixing, followed by triple roll mill operation to form thick paste-like consistency and subsequently vacuum bagging and compression moulding in an autoclave under a predetermined temperature programme for curing cycle. The absorbance in X-band was above 90% for all compositions, with maximum of 98% at the centre frequency (about 10 GHz).

Graphene nanosheets are potential microwave lossy material used in polymer matrix. The introduction of graphene in MW absorber laminate design is relatively new, and the research accelerated only after 2010. FRP-based RACS with graphene-loaded resins are the addition in radar stealth technology. The advantages of structure of graphene and residual groups present in rGO facilitate MW absorption capability of graphene better than carbon nanotubes. The residual groups in reduced graphene nanoplatelets enhance the polymer-filler interface and relaxational energy loss more than other forms of carbon nanoparticles. Moreover, plate-type morphology of graphene imparts some more advantages such as better dissipation of heat, mechanical strength and better thermal stability. Reduced graphene oxide (rGO) nanocomposites of PEO [78] and PVA [79] showed very high MW due to the better interface, and hence better distribution, which led to high complex permittivity and consequently high MW absorption due to high dielectric loss. MW absorption of graphene/polymer composites is due to the combined effects of conductive loss, polarization, interfacial scattering and multiple scattering.

Meng et al. [80] reviewed the MW-absorbing composites with graphene nano-plates. The authors discussed the mechanism of electrical and dielectric properties of graphene nanocomposites in relation to the molecular structure, presence of residual groups such as oxygen in rGO, energy levels and morphology. It has been shown that a 2.1 mm-thick composite containing only 7 wt% ultrathin graphene nanosheets can show 35 dB reflection loss at 12 GHz frequency, and for the entire X-band, the composite showed >7 dB reflection loss, measured at 140 °C. Moreover, the graphene composite showed quite high reflection loss (>24 dB) for a wide temperature range of 50–200 °C, having maximum loss at 140 °C. The authors discussed various combinations of graphene with conducting polymer (PANI), magnetic iron oxides, graphene-Ni nanocomposites, and combinations with alloys FeCo and NiCo, which have large saturation magnetization, high conductivity and high Curie temperature reported by various researchers, which all show high MW absorption properties as polymer-based composites/nanocomposites.

Pisu et al. developed broadband radar-absorbing polymeric composite laminate for aircraft application using graphene nanoplates, as described in a WIPO patent [81]. The inventors described the method of manufacturing the multilayer laminate composite using graphene-containing conducting layers alternately with non-conducting layers. The laminates are prepared by infusion of liquid resin (resin transfer moulding) technique containing graphene in a mould of vacuum bag system, where fibre preforms (plies) are kept stacked. Due to the pressure difference, the resin with graphene is forced through the plies and excess resin is removed. Alternately, the resin is injected by pressure. Finally, the laminate is formed by heat and pressure. The layered laminate basically is a backing of reflecting composite (CFRP) followed by a middle absorber layer containing graphene-loaded GFRP and front face of a non-conducting layer of a simple GFRP. The MW absorption in 6–17 GHz was >5 dB, with two distinct peak losses near 9.1 GHz (33 dB) and 13 GHz (34 dB), which is to some extent broad band absorber.

The challenges in development of MW-absorbing composite are extension of band width, reduction of thickness as well as weight penalty. In airborne structures, due to the restriction of weight, many high-absorbing magnetic particles cannot be used, because of reasons such as high density and high loading required in the polymer matrix for satisfactory RCS reduction; and consequent to high loading, these composites show lower toughness and mechanical properties compared to the pristine FRPs. A worse complication is that composite aircrafts are better made by carbon fibre-epoxy composites (CFRP) because of lighter weight and very high strength compared to glass-epoxy, but in case of RACS, carbon fibre can only be used as the backing, and not in the absorbing layers. This increases weight and decreases strength.

The pay load management is better done by carbon nanoparticles like graphene or CNT, since the density of any form of carbon is not more than 2.3 g/cc, almost one-third of iron, and small quantities are sufficient for desired RCS reduction, and due to much less loading, the mechanical property of the composite is not compromised, rather these fillers act as reinforcement for the resin even at very low concentrations.

The absorbing layers and the non-conducting front layer have to be with glass or glass/Kevlar hybrid fibre. A multilayer laminate having a thick CFRP backing with optimum thicknesses of absorbing and front dielectric layers, which are much less than the backing thickness, can be designed to achieve as close strength as possible compared to a pristine CFRP laminate. Another consideration is that radar frequencies are from 1–40 GHz for all military applications. High absorption in this wide band of frequency is difficult to realise, since apart from complex permittivity and permeability, the MW loss also depends on internal scattering and multiple reflections as well as phase cancellation and these are not basic properties of the composite. The losses due to scattering, multiple reflection and phase cancellation are dependent on the relation between MW wavelength and thickness of the absorbing-dielectric part of the composite. For a stealth aircraft, the design of the different portions can be done with variation of such composite layers. However, in ground objects and ships, the RACS structural requirements are easily met with even ferrites and combination of conducting composites, nanocarbons and ferrites. The superstructure of a ship with such RACS can have a thickness of even 20 mm for a wide band absorbing capability (2–40 GHz). Several layers of alternate lossy and non-conducting (pristine FRP) with graded conductivity of the lossy layers can be designed with a CFRP backing in such applications.

There is very limited scope for radar absorbing flexible sheet in aircraft application, but it has very good potential for ships and ground installations. The best advantages of such sheets are wide choice of dielectric and magnetic lossy fillers, easy application and installation when needed, using a pressure-sensitive adhesive on an MW-reflecting surface. The sheets can be applied on any contours. The weight penalty for ships and ground objects are not so critical and moving targets may not experience significant increase in energy requirement (fuel consumption, for example) due to add-on absorber sheets. For a ground object or ship, the design of multilayer absorber with wide band width covering 2–40 GHz, is more viable, even with hybrid MW active materials such as combination of graphene and ferrites, conducting polymer- ferrite, CNT-ferrite and many possible combinations. An additional advantage of the flexible sheets is reduction of structural vibration for moving objects. The reduction of vibration intensity is of higher importance for ship superstructures as improvement of acoustic stealth measure, since the vibrations are transmitted to the water through the hull and can be detected by passive acoustic sensors, especially underwater mines.

5.4 Application of smart composites

Smart composites based on polymer as matrix are being used in many areas in civil sectors such as construction industries, electronics, medical field, packaging, transportation and in defence sectors. The smart composites are meant to enhance safety and reliability. The smart response of the composites can be categorized into two functions, viz. (i) sensing a change in the environment in which it is deployed and (ii) specific and quantitative response to manage/mitigate the change caused by an external stimulation.

An example of the first kind is an electrically conducting composite that can detect humidity or presence of a gas in the atmosphere. The usefulness of a smart composite sensor depends on several characteristics such as limits of sensing, for example, lower and upper limit of humidity, secondly, accuracy in sensing and thirdly, repeatability in use. All these factors add to ensure quality of an environment. A smart sensor not only detects a parameter but should also accurately and repeatedly sense the change in the parameter. A minute change in gas concentration is sometimes very critical, for example, a toxic gas contamination in a chemical plant, or in the case of chemical warfare. Likewise, in maintaining a dehumidified room, smart humidity sensors are required to perform with accuracy. A smart composite is needed to be designed for very minute changes in such critical applications.

A similar serious situation can be structural health monitoring by smart composite sensors. A very minute crack should be reflected in the vibration signature to avoid a catastrophic failure of the structure, for example a wing of an aircraft. Repeatability is another important, though not essential requirement. The sensors can be repeatedly used, where the physical properties are changed, so that the change is reversible. Secondly, the repeated sensing has to be at the same quantitative value with same accuracy after every cycle of sensing and recovery. In many smart sensing scenarios where absorption/adsorption is involved, the sensor response changes for the first few cycles because forward and reverse isotherms are not the same, and the residual concentration after a few cycles attains an equilibrium point, after which the sensing might be reliably repeatable. However, in some cases, the sensing is by an irreversible chemical reaction and hence, the repeatability is not applicable, for example, a corrosion-sensing composite coating on steel surface. The change in colour in the smart coating is an irreversible chemical reaction due to corrosion of the steel is and hence the coating cannot be used again for sensing the corrosion.

The second category of smart composites used for mitigating a change due to external stimulation should be either passive or active in response to the stimulation. In both cases, the mitigation of the change is without human intervention and fully automatic. In passive response, the composite has its own internal mechanism for mitigation, for example, a passive vibration damping, where the composite can damp a vibration with internal viscoelastic loss as well as by frictional damping by anisotropic filler loaded in it. In an active mitigation, a quantitative response is provided by a sensing-actuating system. For example, in an active-passive CLD damping, the vibration amplitude and phase are sensed to activate an equal but reverse signal using a piezoelectric or magnetic cum viscoelastic composite damping arrangement. In the case of such smart composites, the sensing and actuating both are required to be accurate and repeatable. The smart damping is perfectly repeatable as there is no physical or chemical change in the composite is involved, even in the case of magnetic CLD. EMI shielding and MW absorbers act due to their intrinsic MW properties and configuration, similar to what is adopted or mechanical vibration damping by CLD. However, in case of some smart composites in this category, the process of mitigation may be due to chemical reaction, which could be irreversible. For example, in the case of a self-healing composite, which works on release of resin and hardener from capsules, the healing is due to irreversible curing reaction of these two reactants. Therefore, the next crack appearance on that location cannot be healed. However, repeatable self-healing composites are possible, if the reaction is reversible like reversible covalent bond formation and bond exchange-type self-healing composites. Other types of physical changes to mitigate an external stress are phase change composites, which are used to control temperature by absorbing large heat of melting/phase change at a set temperature, thermochromic composite coating and textile, which changes colour due to change in temperature and likewise, electrochromic composite coating, which changes colour due to change in electric field.

Some special applications of smart composites described so far are discussed below.

5.4.1 Biomedical application

The field of biomedical science is the fastest growing field for novel technological innovation and commercial deployment. There are a number of smart technologies required for surgical treatment, organ replacement, repair and implants in human body. The most important requirement for a smart composite to be used in the medical field is biocompatibility and secondly, degradation in human body and useful life as replaced/repaired or implanted composite. As a common example, self-healing implants are used in bone repair and tissue engineering, where a self-healing composite is fixed in a human body. The matrix polymer, all additives including healing components and product after healing reaction should be compatible with the human body. A typical example is the development of self-healing polyurethane elastomer for multiple biomedical applications by Jiang et al. [82]. The authors developed a series of biodegradable and biocompatible self-healing polyurethane elastomers (SHEs) with tunable mechanical properties and applied them to various disease models in vivo and tested their reparative potential in multiple tissues and at physiological conditions. They also validated the effectiveness of SHEs as promising therapies for aortic aneurysm, nerve coaptation and bone immobilization in animal models. In a survey, it was observed that globally, the market share of biomedical applications of self-healing materials is the highest among all areas of application, followed by construction industry [83].

One common self-healing material is dental composite [84–86]. In dental composites, the fracture toughness and fatigue crack growth resistance are very important for durability and self-healing efficiency. Huyang et al. [84] addressed the problem of crack development in artificial dental parts using conventional resin, by developing a self-healing dental composite (SHDC), which uses a healing powder (HP, strontium fluoroaluminosilicate particles) and a healing liquid (HL, aqueous polyacrylic acid) enclosed in silica microcapsules. On release due to a crack, the HP reacts with HL to form an insoluble solid product as the healing material filling the cracks. This developed dental composite can provide a longer service life compared to conventional resin based dental restoration. Yahyazadehfar et al. [85] studied similar healing capsules in a dental composite (the matrix resin being a mixture of two methacrylate monomers) for their performance in fracture toughness and fatigue crack growth resistance under monotonic and dynamic loading using single V notch specimen and compact tension samples. The authors used 1–5N for cyclic load, depending on compositions, with the stress ratio $R = P_{min}/P_{max} = 0.5$ for the initiation of the crack, and thereafter, a dynamic load (1–5N) was applied at 5 Hz frequency using $R = 0.1$. The cyclic loading was conducted in increments of 20,000 cycles. One particular result showed that the critical stress intensity factor (K_{ic}) for crack initiation and after healing were 0.91 and 0.22 for a typical healing composite with 5 wt% of microcapsules in the composite, and the healing efficiency was average 24%. Ning and co-workers [86] used a poly(urea-formaldehyde) microcapsule and a healing liquid as a mixture of triethyleneglycol dimethacrylate (TEGDMA) and N,N-dimethyl-p-toluidine (DMPT). The composite resin was a commercially available flowable monomer, polymerized using varying initiator concentrations to observe the effect of initiator and the size of microcapsules on fracture toughness (K_{ic}) and healing efficiency. It was observed that higher concentration of the initiator and larger microcapsules had higher fracture toughness after healing. However, the healing period was 72 h at room temperature.

Electrically conducting polymer bio-nanocomposites (CBNCs) are being used in tissue engineering, organ transplant and drug delivery. Conducting polymers such as polyaniline (PANI), polypyrrole (Ppy) and polythiophene (PTH) are synthesized as nanosized particles and this results in a very high conductivity, which cannot be achieved by micron size particles of these polymers. These conducting polymers are not biodegradable and have some toxic effects too. These difficulties are solved by adding small quantities (up to 5%) of the nanoparticles of these polymers in some biodegradable and biocompatible polymers such as polycaprolactone, polylactide, polylactic acid, poly (lactide-co-glycolide), chitosan, gelatin, collagen, poly(lactide co-polycaprolactone), etc. Flexible polymeric bio-nanocomposites such as polyurethane-Ppy composite and poly (glycerol sebcate)/PANI composite can be used for regeneration of skin, soft skeletal muscle and blood vessels [43].

5.4.2 Industrial application

Self-healing composites are being used as immediate repairable structure, which prevents catastrophic failure resulting from propagation of a crack due to high stress, for example, in interlaminar shear. The large structures remain unattended and such automatic repair of a minute crack is essential. Self-healing concrete structure may contain a large variety of healing microcapsules and healing agents in a polymeric resin. One important feature of the healing is that the presence of water during healing is mandatory. Typical examples are urea-formaldehyde resin capsule containing epoxy resin and a diamine as healing agents, silica microcapsule containing methyl methacrylate as healing agent, poly (styrene-divinyl benzene) microcapsules with epoxy resin as healing agent, and many such polymer-based self-healing composites are reported. Considering the different polymerization or curing rates at conditions of initial curing and curing at the healing stage, the performance of the self-healing system varies [87]. Self-healing composite coatings are used as paints on concrete buildings for crack repair due to damage by mechanical impact or climatic causes. Thakur et al. reviewed the various self-healing coatings where nanoparticles are used in coatings [88]. The authors discussed the nanocontainers used to manufacture self-healing anticorrosive coatings with the self-healing mechanism. The method of making nanocontainers such as layer-by-layer assembly of layered double hydroxide has been discussed. The review deals with manufacture of nanocontainer-facilitated self-healing corrosion coatings used in different construction industries.

Conducting composites, especially flexible PVC-based carbon composites of various carbon and conducting fillers, as flooring sheets and similar rubber-based composites are widely used as anti-static floor cover and wrist bands. Industries manufacturing various electronic/electrical gadgets using electrical/electronic power sources essentially use such floor mats of the lab and wrist bands for technical personnel handling even small power circuits, which are connected to main electrical lines of 220/440 volts. Static dissipative mats can be found in many places, such as: computer repair shops, assembly lines, healthcare offices, computer server rooms, auto shops and any place where flammable materials are used like chemical laboratories or manufacturing facilities. Such mats are termed as ESD (electrostatic discharge) mats and it is well established that ESD has proven to be the best method of protection against accidents caused by electrostatic discharge.

Conducting composites are now being used for lightning strike protection (LSP) of passenger aircrafts for prevention of damage due to possible lightning. Till recent times, carbon fibre-reinforced epoxy composite (CFRP) was used in conjunction with thin metallic foils. The performance of such LSP was best, comparing the low weight penalty and protection efficiency, and was definitely better than CFRP without a foil attached [89]. However, the manufacturing difficulties, improper metal-to-CFRP adhesion and associated cost of manufacturing are some disadvantages [90]. To overcome these problems, new LSPs are being made with inclusion of conducting nanoparticles in the resin matrix to manufacture the CFRP lightning arrestor. The epoxy matrix was mixed with 30% doped PANI, as described by Katunin et al. [37] or carbon nanofibre as described by Gou et al. [91] or nanotubes as described by Han et al. [92].

Another important application of smart composites in the electrical/electronic industry is EMI shielding of equipment to protect damage due to electromagnetic conduction/radiation. Industrially, the best EMI shields are definitely metal sheets because of their highest conductivity and hence, near-perfect reflectivity of EM wave. However, metals suffer from the possibility of corrosion, high weight penalty and cost of manufacturing in different shapes. Domestic, consumer markets of electrical and electronic consoles and gadgets require the traits of being light, durable, fashionable and competitive in manufacturing cost. Typical examples are laptop, PC tablets and mobile phones. Polymeric conducting nanocomposites are the most promising materials for EMI shielding at present and have replaced almost all metallic versions due to all advantages that metals do not offer. Among the polymer matrices, epoxy resin is most preferred, followed by polyurethane (PU), polyethylene (PE), polycarbonate (PC) and polytetrafluoroethylene (PTFE). Carbon nanotubes, graphene, exfoliated graphite and nanocarbon powders are used to make conducting composites with the polymer matrix [93, 94]. The balance between conductivity and complex permittivity of such polymer nanocomposites make them efficient absorber-type shields. On the other hand, epoxy-carbon fibre composite (CFRP) is another conducting FRP composite used in EMI shielding very effectively, since the conductivity of CFRP is very high, and hence it is a reflecting type of shield. CFRP is used as such [95] or with conducting nanoparticles such as carbon nanotube and continuous graphene oxide fibre [96]. As a new field of implementation, these modified epoxy composites with a high strength-to-weight ratio are gaining considerable interest as electromagnetic interference (EMI) shielding materials, primarily in consumer electronics, aircraft and automobiles.

5.4.3 Application in defence

All types of smart composites are used in many critical and general areas of defence vehicles, aircrafts, ships, submarines and installations. Self-healing composites have good potential to be used on FRP superstructure of ships, since large areas of the superstructure cannot be monitored for minute crack development on small impacts and self-healing process at initial crack generation can prevent sudden structural failure. Moreover, self-healing corrosion-resistant underwater paint would be another advantage, since the crack generated can lead to very severe corrosion of the underwater hull in the sea water. Similarly, a smart composite coating with microcapsule resin and/or bond exchange-type elastomeric smart cover mats can be used in battery pit of submarines. The cracks can be filled immediately to prevent corrosion of the steel pit surface by dilute sulphuric acid (battery acid).

Smart damping composites are potential acoustic stealth items. Vibration of ship structures due to rotating machines and the propeller results in hull vibration and the same is transferred to seawater as radiated underwater noise. The u/w acoustic signal is thus of low frequency, since the revolution frequencies of the machines and the propeller are quite low, typically 2–120 Hz, with some harmonics and higher modes extending up to 2 kHz. The intensity of the vibrations is typically in the range of 65–125 dB depending on machine power and mounting system of machines. The radiated u/w noise can be picked up by a small acoustic sensor embedded in a u/w mine. Additionally, the low frequency noise can travel longer distances in the sea and can be detected by a passive SONAR. Thus, the vibrations are needed to be damped at the base frame of the large machines such as engine, pumps, etc. apart from the elastomeric shock mounts. A simple passive CLD can reduce vibrations to the extent of 10–15 dB in mid frequency range (100–3,000 Hz) only, and lower frequencies are not damped significantly. Smart CLDs would very effectively minimize low frequency (5–100 Hz) vibration intensity by active electrical control or magnetic force control. In electrical control, the damping is independent of VEM damping mechanism, but augments the viscoelastic damping only in low frequency range and is not effective beyond 400 Hz. In a passive VEM-piezoelectric system, the vibration acceleration drives the piezo-effect and the transformation from dynamic strain to electric charge is proportional to the strain rate. In case of magnetic CLD, the shear strain of the VEM is controlled (increased) by the magnetic force and hence the damping is enhanced since shear strain energy loss would be higher. Therefore, the magnetic force is an augmentation of viscoelastic damping. The magnetic force and its effect are independent of frequency since there is no active control. However, active control by electrical sensor (actuator) cannot be installed in large areas such as base frame of engines, internal hull structures near vibrating machines, etc. To date, due to large areas to be covered, passive CLDs are being used in ship structures. Active control is possible in places like helicopter rotor assembly [97, 98] and military ground vehicles [99].

EMI shielding of equipment in defence is compulsorily required for safety of the electrical systems and personnel. Apart from EMI, the equipment also needs to comply with electromagnetic compatibility (EMC) with the electromagnetic environment in which it is to be used. Each equipment to be installed on board any vehicle, ground, sea or air, has to qualify the EMI-EMC limits specified in MIL standard MIL-STD 461, by the Department of Defence, USA. The standard describes the EMI requirements for a wide range of applications, from ground vehicles and ships to aircraft and ground installations, as well as the various requirements within an application (e.g. SONAR room and engine room on a naval ship). There is also scope to tailor the requirements to particular applications. The tests are for conducted and radiated emissions and susceptibility of the equipment. In a document, EMI prediction and protective margin calculations are given in details in published literature [100, 101] for naval equipment. A quantitative example is a composite element of a ship that may need EMI shielding of minimum 25 dB in the frequency range of 1 GHz to 40 GHz. The actual limits of the EMI/EMC are critical for most defence systems and cannot be generalised.

Radar-absorbing composites (RACS) and elastomeric flexible composites (RASH) as stealth materials, the performance and efficacy in reduction of RCS of a defence target, and a general attenuation characteristic as research documents are available in open literature in abundance as already discussed in Section 5.3.2. However, broad band radar flexible composite sheets are required to be used on superstructures and for X- and Ku-band for specific areas such as radar mast, and also in some portions of submarines. The attenuation of a surface depends on detection range. A 10-dB lossy RACS would reduce the distance of detection to 56% of uncoated metal target. Therefore, when a 2-GHz radar frequency is used, the detection distance of metal target is more, assuming 100 km for an aircraft. A 10-dB attenuation means the target can be detected only at 56 km; whereas, when the same target at a distance of 50 km (attained by it within 30 s), is being tracked by 10 GHz (X-band) for exact coordinates to activate fire control system, the same 10-dB lossy RACS would reduce the range to 28 km. Now, at the same speed, the effective time to react is just half with identical attenuation in high frequency than at low frequency. This indicates that at high frequency, the attenuation must be higher. However, the same logic does not apply to slower vehicles such as ship, battle tank, truck, etc. In this case, the speed of the vehicle is almost immaterial compared to the reaction time available. For ships, the choice of RASH and RACS are much wider than aircrafts. For aircrafts, use of RACS as the main skin, leading edges, air intake and other places need different specific properties. For example, thermal and structural stability requirement in air intake is 350 °C, while for the leading edge (wing tip, etc.), the requirement is a thermal and structural stability in several thermal cycles of −40 to +150 °C. However, known stealth aircrafts of all developed and developing countries such as F117A, F-22 raptor or B-2 of US Airforce, Sukhoi-57 of Russia, F-35B Lightning of UK Royal Airforce and J-20A of China, do not reveal the radar stealth materials used on these aircrafts.

For a ship, the requirement of chemical stability in the presence of marine environment at ambient temperature is most important. Several metal nanopowders being reported now are not suitable in this case because of corrosion in marine conditions, and nanocarbons are a better solution. Conducting polymers also suffer from de-doping in such atmosphere and the effective conductivity drastically reduces. As a general observation, a water barrier coating containing a polymeric resin-clay nanocomposite as matrix with neutral pigments (for Admiralty grey or light grey colour requirement) provides a very effective protection against corrosion or degradation of radar stealth items for naval ships. Polyurethane resins with low isocyanate contents (<3.5%) and Cloisite 30B clay are most suitable for such clay nanocomposites for long-life protective coating. The coating being purely dielectric, transmission of MW is almost 100% through the coating. Examples of stealth ships are the famous La Fayette class French Frigate, destroyer of US Navy USS Zumwalt (which is an RCS similar to a fishing boat, despite large size) and Shivalik class Frigates of Indian Navy. However, the radar stealth materials used in these strategic ships are closely guarded secret.

5.5 Conclusions

Research on smart polymeric composites has gained momentum with the demand of high-performing and energy-efficient systems. Lighter and robust materials with desired performance are aimed for the development of smart materials. Application of smart composites is rapidly increasing compared to homogeneous smart polymers, because of the many advantages that a composite provides. For example, in a composite, the physical and chemical nature of the individual components is retained, and some properties could be synergistically enhanced, such as strength and vibration damping of rubber by carbon nanotubes. Secondly, a component in a composite can enhance structural integrity far more than a simple smart polymer. Thirdly, a component might be providing the functional property while the polymer is only a binder for developing a shaped item with structural integrity. Moreover, smart composites are the preferred choice for lighter weight at competitive cost of production. Invention of different nanoparticles and the technologies to integrate them in polymeric materials have increased the use of polymer nanocomposites in almost all fields from automobile to biomedical engineering. Nanotechnology also allows developing lighter composites than conventional polymer matrix composites and the nano-effect increases the functional properties to great extent. The only disadvantage of nanoparticles is handling hazard, both in terms of possibility of material loss and human health. In modern industrial and laboratory practices, particularly automation with robotic operation and clean room system, these problems can be eliminated with a hygienic environment and better quality of smart composites.

References

[1]   Wang, Y., Pham, D. T., Ji, C. Cogent Eng. 2015, 2, 1075686.
[2]   Islam, S., Bhat, G. Mater. Adv. 2021, 2, 1896–1926.
[3]   Bleay, S. M., Loader, C. B., Hawyes, V. J., Humberstone, L., Curtis, P. T. Compos. Part A. 2001, 32, 1767–1776.
[4]   Pang, J. W. C., Bond, I. P. Compos. Sci. Technol. 2005, 65, 1791–1799.
[5]   Trask, R. S., Bond, I. P. Smart Mater. Struct. 2006, 15, 704–710.
[6]   Zhu, Y., Ye, X. J., Rong, M. Z., Zhang, M. Q. Compos. Sci. Technol. 2016, 135, 135146–135152.
[7]   Lee, J., Bhattacharya, D., Zhang, M. Q., Yuan, Y. C. Compos. Part B Eng. 2015, 78, 515–519.
[8]   Yuan, Y. C., Ye, Y., Rong, M. Z., Chen, H., Wu, J., Zhang, M. Q., Qin, S. X., Yang, G. C. Smart Mater. Struct. 2010, 20, 1–11.
[9]   Jhanji, K. P., Asokan, R., Kumar, R. A., Sarkar, S. Mater. Sci. Eng. 2020, 912, 1–8.
[10]   Suryanarayana, C., Rao, K. C., Kumar, D. ProgOrgCoat. 2008, 63, 72–78.
[11]   Lin, C., Ge, H, Wang, T., Huang, M., Ying, P., Zhang, P., Wu, J., Ren, S., Levchenko, V. A. Polymer. 2020, 206, 122894.
[12]   Jia, H., Gu, S. Y. J. Polym. Res. 2020, 27, 298.
[13]   Nokes, D. S., Nelson, F. C. Shock Vib. Bull. 1968, 38, 5–12.
[14]   Mead, D. J., Markus, S. J. Sound Vib. 1969, 10, 163–175.
[15]   Plunkett, R., Lee, C. T. JASA. 1970, 48, 150–161.
[16]   Rao, Y. V. K. S., Nakra, B. C. J. Sound Vib. 1974, 34, 309–326.
[17]   Rao, D. K. J. Mech. Eng. Sci. 1978, 20, 271–282.
[18]   Torvik, P. J. Analysis and Design of Constrained Layer Damping Treatments. Report AFIT VR 80-4, Airforce Institute of Technology, Ohio, USA, July 1980.

[19]   Murali, D. M., Chakraborty, B. C., Begum, S. S., Natarajan, R., Chandramohan, S. Polym. J. 2022, 31, 1247–1261.

[20]   Plump, J. M., Hubbard Jr., J. E. Modeling of an Active Constrained Layer Damper, In 12th International Congress on Acoustics. Toronto, Canada, 1986, #D4–1.

[21]   Lam, M. J., Inman, D. J., Saunders, W. R. J. Intel. Mat. Syst. Str. 1997, 8, 663–677.

[22]   Shen, I. Y. J. Vib. Acoust. 1994, 116, 341–349.

[23]   Ba, A. J. Sound Vib. 1998, 211, 467–489.

[24]   Trindade, M. A., Benjeddou, A., Ohayon, R. Int. J. Numer. Methods Eng. 2001, 51, 835–864.

[25]   Van Nostrand, W. C., Inman, D. J. Finite Element Model for Active Constrained Layer Damping. In: Anderson, G. L., Lagoudas, D. C., eds. Active Materials & Smart Structures, Vol. 2427. SPIE, Bellingham, WA,USA, 1995, 124–139.

[26]   Varadan, V. V., Lim, Y.-H., Varadan, V. K. SmartMaterStruct. 1996, 5, 685–694.

[27]   Veley, D. E., Rao, S. S. Smart Mater. Struct. 1996, 5, 660–671.

[28]   Trindade, M. A., Benjeddou, A. J. Vib. Control. 2002, 8, 699–745.

[29]   Zheng, H., Li, M., He, Z. Int. J. Solids Struct. 2003, 40, 6767–6779.

[30]   Hansaka, M., Mifune, N., Sato, H., Takinosawa, H., Nishimoto, K. Magnetic-vibration-damper of composite type and damping construction method. European Patent No. EP 0849495A2, dt.24 June 1998, assigned to CJ Kasei Co. (Ltd), Chou-ku, Tokyo, JP, Nichias Corporation, Tokyo JP, Railway Technical Research Institute, Kkubunji-shi, Tokyo JP.

[31]   Hansaka, M., Hayashi, I., Iwatsuki, N., Mifune, N., Morikawa, K., Kagawa, Y. Jpn. Soc. Mech. Eng. C Part C. 2022, 68, 374–382.

[32]   Ochôa, P. A., Groves, R. M., Benedictus, R. J. Sound Vib. 2020, 475, 115289.

[33]   Ko, J. M., Ni, Y. Q. J. Eng. Struct. 2005, 27, 1715–1725.

[34]   Montalvão, D., Maia, N. M. M., Ribeiro, A. M. R. Shock Vib. Dig. 2006, 38, 295–324.

[35]   Singh, T., Sehgal, S. Arch. Comput. Methods Eng. 2022, 29, 1997–2017.

[36]   Mansor, M. R., Fadzullah, S. H. S. M., Masripan, N. A. B., Omar, G., Akop, M. Z. Synthesis, Processing and Applications. Elsevier Inc, Vol. 9, 2019, 177–204.

[37]   Katunin, A., Krukiewicz, K., Turczyn, R., Sul, P., Bilewicz, M., Conf, I. O. P. Ser. Mater. Sci. Eng.2017, 201, 012008.

[38]   Chavan, J. G., Chandrasekhar, L., Samui, A. B., Chakraborty, B. C. Polym. Mat. 2004, 21, 293–304.

[39]   Jain, S., Samui, A. B., Patri, M., Hande, V. R., Bhoraskar., S. V. Sens. Actuators B Chem. 2005, 106, 609–613.

[40]   Jain, S., Chakane, S., Samui, A. B., Krishnamurthy, V. N., Bhoraskar, S. V. Sens. Actuators B Chem. 2003, 96, 124–129.

[41]   Shukla, S. K., Govender, P. P., Agoru, E. S. Microchim. Acta. 2016, 183, 573–580.

[42]   Sivaraman, P., Kushwaha, R. K., Shashidhara, K., Hande, V. R., Thakur, A. P., Samui, A. B., Khandpekar, M. M. Electrochim. Acta. 2010, 55, 2451–2456.

[43]   Dubey, N., Kushwaha, C. S., Shukla, S. K. Int. J. Polym. Mater. Polym. Biomat. 2020, 69, 709–727.

[44]   Cao, Q., Yu, Q., Connell, D. W., Yu., G. Clean Technol. Environ. Pol. 2013, 15, 871–880.

[45]   Lekawa-Raus, A., Patmore, J., Kurzepa, L., Bulmer, J., Koziol, K. Adv. Funct. Mater. 2014, 24, 3661–3682.

[46]   Earp, B., Dunn, D., Phillips, J., Agrawal, R., Ansella, T. Aceves, P., Igor De Rosa, I. D., Xin, W., Luhrs, C. Mater. Res. Bull. 2020, 131, 110969.

[47]   Fang, X.-Y., Yu, X.-X., Zheng, H.-M., Jin, H.-B., Wang, L. Phys. Lett A. 2015, 379, 2245–2251.

[48]   Hu, D., Yi, X., Jiang, M., Li, G., Cong, X., Liu, X., Rudd, C. Aerosp. Sci. Technol. 2020, 98, 105669.

[49]   Ma, X., Scarpa, F., Peng, H., Allgeri, G., Yuan, J., Ciobanu, R. Aerosp. Sci. Technol. 2015, 47, 367–377.

[50]   Kim, M., Jeong, J. H., Lee, J. Y., Capasso, A., Bonaccoroso, F., Kang, S. H., Lee, Y. K., Lee, G. H. ACS Appl. Mater. Interfaces. 2019, 11, 11841–11848.

[51] Park, S. H., Hwang, J., Park, G. S., Ha, J. H., Zhang, M., Kim, D., Yun, D. J., Lee, S., Lee, S. H. Nat. Commun. 2019, 10, 2537.

[52] Kondawar, S. B., Modak, P. R. Theory of EMI Shielding. In: Kurivilla, J., Wilson, R., George, G., eds. Materials for Potential EMI Shielding Applications -Processing, Properties and Current Trends. Elsevier Inc, Vol. Chapter-2, 2020, 9–25.

[53] Skin depth of electromagnetic waves in conductors. Physics Page, http://physicspages.com/pdf/ Electrodynamics/Skin%20depth%20of%20electromagnetic%20waves%20in%20conductors.pdf. (accessed on 02 May, 2023)

[54] Kurivilla, J., Wilson, R., George, G. Materials for Potential EMI Shielding Applications -Processing, Properties and Current Trends. Elsevier Inc, 2020.

[55] Dhawan, S. K., Singh, A. P., Anil Ohlan, A., Kakran, K. S., Sambyal, P. Smart Materials Design for Electromagnetic Interference Shielding Applications. Bentham Science Publishers, Sharjah, 2022.

[56] Saritha, A., Thomas, B., George, G., Wilson, R., Kuruvilla, J. Elastomer-Based Materials for EMI Shielding Applications. In: Joseph, K., Wilson, R., George, G., eds. Materials for Potential EMI Shielding Applications. Elsevier, Vol. Chapter 8, 2020, 121–143.

[57] Lu, S., Bai, Y., Wang, J., Chen, D. Flexible GnPs/EPDM with excellent thermal conductivity and electromagnetic interference shielding properties. Nano Brief Rep. Rev. 2019, 14(6), 1950075. 10.1142/S1793292019500759.

[58] Iqbal, S., Ahmad, S. Conducting Polymer Composites: An efficient EMI Shielding Material. In: Kurivilla, J., Wilson, R., George, G., eds. Materials for Potential EMI Shielding Applications - Processing, Properties and Current Trends. Elsevier Inc, Vol. Chapter-16, 2020, 257–266.

[59] Sun, X., Lv, X., Li, X., Yuan, X., Li, L., Gu, G. Mater. Lett. 2018, 221, 93–96.

[60] Gupta, T. K., Singh, B. P., Mathur, R. B., Dhakate, S. R. Nanoscale. 2014, 6, 842–851.

[61] Guo, J., Song, H., Liu, H., Luo, C., Ren, Y., Ding, T., Khan, M. A., Young, D. P., Liu, X., Zhang, X., Kong, J., Guo, Z. J. Mater. Chem. C. 2017, 5, 5334–5344.

[62] Sun, J., Shen, Y., Hu, X. S. Polym. Bull. 2018, 75, 653–667.

[63] Liang, J., Bai, M., Gu, Y., Wang, S., Li, M., Zhang, Z. Compos. Part A Appl. Sci. Manuf. 2021, 149, 106481.

[64] Maiti, S., Shrivastava, N. K., Suin, S., Khatua, B. B. ACS Appl. Mater. Interfaces. 2013, 5, 4712–4724.

[65] Skolnik, M. I. Radar. In: Middleton, W. M., Van, V. M. E., eds. Reference Data for Engineers-radio, Electronics, Computer, and Communications. Elsevier Inc, 2002, 1–22, Chapter 36.

[66] Atay, H. Y. J. Phys. Conf. Ser. 2022, 2413, 012005.

[67] Zhang, L., Zhou, P., Zhang, H., Lu, L., Zhang, G., Chen, H., Lu, H., Xie, J., Deng, L. IEEE Trans. Magn. 2014, 50, 1–5.

[68] Kim, S. Y., Kim, S. S. PolymPolymCompos. 2018, 26, 106–110.

[69] Kartal, I., Demirer, H. Acta Phys. Pol. 2017, 131, 559–561.

[70] Go, J.-I., Lee, W.-J., Kim, S.-Y., Baek, S.-M., Choi, W.-H. Compos. Sci. Technol. 2020, 199, 108366.

[71] Peymanfar, R., Javanshir, S., Naimi-Jamal, M. R., Cheldavi, A. Mater. Res. Exp. 2019, 6, 0850–9.

[72] Oyharcabal, M., Olinga, T., Foulc, M.-P., Lacomme, S., Gontier, E., Vigneras, V. Compos. Sci. Technol. 2013, 74, 107–112.

[73] Jang, H. K., Shin, J.-H., Kim, C.-G., Shin, S.-H., Kim, J. B. Adv. Compos. Mater. 2011, 20, 215–229.

[74] Potphode, D. D., Sivaraman, P., Mishra, S. P., Patri, M. Electrochim. Acta. 2015, 155, 402–410.

[75] Sivaraman, P., Bhattacharrya, A. R., Mishra, S. P., Thakur, A. P., Shashidhara, K., Samui, A. B. Electrochim. Acta. 2013, 94, 182–191.

[76] Hu, W., Yin, H., Yuan, H., Tang, Y., Ren, X., Wei, Y. Compos. Sci. Technol. 2023, 233, 109927.

[77] Choi, I., Lee, D., Lee, D. G. Compos. Struct. 2015, 122, 23–30.

[78] Bai, X., Zhai, Y., Zhang, Y. J. Phys. Chem. C. 2011, 115, 11673–11677.

[79] Wang, T. H., Li, Y. F., Geng, S., Zhou, C., Kia, X. L., Yang, F., et al. RSC Adv. 2015, 5, 88958–88964.

[80] Meng, F., Wang, H., Huang, F., Guo, Y., Wang, Z., Hui, D., Zhou, Z. Compos. Part B Eng. 2018, 137, 260–277.

[81] Pisu, L., Iagulli, G., Sarto, M. S., Fabrizio, M., Lecini, J., Tamburrano, A. Multilayer radar-absorbing laminate for aircraft made of polymer matrix composite material with graphene nanoplates, and method of manufacturing same. Patent WO2019/167009 A1, assigned to Leonardo SPA, Piaza Monte Grappa 4, Roma (Italy) dated 06 September, 2019.

[82] Jiang, C., Zhang, L., Yang, Q., Huang, S., Shi, H., Long, Q., Qian, B., Liu, Z., Guan, Q., Liu, M., Yang, R., Zhao, Q., You, Z., Ye, X. Nat. Commun. 2021, 12, 4395.

[83] Idumah, C. I. PolymPolymCompos. 2021, 29, 246–258.

[84] Huyang, G., Debertin, A. E., Sun, J. Mater. Des. 2016, 94, 295–302.

[85] Yahyazadehfar, M., Huyang, G., Wang, X., Fan, Y., Arola, D., Sun, J. Mater. Sci. Eng. C. 2018, 93, 1020–1026.

[86] Ning, K., Loomans, B., Yeung, C., Li, J., Yang, F., Leeuwenburgh, S. Dent. Mater. 2021, 37, 403–412.

[87] Zhang, W., Zheng, Q., Ashour, A., Han, B. Compos. Part B. 2020, 189, 107892.

[88] Thakur, A., Savas, K., Kumar, A. CurrNanosci. 2022, 18, 203–216.

[89] Feraboli, P., Miller, M. Compos. Part A. 2009, 40, 954–967.

[90] Gagne, M., Therriault, D. Prog. Aerosp. Sci. 2014, 64, 1–16.

[91] Gou, J., Tang, Y., Liang, F., Zhao, Z., Firsich, D., Fielding, J. Compos B. 2010, 41, 192–198.

[92] Han, J. H., Zhang, H., Chen, M. J., Wang, D., Liu, Q., Wu, Q. L., Zhang, Z. Carbon. 2015, 94, 101–113.

[93] Banerjee, P., Bhattacharjee, Y., Bose S, J. Electron. Mater. 2020, 49, 1702–1720.

[94] Aal, N. A., El-Tantawy, F., Al-Hajry, A., Bououdina, M. Polym. Compos. 2008, 29, 125–132.

[95] Ramadin, Y., Jawad, S. A., Musameh, S. M., Ahmad, M., Zihlif, A. M., Paesano, A., Martuscelli, E., Ragosta, G. Polym. Int. 1994, 34, 145–150.

[96] Ucar, N., Kayaoglu, B. K., Bilge, A., Gurel, G., Sencandan, P., Paker, S. J. Compos. Mater. 2018, 52, 3341–3350.

[97] Shevtsov, S., Soloviev, A., Acopyan, V., Samochenko, I. PHYSCON 2009, Catania, Italy, September, 1–4.

[98] Konstanzer, P., Enenkl, B., Aubourg, P.-A., Cranga,, P. Recent advances in Eurocopter's passive and active vibration control. Presented at the American Helicopter Society 64th Annual Forum, Montréal, Canada, April 29 – May 1, 2008.

[99] Miller, L. R., Nobles, C. M. The design and development of a semi-active suspension for military tank, SAE, 1988, Paper No. 881133.

[100] Dixon, D. S., Sniegoski, R. J. Low frequency EM1 modeling and EMC prediction program for shipboard electrical/electronic equipments. IEEE International Symposium on EMC. 1982.

[101] Kowalczyk, K. R., Schlachter, F. An EMC prediction and protective margin calculation program for the Dec Vax computer series. IEEE 1988 International Symposium on Electromagnetic Compatibility 2–4 Aug. 1988. 10.1109/isemc.1988

Chapter 6
Experimental techniques

Description of abbreviations

Abbreviation	Description
FTIR	Fourier-transform infrared
ESR	Electron spin resonance
FMR	Ferromagnetic resonance
SEM	Scanning electron microscop
TEM	Transmission electron microscopy
XRD	X-ray diffractometry
AFM	Atomic force microscopy
EMC	Electromagnetic compatibility
UAV	Unmanned aerial vehicle
UUVs	Unmanned underwater vehicles
UTM	Universal testing machine
RTM	Tesin transfer moulding
DMA	Dynamic mechanical analyser
DTA	Differential thermal analysis
TGA	Thermogravimetric analysis
TMA	Thermomechanical analysis
DSC	Differential scanning calorimetry
DEA	Dielectric analysis
DC	Dielectric constant
DF	Dissipation factor
LM	Light microscope
PEO	Poly(ethylene oxide)
EDS	Energy-dispersive spectrometer
CBS	N-Cyclohexyl-2-benzothiazolesulphenamide
IPN	Interpenetrating polymer network
MBTS	Mercaptobenzothiazyl disulphide
MPa	Mega-Pascal
PE	Poly(ethylene)
PP	Poly(propylene)
phr	Parts per hundred rubber (by weight)
SEM	Scanning electron microscope
TEM	Transmission electron microscope
TMTD	Tetramethylthiuram disulphide
FRP	Fibre-reinforced plastic

https://doi.org/10.1515/9783110781571-006

6.1 Introduction

The development and Manufacturing of polymer composite items need reproducible and accurate evaluation of relevant properties. The relevance is dependent on conditions of use of the composite item. An underwater fibre-reinforced plastic (FRP) pipeline needs more attention on creep, vibration and effect of interaction with seawater, biocorrosion and biofouling, rather than fire rating and electrical property. Likewise, a cooling tower requires evaluation of stresses and strains due to static and dynamic loads and UV/ozone ageing effects rather than dielectric strength or fire rating. Since there are a large number of properties that are required to understand a material, it becomes difficult and impractical to study all properties elaborately. The identification of the properties relevant to the application is of foremost importance. Subsequently, the evaluation is to be divided in three distinct categories, stage-wise as follows:

Stage 1: Tests for quality building of polymer-reinforcement combination: molecular level, in the context of physical blending or chemical kinetics, structure and morphology of final form. This includes physical homogeneity, reaction data, identification of chemical structure, etc. by Fourier-transform infrared (FTIR), NMR, electron spin resonance (ESR), ferromagnetic resonance (FMR) spectroscopy, morphology by scanning electron microscopy (SEM), transmission electron microscopy (TEM), X-ray diffractometry (XRD) especially for nanocomposites, surface properties by atomic force microscopy (AFM) and any other basic characteristics, if required.

Stage 2: Macroscopic properties to characterize the formed material. This includes mechanical, dynamic mechanical, thermal and electrical property evaluation, which is restricted to the material behaviour. However, thermal and dielectric studies are also very useful to determine some molecular level properties and reaction dynamics. These studies would reveal the nature of polymer-reinforcement interaction, efficacy of the reinforcement, failure process of the composite, degradation, estimation of service life, etc. The evaluation results would provide input parameters to design an item using the composite so developed.

Stage 3: Functional and structural evaluation of the fabricated item and prototypes. This involves acoustic, electromagnetic, electrical, i.e. electromagnetic induction (EMI) and electromagnetic compatibility (EMC), vibration, hydrostatic, fire rating, load bearing properties for systems such as machinery base frame, hydraulic piping, valve, storage vessels, reactor, cooling tower, wind mill, furniture, home appliances, electronic gadgets, printed circuit boards, building construction elements, automobile parts, unmanned under water vehicles (UUVs), unmanned aerial vehicle (UAVs), and others according to the need in specific application.

The experimental techniques of all the above are adequately sited in literature and various standards. However, there are many salient features of common experimental techniques which are discussed here, without elaborating the complete test procedure. Brief description of the machines with some criticalities in experiments is also included. The laboratory evaluation as in stage 2 are discussed here, because stage 1 evaluations come under analytical methods, and stage 3 is very wide in spectrum, since there are enormous types of application. Stage 2 tests are mostly required for physical and mathematical modelling and design of components or subsystems and systems.

One of the main aspects of experimental techniques in the laboratory is the sample preparation of composites and conditioning of the samples prior to test. As far as possible, moulding of the thermoplastic polymer composites should be thermoformed in a mould in case of intricate or no uniform shapes. Extrusion process is suitable for uniform cross section. However, where a fabricated sheet is provided, sample has to be cut out. The dimensions are to be uniform and be measured accurately since the modulus value much depends on original dimensions. A micrometre with 0.01 mm accuracy will be adequate for thickness measurement, while a slide calliper with accuracy of 0.1 mm will be good for lateral dimensions. In order to avoid human error, repeated measurements should be done. For FRP composites, forming in a mould will be difficult with fibre, etc. In such case, samples may be cut out from sheets by a high-speed machine and the cut surfaces should be polished by surface grinder and polishing machine. There is possibility of fibre-polymer separation at the cut edges, which can be covered by the thermoset resin as a fine coating of 50–70 μm. For elastomeric composites with particulate and nanomaterials, standard sample cutters of dumbbell shapes according to ASTM standards can be used. For each type of composites, the sample sizes are precisely described in standards such as ASTM, ISO, BS, DIN and BIS. However, the researcher can use other sizes not described in these standards, if the entire study and repetition, if any, are done using same dimensions. In addition, in situations where samples are to be cut from an existing item, there could be a sample dimension which might not be as per the standards. That might be an exception, but at least a comparative assessment can be done with similar composition made in the lab under ideal condition and dimension.

Conditioning of the samples is another important part of testing. Conditioning is required to allow the samples to be in exactly similar conditions prior to test. Composites, in general, are bad conductors of heat, and requires prolong time to equilibrate at the room temperature. Secondly, composites may absorb moisture to different extent if not kept in a constant humidity environment. Polar polymers absorb more moisture than nonpolar ones. For example, polycarbonate absorbs about 0.2% moisture and needs to be dried in oven for 24 h at 100° to remove moisture before melt processing, while polypropylene does not need such elaborate drying. The moisture acts as a plasticizer at a low concentration and may reduce the mechanical properties and also lower the glass transition temperature of polar thermosets and their composites, depending on extent of moisture absorbed. Therefore, composite samples are to be conditioned. For example, all samples are to be kept in a room with less than 45%

relative humidity and temperature of 25 ± 1 °C for 48 h before test. Some standards specify 24 h of conditioning. In tropical countries, the humidity in most places is higher than temperate climates. It is recommended that the duration of conditioning the samples should be decided considering ambient atmospheric temperature and humidity of the place of study.

Sampling is the other important aspect of testing in a regular production unit, but not so in research, since during a development of composite, every experimental batch has to be evaluated completely. The intension to test all the batches and the variants is to investigate the causes of possible quality variation and attempt to reduce the variations as far as possible in the development stage.

Instruments for any test should be well calibrated. Measuring instruments of high accuracy and high precision are to be used. Internationally reputed manufacturers generally make instruments which are suitable for measurements according to international standards such as ASTM or ISO. The accuracy depends on the allowable tolerance at the minimum value to be measured. Precision depends on the variation in repetitive measured results. Sometimes dimensional tolerances are very critical such as for high modulus FRPs.

Different methods are applicable for different type of samples depending on rigidity, plastic flow, thermodynamic state, type of reinforcements, extent of non-linearity, etc.

6.2 Mechanical testing

A fundamental property of a real solid body is Young's modulus, which is measured in tension. Ideally, it should be same as compression and flexure too. But in practice, the real solids do not follow ideal elastic behaviour, and there are considerable differences. However, there is a well-established relationship of Young's modulus and shear modulus through Poisson's ratio. Subsequent paragraphs give brief descriptions of some of these mechanical properties and impact strength with type of instruments, methods and international standards.

Various types of mechanical testing depending on the mode of force application, such as tension, compression, shear or bending. The corresponding properties are defined by modulus of elasticity, ultimate strength and ultimate strain. All static mechanical properties can be evaluated using a universal testing machine (UTM).

6.2.1 Universal testing machine (UTM)

The UTM has a rigid base and a top head, fixed in a frame supported by two rigid columns, acting as guide for moving a cross-head. The daylight of a standard UTM is about 1 metre and the movement of the cross-head is controlled by a hydraulic or pneumatic system. A typical UTM is shown in Fig. 6.1. The movement is sensed by calibrated sensors, either laser or other non-contact extensometer. Since most composites are low strain, high modulus materials, except elastomeric composites, the movement measurement has to be very precise. The upper cross-head houses the load cell, which measures the force of deformation. The brief specification of an UTM machine is given as Tab. 6.1, where only main functional parameters are listed.

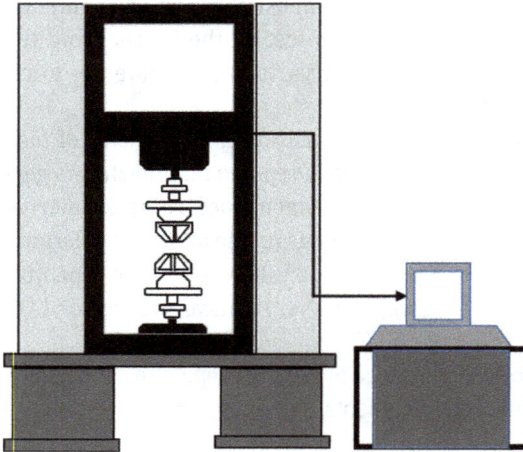

Fig. 6.1: A sketch of functional pats of a UTM machine.

Tab. 6.1: A brief specification of a universal testing machine (UTM).

Parameter	Specification
Load ranges	0.1 N to 50 kN with different load cells. Load accuracy: better than ±0.5% of reading from 2% to 100% of full-scale load. Load cells: 1, 5 and 50 kN
Maximum cross-head travel	1,000 mm
Testing speed range	0.001–500 mm/min speed setting increments of 0.001 mm/min
Accuracy of test speed	0.1% of set speed
Sample grips	Tensile, compression, shear and flexure in three-point bending and four-point bending. Flat specimen width up to 40 mm and thickness up to 25 mm. Circular specimen up to 18 mm diameter. Three-point bending fixture typically for specimen span length from 8 to 250 mm and optionally up to 450 mm span. Compression test platens of top and bottom plate diameter 150 mm for width of specimens up to 100 mm
Drive system	Electromechanical/pneumatic/servo-hydraulic

Tab. 6.1 (continued)

Parameter	Specification
Data sampling rate	400 kHz or better
Post-processing functions	Young's, secant and shear modulus, stiffness, Poisson's ratio proportional limit, yield stress and strain, breaking stress/strain/force/deflection, ultimate stress/strain/force/deflection, work to break and/or total work. Modulus, creep, stress relaxation and hysteresis
Computer	Programming of runs, data storage, processing, and output display as tables and graphs
Safety	No release of load under power failure, emergency stop control and mechanical limit switches, automatic stop or return following sample break, automatic return of cross-head

6.2.2 Tensile test

Tensile strength of common FRP composites are generally very high compared to rubber based composites, and some thermoplastic composites have intermediate strengths. Therefore, tensile test of such variety of composites requires samples of different size and configuration according to the stiffness and ultimate elongation. The standards of the tensile test of various composites are listed in Tab. 6.2.

Tab. 6.2: Standards for tensile testing of composites based on matrix polymer.

Composite types	ASTM standard	ISO standard
Elastomers and thermoplastic elastomers	D412 type C	37:2017
Thermoplastics	D638	527-1:2019
Thermoset	D3039	527-4:1997

The criticality in the test of high strength composites such as FRPs is the grip of the samples. Irrespective of the type, all grips are self-tightening, which means that the gripping is maintained even as the force of deformation increases. All grips should have serrated faces for proper gripping of the sample and to avoid slippage during the test. Dimension of FRP samples recommended are 250 mm in length, 25 mm in width and about 2 mm thickness. Pneumatic grips for soft samples such as elastomers and thermoplastic elastomers are good enough. Dumbbell-shaped specimens are used for such soft samples. The cross-head speed is another important test parameter. In case of FRPs, since the ultimate elongation is generally very low (<10%), the cross-head speed recommended in ASTM standard is 2 mm/min, while for an elastomer composite, the user can select from 50 to 500 mm/min. For composites of engineering plastics, 5–500 mm/min can be selected depending on the stiffness. To avoid tear at the grips, aluminium tabs can be boned at both ends of the sample. Sensing the deformation is very critical for high modulus composites, and may be very accurately measured by universal video extensometer or a laser extensometer. Clip-on extensometers previously used for rubber samples are not recommended now.

A typical tensile test result in terms of stress-strain plot of an unfilled elastomer and 20 phr graphene containing elastomer nanocomposite is shown in Fig. 6.2. The test result shows uniformity of the behaviour of the elastomers and their nanocomposite with progressive stretching. The ultimate strength is almost same for both, about 18 MPa, while the ultimate elongation is reduced from 690% to 580%, but the Young's modulus (at ≈ 1% elongation) is much enhanced from 3.5 to 11.5 MPa, upon formation of graphene nanocomposite. The test result depicts a typically strong elastomer composite behaviour. The difference in the rate of increase in stress due to initial stretching shows the importance of the stress-strain study to understand the reinforcing efficacy of a nanofiller. There is a significant improvement in initial stiffness without much reduction of rubbery flexibility.

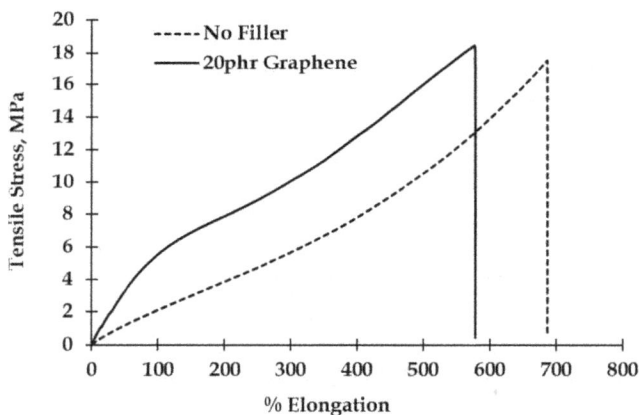

Fig. 6.2: Result of a tensile experiment of an elastomer and its graphene nanocomposite.

6.2.3 Flexure test

FRP composites and high modulus plastics are better studied in bending mode rather than tensile mode. The bending experiment, also known a flexure experiment, is easier to perform, require no rigid gripping and less force compared to pulling apart. The bending or flexural test can be done using a three-point bending or a four-point bending method. Figures 6.3 and 6.4 show these two typical arrangements. The test can be performed in UTM using appropriate fixtures. The moving cross-head is moved downwards in this experiment to cause a bending deformation. Depending on the geometrical configuration and support, different mathematical expressions for stress, strain and flexural modulus are used.

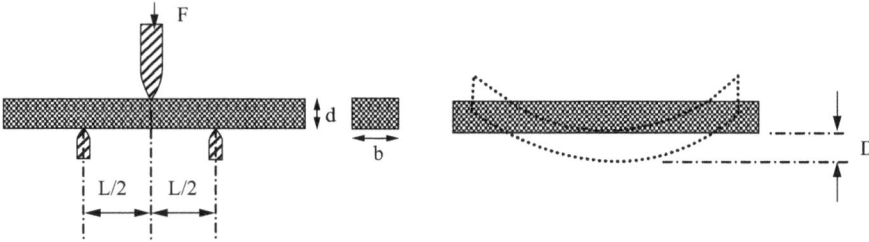

Fig. 6.3: Three-point bending test arrangement.

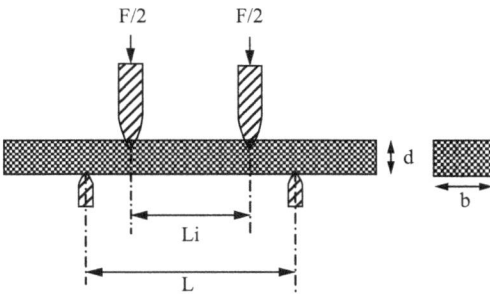

Fig. 6.4: Four-point bending test arrangement.

Commonly, three-point bending test is done for FRP and other high modulus composites. The span length between two supports can be a short span of 60 mm with 10% overhang (the length beyond the support) or a long span of maximum 100 mm with 16–20% percent overhang. Long span is required for very high modulus composites such as carbon fibre-epoxy composites.

6.2.3.1 Stress and strain in a three-point bending test
Typical short span flexure parameters are described:
    Width: 12 mm, thickness (depth: 2.5 mm, span length: 40 mm)
    Overhang: 4 mm each side
    Total sample length: 48 mm

$$\text{Bending strain: } \varepsilon_f = \frac{6dD}{L^2} \tag{6.1}$$

where D is the vertical deflection of the outer surface (convex) of the beam at the centre, L is the span length and d is the depth of the beam:

$$\text{Corresponding stress: } \sigma_f = \frac{3FL}{2bd^2} \tag{6.2}$$

where F is the force applied, b is the width of the beam. The flexural modulus is the ratio of the stress to strain and is given by

$$E_s = \frac{\text{Stress}}{\text{Strain}} = \frac{FL^3}{4bDd^3} \tag{6.3}$$

In case of large span test, typical parameters are described as follows:
Width: 12.5 mm, depth: 3 mm and span length: 100 mm.
Overhang: 16 mm each side.
Total sample length: 132 mm.

If deflections in excess of 10% of the support span occur, the stress in the outer surface of the specimen for a simple beam can be approximately given by

$$\sigma_f = \frac{3FL}{2bd^2}\left[1 + 6\left(\frac{D}{L}\right)^2 - 4\left(\frac{d}{L}\right)\left(\frac{D}{L}\right)\right] \tag{6.4}$$

Strain is given by the same expression as short span and the flexural modulus is given by

$$E_L = \frac{FL^3}{4bDd^3}\left[1 + 6\left(\frac{D}{L}\right)^2 - 4\left(\frac{d}{L}\right)\left(\frac{D}{L}\right)\right] \tag{6.5}$$

6.2.3.2 Stress and strain in four-point bending test

For the four-point bend setup, as in Fig. 6.4, if the loading span is 1/2 of the support span (i.e. $L_i = 1/2\ L$), the stress is

$$\sigma = \frac{3FL}{4bd^2} \tag{6.6}$$

If the loading span is neither 1/3 nor 1/2, the support span for the four-point bend setup, the stress is

$$\sigma = \frac{3F(L - L_i)}{2bd^2} \tag{6.7}$$

Methods of measurement in flexure mode are described in detail in many standards, and a few important methods are given from ASTM and ISO in Tab. 6.3.

Tab. 6.3: Standards of flexural tests for polymeric composites.

Standard	Description
ASTM D790-03	Standard Test Methods for Flexural Properties of Unreinforced and Reinforced Plastics and Electrical Insulating Materials – Procedure A: Material that breaks at relatively low deformation and Procedure B: Material that Fails at relatively large deformation
ASTM D 7264/ D 7264 M-21	Standard Test Method for Flexural Properties of Polymer Matrix Composite Materials
DIN EN ISO 14125:1998	FRP Composite – Determination of Flexural Properties
ISO 14125 (Method B)	FRP Composites – Determination of Flexural Properties – Measurements in four-point bending.
DIN EN ISO 178:2019	Plastics – Determination of Flexural Properties
ISO 20975-2:2018	Fibre-reinforced plastic composites— Determination of laminate through-thickness properties — Part 2: Determination of the elastic modulus, the strength and the Weibull size effects by flexural test of unidirectional laminate, for carbon-fibre based systems.
ASTM D4476	Flexural Properties Testing of Fiber Reinforced Pultruded Plastic Rods
ASTM D6272	Flexural Properties Testing of Unreinforced and Reinforced Plastics by Four-Point Bending

Flexure test is not suitable for soft materials with large extensional strain such as composites of elastomers and thermoplastic elastomers. For these soft materials, tensile test is appropriate. On the other hand, tensile test for very high modulus composite is not suitable, and flexure test is more preferable. As a thumb rule, any composite having more than 30% elongation at break, can be tested by tensile method.

Thickness of flexure test is important and should be restricted to 1/16 of span length in case of short span and 1/40 to maximum 1/32 of span length for large span. The high thickness leads to shear strain in between the outer/inner surface compared to the neutral axis of the beam, and large difference in strain and stress across the thickness, causing considerable error in calculation of modulus.

A three-point bending flexural test of carbon fibre reinforced epoxy composite (45 mm x 12.5 mm × 2.5 mm) is taken here as an example from ref [1]. The load-deflection data extracted from a figure in the reference was used to calculate strain, stress and secant modulus at various strains.

The ultimate bending strength is 190 MPa and % strain at break is 10%. The calculated stress and secant Flexural modulus are plotted against strain as shown in Fig. 6.5

to show that the composite has decreasing modulus from 14.6 GPa at 0.15% strain to 2 GPa at 9% strain, typical of a viscoelastic material, despite 60% carbon fibre loading by volume.

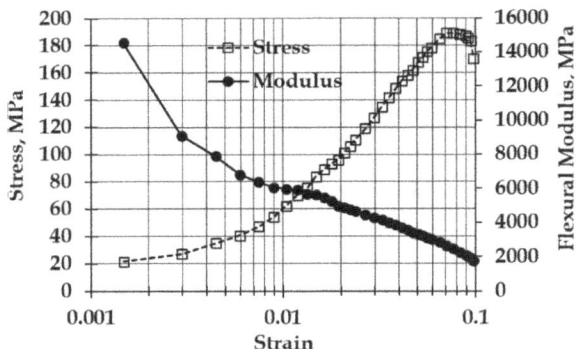

Fig. 6.5: A typical flexural test result for a CFRP composite using three-point bending method (reprinted with permission from Ben Wang [1]. © 2020, De Gruyter, Publisher Publishing Company).

6.2.4 Compression test

Compression properties are used for structural components of composites in most load-bearing application such as machinery foundation, vibration mounts and columns. Compression experiment of high strength composite is difficult due to three reasons [2, 3] such as end splitting of the surface of the sample, possibility of buckling of sample under force and perfect alignment along the axis of force without slippage of the sample. A careful arrangement can eliminate these problems to large extent. All standards suggest some suitable fixtures and gripping arrangement to avoid non-alignment, slippage and buckling. Table 6.4 lists the standards of compression test where anti-buckling guides are provided for correctness of compression strength.

Tab. 6.4: Standards for compression test of polymer matrix composites.

Standard	Description
ASTM D695-15	Standard Test Method for Compressive Properties of Rigid Plastics
ASTM D3410/D3410M or D6641/D6641M	For compressive properties of resin-matrix, composites reinforced with oriented continuous, discontinuous, or cross-ply reinforcements
DIN EN ISO 14126:1999	Fibre-reinforced plastic composites – determination of compressive properties in the in-plane direction
ISO 604:2002	Plastics — determination of compressive properties

Compressive properties can be measured for laminates using through the thickness mode where the force is applied transverse to the fibre lay-up. It has been observed [4] that the cross-ply arrangement (0°/90°) of fibre has higher compressive modulus compared to the unidirectional orientation. Compressive strength of cylindrical samples of FRPs are reported to be lower than their tensile strength because of the fact that in tension, failure took place by fibre rupture while in compression, the failure was by separation of longitudinally aligned fibres because of matrix rupture, rather than buckling [5]. Effect of buckling, however, has also been reported by Deitz et al. [6] while measuring compression properties of FRP bars of 15 mm diameter and 210 mm buckling length in the direction of fibre orientation.

The UTM can be used for measurement of compression properties with special arrangements to prevent buckling. Both ASTM and ISO standards recommend anti-buckling fixtures for compressive properties. Figure 6.6 shows one typical arrangement of sample holder for a cylindrical sample for compression test and one for prismatic sample for through the thickness measurement.

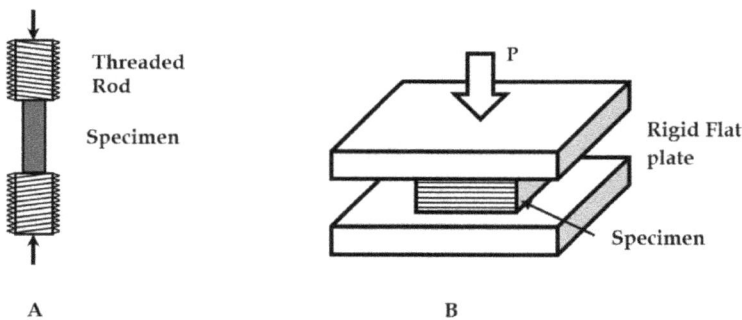

Fig. 6.6: Compression specimen arrangement: (A) cylindrical sample and (B) prismatic sample.

Figure 6.7 shows a compression test result as dependence of stress and Compressive modulus on compressive strain for a sample of carbon fabric-vinyl ester-epoxy hybrid resin system intended for use in making a high-pressure ball valve body for naval application. The sample was prismatic in shape with 10 mm × 10 mm cross section and 15 mm height, tested in through the thickness mode as shown in Fig. 6.6(B). The maximum strength is seen to be about 1.5 GPa, and the yielding starts at 18% strain. Maximum compressive modulus was 8.3 GPa at 16% strain. The stress and modulus range were important to decide its suitability.

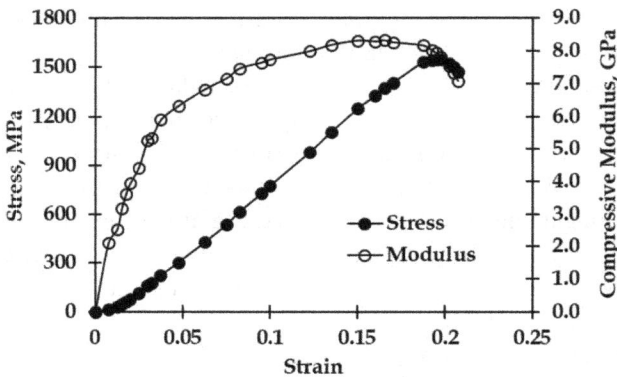

Fig. 6.7: Compression test result of a CFRP in through the thickness test of a prismatic sample.

6.2.5 Shear test

Shear test is somewhat more complicated since there should be a force couple acting on two surfaces to shear a solid. The sample is to be held between two jigs in a UTM and the force is applied to one fixer jig to shear the specimen. This is difficult for high modulus composites, but simple for elastomeric composites. However, using some appropriate fixing method and using two strain gauges, one for in-axis and the other for transverse strain, and the shear stress and modulus is calculated from the strain gauge readings. A simple method almost similar to three-point bending is also done for shear properties of composites. The standards for shear tests are given in Tab. 6.5, and different methods are also described in brief.

Tab. 6.5: Standards of shear test for polymer matrix composites.

Standards	Description
ASTM D-2344	Short beam shear testing for interlaminar parallel fibers
ASTM C 273	Standard test method for shear properties of sandwich core materials
ISO 14130:1998	Fibre-reinforced plastic composites. Determination of apparent interlaminar shear strength by short-beam method. Applicable also for reinforced pultruded thermoset products
ASTM D3846-08(2015)	Standard test method for in-plane shear strength of reinforced plastics in-plane shear strength of reinforced thermosetting plastics in flat sheet form in thicknesses ranging from 2.54 to 6.60 mm
ASTM D732-17	Standard test method for shear strength of plastics by punch tool
ISO 20337: 2018	Fibre-reinforced plastic composites — shear test method using a shear frame for the determination of the in-plane shear stress/shear strain response and shear modulus

Tab. 6.5 (continued)

Standards	Description
ISO 1827:2016	For rubber, vulcanized or thermoplastic – determination of shear modulus and adhesion to rigid plates – quadruple-shear methods
DIN EN 2563	Carbon fibre reinforced plastics – unidirectional laminates. Determination of the apparent interlaminar shear strength (aerospace)
DIN 65148	Fibre-reinforced plastics – determination of interlaminar shear strength by tensile test
DIN 53294	Testing of sandwiches – shear test

6.2.5.1 The short beam shear test for composite material

Different standards (e.g. ASTM D2344 and BS EN ISO 14130:1998) describe slightly different versions of the short beam shear test. However, the underlying principles are the same. A specimen with a rectangular cross-section is placed upon two lower supports that are a fixed distance apart (the "span"). A central load is then applied to the specimen from above. The geometry of the specimen is such that its span-to-thickness ratio is relatively low. This ensures that the specimen is predominantly loaded in shear. Figure 6.8 shows a typical short beam shear test arrangement with the semi-circular supports and the ram with semi-circular tip for force application.

6.2.5.2 In-plane shear testing of FRP composites

In-plane shear properties of composites are measured by a tensile test of +45/−45 ply in resin matrix with a notch at the middle of the FRP with two strain gauges. It is known that shear at 45° is uncoupled from axial stress in tension. Hence, in-plane shear modulus can be determined by a simple tensile test by monitoring strains in ±45° ply oriented FRP. Figure 6.9 shows a typical sample of FRP with 45° cross-ply prepared for in-plane shear test, with two orthogonal (0°/90°) degree strain gauges fixed onto the sample beam.

6.2.5.3 Shear and bond strength test for plastics and rubber composites

For particulate and nanocomposites of plastics and rubber, two types of shear test are dual shear arrangement as shown in Fig. 6.10 which is especially applicable for softer plastics such as polyethylene and polypropylenes and rubbers. The downward push of the two samples also gives the adhesion bond strength with the two metal fixers as seen in the figure. For exclusively rubber-based composites, a quadruple sample arrangement with two rigid end plates adhered by a strong glue is used for shear test as shown in Fig. 6.11. This test is also used to determine the strength of the adhesive bond between the polymer and metal.

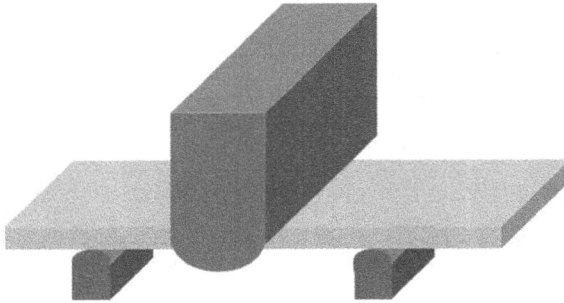

Fig. 6.8: Short beam shear test set-up.

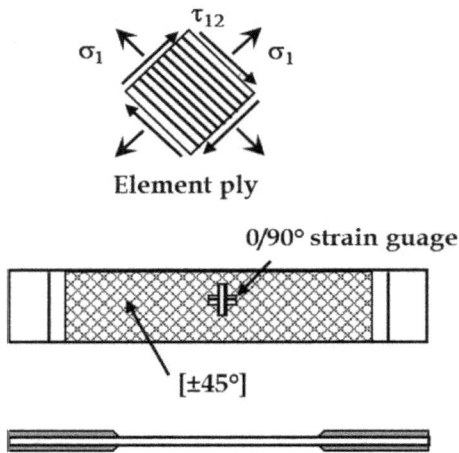

Fig. 6.9: In-plane shear test sample of FRP.

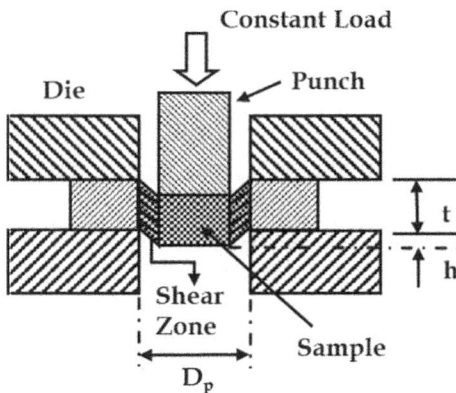

Fig. 6.10: Dual shear test for plastic- and rubber-based composites.

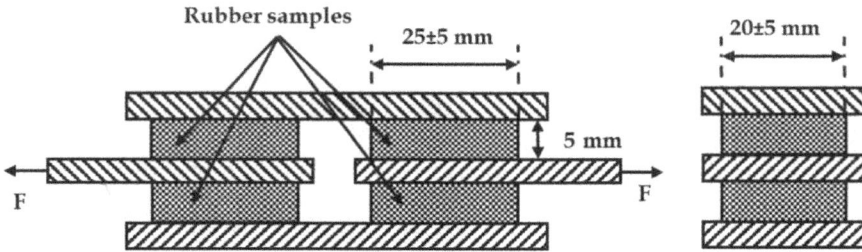

Fig. 6.11: Set-up for quadruple shear test for rubber composites.

6.2.6 Stress, strain and modulus in shear

(1) *Short beam test*
This is not a test for determination of fundamental properties such as shear modulus and its variation with strain, etc., but to qualify a material for a shear application to decide the ultimate shear strength of a composite. The ultimate "short beam strength" (ASTM D2344) is calculated from the test parameters as

$$\text{Shear strength: } \sigma_s = 0.75 \frac{F_s}{bh} \tag{6.8}$$

where b and h are width and depth of the specimen and F_s is the force applied to the specimen at the centre to rupture.

(2) *In-plane shear*
The details of the mathematical equations for this type of in-plane test are described in reference [7]. According to the sample arrangement in Fig. 6.9,

$$\sigma_1 = B\bar{\sigma}_{xx}$$
$$\sigma_2 = (1-B)\bar{\sigma}_{xx} \tag{6.9}$$

where $\bar{\sigma}_{xx}$ is the axial stress along the sample axis and σ_1 and σ_2 are the stresses at normal to the ply orientation as shown in Fig. 6.9, and B is given by

$$B = \left[\frac{m^2(2m^2-1) + 4m^2n^2 \frac{G_{12}}{E_2}\left(\frac{E_2}{E_1}\nu_{12}+1\right)}{4m^2n^2\frac{G_{12}}{E_2}\left(\frac{E_2}{E_1}+2\frac{E_2}{E_1}\nu_{12}+1\right) + (m^2-n^2)(2m^2-1)} \right] \tag{6.10}$$

and the in-plane shear stress is given by

$$\tau_{12} = \frac{-1}{2mn} \left[B\left(1 - 2m^2\right) + m^2 \right] \bar{\sigma}_{xx} \tag{6.11}$$

For element orientation is 45° with the applied force axis, as in Fig. 6.9, the in-plane shear stress is simplified as

$$\tau_{12} = \mp \frac{\bar{\sigma}_{xx}}{2} \tag{6.12}$$

The shear strain in the principal material direction is

$$\gamma_{12} = -\left(\varepsilon_{xx} - \varepsilon_{yy}\right) \tag{6.13}$$

where ε_{xx} and ε_{yy} are strains in the orthogonal directions of the element. The shear modulus is given by

$$G_{12} = \frac{\bar{\sigma}_{xx}}{2\left(\varepsilon_{xx} - \varepsilon_{yy}\right)} \tag{6.14}$$

(3) **Dual shear test**

According to Fig. 6.10, the shear stress is given by

$$\sigma_s = \frac{F}{\pi D_{\text{die}} t} \tag{6.15}$$

where F is the force applied to the punch, t and D_{die} are the thickness of the test specimen and diameter of the die punch. Shear strain is given by

$$\gamma = \frac{h}{\left(R_{\text{die}} - R_{\text{sample}}\right)} \tag{6.16}$$

where h is the deformation of the sample as shown in Fig. 6.10.

Therefore, the shear modulus is

$$G = \frac{\sigma_s \left(R_{\text{die}} - R_{\text{sample}}\right)}{h} \tag{6.17}$$

6.3 Impact testing

6.3.1 Izod and Charpy tests

Polymeric materials are sometimes subjected to rapid stress loading or impact loads. A number of test methods are available for assessing impact strength of hard materials and most used methods are so-called *Izod* and *Charpy* (pendulum) impact tests. The Izod test is the more popular method for plastics and composite materials, whereas

Charpy is very common for metals. Table 6.6 lists the standards for Izod pendulum impact energy measurement.

Tab. 6.6: Izod impact test standards.

Standard	Description
ASTM D256	Standard test method for determining the Izod pendulum impact resistance of plastics. The test requires specimens to be made with a milled notch
ISO 180: 2019	Plastics – determination of Izod impact strength

A sketch of a pendulum Izod impact tester is shown in Fig. 6.12. A proper sample preparation with the precise notch and repeated measurements using 5–10 samples is recommended for a good average data. The total impact energy depends on both the size of the test specimen and the notch shape and length. A standard specimen is usually used to allow comparison between different materials.

The pendulum impact test involves the measurement of the energy required to break a test specimen that is clamped at the ends and then struck in the centre by a pendulum weight. The energy required to break the specimen is obtained from the loss in energy of the pendulum. This energy is simply the difference in potential energy of the hammer before and after the impact and given by

$$E_{fr} = mg(h_S - h_E) \tag{6.18}$$

where m is the mass of the hammer, g is the gravitational acceleration (9.81 m/s^2), h is its height, while the suffixes S and E denote start and end.

The values are reported in terms of absorbed energy per unit of thickness at the notch (such as J/m or ft.lb/in). Alternatively, the results may be reported as absorbed energy per unit cross-sectional area at the notch (J/m^2 or ft.lb/in^2). The dimensions of an ASTM D256 standard specimen are 63.5 × 12.7 × 3.2 mm (2.5 × 0.5 × 0.125 in), with a 45° (angle) notch with a depth of 2.5 mm and slightly round at the wedge end of radius 0.25 mm. Due to the intricate shape and small dimensions of the notch, it is best to mould the sample in a polished metallic die rather than cutting from a sheet. Particularly, fibre reinforced composites are difficult to cut to such dimensions, since there

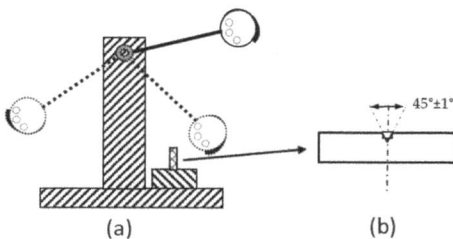

$45°\pm1°$

(a)   (b)

Fig. 6.12: (A) Sketch of an Izod impact tester and (B) sample with notch.

can be fibre pull out at the cutting edge. Injection moulded samples for thermoplastic-based particulate and short fibre composites would be good for this test.

6.3.2 Drop weight impact test

In drop weight impact test, a dart is allowed to fall on the sample and impact energy along with failure characteristics is evaluated. The test is carried out as per ASTM D7136. A photograph of an instrumented drop weight impact tester used by Çallıoğlu and Ergun [8], i.e. Instron-Dynatup 9250 HV model is shown in Fig. 6.13. The instrumented drop weight impact tester comprises a dropping crosshead with its accessories, a pneumatic clamping fixture, a pneumatic rebound brake, and Dynatup 930-I impulse data acquisition system. Testing machine can apply an impact energy ranging from 2.6 to 826 J, with the support of a spring without adding an extra weight to the weight box. Maximum physical drop height at which weight can be dropped is 1.25 m. Similarly, the maximum impact velocity is 5 m/s. The top of the impactor has a 12.5 mm diameter hemispherical tip and pneumatic rebound brake system which prevents the repeated impact on specimens. Impulse data acquisition system records the electronic signals and converts them into the impact parameters, such velocity, and energy. The specimens are clamped with the pneumatic fixture with an impact window of 30 mm diameter and the desired impact energy is applied. A schematic diagram of standard test piece size and impact position is presented [8] in Fig. 6.14. A typical load vs. deformation plot of 12-layer glass/epoxy composite is shown [8] in Fig. 6.15.

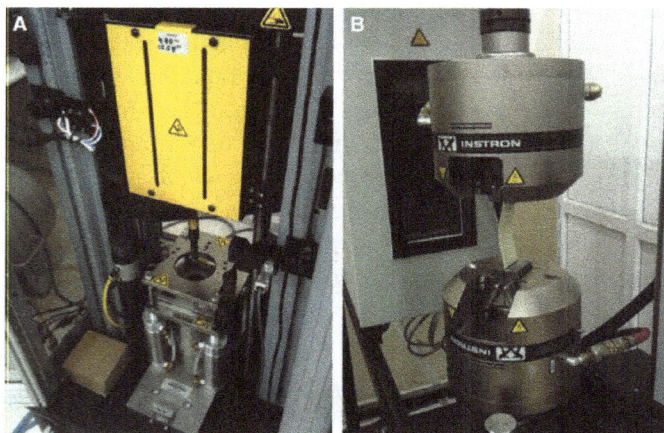

Fig. 6.13: Test equipment: (A) Dynatup 9250 HV impact test (B) Instron 8801 compression test (reprinted with permission from Hasan Çallıoğlu [8]. © 2014, De Gruyter Publishing Company).

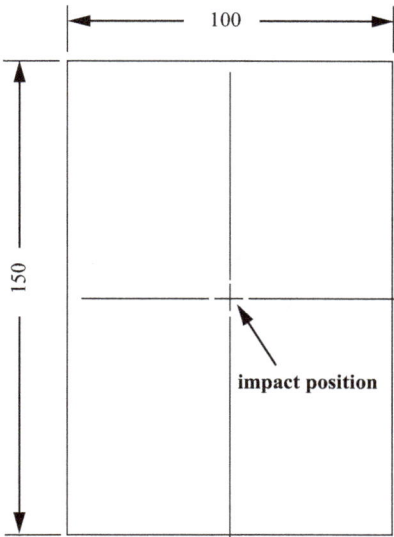

Fig. 6.14: Schematic diagram of standard test piece size and impact position (Reprinted with permission from Kang Yang [9]. © 2022, De Gruyter Publishing Company).

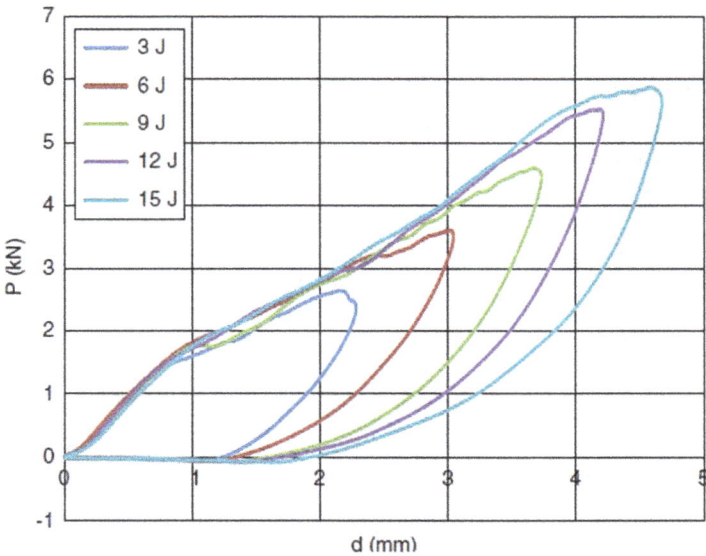

Fig. 6.15: Impact load-deflection curve for 12 layers (reprinted with permission from Hasan Çallioğlu [8]. © 2020, De Gruyter Publishing Company).

6.3.3 Post-impact testing

6.3.3.1 Compression test

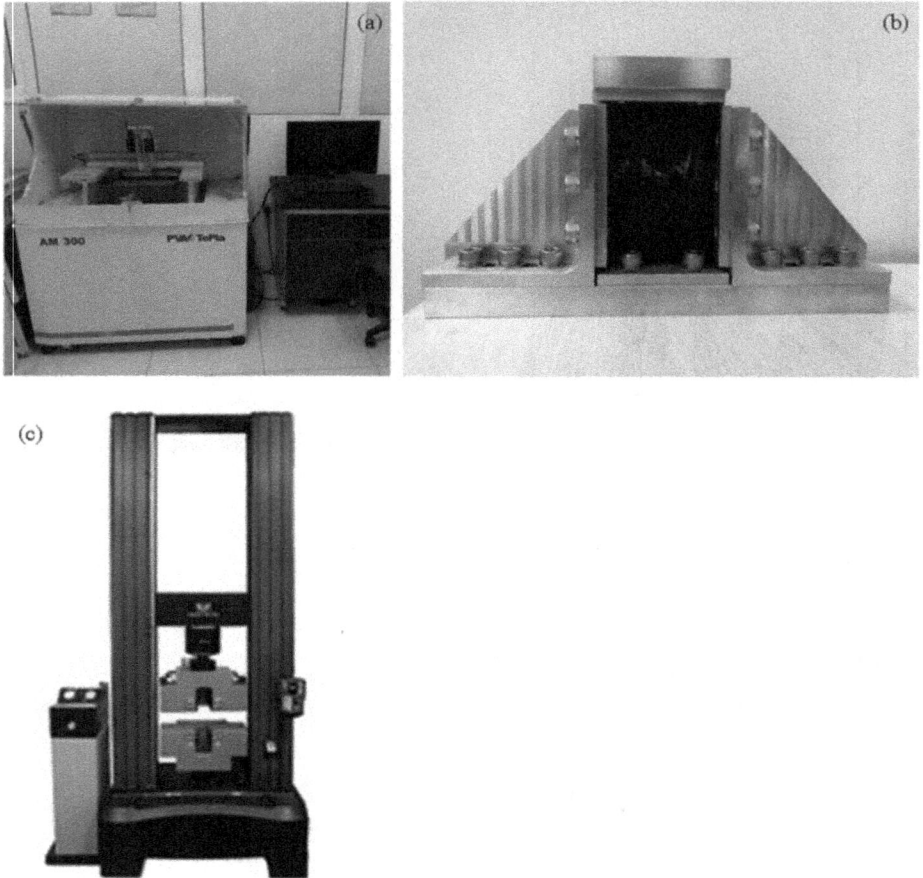

Fig. 6.16: Test equipment: (a) the ultrasonic C-scan, (b) the post impact compression test fixture and (c) the static universal testing machine (reprinted with permission from Kang Kang Yang [9].
© 2022, De Gruyter Publishing Company).

Unlike Izod or Charpy test, in drop weight impact test, complete failure may not take place. Therefore post impact analysis and testing is very important. Yang et al. [9] carried out detail post impact testing of three types of carbon fibre composite laminate prepared by resin transfer moulding (RTM) technique with different types of fibres: unidirectional carbon fibre prepreg (UIN23100), carbon fibre woven fabric (W-3021FF) and carbon fibre-aramid fibre-blended fabric (W-38211). They have studied the compression and ultrasonic C-scan tests. The photographs of the ultrasonic C-scan, the post impact compression test fixture and the static universal testing machine are shown in Fig. 6.16. The impact generated dents on the front side of the laminate as we can see the typical impact damage indentation in Fig. 6.17. After the impact test, the three-material composite laminates were visually inspected, and no visible damage was found on the reverse side of the impact. Then, the laminates were subjected to ultrasonic C-scan to detect the damage of the specimen under the same impact condition. Figure 6.18 shows the typical results of ultrasonic C-scan after the impact. It can be seen that the image after the impact shows failure morphology similar to the "top hat shape". Similar results were reported by Talreja et al. [10].

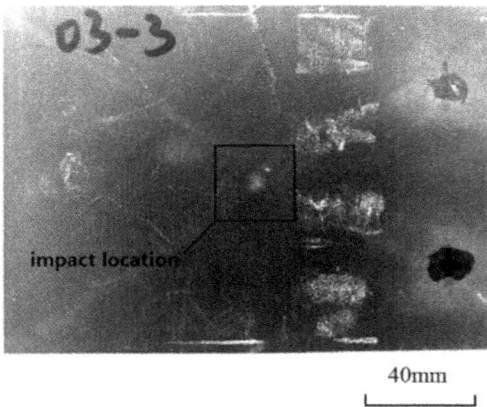

Fig. 6.17: Typical impact damage indentation (reprinted with permission from Kang Yang [9]. © 2022, De Gruyter Publishing Company).

In addition, the dent depth of the UIN23100/9314 specimen is smaller than that of the woven fabric specimen; therefore, the ability of UIN23100/9314 specimen to absorb and transfer impact energy is greater than that of woven fabric specimen. According to the ultrasonic C-scan results of the woven fabric specimen (Fig. 6.17b and c) the dent diameter of the W-38211 specimen is larger and the ability of absorbing and transferring impact energy is improved compared to that of the W-3021FF specimen. The load-bearing compressive strength values of UIN23100, UIN23100 and W-3021FF laminates are 147, 175 and 191 MPa, respectively. The results show that the compression bearing capacity of the woven fabric specimen after impact is higher than that of the carbon fibre unidirectional prepreg specimen, while for the resin-based woven fabric material; the compression bearing capacity of the W-38211 laminate is higher than that of the W-3021FF laminate. Failure of cracks of three kinds of materials extends along the width direction, but not along the length direction. Because the impact on the carbon fibre unidirectional prepreg specimen caused less damage, the damage modes of the front and back of the impact are consistent, and the damage of the speci-

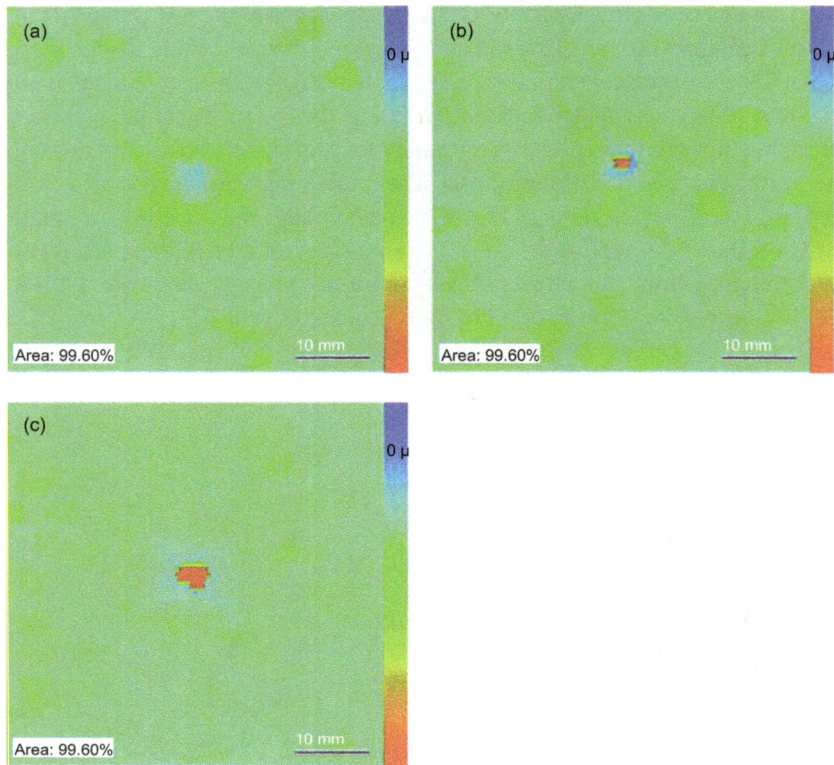

Fig. 6.18: Typical results of ultrasonic C-scan after impact: (a) UIN23100/9314, (b) W-3021FF/285/287 and (c) W-38211/285/287 (reprinted with permission from Kang Yang [9]. © 2022, De Gruyter Publishing Company).

men is caused mainly by the shear failure between the fibres. In the woven fabric laminates, the front of the impact presents a failure morphology characterized by shear failure, with less delamination failure. On the back of the impact, a large number of spalling occurred, and the local deformation presented a "C"-shaped bulge in the same direction as the dent.

6.3.3.2 Buckling tests

Ergun and coworkers [8] investigated the buckling behaviours of impacted and non impacted specimens at different impact points and energy levels, and having different thicknesses using a universal testing machine (UTM) of 50 kN loading capacity. Both the ends of the specimen are clamped into the machine's jaws, one of which is movable toward and away from the other. Axial Servo hydraulic testing machine has an adjustable crosswise head and its maximum hydraulic pressure is 207 bars. The axial compression tests are performed using displacement control with a speed of 0.3 mm/

min. During the test, the load versus displacement (contraction) curve is measured and recorded automatically by test machine. The hydraulic pressure in the head of machine during buckling tests was almost 40 bars. It was concluded from the above mentioned study that when a thin laminate is impacted at the centre the buckling loads are dramatically decreased for lower energy level. On the other hand a gradual decrease in buckling load is observed for a thicker composite. The change in the critical buckling loads decreases with increasing energy levels. As the distance between the point of impact and the centre of the specimen increases, the change in the critical buckling load values tend to reduce rapidly.

6.4 Dynamic mechanical analysis

Dynamic mechanical analysis is done to study the viscoelastic properties and fatigue of composite materials under time-varying force. In order to calculate the properties in time domain and frequency domain, the dynamic part is a pure sign wave so that mathematical equations can be used without complications. Secondly, the frequency of force cycle is very limited, so that the data acquisition is accurate enough to capture the locus of force and deformation (stress and strain) and also, to precisely measure the difference in phase angle between the stress and strain. The test is done as a temperature sweep or isothermal, at fixed frequencies and not in a frequency sweep method. The fundamental properties revealed by this analysis are viscoelastic parameters, mechanical energy loss, several transitions of a polymer such as a-, b- and g-transition temperatures, composition of a polymer mix, process and kinetics of curing, effect of moisture and plasticizer on modulus of elasticity, creep, relaxation and temperature limits of practical application for a structural material, such as a composite. The elastic modulus can be measured in all modes such as tension, compression, bending and shear. Table 6.7 lists the standards of measurement procedures. The standards elaborately describe the sample preparation, and testing methods by all modes of operation in DMA.

Tab. 6.7: Standards for analysis of dynamic mechanical properties.

Standard	Description
ASTM D4065, D4440, D5279	Dynamic mechanical analysis
ISO 6721 (Parts 1 to 12)	Plastics – determination of dynamic mechanical property

The machine, commonly known as dynamic mechanical analyser (DMA), has also some synonyms such as DMTA and DMS. The machine uses the basic principle of UTM, but with very low force because the moving ram has to oscillate quite fast even at 10 Hz (cycle time = 0.1 s), without much inertial resistance. The brief functional specification of a typical DMA is listed in Tab. 6.8.

Tab. 6.8: Brief functional specification of a DMA.

Parameters	Typical specification
Elastic modulus range	$10^3 - 10^{12}$ N/m^2 (precision: ±1%)
Temperature range and accuracy	−170 °C to + 500 °C (controllable using liquid nitrogen and heater) ±0.1 °C
Temperature ramp	0.1 °C/min to 10 °C/min
Fixed frequencies	0.01, 0.03, 0.05, 0.1, 0.3, 0.5, 1, 2, 3, 5, 10, 20, 30, 50.
Maximum force	100 N (static maximum 30 N)
Deformation precision	0.01 micron
Deformation range(amplitude)	0.01 micron to 10 mm
Loss factor range and accuracy	0.01–10. Accuracy: ±0.0001
Modes of measurement	(i) Tension, (ii) Single and double shear, (iii) single cantilever, (iv) dual cantilever, (v) three-point bending (vi) Compression
Programs	(i) Isothermal run, at one or several fixed frequencies (ii) Temperature sweep at a fixed frequency (iii) Creep and stress relaxation
Type of samples	Soft and hard composites including FRP
Sample geometry	Should be defined by manufacturer of the DMA
Software	(i) For all operations, data analysis, graphical and tabular representation of storage modulus, loss modulus, complex modulus, loss factor, complex viscosity (ii) Arrhenius and WLF software for master curve and activation energy determination All data should be available in Excel spreadsheet.

Details of criticalities in measurement regarding preferred frequency limit, sample dimensions and data analysis including time-temperature transformation are given in ref. [11]. A typical result of variation of dynamic (storage) modulus with temperature of an elastomer-graphene nanocomposite is shown in Fig. 6.19 at various fixed frequencies and subsequent frequency scale master curve at a reference temperature of 30 °C is shown in Fig. 6.20.

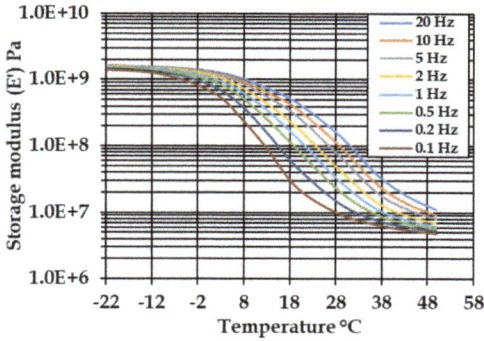

Fig. 6.19: Storage modulus of a rubber-graphene nanocomposite converted in frequency scale at 30 °C reference temperature.

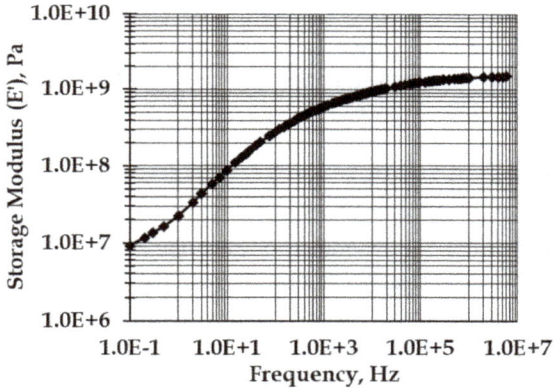

Fig. 6.20: Frequency-scale master curve at a reference temperature of 30 °C.

6.5 Thermal analysis

Physical and chemical changes in a polymer due to heat are studied for several investigative and application-oriented requirements. Behaviour of a polymer from cryogenic temperature to its decomposition temperature reveals very important information about the polymer. At extremely low temperature, the oscillatory motion and relaxation of even the smallest group or molecules attached in a polymer remains frozen. On increasing the temperature gradually, the γ-relaxation of an end group or defect is observed, followed by onset of movement by side groups and a crank-shaft-type motion by 4 carbon atoms in the chain, resulting in β-relaxation and finally, segmental motion, as α-relaxation, sets in, which is a glass to rubber transition. The segments of a linear polymer chain can be made of minimum 8 carbon atoms to maximum about 50 carbon atoms. At this point, the free volume increase, the specific heat changes to a higher value and the coefficient of thermal expansion also increases. However, the phenomenon of glass to rubber transition only exhibited by amorphous phase in the polymer. Further heating results in melting of the crystalline phase in the polymer to a flowable consistency. However, for fully amorphous polymers and cross-linked polymers, the fluidity is not observed. On further heating, the decomposition takes place, rather than volatilization, since the secondary valence forces are much stronger than primary bonds in polymers. The decomposition can be due to chain scission, fragmentation, oxidation, release of side groups, etc., and finally charring. Heat can also be used to study volatile matter content such as moisture, or, crystalline melting point and heat of melting, and also to study reactions such as curing of resins. For composites, all the thermal behaviours are similar to their parent matrix, with change in the intensity of the properties. Therefore, the study of change in the thermal properties and of reaction dynamics with variation of the content of reinforcements is important to design appropriate composite material.

Various thermal analysis instruments are used to study different types of thermal effects on polymers and composites. Some of the important instruments and method of thermal analysis with few relevant international standards are described below.

6.5.1 Differential thermal analysis (DTA)

In this analytical method, the polymer sample in a metallic/ceramic pan is heated at a constant heating rate along with another blank pan as reference. Both the sample and the reference are heated by the same heat source and the difference in temperature ΔT between the two is recorded. The plot of ΔT vs. temperature will be a straight line with a steady slope, if the heat capacity is independent of temperature. When a transition occurs in a sample, a temperature difference will show as endothermic or exothermic depending on the transition. Therefore, the DTA can reveal a phase change or decomposition, which is either exothermic or endothermic, with a peak which is indicative of the temperature of the phase change or maximum decomposition temperature. Table 6.9 lists the standards for measurement procedures by DTA.

Tab. 6.9: Standards for measurement of thermal property by DTA.

Standard	Description
ASTM E794-06(2018)	Standard test method for melting and crystallization temperatures by thermal analysis
DIN 51004	Thermal analysis; determination of melting temperatures of crystalline materials by differential thermal analysis
DIN 51007	Thermal analysis – differential thermal analysis (DTA) and differential scanning calorimetry (DSC) – general principles

The analyser consists of two sample holders fitted with two thermocouples for temperature measurement of sample and reference, while another thermocouple is used controlling the heating rate by adjustment of the current flow through the heaters of the furnace. The sample and reference pan with the base holder is enclosed in a furnace which can be cooled or heated by programmable rate using liquid nitrogen and electric heaters. Nitrogen gas flow at a rate not more than 50 mL/min is used to provide inert atmosphere to avoid oxidation. A typical sketch of the functional part of a DTA is shown in Fig. 6.21.

A schematic plot in Fig. 6.22 shows the nature of the plot of a polymer showing T_g, Crystallization, crystalline melting, oxidation/decomposition. During crystallization or melting, the temperature of the sample remains constant, but the reference temperature changes, hence ΔT changes, and on completion of the event, the base line again shows a linear nature, till oxidation. Oxidation is associated with exothermic heat, therefore the positive peak in y-axis (ΔT) indicates oxidation. However, if the measurement is done in inert atmosphere, then the dotted line shows linear change in temperature instead of the peak. Decomposition, at inert atmosphere, for a polymer is endothermic.

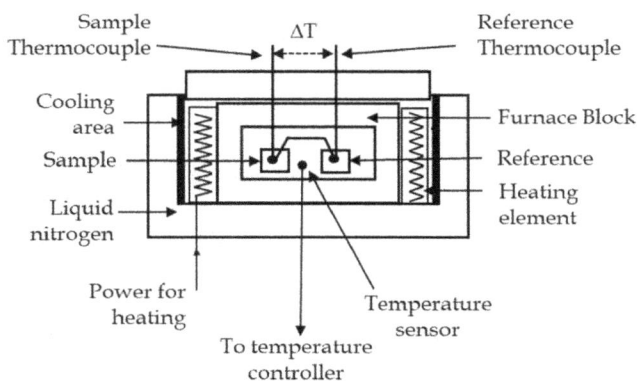

Fig. 6.21: Sketch of a DTA functional part only.

6.5.2 Thermogravimetric analysis (TGA)

A polymer composite, on heating, may release volatile substances physically mixed in it, and/or decompose to produce volatile molecules and oxidized gases and finally, the residue may be the charred carbon and metal oxides and silica (if any). Each step of mass loss occurring at specific temperature zones is recorded by the TGA instrument. The mass of the polymer composite sample is continuously recorded against temperature in either isothermal mode or under constant hating rate. The mass loss upon heating, gives the valuable information on the initial volatile (such as moisture) content, decomposition

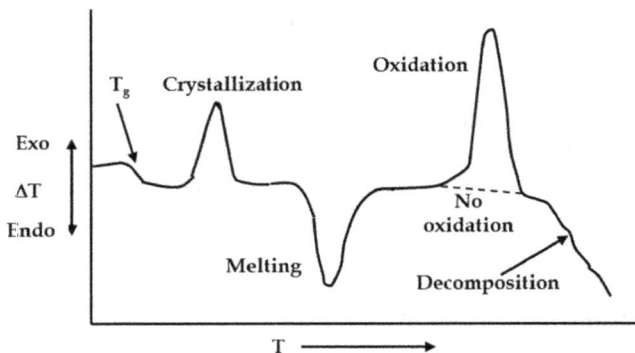

Fig. 6.22: A typical DSC plot showing T_g, T_m and T_c of a polymer sample.

stages, and thermal stability of a material. On coupling with a gas chromatograph-mass spectrometer (GC-MS) or Fourier-transform infrared spectrophotometer (FTIR), composition of the evolved gases can be analysed, which is used in composition analysis of a sample. Additionally, the first derivative of the weight loss (dw/dt), commonly known as DTG, gives the maximum decomposition rate and temperature at which it occurred. The residue helps in analysis of the mineral/silica content. Hence, the TGA allows one to study the composition of the material. Table 6.10 lists the standards for measuring such properties and analysis by TGA. Generally, a TGA for polymeric materials operate in the temperature range of cryogenic to maximum 1,000 °C and heating rate can be 0.1–100 °C/min, with flexible options of start and end temperatures, isothermal, dynamic run, and also different programmed zones in a single run. For low temperature, liquid nitrogen envelop is used with special flow control system. Figure 6.23 shows a functional part of a typical TGA instrument.

Tab. 6.10: Standards for analysis by TGA.

Standard	Description
ASTM E1131	A method of measuring volatile components, combustible matter, and ash content by TGA
ISO 11358-1:2014	Thermogravimetry (TG) of polymers – Part 1: general principles
ISO 11358-2:2014	Thermogravimetry (TG) of polymers – Part 2: determination of activation energy
DIN 51006	Thermal analysis (TA) – thermogravimetry (TG) – principles

Present-day TGA uses a modulated temperature scan, where, the steady state constant heating ram is continuously superimposed by a dynamic sinusoidal heating power, which is very effective in enhancing the minute event, e.g. small rate of mass change. *Modulated TGA* is now mostly used for thermal analysis of all critical materials including nanocomposites. TGA study can be done under a selected environment such as blanket of nitrogen/air/helium/argon. In certain cases, oxygen or nitrogen addition increases the weight, hence weight gain in such study is used to analyse the formation of nitride or oxide in a composite sample. A good example is microwave absorbing composite with nanoiron particles, where weight gain is observed in the residue compared to added weight of the iron in the initial composite. For a polymer, the decomposition under inert atmosphere shows the inherent stability of the polymer, while under air, it shows actual stability in most application. TGA is also a powerful tool to study Kinetics of decomposition of composites, accelerated thermal ageing and subsequent life time estimation at service conditions. A typical TGA-DTG (dw/dt) plot for an epoxy thermoset is shown in Fig. 6.24.

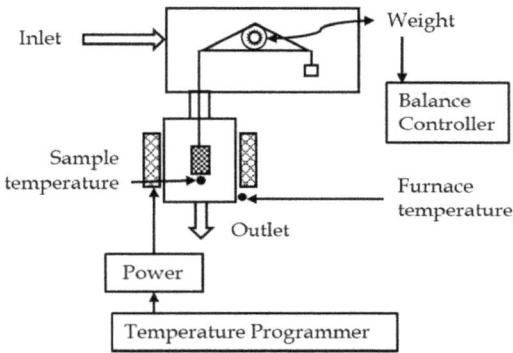

Fig. 6.23: Sketch of functional parts of a TGA instrument.

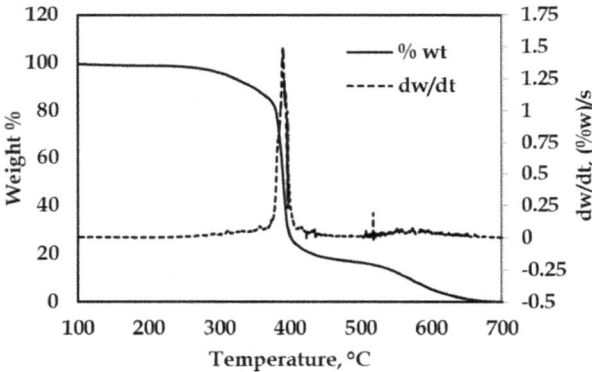

Fig. 6.24: A typical TGA-DTG plot of an epoxy thermoset.

6.5.3 Thermomechanical analysis (TMA)

Polymers exhibit different rates of dimensional and volume change with change in temperature. Below the glass transition, the expansion coefficient is lower than that at above the transition, due to increase in free volume as a result of onset of segmental mobility beyond the glass transition. In addition, the transition is characterized by drastic reduction in hardness, which can be observed as an increase in indentation of a probe onto the surface of a polymer sample. The behaviour of the composites is no exception. Therefore, the linear and volume expansion coefficients, glass transition and heat deflection temperature of a composite can be measured with change in temperature using appropriate probes, dilatometer and indentation. Table 6.11 lists the various standards related to different modes of TMA measurement procedures.

Tab. 6.11: Standards for measurement procedures by TMA.

Standard	Description
ASTM E1545-11(2016	Standard test method for assignment of the glass transition temperature by thermomechanical analysis
ASTM E 2347	Standard test method for indentation softening temperature by thermomechanical analysis
ASTM E 2092	Standard test method for distortion temperature in three-point bending by thermomechanical analysis
ISO 11359-1:2014	General conditions for the thermomechanical analysis of thermoplastics and thermosetting materials, filled or unfilled, in the form of sheet or moulded parts
ISO 11359-2:2021	Thermomechanical analysis (TMA) – Part 2: determination of coefficient of linear thermal expansion and glass transition temperature
DIN 53752	Testing of plastics; determination of the coefficient of linear thermal expansion coefficient
DIN 52545	Testing of rubber – determination of low-temperature behaviour of elastomers – principles and test methods

TMA is used to study these characteristics of polymers and composites. The instrument consists of a sample holder, where quartz vials containing the sample is placed in case of volume expansion measurement. The vial, fitted with a quartz rod probe is actually a dilatometer. The other end of the probe is attached to a system of non-contact motor for force application to the probe. Initial force can be adjusted so that the probe end in the dilatometer is in good contact with the sample. For volume expansion, the polymer should be powder or small granules to fill up the vial properly. For linear expansion, the regular shaped sample can be placed on a flat base of the sample holder and a quartz probe tip rests on the sample. A small initial force can be applied for firm contact of the probe tip to the surface of the sample. Present day TMA can apply a wide range of force, from 0.001 to 2 N and samples of 25 mm length can be accommodated with a measuring range of ±2.5 mm. The movement of the probe is measured by an LVDT with an accuracy of about 20 nm. Samples of elastomeric to stiff composites can be used for TMA study. The furnace, housing the sample holder and temperature sensor, is fitted with both controlled heating and cryogenic cooling by liquid nitrogen. The measurements are done in nitrogen blanket to avoid oxidation. A sketch of functional parts of the TMA and some probes are shown in Fig. 6.25.

A linear expansion study of a cured phenolic resin was done in a TMA to determine the glass transition temperature. Figure 6.26 shows the linear expansion curve with respect to rise in temperature. The thermal expansion curve showed two distinct regions, below and above the glass transition (T_g = 150 °C), characterized by a sudden increase in the slope.

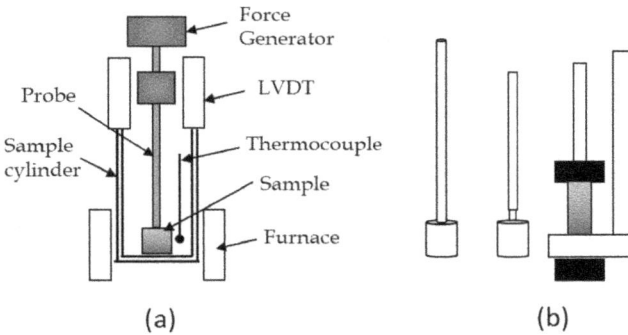

Fig. 6.25: (A) Sketch of functional parts of a TMA and (B) different probes.

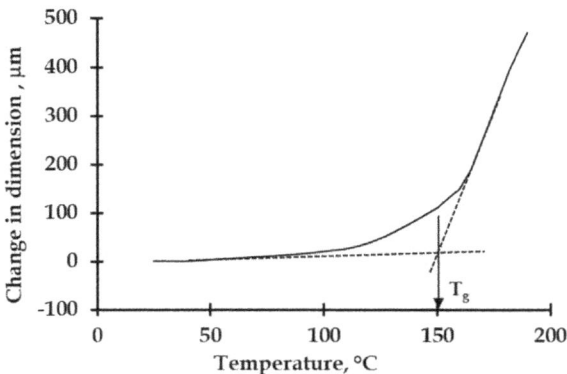

Fig. 6.26: Thermomechanical analysis of a cured phenolic resin: linear expansion mode.

6.5.4 Differential scanning calorimetry (DSC)

The name implies that it is basically a calorimeter, which measures heat as a function of temperature and time. When a certain mass of a polymer is heated from below its glass transition temperature, it absorbs heat and the temperature increases according to its heat capacity. The enthalpy will be linearly varying with temperature due to the constant heat capacity. At the glass transition, the mobility of the chain segments set in, thereby causing an increase in the heat capacity, and hence, there will be a change in the base line of the enthalpy vs temperature plot. Similarly, during a crystalline melting, there will be an endothermic peak as the material will absorb energy without change in temperature. Hence, crystallization phenomenon is characterized by an exothermic peak when the molten mass is cooled down. In a chemical reaction such as curing of thermoset, the reaction is indicated by an exothermic peak. The physical and chemical changes including reactions of a polymeric composite takes place within about 350 °C, beyond which, most polymers decompose. There can be quite large number of parameters and study by DSC. However, some standard methods of measurement are listed in Tab. 6.12.

Tab. 6.12: Standards for some measurements by DSC.

Standard	Description
ASTM D3418-21 and E1356	Heat of fusion, crystallization, melting point and glass transition by DSC
ISO 11357	General DSC principles, method of determination of temperature and enthalpy of melting and crystallization, and determination of glass transition temperature
DIN 53765	Testing of plastics and elastomers; thermal analysis; DSC-method
DIN 65467 Aerospace	Testing of thermosetting resin systems with and without reinforcement – DSC method

These phenomena are studied using a Differential Scanning Calorimeter (DSC) which compares the change in enthalpy (endo or exo) of two identical metallic/ceramic pans, one blank as reference and the other with the polymeric sample. Generally, a DSC works on either temperature compensation mode or power compensation mode, to compare the difference in consumed power between reference and sample. Any change from the base line during the run is due to the thermal effect on the polymer, since the metallic or ceramic pan do not show any physical or chemical change in the DSC temperature range (cryogenic to 600 °C) under inert gas blanket. A sketch of functional parts of a DSC instrument is shown in Fig. 6.27.

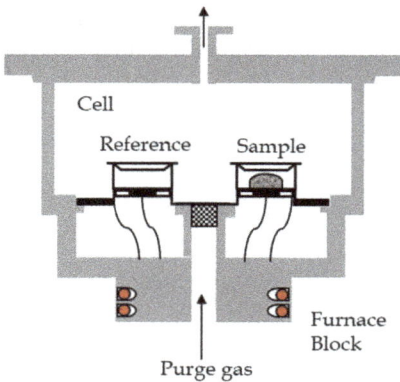

Fig. 6.27: Sketch of functional parts of a DSC instrument.

It is known that thermosets do not have crystalline phase, hence on heating, they exhibit glass transition and on further heating, decompose. Similarly, all amorphous polymers and their composites do not exhibit melting phenomenon.

Crystalline melting is only shown by thermoplastics with crystalline phase. Examples of glass transition and crystal melting can be seen for a thermoplastic material as shown in Fig. 6.28. The forward run (solid line) shows the Glass transition temperature at about 117 °C, indicated by a change in the base line, followed by an endothermic peak for melting at 162 °C. Upon cooling (dotted line), crystallization takes place, indicated by an exothermic peak (heat is released upon crystallization) at 115 °C. The crystallization temperature is the temperature where the molecules rearrange into ordered formation (crystallize, and an exothermic process). The melting temperature is where the polymer chains can freely move and ordered arrangements are disrupted (this annihilates any prior crystallization; an endothermic process). This explains the reason for different temperatures for melting and crystallization.

The heat flow in the Y-axis is actually the rate of flow of heat (dH/dt), either in mW or in W/g, if the mass of the polymer sample is known. Therefore, the area under the curve of melting or crystallization peak (in dH/dt vs time) would give enthalpy of melting or crystallization in J/g. The enthalpy of melting is additive for polymer blends and DSC study can be useful to find a blend composition, if the pure components are both crystalline/semi-crystalline and individual melting values are known or measured by DSC. Similarly, polymer content in an unknown composition of a composite can also be determined, if cry thermoplastic semi-crystalline polymer is identified.

DSC is also used to study reaction kinetics and very widely used for quantitative studies on enthalpy of reaction in vulcanization of elastomer composites, curing of thermoset composites including epoxy, phenolic, polyurethanes, etc. Various kinetic expressions are derived using the heat flow rate (dH/dt) vs time and temperature data, and this is discussed in detail in Chapter 3 under Curing of Thermoset-Based Composites. A typical curing plot of an epoxy resin with an amine hardener is shown in Fig. 6.29. The area under the curve gives the enthalpy of reaction in J/g. The peak of the curve shows maximum rate of the curve, and the corresponding temperature from the figure is 157.4 °C.

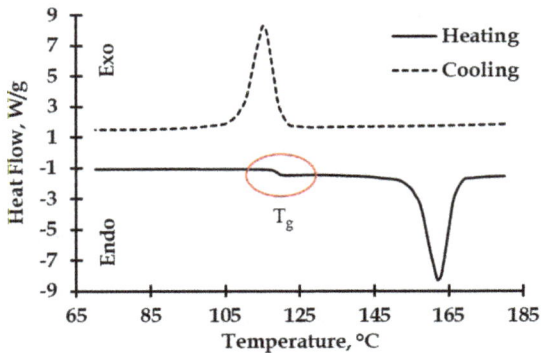

Fig. 6.28: DSC plot of a semi-crystalline thermoplastic: heating and cooling runs.

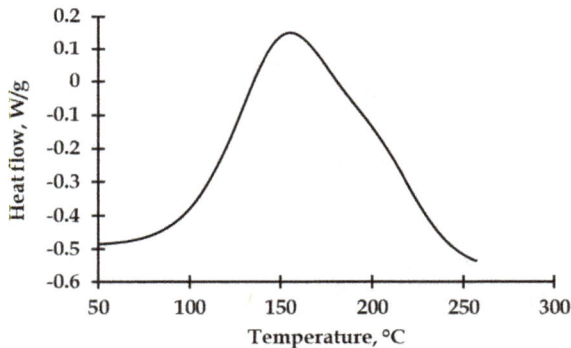

Fig. 6.29: A typical curing curve of an epoxy resin cured by a polyetheramine.

6.5.5 Dielectric analysis (DEA)

Dielectric properties of composites are very important for application in electrical gadgets, casings, printed circuit boards, etc. Most thermosets are polar in nature, and so are their composites. Dielectric strength (permittivity) of a polar polymer undergoes drastic change in frequency and temperature domain with loss of the property due to relaxation of the dipoles. In addition, ionic translations towards opposite electrodes also changes in similar manner. Therefore, like dynamic mechanical properties, dielectric properties are characterized by relaxation and are complex like the complex elastic modulus. The complex dielectric permittivity (e^*) has, therefore, a real part (ε') and an imaginary or loss part (ε''), widely varying with temperature and frequency. The dielectric relaxation time is the inverse of the angular frequency at which the loss (ε'') attains a peak value, just as loss modulus peak.

There is no specific standard for complex permittivity measurement below 0.50 GHz. Only, ASTM D150 and IEC 60250 are the standards for measurement of dielectric constant (DC) and dissipation factor (DF) in the general frequency range of 10 Hz to 2 MHz. However, these standards do not specify temperature sweep, nor mentions the necessity to use the DEA instrument. ASTM D2520–21 is the relevant standard of test method for Complex Permittivity of Solid Electrical Insulating Materials at Microwave Frequencies (0.5–50 GHz) and Temperatures up to 1,650 °C. The test methods cover the determination of relative complex permittivity (dielectric constant and dissipation factor) of nonmagnetic solid dielectric materials.

Dielectric analysis provides the most sensitive method to study local motions along the chains of polar polymers, since polar bonds (such as >C=OH, -COOH, -C $\equiv$ N and >N–H) are directly affected by the electric stimulus. Therefore, DEA can be used to study complex permittivity even beyond GHz range, and there are different instruments for such high frequency ranges, such as vector network analyser. The GHz range is very useful in strategic defence subjects such as radar stealth coatings and composites. Details of dielectric analysis are described in ref. [12].

A sketch of the functional parts of a typical DEA is shown in Fig. 6.30. The instrument houses a source of alternating electric field applied through gold coated electrodes across the thin disc type sample. The field can have sweep of frequency and fixed as well, at fixed temperatures and temperature sweep at fixed frequencies. The DEA cell is housed in a furnace with controlled electrical heating arrangements and also cooling by liquid nitrogen and a system of purging inert gas such as nitrogen to avoid oxidation of the test sample. Common DEA instruments are very widely used in study of curing thermosets in research and production as well. As the curing proceeds, the loss part reduces due to the reduction of mobility of the dipoles while the real part of complex permittivity (ε') and ion viscosity rises sharply at gel point.

An epoxy thermoset was studied for complex dielectric property in temperature scale at fixed frequencies of 10^3, 10^4 and 10^6 Hz. The real part of permittivity (ε') and the tanδ is shown in Fig. 6.31. At higher frequency, the peak loss shifts towards higher temperature, indicating faster relaxation, which is the general behaviour of a polymer. Also, the value of tanδ is progressively higher for higher frequency, which is a typical property of dielectrics.

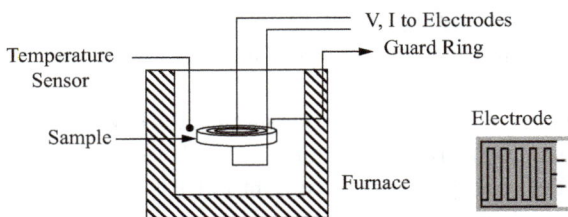

Fig. 6.30: A sketch of a DEA instrument and the electrode design.

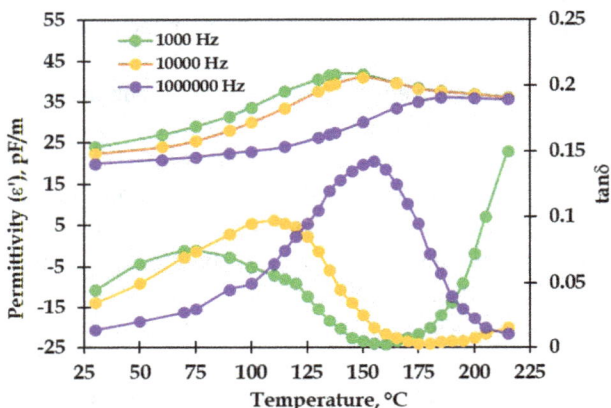

Fig. 6.31: A DEA experiment of an engineering plastic sample at fixed frequencies.

6.6 Morphological study

6.6.1 Light microscopy

Light microscope (LM) is utilized to generate magnified image of an object by using of a series of glass lenses so that one can visualize a fine details of the same. In LM a beam of light is focused onto or through an object, and the image formed is enlarged by using convex objective lenses. In most of the commercially available LM, the image is observed directly through a binocular eyepiece, which acts as a secondary lens in the form of a magnifying glass in order to observe the projected image. These are known as 'compound microscopes,' in which the total magnification is calculated from the sum of the objective magnification and the eyepiece magnification. The magnification range varies from ×10 to ×1,000, with a resolving power of the order of 0.2 µm which again depends on the type and numerical aperture of the objective lenses. Light microscope is an extremely versatile instrument and allows to examine specimens of any size, whole or sectioned, wet or dry, hot or cold, and static or fast-moving. When in LM heating arrangement for analysing the sample at a temperature higher than room temperature is available then it is called hot-stage microscope. In order to produce better resolution polarized light microscopy is used. In a polarized light microscope, a polarizer is placed in between the light source and the sample, which converts the polarized light into plane-polarized light before it hits the sample. The basic difference between unpolarized light and plane-polarized light is that plane-polarized light vibrate in in a single direction whereas normal light vibrate in two mutually perpendicular directions. This polarized light falls on a doubly refracting specimen which generates two wave components that are at right angles to each other. The waves pass through the sample in different phases. An analyser is used in polarized light microscopy, which combines the polarized light using constructive and destructive interference leading to generation of high-contrast image.

LM with is widely used to study the crystallization behaviour of semi-crystalline polymer based composites. It is observed that the incorporation of nanoparticles in a semi-crystalline polymer matrix affect significantly the crystallization behaviour of the polymer. Depending on the polymer/filler interactions, three different types of behaviour are expected during the crystallization process such as (i) development of new crystal structures, (ii) heterogeneous nucleation by fillers and (iii) polymer amorphization by filler [13]. The author has investigated [14] the crystallization behaviour of nanocomposites of poly(ethylene oxide) (PEO) and organoclay using polarized light microscopy. Figure 6.32 compares the POM images of PEO and PEOCN3 (nanocomposite of PEO containing 3 wt% clay) (which were isothermally crystallized at 40 °C. The morphology of the crystals is shown at the beginning and at the final stage of crystallization. For the PEO alone, it can be clearly seen that the spherulites are similar in size. Before impinging upon one another, they appear circular, suggesting a spherical shape. For the intercalated system, the spherulite formation varies considerably. The spherulites are typically much smaller than those seen in the virgin PEO. A similar behaviour was reported by Strawhecker and Manias [15] and also by Ratna et al. [16] for solution intercalated PEOCN systems. It is clear from Fig. 6.32 that the clay particle acts as a nucleating agent and the PEO crystallized in the heterogeneous nucleation mode. As the number of nucleation sites increased the number of the spherulites also increased leading to the formation of more number of smaller spherulite.

It has also been utilized to study [17] the reaction-induced phase separation of a toughened epoxy resin (modified with epoxy-functionalized hyperbranched polymer) for application as a matrix for fibre-reinforced composites. A mixture epoxy and hyperbranched polymer (HBP) is heated from room temperature to 100 C and kept at constant temperature. At various times, photographs were taken on hot-stage optical microscopy and presented in Fig. 6.33. We can see that the HBP is not miscible with epoxy at room temperature. As the temperature increases the HBP gets miscible with epoxy making the mixture homogeneous functionalized HBP is not miscible with epoxy at room temperature but becomes miscible at higher temperatures (100 °C). With the advancement of curing reaction, the HBP undergoes

Fig. 6.32: Cross-polarization optical microscope images of (a) PEO at the beginning of crystallization, (b) PEO after the crystallization and (c) PEOCN3 after the crystallization (reprinted with permission from T.N. Abraham [17]. © 2008 Society of Plastics Engineers Publishing Company).

phase separation and as a result a two-phase microstructure is formed. The reason for phase separation is the decrease in combinatorial entropy due to formation of network. Note that mixing of epoxy with HBP is favoured by the increase in combinatorial entropy of the mixture.

6.6.2 Scanning electron microscopy

Electron Microscopes are capable of much higher magnifications and have a greater resolving power than a light microscope. This is because the wavelength of electron is much smaller than light and resolution is directly proportional to the wave length. In fact electron microscopy like scanning electron microscope (SEM) and transmission electron microscopy (TEM) are able to image much smaller objects at sub cellular, molecular level. Another difference is that LM reveals only in-plain details of morphology whereas SEM image is able to provide some kind of 3-D appearance of the surface under investigation. That is why SEM is widely used to characterize the phase morphology and failure behaviour of FRP composites, i.e. to understand whether the fracture is brittle or ductile one. The working principle of SEM is that a beam of electrons is formed by an electron source and accelerated toward the specimen using a positive electrical potential. The electron beam is confined and focused using metal apertures and magnetic lenses into a thin, focused, monochromatic beam. Electrons in the beam interact with the atoms of

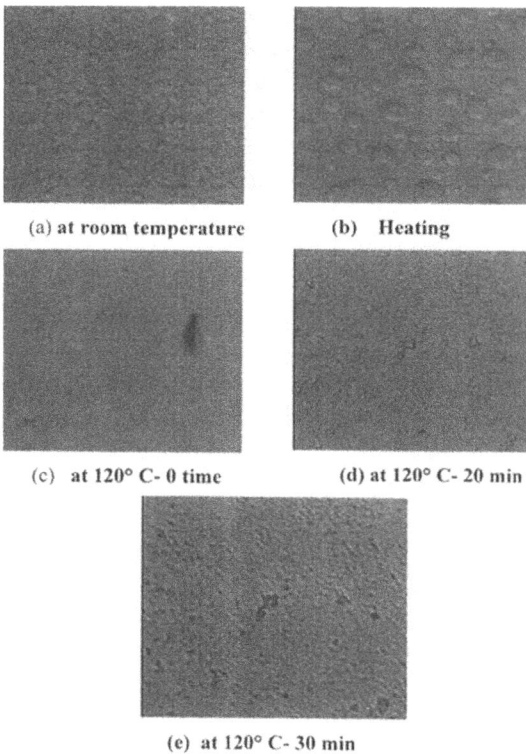

(a) at room temperature   (b) Heating

(c) at 120° C- 0 time   (d) at 120° C- 20 min

(e) at 120° C- 30 min

Fig. 6.33: Optical micrographs of epoxy/HBP/DETDA mixture: (a) at room temperature, (b) heating, (c) at 120 °C in zero time, (d) at 120 °C after 20 min and (e) at 120 °C after 30 min (reprinted with permission from Debdatta Ratna [17]. © 2010 Wiley Periodicals, Inc.).

the specimen, producing signals that contain information about its surface topography, composition and other electrical properties. These interactions and effects are detected and transformed into an image. Images are formed by rastering the electron beam across the specimen using deflection coils inside the objective lens. The electron beam should have a circular cross section when it strikes the specimen.

An SEM machine comprises various components like electron gun which is used as a source of electron in tungsten single crystal, lanthanum hexaboride, zirconium oxide, etc. The electron column is where the electron beam is generated under vacuum, focused to a small diameter, and scanned across the surface of a specimen by electromagnetic deflection coils. After the beam passes the anode it is influenced by two condenser lenses that cause the beam to converge and pass through a focal point. In conjunction with the selected accelerating voltage, the condenser lenses are primarily responsible for determining the intensity of the electron beam when it strikes the specimen. After the beam passes the anode it is influenced by two condenser lenses that cause the beam to converge and pass through a focal point. In conjunction with the selected accelerating voltage, the condenser lenses are primarily responsible for determining the intensity of the electron beam when it strikes the specimen. When the primary electron beam interacts with the sample, the energy exchange between the electron beam and the sample results in the reflection of high-energy back scattered electrons by elastic scattering, emission of low energy secondary, auger electrons by inelastic scattering and the emission of electromagnetic radiation (X-rays and cathodoluminescence), each of which can be detected by respective detectors. Various detectors are used for example Everhart-Thornley detector for secondary electron, solid-state detector for backscattered electrons and energy-dispersive spectrometer (EDS) detector for the produced X-ray.

Generally polymer composite samples are cryogenically fractured and the fracture surface is coated with a gold or carbon to make the surface conducting so that any charging problem does not arise as discussed above, SEM can provide resolution up to 1 nm and as a result widely used for characterization of nanofillers and polymer nanocomposites. For example, Ratna and coworkers [18] studied the morphology of unmodified and a modified MWCNT (with half-neutralized salt of adipic acid) which is widely used as a nanofiller for making polymer nanocomposites. The results are shown in Fig. 6.34. We can clearly see that in the unmodified sample, the nanotubes are highly agglomerated and bundled. On the other hand in the modified MWCNT sample, the tubes are debundled and individual tubes can be seen. Thus the effect of modifier can be clearly established using SEM. It is also used to analyse the FRP composite after subjecting to mechanical test. For example, Wan et al. [19] investigated the fracture behaviour of composite sample after conducting fatigue test at frequencies of 5 Hz for fatigue stresses ranging from 55 to 620 MPa using SEM. The SEM microphotograph is presented in Fig. 6.35. The micrograph shows typical fatigue cracks of a composite occurred under tension-tension loading, in which one can obviously see the crack damage zone. The crack seems to be somewhat blunt, beyond which only little damage is observed. The voids are localized beyond the crack and along the crack lips. This indicates the suitability of SEM for fracture characterization of polymer composites.

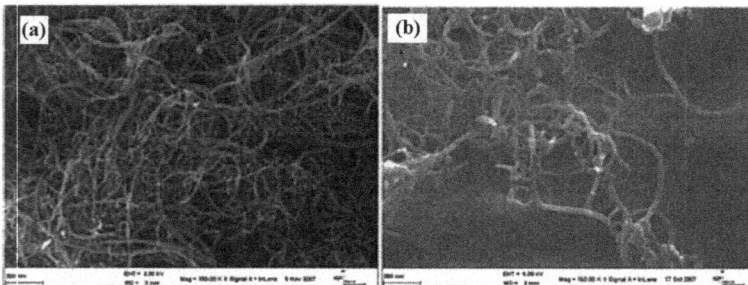

Fig. 6.34: SEM photographs of (a) unmodified and (b) HNSA-modified MWCNT. The nanotubes were dispersed in ethyl alcohol by sonication and cast onto the stub directly and inspected after providing a gold coating (reprinted with permission from Debdatta Ratna [18]. © 2009 Wiley Periodicals, Inc. Publishing Company).

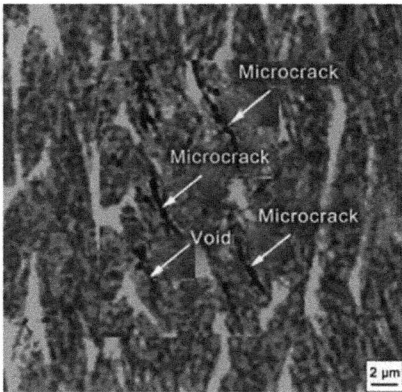

Fig. 6.35: SEM micrograph of the interior of braided composites (reprinted with permission from Zhenkai Wan [19]. © 2017, De Gruyter Publishing Company).

SEM is also used to characterize the two-phase morphology of a toughened polymer system. As it is discussed in previous sections, in order to toughen a brittle polymer system a rubbery modifier is incorporated as a separate phase. The dispersed two-phase morphology can be characterized by SEM. Ratna et al. [20, 21] developed toughened epoxy systems using acrylate based elastomer. In fact in case of rubber-toughened epoxy, the liquid rubber initially remains miscible with the epoxy resin and in the course of cure reaction undergo phase separation generating a two-phase morphology with uniformly dispersed rubber particles. The morphology depends on molecular weight of the liquid rubber. An increase in molecular weight of liquid rubber favours phase separation. Hence to ensure complete phase separation it is necessary to use liquid rubber with sufficiently high molecular weight. However, beyond a certain molecular weight, the liquid rubber undergoes phase separation at a very early stage leading to agglomeration and macrophase separation. The SEM photographs of fracture surfaces of toughened epoxy systems modified with CTPEHA of two different molecular weights, e.g. 3,600 and 9,500 g/mol are shown in Fig. 6.36. The fracture surface of modified epoxy with CTPEHA having molecular weight 3,600 g/mol shows uniform distribution of globular rubber particles whereas the same modified with CTPEHA having molecular weight 9,500 g/mol shows agglomerated particles having a size greater than 10 microns.

6.6.3 Transmission electron microscope

In addition to SEM discussed above, transmission electron microscope (TEM) is another electron microscopic technique which is extensively utilized to characterize polymer composite in general and polymer nanocomposite in particular, which involves nanoscale dispersion of inorganic fillers or other polymeric components in a polymeric matrix. Similar to SEM, a TEM equipment consists of an electron Gun which produces the electron beam, Electromagnetic Lenses, which focuses the beam

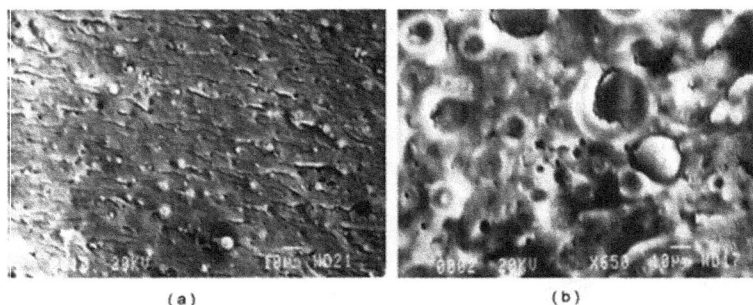

(a)   (b)

Fig. 6.36: SEM microphotographs of fracture surfaces of (a) LR-4 (Mn = 3,600) modified epoxy and (b) LR-1 (Mn = 9,500) modified epoxy network (reprinted with permission from, D. Ratna [20]. © 2000 John Wiley & Sons, Inc Publishing Company).

onto the object, specimen stage and the image-producing system, consisting of the objective lens. The electron gun has to be placed in a vacuum system which can produce and maintain a pressure as low as 10–8 Pa. In SEM the specimens are placed on a stab whereas in TEM, they are placed on a mesh or grid (made of copper, molybdenum, gold, or platinum) having a diameter of about 2.5 mm and is held by a specimen holder. The specimen stage is associated with airlocks to allow for the insertion of the specimen holder into the vacuum so that there is no significant increase in pressure in other areas of the microscope.

Unlike in SEM where image is generated from secondary or backscattered electron, and the transmitted electrons provide the information about the inner micro- or nano-structure, which is captured to produce an image. Therefore, TEM images are produced by focusing a beam of electrons onto a very thin specimen having thickness less than 80 nm. It may be noted that the preparation of samples having such a low thickness is a cumbersome process and is practically difficult for highly flexible polymers. The second issue is related to the poor contrast observed during TEM analysis of many polymer samples due to a weak interaction of electrons with the sample. As a solution to this problem, various heavy metal compound based stains are used which enhances the contrast due to interaction of electron beam with the dense electron clouds of the heavy metal atoms. Third issue is the possibility decomposition of polymers due to exposure of electron beam as a result of increase in temperature. This can be avoided by using cryogenic microscopy (cryo-TEM), which keeps the sample in liquid nitrogen. The modern high-resolution (HR) TEM can provide resolution up to 0.1 nm. Therefore, it can be utilized for characterization of both the microstructure and nanostructure of composite materials.

Figure 6.37 shows TEM images of nanocomposite samples prepared [22] using poly (ethylene oxide) (PEO) as a matrix and modified graphene as nanofiller in which two different concentrations of graphene, i.e. 0.75 wt% and 1.25 wt% were used. We can see that 0.75 wt% containing composite exhibits the exfoliation of graphene into a single sheet. On the other hand, the morphology obtained for 1.25 wt% graphene containing composite, exhibits the presence of agglomeration. This clearly indicates that at a higher concentration of graphene, it is difficult to prevent the agglomeration. Therefore, all nanofillers have to be used in a very low concentration to realize the nano effect. The require concentration has to be optimized for a particular system depending on the nature polymer and nanofiller. TEM analysis is extremely useful for such optimization and successful development of a useful polymer nanocomposite.

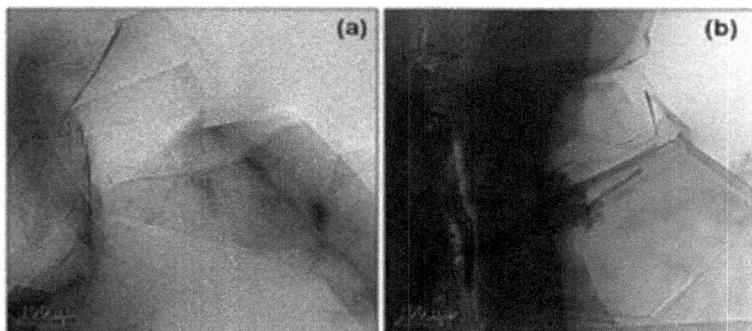

Fig. 6.37: TEM images of (a) PEO/graphene (0.75 wt%)/MC-2 and (b) PEO/graphene (1.25 wt%)/MC-2 nanocomposites.

6.6.4 X-ray diffraction analysis

Apart from the microscopic techniques discussed above, another versatile non-destructive analytical technique used to characterize polymer composite materials in terms of crystal structure and phase composition is X-ray diffraction (XRD). A typical X-ray diffractometer comprises three main parts: an X-ray source, a sample holder and a detection system. The sample to be analysed is illuminated by the X-rays produced by the source. The X-ray is then diffracted by the sample phase and enters the detection system. The diffraction angle (2θ) can be varied by rotating the tube or sample and detector and accurately measured by using a goniometer. The intensity as a function of diffraction angle is then recorded. The wavelength of the X-rays is in the same order of magnitude of the distance between the atoms in a crystalline lattice. Therefore, the diffraction resulted in constructive interference between X-rays and a crystalline specimen. The diffraction pattern produced during the analysis can be analysed in various ways and the most popular is to apply Bragg's law, which is widely used in the measurement of crystals and their phases present in the composite material. The Bragg's relation is given as follows:

$$\lambda = 2d \, \sin\theta \qquad\qquad (5.1)$$

where λ is the wavelength of the X-ray used, d is the spacing between specific diffraction lattice planes and θ is the measured diffraction angle. Since the value of λ is known ($\lambda = 1.5405$ Å), by measuring the value of θ we can determine the value of d. As the value of θ decreases, the value of $\sin\theta$ decreases, since λ is constant, the value of d increases.

The author has used XRD analysis for characterization of polymer nanocomposite material. In the previous sections different types of nanocomosites have been elaborated. The XRD patterns for organically modified clay (MMT-30) and PEOCN6 (nanocomposite containing 6 wt% of clay in PEO) is shown [23] in Fig. 6.38. The clay shows the d_{100} peak at $2\theta = 4.04$, which corresponds to a d-spacing of 2.18 nm. This indicates that modification of clay with organic ions not only makes the clay surface hydrophobic but also resulted in a 1.18 nm increase in d-spacing (as the d-spacing for untreated clay is about 1 nm). This facilitates the penetration of PEO into the interlayer galleries [24, 25]. In case of the composite sample, the basal spacing of clay is increased to 4.32 nm as the diffraction angle shifts from $2\theta = 4.04$ to $2\theta = 2.06$ for the d_{001} peak. Hence, the polymer layer in between the two silicate layers is 3.14 nm. The result is consistent with the other studies in the literature [26–28] on the adsorption of PEO chains onto the clay platelets at low polymer and the clay concentrations using contrast variation methods to separate the contribution from the bulk and adsorbed polymer chain.

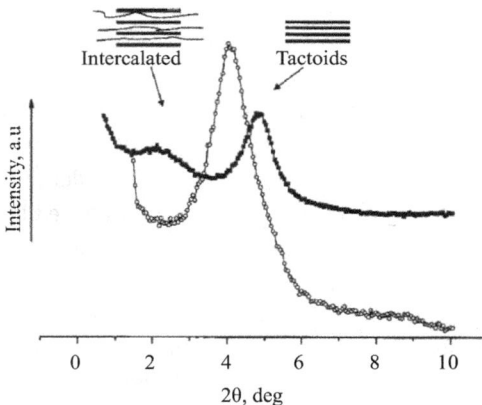

Fig. 6.38: XRD patterns for PEOCN6 (–■–) and Cloisite 30B (–o–) (reprinted with permission from T.N. Abraham [20]. © 2008 Society of Plastics Engineers Publishing Company).

6.7 Conclusion

Mechanical, dynamic viscoelastic and complex dielectric properties are commonly used for design of most composite items in structural elements, electrical application, processing conditions and for determination of optimum reinforcement and very importantly, the thermal and temporal stability. These macroscopic properties of composite material, brief methods of evaluation of polymeric matrix composites are briefly described in this chapter. In each test method, international standards (ASTM, ISO and, in some cases, DIN) are listed. The graphical examples are given only to show how the properties depend on some external stimuli like force, heat and frequency. Basic functional details of the instruments are discussed to introduce the techniques for composite characterization. The instruments are commercially available from very reputed industries in the field of analytical instrumentation, and most of the instruments are designed to perform tests as per international standards. Due to the difference in design by different companies, the operational methods would vary, but the tests must be within the specified domain of parameters prescribed in international standards.

Functional evaluation such as structural stability, dynamic behaviour, fatigue life and environmental degradation of items built with such composites is deliberately avoided, because of the large diversity in application.

The mechanical and thermal properties of the composites at the material level forms the basis of design of components/systems and also used in basic engineering study to develop physical and mathematical models for prediction of functional behaviour as a component. The predictive mathematical models are subsequently validated by experimental evaluation of the fabricated components/systems. Thus, the accuracy and correctness of basic property evaluation is most important step in a system development.

It is a common experience that incorrect analysis of properties leads to very wide differences between predicted and practical result. To eliminate errors in instrumental analysis, few basic steps may be very useful, such as regular calibration of the instrument, uniformity in sample size and shape in repeated experiments, maintaining identical accuracy in dimensional measurements and identical conditioning. It is also suggested that each experiment should be performed following one standard only, such as ASTM or ISO, for consistency in experiments, using same instrument, so that the results of different batches in development and production stage can be compared. In many applications, catastrophic failures of composite items are considered dangerous to human life. In such cases, the test results are more critical requiring instruments of higher quality, consistency and accuracy in addition to consistence in measurement.

References

[1]   Wang, B., Yang, B., Wang, M., Zheng, Y., Hong, X., Zhang, F. Sci. Eng. Compos. Mater. 2019, 26, 394–401.

[2]   Hofer, K. E., Rao, P. N. J. Test. Eval. 1977, 5, 278–283.

[3]   Harper, J. F., Miller, N. A., Yap, S. C. Polym. Test. 1993, 12, 15–29.

[4]   Zhang, C., Ganesan, R., Hoa, S. V. Sci. Eng. Compos. Mater. 2000, 9, 163–176.

[5]   Bakker, A. A., Jones, R., Callinan, R. J. Compos. Struct. 1985, 15, 154.

[6]   Deitz, D. H., Harik, I. E., Gesund, H. J. Compos. Constr. 2003, 7, 363–366.

[7]   Jang, B. Z. Sci. Eng. Compos. Mater. 1991, 2(1), 29.

[8]   Çalhoğlu, H., Emin, E. Sci. Eng. Compos. Mater. 2014, 21, 463–470.

[9]   Yang, K., Yang, L., Gong, P., Zhang, L., Yue, Y., Li, Q. e-Polymers. 2022, 22, 309–317.

[10]   Talreja, R., Phan, N. Compos. Struct. 2019, 219, 1–7.

[11]   Chakraborty, B. C., Ratna, D. Polymers for Vibration Damping Applications. Elsevier, 2020, 121–125.

[12]   Vassilikou-Dova, A., Kalogeras, I. M. Dielectric Analysis. In: Menczel, J. D., Prime, R. B., eds. Thermal Analysis of Polymers: Fundamentals and Applications, 1st ed. Wiley, NY,USA, 2008, 497–613.

[13]   Ratna, D., Abraham, T., Karger-Kocsis, J. Macromol. Chem. Phys. 2008, 209, 723–733.

[14]   Abraham, T., Siengchin, S., Ratna, D., Karger-Kocsis, J. J. Appl. Polym. Sci. 2010, 118, 1297–1305.

[15]   Strawhecker, K. E., Manias, E. Chem. Mater. 2003, 15, 844.

[16]   Ratna, D., Abraham, T. Polym. Compos. 2011, 32, 1210–1217.

[17]   Ratna, D., Simon, G. P. J. Appl. Polym. Sci. 2010, 117, 557–564.

[18]   Ratna, D., Abraham, T., Siengchin, S., Karger-Kocsis, J. J. Polym. Sci. Polym. Phys. 2009, 47, 1156–1165.

[19]   Wan, Z., Guo, J., Jia, M. Sci. Eng. Compos. Mater. 2017, 24, 213–220.

[20]   Ratna, D., Banthia, A. K., Deb, P. C. J. Appl. Polym. Sci. 2000, 78, 716–723.

[21]   Ratna, D. 2001, 50, 179–184.

[22]   Jagtap, S. B., Khushwaha, R., Ratna, D. RSC Adv. 2015, 5, 30555–30563.

[23]   Abraham, T. N., Ratna, D., Siengchin, S., Karger-Kocsis, J. Polym. Eng. Sci. 2009, 49, 379–390.

[24]   Okamoto, M. Rapra Rev. Rep. 2003, 14, 1.

[25]   Nguyen, Q. T., Baird, D. G. Adv. Polym. Technol. 2006, 25, 270.

[26]   Malwitz, N. M., Dundigalla, A., Ferreiro, V., Butler, P. D., Henk, M. C., Schmidt, G. Phys. Chem. Chem. Phys. 2004, 42, 3102.

[27]   Stefanescu, E. A., Dundigalla, A., Ferreiro, V., Loizou, E., Porcar, L., Negulescu, I., Garnoa, J., Schmidt, G. Phys. Chem. Chem. Phys. 2006, 8, 1739.

[28]   Loizou, E., Butler, P., Porcar, L., Schmidt, G. Macromolecules. 2005, 38, 2047.

Chapter 7
Lifetime estimation of polymer matrix composite

Description of abbreviations

Abbreviation	Description	Unit
ATM	Accelerated testing method	
CFRP	Carbon fibre-reinforced plastic	
CNT	Carbon nanotube	
CRTA	Controlled rate thermal analysis	
DMA	Dynamic mechanical analysis	
DSC	Differential scanning calorimetry	
EM	Electromagnetic	
EMI	Electromagnetic interference	Volt
FRP	Fibre-reinforced plastic	
FTIR	Fourier-transform infrared spectrophotometry	
GC-MS	Gas chromatography-mass spectroscopy	–
GPa	Giga Pascal	10^9 N/m^2
HDPE	High-density poly(ethylene)	
Hz	Hertz	Cycle/s
J/mol K	Joule per mole per Kelvin	
kJ/mol	Kilo-Joule per mole	
MMF	Micromechanics of failure	
MPa	Mega Pascal	10^6 N/m^2
NMR	Nuclear magnetic resonance	
NR-CB-VR	Natural rubber-carbon black-Vermiculite	
phr	Parts per hundred (part) rubber	
PMC	Polymer matrix composite	
PMMA	Poly(methyl methacrylate)	
PVC	Poly(vinyl chloride)	
RH	Relative humidity	%
SCTA	Sample-controlled thermal analysis	
TGA	Thermogravimetric analysis	
UTS	Ultimate tensile strength	N/m^2
UV	Ultraviolet	
VFT	Vogel-Fulcher–Tammann	
WLF	Williams Landel Ferry	

7.1 Introduction

Ageing is an obvious process for any mater and living beings. This terminology involves three main mechanisms, namely physical ageing, chemical ageing and biological ageing. To distinguish these three processes, we can take example of a slow change in density of a plastic for a long storage time, several years as a physical pro-

https://doi.org/10.1515/9783110781571-007

cess, while corrosion of a steel structure exposed in sunlight or due to humidity is an example of chemical ageing and decay of a processed food item due to bacteria is a biological process. However, chemical and biological ageing are much faster than physical ageing.

The ageing of polymeric composites is no exception and in fact, polymeric composites are quite vulnerable to temperature, ultraviolet ray, ozone, saline atmosphere and a lot of chemical gases and fluids, just like any other organic molecule. In addition, the polymers undergo physical ageing which is measurable within a reasonable experimental time frame, which is not possible for metallic or ceramic items. Since the process of ageing is natural, it is really difficult to exactly develop a logical method to predict the extent of ageing as several stimuli simultaneously act causing the total ageing. In a typical example of a rubber gasket, let us consider that the service condition of the gasket is 2.0 MPa initial tightening stress in compression at 60 °C, the ageing depends on stress relaxation at that temperature, causing decrease in tightening force, chemical degradation due to heat, thereby reducing inherent strength of the material, and ultimately a combined effect on corresponding gradual loss of effectiveness in sealing by the gasket as it ages.

The life of a material is decided by the desired usefulness in service. Since ageing is a natural process, the lifetime is different in storage than in service. For example, a rubber gasket when stored a controlled temperature at 22 °C and 45% relative humidity, might have 15–20 years life, which indicates that its functional properties are within the limits of the specification for use, while in a service at 60 °C and 65–80% RH and 2 MPa initial tightening stress, the life might be only 5 years. In this case, the ageing might be faster for stress relaxation than thermal degradation, and that decides the useful life of the gasket. While in storage, except heat ageing, there is no other factor for deciding the life of the gasket.

The property which deteriorates at the fastest rate to a limiting value, should be taken to study and consequently, to decide the life. If the design of an item is done considering lowest tensile strength as 75% of initial (unaged) value, and 85% of unaged EMI shielding value, then the ageing mechanism which causes fastest deterioration of both the strength and the EM property should be considered for life estimation. The lower value of these life estimations should be considered as the useful life of the item. Therefore, no common strategy can be applied to decide the life of differently used composite items, rather different methods should be adopted for accelerated ageing study. In general, following steps may be considered for a comprehensive lifetime prediction of a polymeric composite:

1. Define the most important functional property which will be studied for arriving at life of the material under aging.
2. Define at what temperature and environment it has to perform or to be stored.
3. Define the physical or chemical phenomenon like creep, stress relaxation, water ingress, evaporation of small molecules, thermal decomposition, UV or ozone effects which causes degradation of the desired property.

4. Mechanism of degradation to be established from pyrolytic GC-MS, FTIR and NMR in case of chemical change – mechanism must not be different in the test envelop.
5. Study the functional property which will be required to predict the life.
6. Define the lowest acceptable numerical value of the desired property upon aging, for example, Young's modulus minimum 80% of initial value.
7. Resort to time–temperature relation, such as Arrhenius or WLF or graphical superposition where change in mechanical, viscoelastic or other properties are studied in accelerated mode. For thermal degradation study, kinetics of degradation with Arrhenius function is to be used.
8. Extrapolate for storage/service temperature/condition
9. Check the theoretical estimation and experimental results.
10. Ageing temperature selection should be as nearer to service temperature as practically possible to obtain a reasonably correct result.

7.2 Physical ageing and life estimation

Physical ageing means that the ageing takes place due to change in bulk or molecular level properties with storage time, under no influence of stress or any other external factors. The properties can be many like enthalpy, specific volume, mechanical or dielectric response as bulk properties and microstructural properties which can be determined by spectroscopy, light scattering, atomic force microscopy, scanning electron microscopy or transmission electron microscopy.

Amorphous polymers are defined as supercooled liquids, wherein the mobility of the randomly oriented chain segments reduce as the polymer is cooled down. Initially, above the glass to rubber transition temperature (T_g), the packing of the chains increase slowly on cooling, but it becomes rapid near the glass transition. Below the transition, the packing again increases slowly. Physical ageing of a polymer is the increase in packing or decrease in free volume on observation at a fixed temperature, with time, below the glass transition, but above the secondary relaxation (β-transition). This is shown in Fig. 7.1. A polymer is in equilibrium state above the T_g because the segments can rearrange themselves to an equilibrium conformation very fast, as their relaxation time (τ) is about 10^{-2} to 10^{-4}s. The equilibrium line is shown above T_g. However, below the T_g, the segmental motion is highly restricted and the time required to attain the equilibrium becomes large. The volume change curve has lower slope as shown in the figure. Whereas the polymer continuously tends to reach equilibrium (dotted line extended from above T_g). This means that an amorphous polymer below the T_g has a volume which is greater than what it should have in equilibrium state. The dotted line below T_g shows what the volume should have been at equilibrium, and the firm line shows the volume actually occupied instantaneously quenched below T_g. The effect of ageing at that temperature is reduction of the free volume (increase in viscosity and relaxation time) along the vertical line towards the equilibrium point.

7.2.1 Basic features

There are some important features of the concept of physical ageing:

(1) Physical ageing occurs for amorphous polymers at any temperature below the glass transition (T_g), but above the β-transition ($T_β$).

(2) There is no chemical reaction due to environment or chemical degradation of the polymer during physical ageing process and also during experimental determination of the effect of physical ageing, such as creep experiment.

(3) Physical ageing process is reversible in the sense that if the polymer is heated above T_g again, all the previous history of relative changes in free volume and related properties would be removed and the polymer would again be in equilibrium state.

(4) Physical ageing of amorphous materials is universal. All amorphous polymers age similarly. For temperatures in the ageing range (T_g to $T_β$), time dependence of the low strain mechanical properties of glassy materials was found to be similar and independent of their chemical structure. Thus, poly (vinyl chloride) will age similarly as PMMA or polystyrene at ambient conditions since their T_gs are far above ambient. This is important in mechanical behaviour of the engineering plastics, thermosets and composites since conventional evaluation can lead to errors.

(5) Normally elastomers are all at equilibrium state at ambient temperature, since their glass transitions take place at sub-ambient temperatures, and physical ageing is not applicable to them.

(6) In physical ageing, since the relaxation time is large below T_g, there will be considerable effect due to the phenomenon of Boltzmann superposition principle on repeated creep experiment, if the repeating interval is near about the relaxation time. The change in properties strongly depends on ageing time at ageing temperatures.

(7) In the ageing range ($T_β$ to T_g), the ageing time is the most important material parameter. It means that the creep rate of a polymer is faster after a particular ageing duration below T_g than the creep rate at a higher temperature (but below T_g) aged for more duration.

(8) In most cases, the physical ageing can be studied quantitatively by simple experiments of creep, relaxation, enthalpy change, dilatometry and microlevel tests such as atomic force microscopy and positron annihilation lifetime spectroscopy.

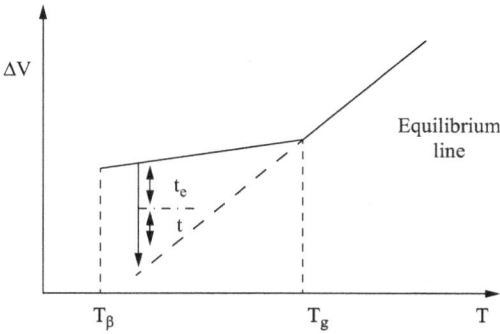

Fig. 7.1: Physical ageing of a polymer: a schematic representation of approach towards equilibrium by segmental relaxation below glass transition.

If a creep experiment is performed on a polymer which has been brought to a temperature T_e below T_g immediately after quenching, then the creep curve will have a particular creep function $\psi(t)$ corresponding to the relaxation time (τ). In this case, the creep is a momentary creep. After a time t_e of ageing at the temperature (T_e), the relaxation time increases as the polymer approaches equilibrium, and the polymer hardens. Now, if the creep experiment is performed, the creep curve deviates from the momentary curve, since relaxation time continuously changes, irrespective of the experiment. Hence, the creep after long ageing must be mapped with the momentary creep as described above. To study the effect of physical ageing, the creep experiment has to be performed at a fixed temperature T_e, after several ageing times t_e (1 to n) and a master curve can be made with respect to a reference ageing time t_{er} with the definition of a simple horizontal shift factor for both momentary creep and long time creep after ageing time t_e.

7.2.2 Example of creep

To demonstrate the difference of experimental creep study for one year and predictive equation fitting for a short duration, and extrapolated for one year without considering physical ageing for an epoxy thermoset is shown here as an example for importance of study of physical ageing.

The epoxy thermoset had a DMA T_g at 79 °C (temperature of E''_{max} taken at 0.2 Hz frequency), and the activation energy of α-relaxation was 343 kJ/mol., determined form the loss modulus peak temperatures (considering $\omega\,\tau = 1$ at E''_{max}) at different frequencies (0.2–5 Hz) using the Arrhenius equation:

$$\tau = \tau_0 \exp\left[\frac{E}{RT}\right] \quad \text{or} \quad \ln(\tau_1) - \ln(\tau_2) = \frac{E}{R}\left[\frac{1}{T_1} - \frac{1}{T_2}\right] \tag{7.1}$$

Corresponding relaxation time was 0.0133 min at 79 °C, and the relaxation time at 60 °C was calculated as 3.83 min. The cured epoxy thermoset was heated to 90 °C for 30 min and brought down to 60 C, and immediately loaded under a tensile creep, keeping the temperature constant in an oven fitted Creep machine in load control mode. The study is termed as momentary creep study to distinguish from the creep studies after ageing. The momentary creep study was done for a year continuously. The data up to 316 min (5.27 h) was fitted fairly with the following model of usual form [1] such as Findlay's power law model with the regression coefficient $R^2 = 0.976$. Fig. 7.2 shows the short-term data and predictive fit.

Predictive model (0–316 min):

$$\varepsilon(t) = 0.0148 + 0.0081t^{0.29} \tag{7.2}$$

where 0.0148 was the initial strain, $\varepsilon(0)$.

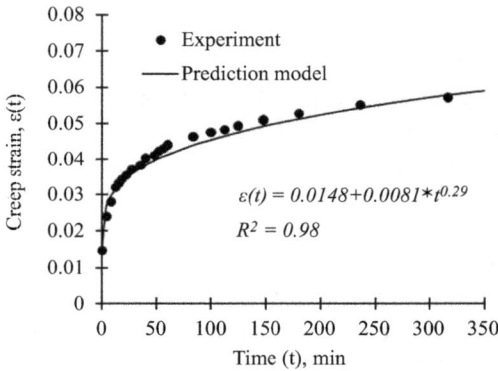

Fig. 7.2: Short-term creep test on an epoxy thermoset at 60 °C.

The same equation was used to predict the long-term creep experimental data (1 year) and found to show large variation. eq. (7.2) predicted much higher strain, compared to experimental data simply because the thermoset aged physically through the experimental time resulting in its considerable increase in relaxation time or decrease in segmental mobility as it approached equilibrium. Fig. 7.3 shows the long-term creep strain experimental data and also by eq. (7.2). A best fit equation with $R^2 = 0.998$ was also obtained for the long-term creep, which is not valid for zero time since it is in logarithmic form with time. The best fit equation is given by

$$\varepsilon(t) = 0.0081 \ln(t) + 0.0111 \text{ for } t > 0 \tag{7.3}$$

and $\varepsilon(t) = 0.0148$ for $t = 0$.

Considering the best fit eq. (7.3), the dependence of creep strain with logarithmic time suggest the delayed creep with progressive lapse of the time of creep experiment.

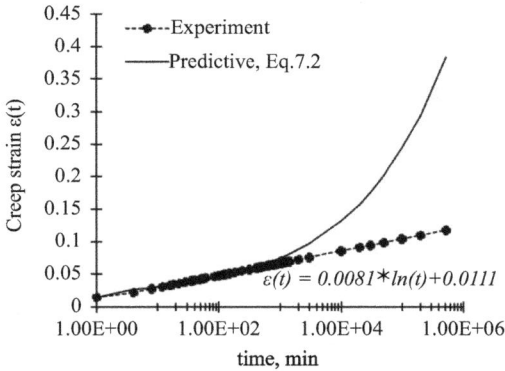

Fig. 7.3: Long-term creep test on an epoxy thermoset at 60 °C.

The relaxation times calculated from the DMA data and using eq. (7.1) resulted in the approximate shift factor $a_M = 18$ for momentary creep study for every 5 °C decrease in temperature, which means that the time required to attain a strain at reference temperature 60 °C is 18 times higher than the time required to attain same strain at 65 °C, and 1/18 times to that at 55 °C. The α-relaxation times at 55, 60 and 65 °C are 1,520, 230, and 37 s, respectively and the mobility of the segments (relaxation time) changes the creep time accordingly. Similar effects are expected due to physical ageing at a constant temperature, as the change in mobility due to both temperature and time of ageing are basically due to change in thermal excitation. Only, the log(ageing time) replaces the temperature, in a reverse order.

7.2.3 Relaxation spectra

Suppose an instant tensile load is applied on an amorphous polymer sample at a temperature T_e which is less than T_g of that polymer. The instantaneous strain is low enough to satisfy linear viscoelasticity. The creep compliance and creep strain are being monitored with time from a zero time immediately after the polymer sample is brought to T_e. The time dependent compliance is

$$J(t) = J_0 + \int_0^\infty \psi(\tau)[1 - e^{-t/\tau}]d\tau \tag{7.4}$$

where $\psi(\tau)$ is the relaxation spectra of the polymer at zero ageing time, J_0 is the creep compliance at zero time, τ is the relaxation time and t is the creep time.

For example, considering the experimental creep data in Fig. 7.3, it is possible to select various relaxation time and sum up the creep strains due to the individual relaxation to obtain the curve of experimental data. The simplest Kelvin-Voigt model as below was used for all creep strain calculation:

$$\phi(\tau) = \Delta\varepsilon[1 - \exp^{-(t/\tau)}] \qquad (7.5)$$

and finally,

$$\varepsilon(t) = \varepsilon(0) + \sum_{i=1}^{4} \phi(\tau)_i \qquad (7.6)$$

In an attempt to do that, several iterations were carried out to finally selecting four relaxation times such as 12, 25, 450 and 1,000 min (across three decades of time) and the integration by simple summing up gave a very accurate fit for the experimental data with regression coefficient $R^2 = 0.991$. Figure 7.4 shows the result.

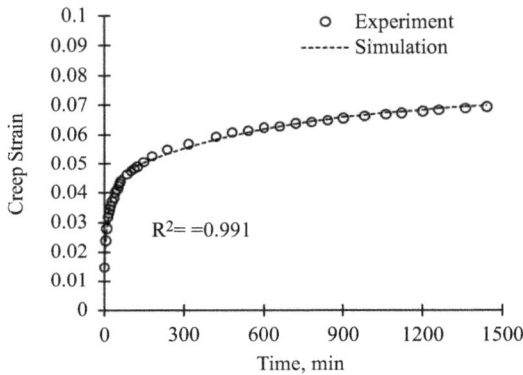

Fig. 7.4: Creep data simulation using four different relaxation experimental data taken from Fig. 7.3.

7.2.4 Theory of physical ageing

If the polymer sample is aged for a time t_e at the temperature T_e and the creep study is done, then the time-dependent creep compliance will be [2]

$$J_e(t) = J_e(0) + \int_0^{\infty} \psi_e(\tau)[1 - e^{-t/\tau}] \qquad (7.7)$$

where the subscript e denotes properties after ageing for t_e duration hence the time t is the creep time measured after the ageing time t_e.

Equations (7.4) and (7.7) are distinguishable since the initial value of compliance after the ageing period (t_e) is different from zero ageing period and also the relaxation spectra are different after ageing. In addition, the actual time elapsed after the polymer is brought to T_e is $t_e + t$. Hence, if one can map the instantaneous or momentary creep onto the aged creep, one needs to validate the following creep compliance with experimental result:

$$J_e(t_e + t) = J(0) + \int_0^\infty \psi(\tau) \left[1 - e^{-(t+t_e)/\tau} \right] d\tau \tag{7.8}$$

In a simpler term, considering a change in relaxation time due to ageing for a duration t_e at a temperature T_e, the relaxation time for an ith molecular process can be expressed as

$$\tau^{(i)}(t_e + t) = \tau_0^{(i)} \phi(t_e + t) \tag{7.9}$$

Here, the relaxation time is continuous from $t_e = 0$ and also applicable to any t_e.

If the creep study is carried out after different ageing periods t_e, t_e', etc., then a master curve of the creep compliance can be constructed with a horizontal shift factor as

$$J_{e'}(t) = J_e(at) \tag{7.10}$$

$$\psi_{e'}(\tau) = a\psi_e(a\tau) \tag{7.11}$$

The shift factor (a) changes due to ageing and $\log(a)$ is linear with $\log(t_e)$. The rate of change of $\log(a)$ with respect to $\log(t_e)$ is designated as double-logarithmic shift-rate (μ), which signifies the intensity of ageing:

$$\mu = \frac{d\log(a)}{d\log(t_e)} \tag{7.12}$$

Above the glass transition, there is no physical ageing, hence, $\mu = 0$. On the other hand, in the ageing range (T_β to T_g), the mobility of the segments depends on the ageing time (at a fixed temperature), because the ageing increases the relaxation time as the molecule tends to attain equilibrium. The ageing is, therefore, a self-delaying process. The mobility is inversely proportional to the ageing time:

$$M \approx \frac{k}{t_e} \tag{7.13}$$

Therefore, the above equation suggests that the ratio of $\log(M)$ and $\log(t_e)$ must be unity, and therefore,

$$\mu = \frac{d\log(a)}{d\log(t_e)} = -\frac{d\log(M)}{d\log(t_e)} \approx 1 \tag{7.14}$$

The negative sign for the double logarithm of M rate is due to the fact that the slope is reverse in this case compared to that of $\log(a)$ vs $\log(t_e)$. This has been reported by Struik [2] and many others for a large number of polymers.

However, the long-term test of creep (or any time-dependent mechanical response) does not follow the short time creep curve. The long and short terminology is expressed in terms of the comparison between the ageing time and creep study duration. For example, a short duration creep is where ageing time is 3,000 s and creep tests are done for 100 s, while long-term is where test time is 1 year for 3,000 s ageing time. In a comparison of short-term and long-term tests, Struik [3] observed that the long-term creep curve is not concave at long-time, but approximately a straight line, while the short-term creep curves are concave upwards, and similar at all temperatures of experiment. To map the momentary creep on all the aged creep curves (done after different ageing times, e.g. $t_e = 1, 3, 10, 30, 100$ h), Zheng and Weng [4] used eq. (7.9) to calculate the relaxation time during the creep experiment after an ageing time of t_e at a constant sub-T_g temperature. The authors used effective time λ which is zero at the start of the creep experiment after an ageing time t_e. The effective time is calculated as

$$\lambda = \int_0^t \frac{d\tau}{\phi(t_e + \tau)} \tag{7.15}$$

The authors stated that the function ϕ for a chrono-rheologically simple polymer must be linear with $(t_e + t)$, and suggested a dimensionless form:

$$\phi(t_e + t) = 1 + \frac{(t_e + t)}{b} \tag{7.16}$$

where b is a material property and has a unit of time. If b tends to infinity, then the material undergoes viscoelastic deformation with time and there is no effect of the physical ageing, and when $b = 0$, then the material undergoes only elastic deformation. The value of b is finite positive between zero to infinity.

The effective time λ becomes (from eq. (7.15))

$$\lambda = b.\ln\left(1 + \frac{t}{b + t_e}\right) \tag{7.17}$$

With the above equation, long-term creep compliance at different ageing time can be horizontally shifted to a reference ageing time curve and it was established by the authors that the master curve would exactly the momentary creep compliance curve ($t_e = 0$) and it is the creep curve with no ageing effect ($\lambda = t$, $\phi = 1$).

For horizontal shifting of the long-term creep curves of different ageing time (t_es), following equation must be valid with a reference ageing time t_e^{ref}:

$$b.\ln\left(1+\frac{a_L t}{b+t_e^{ref}}\right) = b.\ln\left(1+\frac{t}{b+t_e}\right) \tag{7.18}$$

where a_L is the shift factor for the long-term creep experiment for shifting one ageing time curve to the reference ageing time curve (t_e^{ref}), and the double–logarithmic shift rate (μ_L) is also defined as in eq. (7.14):

$$\mu_L = \frac{d\log(a_L)}{d\log(t_e)} \tag{7.19}$$

Therefore,

$$a_L = \frac{b+t_e^{ref}}{b+t_e} \tag{7.20}$$

$$\mu_L = \frac{t_e/b}{1+t_e/b} \tag{7.21}$$

From eq. (7.20), it is seen that the shift factor for long-term creep is simply the ratio of reference (t_e^{ref}) to any ageing time (t_e), when the ageing time is much larger than the material constant b. From eq. (7.21), it is observed that at very long ageing time (t_e), the shift rate tends to be unity.

Further, on mapping the momentary creep at the reference (t_e^{ref})and the momentary creep at any ageing time (t_e), it is shown that the shift factor a_M is also same as the shift factor for long-term creep:

$$a_M = \frac{b+t_e^{ref}}{b+t_e} = a_L \tag{7.22}$$

These two shift factors are identical, independent of loading time (t) (in the creep experiment) but depends on the ageing time (t_e) if the material property (b) is constant.

There are a considerable number of different studies on physical ageing alone and combined effect of physical and chemical ageing reported by many researchers, and few are listed discussed here[2–20].

Struik [3] studied creep of PVC on physical ageing at sub-T_g temperature with different ageing time spanning from 0.03 days to 1,000 days, and it is seen that the time required to attain a creep compliance for a 0.03 days aged sample is a thousandth fraction of the time taken for a 300 days aged sample. The creep curves are all almost parallel, shifts in time scale as the samples are aged for more time. Hence the shift of the curves to a master curve at a reference ageing time was very accurate indicating that the behaviour of the viscoelastic change is uniform. In addition, a different material showed exactly same creep performance as PVC sample when these two are aged for the same period.

Lai [5], studied physical ageing of poly(methyl methacrylate) (PMMA) by free volume change using length contraction method, and observed that the major length contraction occurred in the vicinity of glass transition temperature of PMMA. The author studied the relaxation rate by a thermomechanical experiment of contraction of PMMA under a constant load at various isothermal ageing for PMMA. The relaxation rate is expressed by

$$-r = \left(\frac{1}{2.303}\right)\left(\frac{3}{L_Z}\right)\left[\frac{dL_Z}{d\ln(t_a)}\right] \tag{7.23}$$

and also, can be expressed using modified Doolittle equation:

$$-r = \frac{\Delta f}{\tau_g}\exp\left[\frac{1}{f_g} - \frac{1}{f}\right] \tag{7.24}$$

Relaxation rate was drastically higher at higher ageing temperature, indicating faster physical ageing as the ageing temperature approaches the α-transition (T_g) of the polymer, and beyond T_g, the rate is about 170 times faster than at 60 °C below T_g indicating that at lower ageing temperature, the ageing is slowed down significantly. The DMTA study showed secondary β–relaxation peak temperature and below this temperature, there was no significant physical ageing of PMMA, which confirmed that physical ageing takes place only between T_g and T_β of a polymer. The author calculated the relaxation time and activation energy of relaxation by following equations:

$$\tau = \tau_g\exp\left[\frac{1}{f} - \frac{1}{f_g}\right] \tag{7.25}$$

$$\ln\left(\frac{\tau(T_a)}{\tau_g}\right) = -\frac{E_a}{R}\left(1 - \frac{1}{f}\right)^{-1}\left(\frac{1}{T_a}\cdot\frac{1}{f}\right) + \frac{E_a}{RT_{eq}}\left(1 - \frac{1}{f}\right)^{-1} \tag{7.26}$$

The relaxation time at 40 °C was found to be 10^{203} times higher than at 140 °C, correspondingly, the time to reach equilibrium at 40 °C was almost infinite. The activation energy of ageing (E_a) in the range of 40 °C to T_g was quite high (200–1,200 kJ/mol) signifying the relaxation of the main chain segmental mobility.

The findings on PMMA by the author [5] suggests that a product moulded from PMMA based composite when used at ambient temperature, would change in mechanical and viscoelastic property with time at a very slow rate, as its relaxation time would be almost infinite years. The high activation energy also suggests almost negligible change in relaxation, even when the ambient temperature is enhanced by 10 °C, like in summer of a tropical country. In this case, the life assessment is not dependent on physical ageing, since chemical ageing due to other factors such as heat, pollutant gases, presence of moisture, ozone and UV might be much faster.

Minguez et al. [6] studied polylactide, which is a semi-crystalline polymer, having a T_g = 55 °C for isothermal physical ageing by monitoring the enthalpy change using differential scanning calorimetry (DSC). Polylactide is an important polymer for food packaging, as it is biodegradable and environmentally safe material. The authors suggested that the physical ageing near the T_g is essentially due to a-relaxation (main-chain segmental mobility), but below the T_g, it is β-relaxation due to side group movements. In their experiment, the enthalpy of the β-relaxation was used to quantify the ageing. Upon physical ageing at 50 °C, the endothermic enthalpy associated with change in T_g was increasing with ageing time. The T_g was seen to be progressively increasing with ageing and was about 2–3 °C higher than quenched (unaged) sample after 300 min of ageing. The β_H enthalpy relaxation rate was thus defined as:

$$\beta_H = \left[\frac{\partial \delta_H}{\partial (\log t_a)} \right]_{q_1, q_2, T_a} \tag{7.27}$$

where q_1 and q_2 are heating and cooling rates at the ageing time t_a and isothermal ageing temperature T_a.

The β_H is an indication of bulk relaxation rate, which is related to ageing rate, and a higher value indicates faster ageing in approaching towards the equilibrium state. Measurement of fractional free volume could be another method to study the physical ageing as it is expected that with progress of the ageing, the fractional free volume reduces while the relaxation time increases.

The study of β-relaxation and associated structural relaxation has a quite practical significance in terms of cross-linked flexible thermoset epoxy resins where the relaxational characteristics are designed by changing chain length of the epoxy resin or the hardener amine. Patil et al. [7] studied β–relaxation phenomenon of a series of flexible and rigid epoxy resins and determined fractional free volume by positron annihilation lifetime spectroscopy (PALS).The authors found that the β-relaxation occurred in the range of −56 °C to −80 °C depending on network cross-link density. The fractional free volume ranged from 2.18% to 2.51% for highest cross-linked to least cross-linked epoxy network. Periodic measurement of the fractional free volume at an isothermal ageing study would throw light on the change in loss compliance of the network and corresponding change in damping properties of a structural damping coating.

There is a very comprehensive but tedious method of calculation of loss compliance and storage compliance from creep compliance experiment, as reported by Schwarzl [8]. The simplest equation for calculation of $J'(\omega)$ and $J''(\omega)$ are suggested as

$$J'(\omega) \sim J(t) - 0.86[J(2t) - J(t)] = 1.86J(t) - 0.86J(2t) \tag{7.28}$$

and,

$$J''(\omega) = J'(\omega). \tan \delta \tag{7.29}$$

where

$$t = \frac{1}{\omega} = \frac{1}{2\pi f} \tag{7.30}$$

$J(t)$ and $J(2t)$ are creep compliances at t and $2t$ s, while t is determined from equation (7.30), considering the angular frequency ω is in rad/s.

The above simplistic calculation is not very accurate for high loss factor ($\tan\delta$), but reasonable with maximum 8% error, when the $\tan\delta$ is about 0.6. However, the accuracy is best (upper bound of relative error = 0.8%) for much higher $\tan\delta$ range (0.10–10) corresponding to the most complicated formula for $J'(\omega)$ as:

$$J'(\omega) \sim J(t) + 0.0007[J(32t) - J(16t)] - 0.0185[J(16t) - J(8t)]$$
$$+ 0.197[J(8t) - J(4t)] - 0.778[J(4t) - J(2t)] - 0.181[J(t) - J(t/2)] - 0.049[J(t/4) - J(t/8)] \tag{7.31}$$

Eftekhari and Fatemi [1, 9] studied short-term creep of a number of engineering plastics such as polypropylene (PP), PP/elastomer blend, polyamide 66, polystyrene and polyphenylene ether and their composites with glass fibre and talc and high-density polyethylene (HDPE) and used time temperature superposition principle to predict long-term creep. The authors showed that the experimentally determined long-term creep and predicted long-term creep by TTS based on short-term data were in good agreement.

Chang and Brittain [10] showed that the tensile yield strength of epoxy thermoset increases with ageing time, at temperature below T_g of the epoxy network, while Kong [11] showed that the density, elastic modulus, hardness, damping, moisture absorption capacity and thermal expansion coefficient of neat epoxy and epoxy-carbon fibre composites increases with ageing time, as he studied the properties up to 105 min after annealing at isothermal ageing temperatures 140, 110 and 80 °C. Le've^que et al. [12] studied a carbon fibre reinforced epoxy composite for both physical and chemical ageing. The authors observed the T_g of the resin and the composite increases upon physical ageing for 5,000 h at an isothermal temperature below T_g of the unaged sample, and large decrease in creep strain during the physical ageing. However, no significant change in elastic properties was noticed. The interesting part of the study was a multiscale model to calculate the T_g over ageing time in a combined physical and chemical ageing effects, using a normalized cross-link density τ_R:

$$T_g(\tau_R) = T_{g\infty} - cT_{g\infty}^2(1 - \tau_R) \tag{7.32}$$

following the reaction schemes of structure consolidation and degradation:

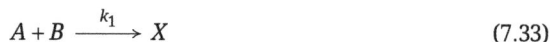

$$A + B \xrightarrow{k_1} X \tag{7.33}$$

$$X \xrightarrow{k_2} C + D \qquad (7.34)$$

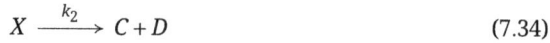

and the rate of reaction was defined as

$$\frac{d[X]}{dt} = \frac{k_1[A_0]}{(k_1[A_0]t_v + 1)^2} - k_2[X] \qquad (7.35)$$

where c is a constant, and $\tau_R = 1$ at the full consolidation, when T_g attains a final value, designated as $T_{g\infty}$.

The consolidation (physical ageing) and degradation are simultaneous process in ageing, and the consolidation is accelerated at elevated temperature, just below T_g of a typical thermoset composite such as CFRP. However, the convex shape of the plot of T_g vs ageing time clearly distinguishes the physical ageing stage (consolidation) in the rising portion and chemical ageing (degradation) as the diminishing portion. The reason for higher ageing time required for significant decrease in T_g is because of the very low rate of degradation of any thermoset, compared to that for physical ageing near T_g. The effect of chemical ageing is more than that of the physical ageing in the long term at that isothermal condition. The unique feature of this theoretical approach of combined ageing using eqs. (7.32)–(7.35) is that the glass transition and related viscoelastic properties can be calculated for any composite, approximately mapping with experimental data.

Frigione et al investigated changes glass transition and relaxation behaviour [13] and mechanical properties [14] of a cold cure epoxy adhesive due to physical ageing for a long period of time. In order to avoid environmental effect (chemical reaction due to UV, moisture, etc.) the epoxy adhesive was studied in dark and dry atmosphere. The authors observed that the T_g of the cold cured epoxy (around 46 °C) did not change significantly in pure physical ageing conditions [13] and behave similarly in physical ageing as conventional heat cured epoxy thermoset reported in literature. The changes due to environmental exposure can be an effect of both physical and chemical ageing, since it was observed that on de–ageing at 50 °C for some period and quenched back to the original ageing temperature, the change in T_g is very small compared to original unaged T_g. However, the mechanical properties such as flexural modulus and strength increased and the strain at break decreased due to pure physical ageing [14], which is in line with other literature reports. The authors distinguished the effect of natural weathering and physical ageing by reporting the properties under both conditions, and shown that the effect of environmental ageing is to degrade all mechanical properties with time. The overall effect of ageing is, however, a combined physical and chemical ageing, notwithstanding the fact that chemical processes results in larger change in properties at isothermal ageing below the glass transition.

The effect of nanoparticles in polymers show remarkable change in glass transition [21–27], which ultimately affects the ageing process. In some cases, the T_g increases due to nanoaddition, while there are reports that the T_g decreases with increase in nanoparticle content. Lee and Lichtenhan [21] showed an 8 °C increase in the mid-point of T_g of epoxy with 10 wt% polyhedral oligomeric silsesquioxanes (POSS) nanoparticle. Consequently, the physical ageing was slowed down compared to the pristine polymer.

The authors stated the reason for increased T_g as the molecular level reinforcement provided by the POSS cages to the epoxy network junctions, slowing down the segmental mobility and hence, the relaxation process. Similar effects are reported by Lu and Nutt [22] for epoxy-10% clay nanocomposite and by Rittigstein and Torkelson [23] for 10% nanoalumina added poly(2-vinylpyridine).

On the other hand, Cangialosi et al. [24] showed that the T_g of Boucher et al showed that a salinized nanosilica composite of PMMA decreased by 3 °C at the maximum content of the silica. The authors explained that the silanized surface of the silica prevented formation of hydrogen bond at the PMMA/silica interface. Similar result was observed for Poly(vinyl acetate)-salinized nanosilica composite by the same author group [25]. Consequently, the physical ageing process is accelerated compared to the pristine polymer.

7.2.5 Example

To observe the nature of the physical ageing, a theoretical exercise was done. An FRP composite of epoxy thermoset reinforced by E-glass fibre was studied for dynamic mechanical properties. Fixed frequencies of 0.2, 0.5, 1, 2 and 5 Hz covering a temperature range of 35–215 °C was used for DMA measurement. The strain was kept at about 0.5% to ensure linear viscoelastic response. The storage (dynamic) flexural modulus was calculated according to the following equations:

$$E_L = \frac{FL^3}{4bDd^3}\left[1 + 6\left(\frac{D}{L}\right)^2 - 4\left(\frac{d}{L}\right)\left(\frac{D}{L}\right)\right] \tag{7.36}$$

The initial glass transition temperature was about 87.6 °C taken at the lowest frequency (0.2 Hz). This DMA T_g increased by 6.33 °C at 5 Hz. The initial dynamic flexural modulus for creep study was measured at 65 °C. The loss factor was very low (<0.1) and T_g being quite high compared to experimental isothermal temperature, the relaxation was very slow. Hence, the storage (dynamic) modulus was approximately same as static modulus.

Subsequently, a hypothetical physical ageing data was generated. Storage modulus was determined by the equivalence of time and temperature since each 5 °C decrease resulted in 1 decade increase (10 times) in time. Accordingly, the relaxation time at 65 °C was about 7,159 s taken for 0.3 days ageing and the relaxation times at various ageing times were calculated by a linear relation of log of ageing time (t_e) vs. relaxation time (τ_{te}).The storage compliance was calculated as inverse of storage modulus. The momentary creep compliance was calculated using constant relaxation time at each ageing period, assuming no physical ageing effect during the creep experiment. The relaxation times at the beginning of each creep study (zero time after each ageing) is shown in Tab. 7.1. The momentary creep compliance at the four ageing period is plotted in Fig. 7.5. The points are generated by a theoretical model of compliance creep as

$$J(t_n) = J(0) + \{J_\infty - J(0)\}\left[1 - e^{(-t_n/\tau)}\right] \tag{7.37}$$

It can be seen from Fig. 7.5 that the creep compliance tends to attain a maxima, which is a property of viscoelastic solid, defined by Kelvin-Voigt model such as eq. (7.37). The maximum compliance, corresponds to a constant minimum elastic modulus is the purely elastic response.

Creep data of continued ageing during the creep study was calculated by a simple consideration of progressive shift factor in horizontal ($a_L = t_{e1}/t_{e2}$, etc.) shift as in eq. (7.10), and also considering vertical shift factor ($b_L = J_1/J_2$, etc.). Tab. 7.1 shows the ageing time, corresponding relaxation time, horizontal and vertical shift factors and the storage compliance for momentary events. This resulted in creep curves with delay due to continued physical ageing during the experiment as shown in Fig. 7.6.

Tab. 7.1: Parameters on ageing for creep study.

t_e, days	a_L	$J(0)$, GPa^{-1}	b_L	τ_{te}, s
0.3	1	0.1218	1	7.159E + 03
1	3.33	0.1134	0.93	1.149E + 05
3	10	0.1082	0.89	2.004E + 06
10	33.33	0.1037	0.85	3.813E + 07

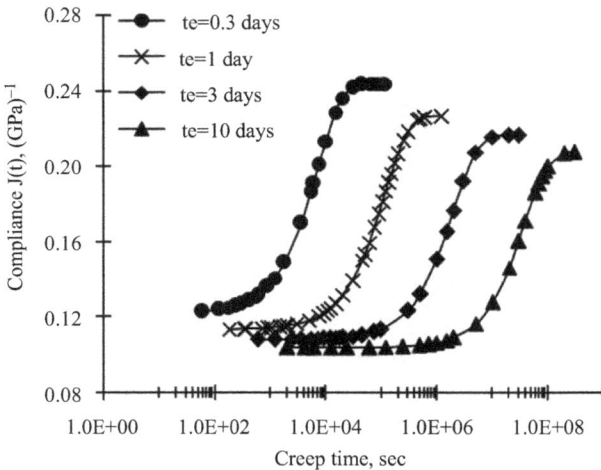

Fig. 7.5: Momentary creep compliance of epoxy thermoset/E-glass composite after various isothermal ageing periods at 65 °C. No delayed creep considered during creep experiment.

For the first ageing time of 0.3 days, the relaxation time at each creep time is calculated as

$$\tau_n = \tau_{n-1}\left(1 + \log\frac{t_n}{t_{n-1}}\right) \tag{7.38}$$

Also, eq. (7.37) is modified by taking progressive difference of compliance from the maximum value (0.2436 GPa^{-1}):

$$J(t_n) = J(0) + \{J_\infty - J(t_{n-1})\}\left[1 - e^{(-t_n/\tau_n)}\right] \tag{7.39}$$

The nature of the delayed creep curve is not the same as the momentary creep curve for some obvious reasons. Firstly, the data is generated by the model of creep as described by eq. (7.37). The equation clearly states that there would be a maximum value of the compliance at $t \gg \tau$, when $\exp(-t/\tau) \to 0$. Secondly, the relaxation time continuously changes during the creep study, and simultaneously, the factor $\{J_\infty - J(t_{n-1})\}$ continuously reduces with time. Therefore, it can be seen from eq. (7.39) that the creep compliance does not attain the maxima at the instant as in momentary creep, and is delayed progressively with time. Hence, there is progressive delay in attaining structural equilibrium for the polymer (Fig. 7.1). Needless to say, that the delay is much reduced, or the time to attain equilibrium is less if the isothermal ageing is near the T_g, and very slow, when far away from T_g. At the T_g or above, there is no physical ageing since the polymer is already in equilibrium as the segmental motion sets in.

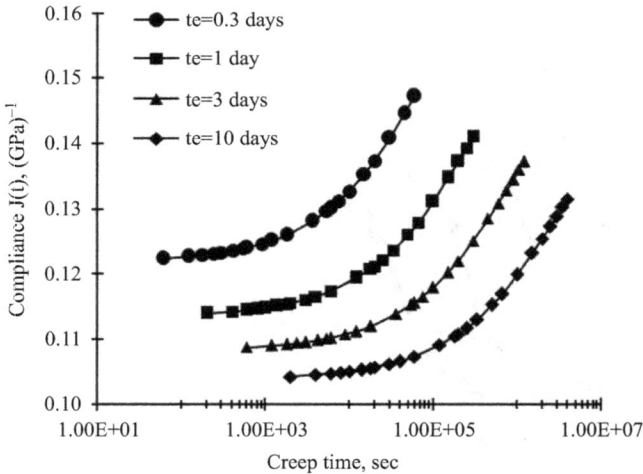

Fig. 7.6: Compliance creep on physical ageing: delayed creep due to continuous ageing during creep experiment.

It may be noted that since this example is shown purely as a hypothetical exercise, the real experimental data may not follow the above equations at all, since relaxation in polymer composites are much more complicated and may need semi-empirical equations to fit the experimental data. Secondly, the life prediction after a period of 10 days ageing at the isothermal temperature is simply the curve for 10 days creep data, since in this example, the shift factors were used to generate the data for all the three ageing (1-, 3- and 10 days ageing) form initial 0.3 days data. If the life prediction is to be done at a different isothermal temperature, then we need to calculate the relaxation time from DMA data for that temperature using $1/T$ vs. $\ln(\tau)$ relationship according to the Arrhenius expression or other time–temperature superposition equation such as WLF or VFT.

7.2.6 Conclusion

It can be concluded that purely physical ageing in a practical operating situation is very rare, unless the use of the polymer composite is in absence of light and atmospheric pollutants, including ozone, humidity, oxides of carbon and nitrogen, etc. However, as the studies reveal, there is a need to systematically study pure physical ageing of composites which are relatively inert to moisture and air pollutants (at minute concentration), since the oxidative degradation of many thermosets are very slow process at ambient, or even 20–30 °C below the glass transition, where the physical ageing could be significant.

7.3 Chemical ageing

The process through which the polymer molecular weight reduces is called degradation of the polymer. Thermal degradation takes place when the vibrational energy of the molecule exceeds the primary bond energy among the atoms. The degradation can be by depolymerization or random chain scission. Influencing factors for degradation can be heat, light, oxygen, ozone, mechanical force, chemical reagents, ultrasonic waves and microorganisms. Since chemical structure of a material decides all the properties, destruction of the structure must be the most important aspect of study for estimation of life of a polymer under storage or service.

Chemical ageing of polymer matrix composites (PMC) essentially means degradation of the basic molecular structure of the polymer and associated reduction of composite properties. The rate at which the chemical ageing takes place, depend on temperature of use and certainly on concentration of the reagent causing degradation. The rate in fact, decides the useful life of the composite. However, composites are of enormous variety and their application is ever increasing in all fields. Therefore, the useful life is to be determined by the deterioration of the functional property of the composite. The life estimation mainly requires studies of thermal and thermo-oxidative degradation and deterioration of important properties. There can be combination of applicable degradation parameters, and it is important to determine the most degrading parameter to decide the useful life. Thermal degradation and corresponding deterioration of other properties can be mapped to determine the acceptable degradation that leads to the minimum acceptable value for deciding the lifetime. Kinetics of thermal and thermo-oxidative degradation is therefore a most basic study. For example, if for an FRP, 5% weight loss corresponds to maximum allowable 10% reduction in Young's modulus compared to unaged state, then taking 5% weight loss as the basis of study, a detailed TGA kinetic analysis can be done to determine the time required to attain 5% weight loss by a time–temperature transformation method.

7.3.1 Thermal degradation kinetics

There are many thermal degradation studies of polymers and composites with kinetic analysis from late fifties onward [28–30] and only few of them, mostly recent reports, are briefly described here. Nikolaidis and Achilias [31] studied thermal composite degradation kinetics of PMMA-organo-montmorillonite clay nanocomposite. Vyazovkin et al. [32] studied kinetics of the thermal and thermo-oxidative degradation of a polystyrene–clay nanocomposite, Menezes et al. [33] carried out TGA kinetic modelling of semi-interpenetrating polymer network protonic conductive membranes. Qiu et al. [34] studied thermal degradation kinetics of a flame-retardant polyester-glass fibre composite. Raveshtian and co-workers [35] studied thermal degradation kinetics of nanoalumina-natural rubber composite. Zhang and Huang [36] studied thermal degradation kinetics of basal fibre-HDPE composite. Yousef and co-workers [37] studied thermal degradation using TG-FTIR, Py-GC/MS, applied to linear and nonlinear isoconversional models of glass fibre reinforced thermoplastic PMMA. Tranchard et al. [38] studied the degradation kinetics using TGA for carbon fibre reinforced epoxy composites. Narayanankutty [39] reported first-order thermal degradation of thermoplastic polyurethane-kevlar short fibre composite using TGA and DSC.

Kinetics of thermal degradation and estimation of life from kinetic parameters are best done by thermogravimetric method using a TGA instrument, while the enthalpy of the decomposition can be directly determined by differential scanning calorimetry. TGA studies can be done using temperature programming with time to navigate a thermal run. In an isothermal study, the temperature of the sample cell is brought to the desired temperature at a very high heating rate, and brought quickly to stabilize at the isothermal temperature of study, followed by carrying out the run for a preset time period. For kinetic analysis, several isothermal runs are taken with fresh samples at various isothermal temperatures. In a non-isothermal study, the temperature is increased at a steady heating rate up to a final temperature. To study at different heating rates, same experiments are repeated with fresh samples. It should be noted that the sample size, shape and other relevant conditions must be uniform for all runs, to avoid errors in experimental results.

The fractional weight loss with time is used as conversion. Mathematical expressions on TGA kinetics and lifetime estimation using the kinetic data are given below.

A general decomposition reaction of a solid, such as a PMC, can be written as

$$-\frac{dC}{dt} = k(T)f(C) \tag{7.40}$$

Introducing Arrhenius expression of dependence on temperature, eq. (7.1) becomes

$$-\frac{dC}{dt} = A \exp\left[-\frac{E}{RT}\right]f(C) \tag{7.41}$$

where C is the fractional weight remaining at any instant t during decomposition, and hence $(1 - C)$ represents faction decomposed, which is fractional weight loss or, more appropriately, the fractional conversion "α", and is calculated as

$$C = \frac{w}{w_0} \text{ and } \alpha = (1 - C) = \frac{w_0 - w}{w_0} \tag{7.42}$$

For convenience, we shall use the following form of eq. (7.41) with fractional conversion α using eq. (7.42):

$$\frac{d\alpha}{dt} = A \exp\left[-\frac{E}{RT}\right]f(\alpha) \tag{7.43}$$

Figure 7.7 shows an example of TGA thermogram with multiple heating rates of a thermoplastic nanocomposite. The horizontal solid straight line in Y-axis shows the 80% weight line and the four vertical lines are corresponding temperatures at four heating rates. The char yield is very low about 0.24% which could be remains of the organically modified nanoclay (originally 0.5% of the thermoplastic). Figure 7.8 shows the calculated fractional conversion (fractional weight loss) "α" against temperature, T °C for all heating rates: 2.5, 5, 10 and 20 °C/min. The solid line at $\alpha= 0.2$ in Fig. 7.8 corresponds to 20% degradation corresponding to 80% weight in Fig. 7.7. The experiment was done in nitrogen atmosphere to show the inherent thermal stability.

It may be noted that the onset of decomposition is progressively delayed for higher heating rates, as shown in Fig. 7.7. Therefore, one important assumption in using the uniform equations of decomposition is that the mechanism of degradation does not change, or any new reaction does not become dominant at the highest temperature for the same extent of conversion (α).

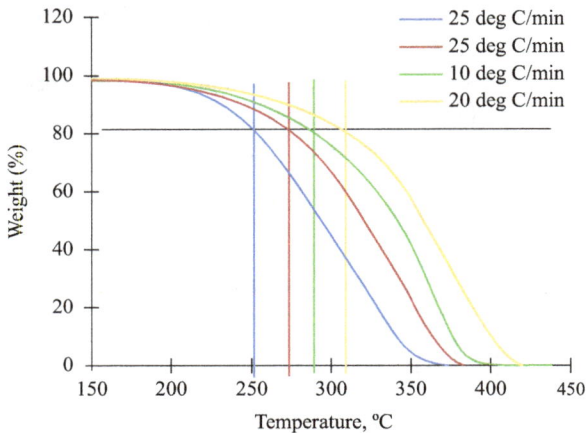

Fig. 7.7: TGA thermogram of a thermoplastic nanocomposite showing weight loss at four heating rates.

7.3.2 Isoconversion kinetics

For a thermal degradation reaction which may be complex and may have a multi-step mechanism, the activation energy (E), Arrhenius pre-exponent (A) and the conversion model $f(\alpha)$ are not independent of each other and are conversion-dependent. Therefore, a single data of E and A throughout the reaction would not correctly describe the kinetics. It is better to study isoconversion kinetics, where the values of E and A are determined as a function of conversion and does not need a model $f(\alpha)$ to estimate E and A. In such model-free kinetics, a more realistic estimation of lifetime at different extent of conversion based on the corresponding E value can be done. The derivation of a model-free kinetic expression is quite simple and effective. This method is described here briefly. There are many attempts to form the equation of model-free expression, both in differential and integral forms.

7.3.3 Differential form of isoconversion kinetics

7.3.3.1 Friedman method

Friedman [40] used the simplest differential method for determination of activation energy. Let a constant extent of degradation α is chosen at temperatures T_1, T_2, T_3, etc. at heating rates β_1, β_2, β_3, etc. for a non-isothermal TGA scan. We can then write from eq. (7.43) after introducing heating rate (β) as

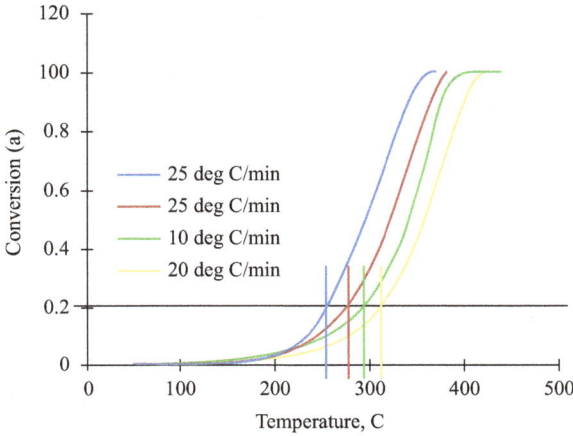

Fig. 7.8: Plot of conversion, a, vs temperature from Fig. 7.7.

$$\left(\frac{da}{dT}\right)\left(\frac{dT}{dt}\right) = A \exp\left[-\frac{E_a}{RT_a}\right]f(\alpha) \tag{7.44}$$

Taking logarithm and noting that $\beta = \dfrac{dT}{dt}$ (constant heating rate), we get:

$$\ln\left(\frac{da}{dT}\right)_{a,i} = \ln\left[\frac{A}{\beta}f(\alpha)\right] - \frac{E_a}{RT_{a,i}} \tag{7.45}$$

The subscript i denotes various thermal runs with different heating rates ($i = 1,2,3$, etc.). The activation energy is calculated from the slope of the plot $\ln(da/dT)$ vs $1/T$.

Since the data of (da/dT) and $(1/T)$ are taken at same conversion (α) from various heating rates, the activation energy thus obtained is conversion dependent. This method of estimating the activation energy is independent of the reaction model $f(\alpha)$. At a particular conversion, the first term of the right-hand side is constant. The plot is essentially a straight line, with a negative slope if $(1/T)$ is taken and positive slope if $(-1/T)$ is taken in X-axis. In any case, the activation energy is calculated from the slope as

$$E_a = slope \times R \tag{7.46}$$

where R is a universal gas constant ($R = 8.314$ J/mol K). The activation energy can be found out at any conversion and while estimating lifetime, the activation energy corresponding to the conversion should be used.

The advantage of this method is that if a different mechanism of degradation becomes prominent at a temperature which may be within T_1 to T_i (corresponding to heating rates β_1 to β_i) at the selected conversion α, then it will reflect in the $\ln(da/dT)$ data and a scatter in the kinetic plot will result. Therefore, activation energy at a series of conversion can be tested by this method to ascertain the validity of the degradation study by kinetics.

The differential method by Friedman does not need any assumption; hence, the activation energy obtained by this method is quite reliable for all conversions.

7.3.3.2 Freeman-Carroll method

Freeman and Carroll [41] used a modification of the Friedman method that it assumed an nth order kinetics of degradation:

$$\frac{d\alpha}{dt} = A \exp\left[-\frac{E}{RT}\right](1-\alpha)^n \tag{7.47}$$

If the logarithmic form of the eq. (7.47) is differentiated and again integrated with respect to T, the following equation is obtained:

$$\frac{\Delta \ln(d\alpha/dt)}{\Delta \ln(1-\alpha)} = n - \frac{E_a}{R}\left[\frac{\Delta(1/T_a)}{\Delta \ln(1-\alpha)}\right] \tag{7.48}$$

The plot between the left-hand side and $\Delta(1/T_a)/\Delta\ln(1-\alpha)$ would give a straight line with slope of Ea/R and intercept n (order of reaction).

This method is, of course, not a model-free method, since it assumes an nth-order model.

7.3.4 Peak rate method

7.3.4.1 Kissinger method

Considering eq. (7.44), Kissinger [28] used the condition of the conversion rate maxima by taking derivative of the rate and equating to zero:

$$\frac{d}{dt}\left(\frac{d\alpha}{dt}\right) = 0 = \frac{d}{dt}\left[Ae^{-E/RT}f(\alpha)\right] \tag{7.49}$$

Hence,

$$\frac{E\beta}{RT_p^2} = A\left[\frac{d}{dt}f(\alpha)\right]_p \exp\left(-\frac{E}{RT_P}\right) \tag{7.50}$$

The Arrhenius pre-exponent A and the derivative of $f(\alpha)$ at peak are constant due to the fixed conversion at the peak, Therefore, taking logarithm of eq. (7.50), we get:

$$\ln\left(\frac{\beta}{T_p^2}\right) = \ln\left(A\left[\frac{d}{dt}f(\alpha)\right]_p\right) - \ln\left(\frac{E_p}{R}\right) - \frac{E_p}{RT_p} \tag{7.51}$$

A plot of the left-hand side against $1/T_p$ would result in a straight line and the activation energy can be calculated from the slope of the line.

Since the activation energy is calculated without $f(a)$, it is also a model-free method. Generally, if the mechanism of degradation is same for the envelop of the temperature considered, then it has been observed that the peak rate corresponds to a particular conversion irrespective of heating rate (β). However, the activation energy thus calculated is not applicable to conversions other than at peak rate. Otherwise, being a differential method, it should yield same value of activation energy as Friedman method at the conversion corresponding to the peak rate.

7.3.4.2 Kim and Park method

Kim and Park [42] used a differential form of kinetic expression for peak decomposition rate. The general kinetic equation of nth order using heating rate (β) is given by

$$\frac{da}{dT} = \frac{A}{\beta} \exp\left[-\frac{E}{RT}\right](1-\alpha)^n \tag{7.52}$$

Differentiation with respect to T and equating to zero for peak rate temperature yields:

$$\frac{E}{nRT_p^2(1-\alpha_p)^{n-1}} = \frac{A}{\beta} \exp\left[-\frac{E}{RT_p}\right] \tag{7.53}$$

The peak rate temperature and height can be utilized to estimate the activation energy and pre-exponential factor of a reaction. Substituting eq. (7.53) into eq. (7.52) yields the expression for the activation energy:

$$E = \frac{nRT_p^2 H_p}{(1-\alpha_p)} \tag{7.54}$$

where H_p denotes the peak height. Subsequently, the pre-exponential factor is given by

$$A = \frac{H_p \beta \exp\left[\dfrac{E}{RT_p}\right]}{(1-\alpha_p)^n} \tag{7.55}$$

The activation energy can also be calculated from comparison of two peak rate temperatures (two thermograms), considering that the peak rate is almost same irrespective of heating rates.

$$E = R\left(\frac{T_{p1}T_{p2}}{T_{p1}-T_{p2}}\right) \ln\left[\left(\frac{\beta_1}{\beta_2}\right)\left(\frac{T_{p2}}{T_{p1}}\right)^2\right] \tag{7.56}$$

The pre-exponential factor A can be derived from any peak rate temperature as

$$A = \left(\frac{E\beta}{nR}\right) \frac{(1-\alpha)^{1-n}}{T_p^2 \exp\left(-E/RT_p\right)} \tag{7.57}$$

Again, this method is not a model-free method as it assumes an nth-order decomposition reaction. The integrated form of Kim and Park [42,43] is given by

$$\ln(\beta) = \ln(A) + \ln\left(\frac{E_a}{R}\right) + \ln\left[1 - n + \frac{n}{0.994}\right] - 5.3305 - 1.0516\left(\frac{E_a}{RT_p}\right) \tag{7.56}$$

where

$$n = \frac{1 - \alpha_p}{RT_p^2(d\alpha/dt)} \tag{7.57}$$

7.3.5 Integral form of isoconversion model-free kinetics

The model-free kinetics of isoconversion can be in the integrated form of eq. (7.4). The integration of eq. (7.43) gives

$$g(\alpha) = \int_0^a \frac{d\alpha}{f(\alpha)} = \frac{A}{\beta} \int_{T_0}^T \exp\left[-\frac{E_a}{RT}\right] dT \tag{7.58}$$

or

$$g(\alpha) = \left(\frac{AE_a}{\beta R}\right) p\left(\frac{E_a}{RT}\right) = \left(\frac{AE_a}{\beta R}\right) p(x) \tag{7.59}$$

where $x = \dfrac{E}{RT}$ and $p(x)$ is the integral defined as

$$p(x) = \left[\frac{e^{-E/RT}}{E/RT} + \int_{-\infty}^{-E/RT} (e^{-x}/x)dx\right] \tag{7.60}$$

The integral in eq. (7.58) cannot be analytically solved. However, there can be either a numerical integration or an approximate value of the integral may be used. Doyle [44], Murray and White [45] gave approximate values. A rational approximation up to four degrees of x is also used which is given in the literature [46,47]. Various solutions to eq. (7.58) based on approximations give minor variations in the value of activation energy.

From eq. (7.59), taking logarithm,

$$\log[g(\alpha)] = \log(AE/R) - \log\beta + \log p[x] \tag{7.61}$$

A table of x ($=E/RT$) and corresponding $-\log p(x)$ is given by Toop [48] for a first approximate value. With Doyle's approximation [44] for the value of $\log p(x)$:

$$\log p\left(\frac{E}{RT_i}\right) \cong -2.315 - 0.457\frac{E}{RT_i} \tag{7.62}$$

and value of $g(\alpha)$:

$$g(\alpha) = 7.03\frac{AE_a}{\beta R}(10^{-3})\exp(-1.052x) \tag{7.63}$$

7.3.5.1 Flynn-Wall-Ozawa method

Flynn-Wall-Ozawa (FWO) method [49, 50] is based on Doyle's approximation as shown in eqs. (7.62) and (7.63). Taking logarithm of eq. (7.63), we have:

$$\ln(\beta) = \ln\left(\frac{AE_a}{R}\right) - \ln[g(\alpha)] - 4.9575 - 1.052\frac{E_a}{RT} \tag{7.64}$$

At a constant conversion α, the function $g(\alpha)$, A, and E_a are constant. Therefore, eq. (7.64) can be written as

$$\ln(\beta) = \text{Cons.} - 1.052\frac{E_a}{RT} \tag{7.65}$$

Therefore, when a particular extent of decomposition is to be taken for lifetime assessment, the TGA runs for several heating rates β_1, β_2, etc. should be taken and corresponding temperatures T_1, T_2, etc. should be noted at that constant extent of conversion (weight loss). Subsequently, a plot of log (β) vs $1/T$ should yield activation energy E_a from the slope of the curve. If more than one weight loss is taken, then we should ideally have parallel lines for each weight loss otherwise the data scattering will be too high to determine a reliable activation of energy at various conversions.

FWO method is not so accurate because of the Doyle's approximation. However, it gives a comparatively acceptable result.

7.3.5.2 Kissinger-Akahira-Sunose (KAS) method

Integral form of KAS method [51] uses the approximate integral value of $p(x)$ from Murray and White [45] as

$$p(x) \cong \frac{e^{-x}}{x^2} \tag{7.66}$$

The KAS equation is supposed to give more accurate estimation of activation energy:

$$\ln\left(\frac{\beta}{T^2}\right) = \ln\left(\frac{AR}{E_a}\right) - \ln[g(a)] - \frac{E_a}{RT} \tag{7.67}$$

$$\ln\left(\frac{\beta}{T^2}\right) = \text{Cons.} - \frac{E_a}{RT} \tag{7.68}$$

A plot of left side of the equation against $1/T$ will be a straight line with a slope E_a/R and the value of E_a is calculated therefrom.

7.3.5.3 Starink method

A comparison of all approximation of $p(x)$ was done by Starink [52], where it is seen that all the approximations are derived from the series expansion of $p(x)$ by different researchers. The following approximation given by Starink for $p(x)$ is supposed to be more accurate:

$$p(x) \cong \frac{\exp(-1.0008x - 0.312)}{x^{1.92}} \tag{7.69}$$

and corresponding expression for estimation of activation energy:

$$\ln\left(\frac{\beta}{T^{1.92}}\right) = \text{Cons.} - 1.0008\frac{E_a}{RT} \tag{7.70}$$

The plot of the left side against $(1/T)$ would give a straight line with the slope $(1.008E_a/R)$. It is seen that the Starink method is almost similar to KAS method, with a small difference in the index of T on the left side expression and also a minute difference for the slope.

7.3.5.4 Coats and Redfern method

Coats and Redfern [53] derived an expression of integral form of kinetic equation taking an nth-order simple model:

$$g(a) = \int_0^a \frac{da}{(1-a)^n} = \frac{1-(1-a)^{1-n}}{1-n} \tag{7.71}$$

and $g(a)$ was substituted in the basic eq. (7.58) to obtain following expression for a non-isothermal TGA:

$$\ln\left[\frac{g(a)}{T^2}\right] = \ln\left[\frac{AR}{\beta E}\left(1-\frac{2RT}{E}\right)\right] - \frac{E}{RT} \tag{7.72}$$

where $\ln\left[\frac{AR}{\beta E}\left(1-\frac{2RT}{E}\right)\right]$ is constant at the same conversion.

Therefore, the authors suggested that for most values of E and the temperature range of usual reactions, the value of the left side in eq. (7.72) against $(1/T)$ would give the activation energy E. However, the Arrhenius pre-factor can be calculated once the activation energy is known. The assumption of the model $f(\alpha)$ is then verified with the theoretical calculation of α vs. T compared to the experimental thermogram.

7.3.5.5 Phadnis-Deshpande method

Phadnis and Deshpande [54] used a combination of $f(\alpha)$ and $g(\alpha)$ to obtain the following expression, which do not require the evaluation of $p(x)$:

$$f(\alpha)g(\alpha) = \frac{RT^2}{E}\left[1 - \frac{2RT}{E}\right]\frac{d\alpha}{dT} \tag{7.73}$$

Neglecting the small term $(2R^2T^3/E^2)$, eq. (7.73) becomes

$$f(\alpha)g(\alpha) = \frac{RT^2}{E}\frac{d\alpha}{dT} \tag{7.74}$$

The products $f(\alpha)$ and $g(\alpha)$ have to be tried out from generally used solid-state reaction models. The left side, when plotted against T^2 $(d\alpha/dT)$ should yield the activation energy from the slope. The method is not a model-free method and may require multiple iterations from the large number of possible $f(\alpha)$ from the literature.

7.3.6 Advanced isoconversion kinetics

In all the above integral methods, one common problem is integration over a large range of temperature (or time), which results in so-called undue "averaging" and thus, with progressive integration (for example, α=0 to 0.2, 0 to 0.4), the discrete errors are added up, resulting in large errors. To circumvent this problem of most integral methods, an advanced nonlinear integral isoconversional method has been proposed by Vyazovkin [55,56].

7.3.6.1 Vyazovkin method

According to this method, for two experiments carried out at two constant heating rates of β_1 and β_2, the E_a value is determined as a value that minimizes the function $\phi(E_a)$

$$\phi(E_a) = \frac{I(E_a, T_{a,1})\beta_2}{I(E_a, T_{a,2})\beta_1} + \frac{I(E_a, T_{a,2})}{I(E_a, T_{a,1})\beta_2} \tag{7.75}$$

where

$$I(E_a, T_{a,i}) = \int_{T_{a-\Delta a,i}}^{T_{a,i}} \exp\left[-\frac{E_a}{RT}\right] dT, \quad i = 1, 2 \tag{7.76}$$

In the above equations, T_{a1} and T_{a2} are the temperatures corresponding to heating rates β_1 and β_2 for the identical conversion a, while $I(E_a, T_{a,i})$ is the temperature integral which can be numerically calculated for the small interval of Δa. In general, a set of n heating programs are carried out (mostly $n = 4$ or 5) for one reaction analysis and the activation energy at a particular conversion (a) is found by fitting a value of E_a which minimizes the function [57]:

$$\phi(E_a) = \sum_{i=1}^{n} \sum_{j\neq 1}^{n} \frac{J[E_a, T_i(t_a)]}{J[E_a, T_j(t_a)]} \tag{7.77}$$

where the integral J is given by

$$J[E_a, T_i(t_a)] = \int_{t_a-\Delta a}^{t_a} \exp\left[\frac{E_a}{RT_i(t)}\right] dt \tag{7.78}$$

The process, is therefore, iterative, and proceeds with assuming a value of E_a to begin with. The initial value can be taken from a previously determined activation energy of the same study calculated by any other method, either differential or integral. For better accuracy, Δa should be chosen as a small value, for example 0.01 as used by Vafayan et al. [58].

7.3.6.2 Cai and Chen method

Cai and Chen [59] developed a new iterative linear integral isoconversion method and claimed that the new method is less complicated than the Vyazovkin method and requires lesser number of iterations. The method is based on the integration of the kinetic equation over a small range of conversion, a similar consideration as Vyazovkin. The integral function $g(a, a - \Delta a)$ is given by the following equation:

$$\begin{aligned} g(a, a - \Delta a) &= \frac{A_{a-\Delta a/2}}{\beta} \int_{T_{a-\Delta a}}^{T_a} \exp[-E_{a-\Delta a/2}/RT] dT \\ &= \frac{A_{a-\Delta a/2}}{\beta} \left[\int_0^{T_a} \exp(-E_{a-\Delta a/2}/RT) dT - \int_0^{T_{a-\Delta a/2}} \exp(-E_{a-\Delta a/2}/RT) dT \right] \end{aligned} \tag{7.79}$$

The authors used an integral $h(x)$, which is given by

$$h(x) = x^2 e^x \int\limits_0^\infty \frac{e^{-x}}{x^2} dx \tag{7.80}$$

to solve the integration $I\,(E,T)$ as

$$I(E,T) = \int\limits_0^T e^{(-E/RT)} dT = \frac{RT^2}{E} e^{(-E/RT)} h(x) \tag{7.81}$$

Finally, the equation for determination of activation energy is

$$\ln\left\{ \frac{\beta}{T_\alpha^2\left[h(x_\alpha) - \frac{x_\alpha^2 e^{x_\alpha}}{x_{\alpha-\Delta\alpha}^2 e^{x_{\alpha-\Delta\alpha}} h(x_{\alpha-\Delta\alpha})} \right]} \right\} = \ln\left[\frac{A_{\alpha-\Delta\alpha/2} R}{E_{\alpha-\Delta\alpha/2} g(\alpha, \alpha - \Delta\alpha)} \right] - \frac{E_{\alpha-\Delta\alpha/2}}{RT_\alpha} \tag{7.82}$$

where

$$x_\alpha = \frac{E_{\alpha-\Delta\alpha/2}}{RT_\alpha}, \quad x_{\alpha-\Delta\alpha} = \frac{E_{\alpha-\Delta\alpha/2}}{RT_{\alpha-\Delta\alpha}} \tag{7.83}$$

In the above expression, $h(x)$ does not have a numerical solution, and approximate value of $h(x)$ can be taken from various approaches of the integral, and the following polynomial rational gives quite accurate value as shown by Senum and Yang [46]:

$$h(x) = \frac{x^4 + 18x^3 + 86x^2 + 96x}{x^4 + 20x^3 + 120x^2 + 240x + 120} \tag{7.84}$$

Observation of eq. (7.82) reveals that the equation is almost similar to KAS [51] with a modification that the term T^2 has a coefficient here which is dependent on activation energy. Since the right-hand side has $E_{\alpha-\Delta\alpha/2}$ in both the terms, there should be an iterative calculation until two successive values of E_α are almost same, and the Cai and Chen [59] have suggested that the two successive values of E_α should not have difference more than 0.001 kJ/mol.

To determine the activation energy at a particular conversion, following steps are to be performed using the above equations:

1. Select a conversion α and $\alpha - \Delta\alpha$ and $\alpha - \Delta\alpha/2$ in a thermogram of heating rate β_1. The value of $\Delta\alpha$ should ideally be 0.01, so that $\Delta\alpha/2 = 0.005$.
2. Assume any value of $E_{\alpha-\Delta\alpha/2\alpha}$ to start with. The best assumption could be such that E/RT should be more than 10, so that the error in estimation of $h(x)$ is less than 0.1% [46].
3. Calculate x for conversions α and $\alpha - \Delta\alpha/2$ corresponding to T_α and $T_{\alpha-\Delta\alpha}$.
4. Subsequently, calculate $h(x_\alpha)$ and $h(x_{\alpha-\Delta\alpha})$ using eq. (7.84).

5. Similarly, calculate these terms for other heating rates β_2, β_3, β_4, etc. at the same conversions α and $\alpha - \Delta\alpha/2$. Minimum four heating rates should be used, like other methods.

6. Calculate the left-hand term of eq. (7.82) and plot against $1/T_\alpha$ for all heating rates to obtain a straight line, with the negative slope as $E_{\alpha-\Delta\alpha/2}/R$.

7. Repeat this iteration with all the calculation using the $E_{\alpha-\Delta\alpha/2}$ obtained in step 6.

8. Find the next $E_{\alpha-\Delta\alpha/2}$.

9. Continue till the last two successive values of $E_{\alpha-\Delta\alpha/2}$ are either equal or with a maximum difference of 0.001 kJ/mol. The authors have shown that maximum three to four iterations are enough to achieve this accuracy.

Vafayan et al. [59] showed that the relative error of activation energy by this method is less than −0.02% at low conversion, and maximum −0.3% at $\alpha = 0.90$. The relative errors of FWO [49,50], KAS [51], Vyazovkin-Dollimore [55] methods increased with conversion and is more than +30% beyond a conversion $\alpha = 0.80$, while the relative error of Vyazovkin method [56,57] was less than +1% up to $\alpha = 0.80$, but the error sharply increased to +3% at $\alpha = 0.90$.

Selection of $\Delta\alpha$ is important since if it is 0.01, the relative error in calculation of activation energy is below 0.1%, but with 0.02, the error may be higher (≈1%) at higher conversion ($\alpha > 0.90$). However, up to 50% conversion ($\alpha = 0.50$), there is very little error even with $\Delta\alpha = 0.05$.

7.3.6.3 Budrugeac method

Budrugeac [60] also suggested an iterative isoconversion method instead of model freeway (MFW) method. His method is almost similar to Vyazovkin [57] and Cai and Chen [59] as he suggested following equations for small $\Delta\alpha$ values to estimate the activation energy:

$$g(\alpha_2) - g(\alpha_1) = \frac{A}{\beta} \int_{T_1}^{T_2} \left[\exp\left(-\frac{E}{RT}\right) \right] dT \tag{7.85}$$

where the integral is given by

$$I(E, T_1, T_2) = \int_0^{T_2} \left[\exp\left(-\frac{E}{RT}\right) \right] dT - \int_0^{T_1} \left[\exp\left(-\frac{E}{RT}\right) \right] dT$$

$$= \frac{E}{R} [p(x_2) - p(x_1)] \tag{7.86}$$

Therefore,

$$g(\alpha_2) - g(\alpha_1) = \frac{AE}{R\beta}(T_2 - T_1)\left[\exp\left(-\frac{E}{RT_2}\right)\right]\frac{p(x_2) - p(x_1)}{(T_2 - T_1)\left[\exp\left(-\frac{E}{RT_2}\right)\right]} \qquad (7.87)$$

which is rearranged after taking logarithm as

$$\ln\frac{\beta}{T_2 - T_1} = \ln\frac{A}{g(\alpha_2) - g(\alpha_1)} + \ln R_l - \frac{E}{RT_2} \qquad (7.88)$$

where

$$R_l = \frac{\int_{T_1}^{T_2}[\exp(-E/RT)]dT}{(T_2 - T_1)[\exp(-E/RT_2)]} = \frac{E}{R}\frac{p(x_2) - p(x_1)}{(T_2 - T_1)[\exp(-E/RT_2)]} \qquad (7.89)$$

Plotting the right-hand side of eq. (7.88) against $(1/T_2)$, one should get first value of E and since R_l consists of E, iterative process is applied until the two successive activation energies are equal or have a maximum difference of 0.1 kJ/mol, as suggested by the author. This method also suggests the integral $p(x)$ as a fourth-order polynomial rational as given by Senum and Yang [46].

7.3.7 Accuracy of kinetics analysis

The various methods of estimation of activation energy using thermogravimetric analysis should ideally yield same value at a constant conversion, since the reaction as such and the thermogram are same, only the theoretical methods are different. The difference in numerical values is due to the assumptions, if any, thermal programming and data processing.

Flynn [61] opined that in the present world of high computational capacity, accurate integral can be used instead of approximate integral. In fact, Dubaj et al. [62] used a very accurate method for incremental isoconversional method for kinetic analysis based on the orthogonal distance regression.

Vyazovkin and co-workers [63] reviewed the thermal methods of kinetics analysis and raised several important points on methods of analysis of reaction parameters, and made recommendations to Kinetics Committee of the International Confederation for Thermal Analysis and Calorimetry (ICTAC) for performing kinetic computations on thermal analysis data. A major assumption in the most used kinetics of thermal degradation is that the $f(\alpha)$ is considered to be a single reaction step, which may not be true. In case of multi-step degradation, however, the slowest step can be taken as the rate controlling one, and the analysis is considered for the slowest function $f(\alpha)$, although it is incorrect. Similarly, the activation energy is actually a composite activation energy, consisting of several discrete activation energy barriers for several steps of degradation reactions. Also, the error in temperature measurement leads to inaccuracy in determination of the kinetic parameters E and A, particularly when the error is not uniform throughout the measuring range. A systematic error of only 1–2 °C in measurement at different heating rates may result in quite significant error in the values of E and A. If the kinetics analysis is used for prediction of a physical quantity such as mass loss beyond the experimental temperature measuring range, the error can be very high, rendering the prediction absurd.

Simon and co-workers [64] stated that most serious error takes place due to high heating rate in experimentation with thermally poor conductors, as there would be a temperature gradient in the sample resulting in a combination of different process rates and different heat transfer rates within the sample, causing faulty measurements. The authors suggested a maximum of 10 °C/min heating rate for such insulating materials.

Maqueda and others [65] discussed the problems of evaluation of a reaction by different thermal programs -isothermal, non-isothermal, and newer methods of sample-controlled thermal analysis (SCTA) and constant rate thermal analysis (CRTA), modulated TGA and the application of a generalized time (θ) for generating master plots that can be used simultaneously for any experimental data independent of the heating schedule. For a single heating program, it is known that more than one conversion function can define the reaction rate with high accuracy, but not so with multiple thermal schedules. The authors proposed a combined kinetic analysis of a series of experiments performed under different heating schedule for a wide range of conversion. This suggestion seems to be correct, since a reaction should be unique irrespective of a selection of heating program, since a reaction rate is only governed by temperature and concentration (effect of pressure is neglected here). There can be a unique model $f(\alpha)$, which would satisfactorily be used to describe the reaction carried out irrespective of different heating schedule – isothermal, non-isothermal with varied heating rates, CRTA, etc. To substantiate this postulate, the authors examined a decomposition kinetics using thermal analysis and used are arranged logarithmic form of eq. (7.43):

$$\ln\left(\frac{d\alpha/dt}{f(\alpha)}\right) = \ln(A) - \frac{E}{RT} \tag{7.90}$$

Subsequently, a combined plot (left-hand side vs $1/T$) of all different heating schedules using a few $f(\alpha)$ models were examined. Only one of the models yielded combined plot as a straight line with uniform slope, rather than non-linear curves by all others.

Budrugeac [66] suggested that the kinetic parameters obtained by model-free way can determine the conversion function, which is proportional to the experimentally obtained one, but cannot be unambiguously taken as the correct function, and proposed the relation between activation energy (E_a) and Arrhenius pre–exponent (A_a) (compensation effect) as polynomial instead of linear, and claimed to find a more accurate conversion function. The author suggested that this method could lead to real values of the conversion function even if its algebraic expression is not part of their known theoretical set.

The most important part in the analysis of a kinetics of degradation of solid polymeric composite is accuracy of calculated activation energy and other kinetic parameters for lower range of conversion (fractional degradation), mostly 0.02–0.20, since the life estimation of a composite is decided by maximum 20% molecular degradation, in all structural and functional applications. Therefore, the thermogravimetric analysis should be carried out for a perfectly dry sample, devoid of any other small molecule, completely cured and polymerized, and should have known history of mechanism of degradation up to the maximum temperature, for example temperature at 20% degradation using the selected maximum heating rate. If the chemical degradation mechanism changes within the desired conversion limit, then the same kinetic analysis will not be applicable.

7.3.8 Estimation of lifetime

A very comprehensive review of accelerated ageing and lifetime prediction techniques for polymer materials in general and thermoset composites was discussed by Maxwell et al. [67]. The authors discussed ageing mechanisms of composites, change of glass transition due to moisture absorption, degradation of fibre due to harsh environment, stress corrosion, reduction in mechanical strength and overall thermal degradation.

For ageing of a polymer at a particular temperature, for example the storage or service temperature of a composite, $g(\alpha)$ is given by

$$g(\alpha) = kt_i \tag{7.91}$$

and

$$k = A \exp\left(-\frac{E_a}{RT}\right) \tag{7.92}$$

Therefore, at identical conditions, for the same $g(\alpha)$, the conversion (α) will be the same for both isothermal ageing and also at a constant heating rate in TGA. At the isothermal ageing time t_i at the TGA temperature T_i for same conversion (α), following expression can be written from rearrangement of the integral form:

$$kt_i = \frac{A_\alpha E}{\beta R} p(x_i) \tag{7.93}$$

Substituting eq. (7.87) into eq. (7.88), we get

$$t_i = \frac{E_a}{\beta R} e^{E_a/RT_i} p(x_i) \tag{7.94}$$

where T_i is the ageing temperature (absolute scale), for example, storage or service temperature for the polymer composite. The value of $log\ p(x_i)$ can be found out from the approximations used by different researchers and also from the table by Toop [48]. However, the following "rational" approximation as a fourth-order polynomial by Senum and Yang [46,47] is the best as shown by Starink. The approximation by Starink as in eq. (7.69) gives the same result for $x > 20$ and up to 100:

$$p(x) \cong \frac{\exp(-x)}{x^2} \frac{x^4 + 18x^3 + 86x^2 + 96x}{x^4 + 20x^3 + 120x^2 + 240x + 120} \tag{7.95}$$

Hence t_i gives the life of the polymer composite corresponding to a minimum acceptable value of a functional property at any storage or service temperature (T_i).

A more convenient form of eq. (7.94) is given by Toop [48]:

$$\log t_i = \frac{E_a}{2.303RT_i} + \log\left[\frac{E_a}{\beta_i R}\right] + \log[p(x_i)] \tag{7.96}$$

where E_a is the activation energy of degradation at the selected conversion "α"; T_α is the temperature in Kelvin at a conversion $= \alpha$, taken from the thermogram at a heating rate β_i. It is advisable to take the slowest heating rate thermogram to take the value of T_α, and E_a/RT_α is calculated using this temperature. $\log p(x_i)$ is calculated from eq. (7.95) and directly used in eq. (7.96). T_i is the temperature at which lifetime is to be calculated (storage or service temperature for the polymer or its composite). The term $E_a/(2.303RT_i)$ is calculated accordingly.

The maximum allowable service temperature for a selected service life can also be determined by rearranging eq. (7.96):

$$T_i = \frac{E_a/2.303R}{\log t_i - \log\left[\frac{E_a}{\beta_i R}\right] - \log[p(x_i)]} \tag{7.97}$$

7.3.9 Example

A polyimide thermoset composite is to be used as a high-temperature lamination inside an aircraft. The composite is not subjected to any significant stress other than heat during use. The material is to be qualified on the basis of life due to possible thermal degradation. Maximum 5% degradation is allowed since this is linked to the minimum acceptable structural integrity. The user wants to ascertain the service life at continuous 325 °C and occasional spike up to 350 °C, for maximum 10–20 min.

Solution

A typical TGA kinetics and life estimation is done to ascertain the service life at 325 °C and up to 400 °C. Figure 7.9 shows the TGA thermogram of the polyimide thermoset–carbon nanotube composite carried out in air atmosphere using four heating rates 2.5, 5, 10 and 20 °C/min. The thermogram is shown here up to a weight loss of about 23% to observe that the shape of the curves is uniform, and hence can be assumed that the mechanism of degradation is uniform at least up to 23% weight loss. The aromatic polyimides are generally thermally stable up to minimum 550 °C under nitrogen with uniform degradation pattern and a char yield of about 52–55% [68]. The dashed horizontal line is drawn for constant ordinate at 95% weight (5% weight loss) and vertical dashed lines are drawn to indicate four temperatures at which 5% weight loss takes place for four selected heating rates. With increase in heating rate, the loss of 5% progressively occurs at higher temperature. It is assumed that the mechanism of degradation/decomposition of the composite is same until at least 20% degradation.

The conversion (fractional weight loss) is plotted against absolute temperature in Fig. 7.10. From Figs. 7.9 and 7.10, the kinetic data listed in Tab. 7.2 are obtained for estimation of activation energy by different methods. Conversion is in fractions and heating rate is expressed in K/s. Subsequently, the activation energies are calculated by the methods of Friedman, FWO and Starink using eqs. (7.45), (7.65) and (7.70), respectively, and also by Cai and Chen method using eqs. (7.82)–(7.84). Temperature and time data were taken for each heating schedule at the conversion of 0.05 and the calculated data for first three methods are listed in Tab. 7.2. The kinetic plot for each of these methods is shown in Fig. 7.11(a)–(c). The iterative values of Cai and Chen method is shown in Tab. 7.3.

It is seen from Tab. 7.3 that three iterations are sufficient to obtain a constant value of activation energy according to Cai and Chen method. The initial value is taken from both Friedman method and FWO method, and the final consistent value is same, irrespective of the initial values. The values of E_α calculated from the slopes of the respective plots are listed in Tab. 7.4.

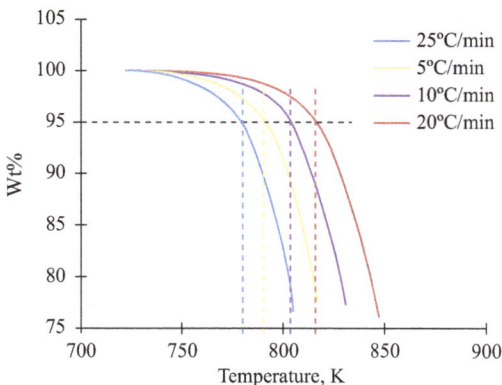

Fig. 7.9: TGA thermogram of the polyimide thermoset-CNT nanocomposite in air at 4 heating rates.

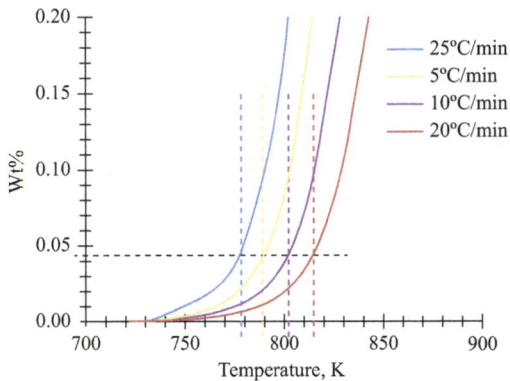

Fig. 7.10: Conversion (weight loss) of polyimide-CNT composite at four heating rates.

Tab. 7.2: TGA data at 5% degradation and calculated values of different kinetic parameters.

β, K/s	T, K	$1/T$, K^{-1}	$\ln(\beta\, da/dT)$	$\ln(\beta)$	$\ln(\beta/T^{1.92})$
0.0416667	779.5	0.0012829	−8.92	−3.1781	−15.96267
0.0833333	791	0.0012642	−8.34	−2.4849	−15.29764
0.1666667	803.5	0.0012446	−7.56	−1.7918	−14.6346
0.3333333	816	0.0012255	−6.92	−1.0986	−13.97109

It is seen from Tab. 7.4 that the activation energies by different methods are different. Particularly, the value for Cai and Chen and Friedman methods are higher than FWO and Starink methods. Further, FWO method yielded almost same value as Starink method. Since KAS method is almost similar to Starink, it is not calculated for this example. The values are higher than that obtained by Ahmad et al. [68], as they reported about 258 kJ/mol at 5% degradation for one aromatic polyimide–SiO_2 nanocomposite. Saha and Bhowmick [69] studied thermal degradation of nanosilica and Cloisite 30B nanoclay filled hydrogenated nitrile rubber and observed that the activation energy of degradation (5%) increased significantly from 218 kJ/mol for HNBR to about 270 kJ/mol for both the nanofiller added composite and the degradation temperature was increased and the rate decreased, consequently the predicted life was increased by 6–7 folds. The authors stated that the degradation resistance enhanced because the nanofillers provide tortuous path for the degraded species of the polymer and shields the underlying polymer. Also, the nanofillers enhanced thermal conductivity of the rubber as measured by the authors, and hence, the heat dissipation is better, and as a result, the degradation was reduced.

In order to calculate the lifetime at a lower temperature than the TGA degradation temperature, it is best to take the slowest heating rate. In this example, we need to predict life at 325 °C, whereas at 5% decomposition, the minimum TGA temperature

(a)

Friedman : $\alpha = 0.05$

$R^2 = 0.9982$

y-axis: $\ln[\beta(d\alpha/dT)]$

x-axis: $1/T$, K^{-1}

(b)

FWO: $\alpha = 0.05$

$R^2 = 0.9999$

y-axis: $\ln(\beta)$

x-axis: $1/T$, K^{-1}

(c)

Starink: $\alpha = 0.05$

$R^2 = 0.9999$

y-axis: $\ln(\beta/T^{1.92})$

x-axis: $1/T$, K^{-1}

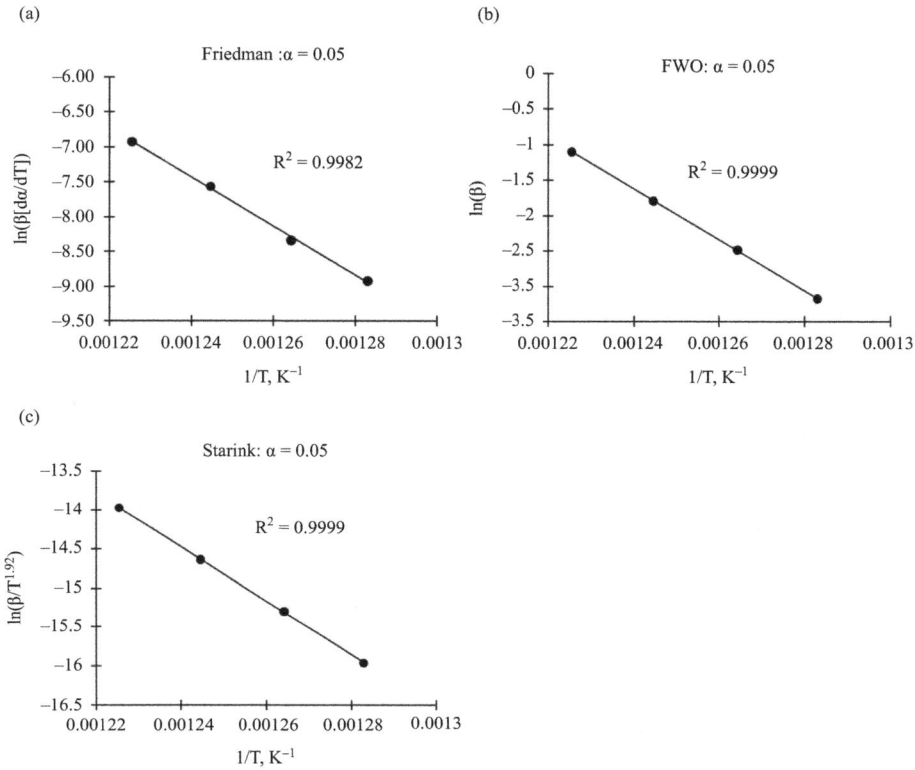

Fig. 7.11: (a) Kinetic plot according to the Friedman method; (b) kinetic plot according to the Flynn-Wall - Ozawa (FWO) method; and (c) kinetic plot according to the Starink method.

Tab. 7.3: Iterations in Cai and Chen method.

Activation energy, E_a at 5% degradation, kJ/mol			
Initial value	Iteration 1	Iteration 2	Iteration 3
294.099	297.142	297.134	297.134
($R^2 = 0.9982$)	($R^2 = 0.9994$)	($R^2 = 0.9994$)	($R^2 = 0.9994$)
285.55	297.176	297.134	297.134
($R^2 = 0.9999$)	($R^2 = 0.9994$)	($R^2 = 0.9994$)	($R^2 = 0.9994$)

is 506.5 °C (from Table 7.2) at 2.5 °C/min heating rate. Hence, the calculations are carried out accordingly.

The value of log $p(x_i)$ is first calculated using eq. (7.95) by Senum and Yang in references [46,47]. The values thus obtained were checked from the table by Toop [48] and was found to be equal, hence the subsequent calculations are done using eq. (7.95).

Tab. 7.4: Activation energy determined at constant degradation (0.05).

Method	Degradation or conversion (a)	Activation energy (E_α), kJ/mol
Friedman	0.05	294.10
Flynn-Wall-Ozawa (FWO)	0.05	285.55
Starink	0.05	287.44
Cai and Chen	0.05	297.13

Using the relevant parameters and the vales of $p(x)$, the lifetime was calculated for each isothermal ageing temperatures 325, 350, 375 and 400 °C.

Table 7.5 shows the calculated parameters of eq. (7.96) and estimated lifetime at the above isothermal ageing temperatures taking all activation energies from Tab. 7.4.

The lifetime estimated by the methods listed in Tab. 7.5 is shown in Fig. 7.12. It is seen from Tab. 7.5 and Fig. 7.12 that at lower ageing temperature, the difference in predicted life is more when activation energies are different, i.e. when the ageing temperature is far below the TGA temperature of degradation. For example, at 325 °C, the lifetime estimated by the activation energy by Cai and Chen method, ($E_a = 297$ kJ/mol) is about 4.28×10^8 s (about 14 years) compared to 2.81×10^8 s (9 years) taking activation energy by Starink ($E_a = 287$ kJ/mol), while at 375 °C, the respective predicted lifetimes are about 4.26×10^6 s (49 days) compared to 3.24×10^6 s (38 days). From Thermogram of 2.5 °C/min, the degradation temperature at 5% conversion (weight loss) is 506.5 °C and the time taken to the 5% weight loss is only 80 min, starting from room temperature of 30 °C.

Tab. 7.5: Lifetime estimation of the polyimide composite by isoconversion kinetics.

Method	T_i, °C	$E_\alpha/2.303RT_i$	$x_i = E_\alpha/RT$	log $p(x_i)$	log(E_α/I)	t_i, s
Friedman	325	25.6856	45.3804	−23.0405	5.9289	3.749E + 08
	350	24.6548	45.3804	−23.0405	5.9289	3.493E + 07
	375	23.7037	45.3804	−23.0405	5.9289	3.908E + 06
	400	22.8231	45.3804	−23.0405	5.9289	5.146E + 05
FWO	325	24.9391	44.0616	−22.4427	5.9161	2.585E + 08
	350	23.9384	44.0616	−22.4427	5.9161	2.581E + 07
	375	23.0148	44.0616	−22.4427	5.9161	3.077E + 06
	400	22.1599	44.0616	−22.4427	5.9161	4.298E + 05
Starink	325	25.1042	44.3532	−22.5750	5.9190	2.81E + 08
	350	24.0968	44.3532	−22.5750	5.9190	2.76E + 07
	375	23.1671	44.3532	−22.5750	5.9190	3.24E + 06
	400	22.3065	44.3532	−22.5750	5.9190	4.47E + 05

Tab. 7.5 (continued)

Method	T_i, °C	$E_a/2.303RT_i$	$x_i = E_a/RT$	log $p(x_i)$	log(E_a/I)	t_i, s
Cai & Chen	325	25.9506	45.8486	−23.2526	5.9334	4.28E + 08
	350	24.9092	45.8486	−23.2526	5.9334	3.89E + 07
	375	23.9482	45.8486	−23.2526	5.9334	4.26E + 06
	400	23.0586	45.8486	−23.2526	5.9334	5.49E + 05

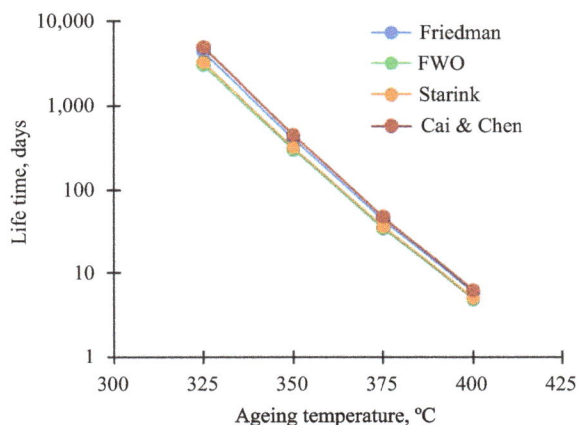

Fig. 7.12: Life prediction by different activation energies on various isothermal ageing temperatures.

As shown in Fig. 7.12, the time logarithmically varies with temperature, as is known for all time-temperature transformations. Also, the lifetime at lower temperatures is widely different for differences in activation energy and the difference reduces as the lifetime is predicted for higher temperature. The estimation of life is closer for all methods as the isothermal aging temperature is chosen closer to TGA degradation temperature. This indicates that for reliable results by interpolation to low temperature, the activation energy must be accurately determined. If, by this method, one has to determine the service life of this composite sample at ambient temperature (25–35 °C), there will be large variation in time determined by different methods, particularly by Friedman or Cai and Chen compared to other integral methods. Generally, experimental data from TGA is the source of error, as there can be many errors in the experiments especially inconsistency in errors of the balance and heating. The other type of error can be in data sampling. It is desired that very small and constant interval of time should be used for data sampling since the first derivative da/dt is calculated by numerical method. Cai and Chen method suggests that time and temperature should be sampled corresponding to half the conversion interval ($\Delta a/2$) and should beat the most 0.005 for consistent kinetic parameters.

7.3.10 Conclusion on predictive Lifetime

The models of thermal degradation kinetics are to be applied to any life prediction where only thermal stress is dominant. In case of composites having quite high degradation temperature than application temperature, the predictive life is infinitely high for prediction at much lower temperature, and is not useful otherwise. This is seen with all the kinetic models considered here. Secondly, the variation from one model to the other is mainly due to considering approximations of the integral value of $p(x)$. The first one can be eliminated by conducting at least two types of heating schedules, one non-isothermal with four or five heating rates and constant conversion decomposition [65]. The integral value is most accurate if it can be computed by any advanced mathematical program as mentioned by Flynn [61]. In addition, the heating rate in non-isothermal study must not exceed 10 °C/min for polymeric composites as suggested by Simon et al [64].

7.4 Lifetime estimation by creep and relaxation

Elastomeric composites do not undergo physical ageing at room temperature because their glass transition is generally much below ambient temperature. This is because the molecular segments of an elastomer are at random motion above glass transition, and hence, the molecules are in equilibrium. Therefore, when an elastomer composite is subjected to constant force or constant deformation, the creep and stress relaxation with time will follow so-called "momentary" creep or relaxation as described in physical ageing. Therefore, to study the long-term effect of creep, etc. and to decide the service life based on limits of strain or stress, accelerated ageing at elevated temperatures can be done and creep or relaxation study can be done immediately after attaining the experimental temperature. However, the experimental elevated temperature should not exceed to such extent that the thermal decomposition becomes prominent. As usual a time-temperature superposition can be applied to extrapolate the life at any isothermal ageing temperature. The reason for using higher temperatures is to accelerate the process, since relaxation time decreases with increasing temperature following an Arrhenius, or WLF or other relationship.

In the case of life of elastomer composite as a structural element under stress, the deterioration in the long time needs to be determined by a minimum elastic modulus or compliance that is a limiting quantity for fail-safe design, and in some cases, the compliance or stress at break of the element is also considered. The time-dependent reduction of mechanical property is either due to stress relaxation or due to creep depending on loading condition. There are studies on life prediction by time-temperature superposition, such as, strength of a polymer melt [70], dynamic modulus and compliance of polymer composites [71], CRPF structures by MMF/ATM method [72], flexural fatigue strength of honeycomb composite structure [73], mechanical properties of rubber O-rings in hydraulic oil [74], creep rupture and life prediction of composites [75], creep failure and life prediction of polymers and composites [76] etc.

A brief description of stress relaxation and creep with the method of life prediction is discussed in this section. It is assumed here that no other external stimuli are prominent for ageing of the elastomeric composite and also the mechanism of chemical degradation reaction is same in the band of test temperature.

7.4.1 Stress relaxation and creep

A structural element under stress experiences either stress relaxation or creep during a long service period. Stress relaxation occurs when the element is subjected to a definite, constant strain in the direction of applied force. The relaxation of the molecules under a constant strain is well known and polymers exhibit strong relaxation phenomenon around their glass transition temperature, since they are basically viscoelastic materials. A simple relaxation function of a polymer can be written mathematically as

$$\frac{\sigma(t)}{\gamma_0} = \varphi(t) + G_\infty \tag{7.93}$$

where $\varphi(t)$ is the relaxation function (modulus) which is changing with time due to change in stress $\sigma(t)$, and G_∞ is the modulus after an infinite time of stress application. The relaxation function can have any mathematical form such as a two-parameter model for viscoelastic liquid by Maxwell, a three-parameter model for viscoelastic solid or liquid by Zenner or Anti-Zenner, a four-parameter model by Burger or semiempirical models such as Kolrusch function. A simple model of stress relaxation can be used in eq. (7.93) to define stress relaxation of a viscoelastic material:

$$\sigma(t) = (\sigma_0 - \sigma_\infty)e^{-t/\tau} \tag{7.94}$$

The time τ is *relaxation time*, which is numerically equal to the time at which the stress reduces to $(1/e)$ or 37% of the initial stress. Putting $\sigma_\infty = 0$, we get Maxwell's equation of a two-parameter viscoelastic liquid.

Creep occurs when the element is under a constant load in a particular direction. Creep can be mathematically expressed by a simple two parameter Kelvin-Voigt model or Zenner or Burger model or also by semi–empirical model such as Kolrusch function. A simple creep function can be defined as

$$\frac{\gamma(t)}{\sigma_0} = J_0 + \psi(t) \tag{7.95}$$

where $\psi(t)$ is the creep function (compliance) changing with time as the strain $\gamma(t)$ changes under constant stress σ_0. The initial compliance is J_0. A simple creep strain can be written as

$$\gamma(t) = \gamma_0 + (\gamma_\infty - \gamma_0)\left[1 - e^{-t/\lambda}\right] \qquad (7.96)$$

where λ is the *retardation time*, similar to relaxation time, but applicable for creep phenomenon.

Retardation time is defined as the time at which the strain becomes (1–1/e) times (63%) the difference of final to initial strain. This difference is ideally the recovery strain when the load is withdrawn. However, theoretically it should take infinite time for complete recovery.

7.4.2 Time-temperature superposition models

The relaxation time is a property of a polymer, which is a result of thermal motion of the chain segments. Thus, relaxation can only take place when the segments are able to move. This condition is the glass transition temperature, above which the segmental motion sets in. Therefore, relaxation time is dependent on temperature of the viscoelastic body. The Arrhenius expression is

$$\tau = \tau_0 e^{E_a/RT} \qquad (7.97)$$

where τ_0 is the relaxation time at infinite temperature, E_a is activation energy of relaxation process, R is universal gas constant and T is the absolute temperature. The expression is based on the assumption that the motion of a segment is independent of the neighbouring segments. This assumption may not be very relevant near the glass transition temperature. A slightly different expression is the Vogel–Fulcher-Tammann (VFT) law [77], which is somewhat valid even below glass transition:

$$\tau = A e^{\frac{B}{T-T_0}} \qquad (7.98)$$

where A, B and T_0 are constants which can be experimentally determined. T_0 is so-called Vogel divergence temperature ($T_0 < T_g$ and $T > T_g$), T_g being the glass transition temperature. Their values would change as the temperature is lowered near T_g when more densification of the molecules would take place.

Equations (7.97) and (7.98) suggest that any viscoelastic property, say, elastic modulus at one temperature for a particular time during relaxation, is same at a different time at a reference temperature. Therefore, there can be a shift factor for shifting the time against change in temperature. For eq. (7.97), let the reference temperature be T_g at which the relaxation time is to be calculated when the value is known at a temperature T. The shift factor is defined as

$$\log(a_T) = \log\frac{\tau}{\tau_g} = \frac{E}{2.303R}\left[\frac{1}{T} - \frac{1}{T_g}\right] \qquad (7.99)$$

According to eq. (7.98), the shift factor will be given by

$$\log(a_T) = \log\left(\frac{\tau}{\tau_g}\right) = \frac{B}{2.303}\left[\frac{T_g - T}{(T - T_0)(T_g - T_0)}\right] \tag{7.100}$$

The temperature dependency of relaxation process can also be expressed by WLF (Williams-Landel-Ferry) equation based on free volume change:

$$\log a_T = \log\left(\frac{\tau}{\tau_g}\right) = \frac{-C_1(T - T_g)}{C_2 + (T - T_g)} \tag{7.101}$$

where C_1 and C_2 are constants, and for most polymers the values when reference temperature is T_g are 17.44 and 51.6, respectively, and valid through $T = T_g + 50$. Modified values of C_1 and C_2 are 8.86 and 101.6 respectively at a reference temperature = $T_g + 50$, valid up to $T_g + 100$. If the reference temperature is different, then modified values of the constants as C'_1 and C'_2 can be either calculated from graphical shift or by following equations given by Ferry [78]:

$$C'_1 = \frac{C_1 C_2}{[C_2 - (T_s - T'_s)]}, \quad C'_2 = [C_2 - (T_s - T'_s)] \tag{7.102}$$

where $T'_s = T_g + 50$, at which $C_1 = 8.86$ and $C_2 = 101.6$ and T_s is the desired reference temperature.

The WLF equation is based on the assumption that the free volume of the polymer does not change below glass transition. Therefore, the constants are not valid below T_g.

For any practical purpose, it is better to use graphical method for better accuracy, since in this method, the experimental data are manually shifted to align with the reference temperature line and there is no mathematical assumption for the shift.

7.4.3 Time-temperature-stress superposition

To accommodate the variation of stress in addition to the temperature for creep or relaxation behaviour of an elastomer composite, there should be a time equivalent of stress to make a master curve, which would include temperature and stress effects. The free volume depends on the combined effect of stress and the temperature as given below [79]:

$$f = f_0 + a_T(T - T_0) + a_\sigma(\sigma - \sigma_0) \tag{7.103}$$

where a_T is the thermal expansion coefficient of free volume fraction, a_σ the stress expansion coefficient of free volume fraction, f_0 is the free volume fraction at a reference temperature of T_0 and reference stress σ_0. There should be a combined shift factor $a_{T\sigma}$ for both temperature and stress, defined as

$$\eta(T, \sigma) = \eta_0(T_0, \sigma_0) a_{T\sigma} \tag{7.104}$$

according to the DoLittle relationship

$$\eta = a \exp(b/f) \tag{7.105}$$

where a and b are constants.

Consequently, the shift factor is given by [79]

$$\log a_{T\sigma} = \frac{-b}{2.303 f_0} \left[\frac{a_T(T - T_0) + a_\sigma(\sigma - \sigma_0)}{f_0 + a_T(T - T_0) + a_\sigma(\sigma - \sigma_0)} \right] \tag{7.106}$$

$$= - C_1 \left[\frac{C_3(T - T_0) + C_2(\sigma - \sigma_0)}{C_2 C_3 + C_3(T - T_0) + C_2(\sigma - \sigma_0)} \right] \tag{7.107}$$

where $C_1 = \dfrac{b}{2.303 f_0}$, $C_2 = \dfrac{f_0}{a_T}$, $C_3 = \dfrac{f_0}{a_\sigma}$

It follows from eq. (7.107) that if there is no change in the stress, $\sigma = \sigma_0$, the equation reduces to WLF equation, eq. (7.101). Also, if there is no change in the temperature, then

$$\log a_{T\sigma} = -C_1 \left[\frac{(\sigma - \sigma_0)}{C_3 + (\sigma - \sigma_0)} \right] \tag{7.108}$$

Pap et al. [80], Emara et al. [81] and Li et al [82] studied creep of epoxy-based adhesive with time–temperature-stress superposition and generated combined master curves, while Jiang et al. [83] applied the combined shift for polycarbonate.

7.4.4 Time-temperature superposition in stress relaxation

Let us consider a stress relaxation phenomenon of a thermoset epoxy-5% Cloisite B clay nanocomposite, examined at temperatures 115, 120, 125, 130, 135 and 140 °C, and at a set of time from (i) 0.1–0.5 min by step of 0.1 min, (ii) 0.5–5 min by step of 1.5 min, (iii) followed by 10 and 20 min at each selected temperature. The relaxation experiment is done above glass transition, and hence, physical ageing is not applicable here. The stress at 1% strain is measured for the study with time. The Young's modulus at ambient conditions is 2.0 GPa. The modulus at this strain is to be predicted at 115 °C after 360 min.

The stress isotherms of above temperatures are plotted against time in Fig. 7.13. The reference temperature is 115 °C. The points of other isotherms are horizontally shifted to align with the line of 115 °C. A few arrows are shown in the plot as shifting of points at various isotherms.

7.4.5 Shift factor

The shift factor is the calculated from the graph as

$$\log(a_T) = \log\left(\frac{X - \text{axis value of time at the end of the arrow}}{X - \text{axis value of time at the begining of the arrow}}\right) \quad (7.109)$$

It may be noted that the shift factor a_T is always a positive value, greater than zero. It can be a fraction, when the shifting is done from a lower temperature to the reference temperature and it is greater than 1, when the temperature for shifting is higher than the reference temperature. The shift factors thus calculated from Fig. 7.13 and also from WLF equation (eq. (7.101)), with coefficients calculated according to the graphical shift factor values are listed in Tab. 7.6.

Tab. 7.6: Shift factors calculated from Fig. 7.13 for test temperatures.

Temperature, °C	$\log(a_T)$: graphical	$\log(a_T)$: WLF $C_1 = 11.1, C_2 = 90.6$
115	0	0
120	0.6989	0.5753
125	1.0969	1.0934
130	1.5229	1.5625
135	1.9031	1.9891
140	2.2844	2.3788

7.4.6 Master curve

The shifted master curve of the stress relaxation is shown in Fig. 7.14. It may be noted that the curve represents the stress values at progressive time at 115 °C when the composite is held at 1% strain, which is a long-time relaxation or the ageing effect of the composite at 115 °C, when it has to perform under a strain of 1% in service. The elastic modulus at ambient is about 2 GPa, which reduces due to relaxation of the stress with time.

Hence it is seen from Fig. 7.8 that at a desired time of 360 min, the stress is about 5100 Pa, and corresponding elastic modulus is about 2.04 MPa. Therefore, the modulus has reduced by three decades of value at this temperature after 360 min.

According to the Arrhenius equation on shift factor obtained from graphical superposition method, a plot of $\log(a_T)$ against $1/T$ should result in a straight line. To examine the validity, Fig. 7.15 was drawn. The linear fit is excellent with $r^2 = 0.99$. The activation energy is calculated from the slopes as

$$E = 2.303 \times R \times \text{Slope} = 271.5 \,\text{kJ/mol}$$

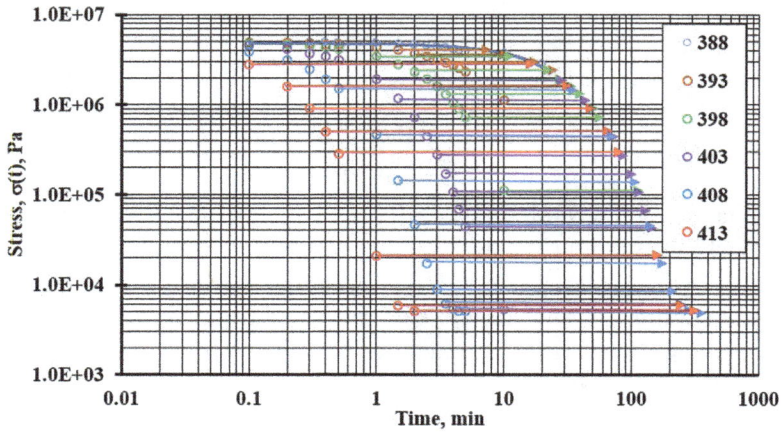

Fig. 7.13: Stress isotherms: graphical shifting of data for time-temperature superposition.

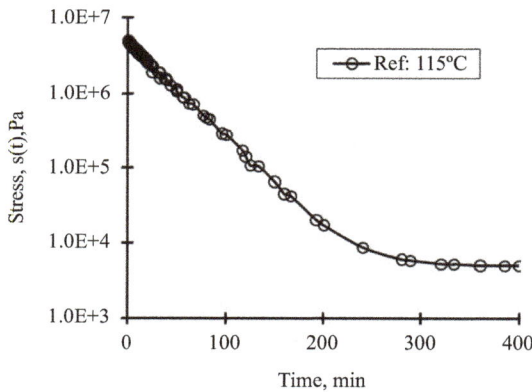

Fig. 7.14: Master curve of stress relaxation at a reference temperature: 115 °C.

Following the Arrhenius equation (7.99) and using the activation energy thus calculated, the shift factor for any temperature can be calculated more or less accurately provided the thermoset does not degrade by some other mechanism. For example, if the time required for the modulus to reduce to 2.04 MPa is to be found out at any ageing temperature, then shift factor at that temperature can be calculated directly from the straight-line equation of Fig. 7.15:

$$\log(a_T) = 14180 \left(\frac{1}{T}\right) - 36.674 \tag{7.110}$$

7.4.7 Lifetime prediction

Subsequently, the shift factors thus obtained are used to calculate the time required to reduce the modulus to the target limit by the following equation:

$$\log(t_{new}) = \log(t_{ref}) + \log(a_T) \tag{7.111}$$

where t_{new} is the shifted time (or reduced time) corresponding to T_i which is any test temperature and t_{ref} is the time at reference temperature.

As an example, from Tab. 7.6, the $\log(a_T)$ for 120 °C is −0.6989. Hence, the time at 115 °C (reference) corresponding to 1 min at 120 °C is calculated as

$$\log(t_{new}) = \log(1) + 0.6989 = 0 + 0.6989 = 0.6989$$

$$\text{Therefore } t_{new} = 10^{0.6989} = 5 \text{ min.}$$

This implies that the stress value after 1 min of loading at 120 °C is same after 5 min of loading at 115 °C. For time shift in case of stress relaxation, it takes lesser time to attain a particular value of stress at a higher temperature, and the reverse is also true.

Using eqs. (7.110) and (7.111), the time required at various temperatures (30–100 °C) is calculated and plotted in Fig. 7.16. The log(time) is perfectly linear with temperature as seen in the figure. The plot can be used as a calibration curve for an approximate prediction of lifetime at any service temperature at the stress condition defined in the problem. As an example, the lifetime is about 2 years at 80 °C and about 32 years at 70 °C, but the life in the said service stress condition is about 2 months at 90 °C.

7.4.8 Critical considerations

Here also, like TGA method of prediction, the accuracy depends on how close to the ageing temperature is selected for the accelerated study. The time required to arrive at the numerical value of the desired property (stress, strain, modulus or compliance) to the set limit, is longer when the accelerated study is closer to the ageing temperature. Practically, it may not be possible to carry out experiment for such a long period. Therefore, the method of time-temperature superposition is employed for an approximate life prediction. The approximation arises due to many reasons such as (1) possible error in experimental determination of the temperature and property, (2) any difference in the sample of different batches and (3) correctness of graphical shift. Further, any other environmental or external factors influencing the property during long-term use is also neglected here.

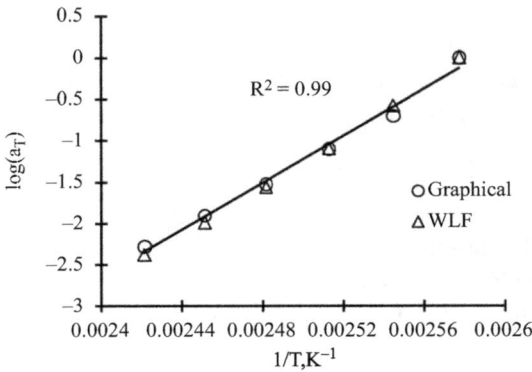

Fig. 7.15: Shift factor vs temperature for the *t-T* superposition.

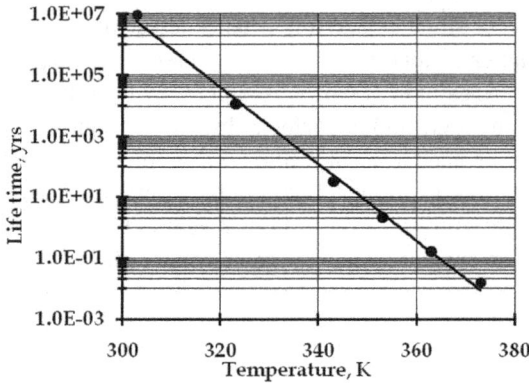

Fig. 7.16: Service life estimation based on stress relaxation.

7.4.9 Creep strain

In a similar manner, creep strain can be studied at various temperature keeping the constant load applied to a sample in any appropriate mode, e.g., tensile, flexure, compression or shear.

The creep is very common for supported beams, cantilevers, ropes, gaskets, machinery base supports, etc. Thermosets and their composites used in such structural elements would experience creep and excessive creep may result in failure of an element or may reduce performance below an acceptable limit. For example, a creep compliance may increase to such an extent that the strength, measured by Young's modulus, may become lower than design value.

7.4.10 Creep study: master curve

A natural rubber–based hybrid composite using 20 phr carbon black (N330) and 125 phr vermiculite designated NR-CB-VR vulcanized with sulphur-MBTS was made for load bearing application. The composite had a glass transition temperature of −40 °C and Shore "A" hardness was 75, tensile strength 8.8 MPa and % elongation at break was 275%. The compression creep of the NR-CB–VR at 3.2 MPa was measured up to 4,300 min (almost 3 days) in steps. Fig. 7.17 shows the isothermal creep strain values at 25, 30, 35 and 40 °C in time scale. There was some scatter in the data and therefore, the creep data were fitted by a similar relationship as in eq. (7.2) with good accuracy ($R^2 = 0.99$). It is decided to predict long time creep at a reference temperature of 30 °C by graphical method similar to stress relaxation example. The shift factors calculated are plotted against $1/T$ as Arrhenius plot, to obtain a perfect straight line with $R^2 = 0.988$. The Arrhenius plot is shown as Fig. 7.18. The equation of the Arrhenius relationship is given by

$$\log(a_T) = \log\left(\frac{t}{t_{\text{ref}}}\right) = -\frac{E_\lambda}{2.303R}\left[\frac{1}{T} - \frac{1}{T_{\text{ref}}}\right] \tag{7.112}$$

From the plot, the equation becomes

$$\log(a_T) = 35.809 - 10,815\left(\frac{1}{T}\right) \tag{7.113}$$

The activation energy calculated as $E_\lambda = 207.07$ kJ/mol.

The agreement of the Arrhenius relationship of shift factor and temperature, it is confirmed that the horizontal shifting by graphical method was nearly perfect. The equation (7.113) can be used to calculate shit factor at any temperature T_n to find out the strain at a reference temperature (T_{ref}) from the value at T_n.

Figure 7.19 shows the master curve of creep strain after the graphical shifting taking reference temperature as 30 °C. The creep strain is thus predicted up to about 77,000 min, or, approximately 54 days. From the figure, it is seen that the strain increased from 0.04 to 0.08 after 2,000 min, but after 50,000 min, the increase was only by 0.02, which suggests that the creep becomes almost negligible after 50,000 min (35 days). The compression modulus reduces from initial 82 MPa to 31 MPa finally after 54 days of creep.

The predicted long-term creep strain as shown in Fig. 7.19 follows a perfect log (time) vs. strain relationship:

$$\varepsilon_c(t) = 0.0064\ln(t) + 0.0322 \tag{7.114}$$

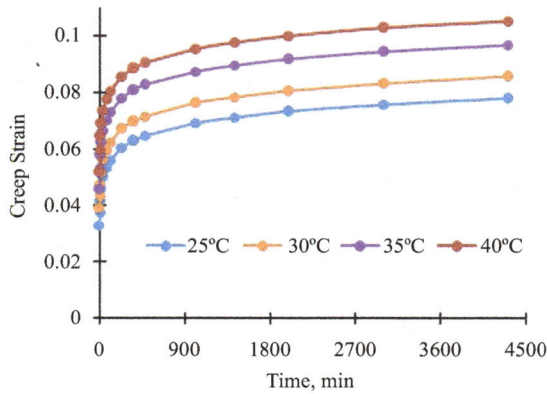

Fig. 7.17: Isothermal compression creep strain of NR-CB-VR composite at various temperatures.

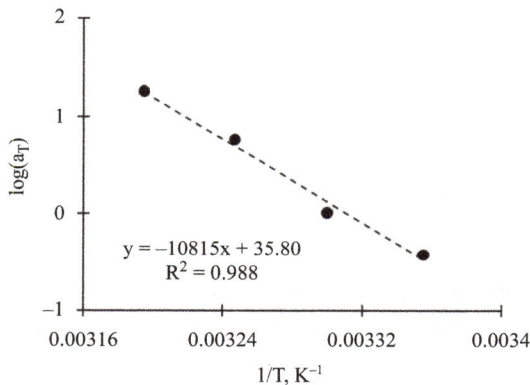

Fig. 7.18: Arrhenius plot of shift factor: creep experiment.

7.4.11 Life estimation from creep

In this example, the NR composite was supposed to be used as a soft capping on hard wooden blocks for drydocking of ships. The composite is designed as a load bearing material under a constant dead weight for a long time, about 2 months, and hence it undergoes creep. It is to be seen that how long the composite can be used under the constant stress of 3.2 MPa so that the deflection is not more than 12% (which is not very high for this example) at most possible coastal temperatures of 30, 35 and 40 °C.

To find the various durations at 30, 35 and 40 °C, the corresponding shift factors are required to be calculated using eq. (7.113):

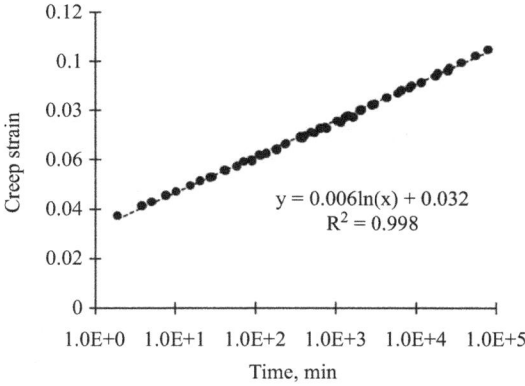

Fig. 7.19: Creep master curve at 30 °C reference temperature.

$\log(a_{30}) = 0$, being the present reference temperature,
$\log(a_{35}) = 35.809 - 10815/(273 + 35) = 0.695364$,
$\log(a_{40}) = 35.809 - 10815/(273 + 40) = 1.256284$.

Now, we need to calculate the time required at 30 °C, using eq. (7.114):
$\varepsilon\,(30) = 0.12 = 0.0064 * \ln(t) + 0.0322 = 907773$ min $= 630$ days.
Hence, the time to attain this strain at 35 °C $= (630/a_{35})$ days $= 121$ days,
and the time to attain this strain at 40 °C $= (630/a_{40})$ days $= 35$ days.

The time limits thus calculated are accurate to the extent the constitutive equation of long-term creep is accurate. This largely depends on strain measurement in the creep machine with accurate time and on the best fit creep equations for all data, if used for master curve generation. Since the time of usage or the lifetime is logarithmic with strain, the deviation from observed strain in the field of use may be quite significant if there is error in basic data and low accuracy in subsequent predictive equations.

7.4.12 Ageing study under stress

The time-temperature-stress transformation method is described in para 7.4.3 with a set of eqs. (7.103)–(7.108), where the dependence on stress is accommodated using modified coefficients for calculation of combined shift factors. It would be more convenient to use a concept of effect of stress on the activation energy of durability following an Arrhenius-type expression.

The effect of stress on the life depends on mode of load application. For example, when the temperature causes a chain scission in a polymer and produce radicals, a tensile stretch can pull the radicals apart and hinder the recombination, thus increasing generation of more new radicals and increase the degradation rate. Compression would bring radicals closer to facilitate the recombination, and discourage new chain reactions, thus delaying the degradation. Stress may have effect on diffusion of reactants such as acid, gas and sea water and subsequent degradation which depends on the diffusion process. The dependence of temperature and stress on failure life of polymers, metals and other solids is expressed by a modified Arrhenius equation known as Boltzmann-Arrhenius–Zhurkov (BAZ) expression [84–86] which includes stress:

$$\tau(\sigma(t)) = \tau_0 \exp\left(\frac{U_0 - \gamma\sigma}{kT}\right) \tag{7.115}$$

In this equation, τ is the time to failure, τ_0 is the time constant, σ is the stress per unit volume and γ is the coefficient of stress, sometimes referred to as *activation volume of bond rupture*, and has the unit of volume. The terms k and T are Boltzmann constant and absolute temperature respectively and (U_0-γ σ) is the activation energy of rupture of interatomic bond due to thermal fluctuation. Arrhenius pre-exponent τ_0 is the period of sub-molecular oscillation at $T \rightarrow \infty$, typically about 10^{-13} s. In absence of any mechanical stress (σ), the equation simplifies to classical Arrhenius expression for time to failure. The failure is due to the thermo-fluctuations, which degrades a material even in absence of stress. The stress only changes the potential barrier E by a quantity $\gamma^*\sigma$ in the stress direction and therefore changes the thermo-fluctuations.

This equation is widely used for prediction of lifetime of polymers under isothermal condition (T) and under a constant stress σ [87,88].

However, for elastomers, time to fracture is given by another expression claimed [89] to be more appropriate:

$$\tau = \tau_0 \left(\frac{\sigma_*}{\sigma}\right)^m \exp\left(\frac{U_0}{kT}\right) \tag{7.116}$$

For elastomers,

$$m = \frac{U_0}{3kT_g} \tag{7.117}$$

where T_g is the glass transition temperature of the polymer.

7.4.13 Bailey Criteria

When the durability (or lifetime) is to be determined for a polymer under time-dependent simple loading conditions and under isothermal ageing, it is possible to use the Bailey criterion of durability [89,90]:

$$\int_0^{t^*} \frac{dt}{\tau(\sigma(t))} = 1 \tag{7.118}$$

Here $\tau\,(\sigma(t))$ is the durability function determined experimentally under constant stress σ and t^* is the durability under a stress $\sigma(t)$. However, the above equation is valid for temperatures below glass transition, and it is assumed that there is negligible Boltzmann superposition effect of stresses.

The above criteria were applied to polyethylene, polyamides, polyoxymethylene, polyethylene terephthalate, polyvinylidene fluoride, copper and aluminium by Vettegren et al. [89].

As an example, a case of any thermoset based composite can be considered, assuming that relaxation time of the thermoset polymer is almost infinity at ambient temperature and the glass transition temperature is very high compared to experimental temperatures. Equation (7.115) is rewritten as:

$$\tau = \tau_0 \exp\left[\frac{U_0}{kT}\right] \exp\left[-\frac{\gamma\sigma}{kT}\right] \tag{7.119}$$

or,

$$\tau = \tau_0 \exp\left[\frac{U_0}{kT}\right] \exp[-a\sigma], \quad \text{where,} \quad a = \frac{\gamma}{kT} \tag{7.120}$$

or,

$$\tau = B \exp[-a\sigma] \tag{7.121}$$

where

$$B = \tau_0 \exp\left[\frac{U_0}{kT}\right] \tag{7.122}$$

The following expression can be assumed with slight approximation:

$$e^{-a\sigma} = b\sigma^{-\kappa} \tag{7.123}$$

or

$$a\sigma = \ln(b) + \kappa \ln(\sigma) \tag{7.124}$$

Introducing eq. (97.124) into eq. (7.121):

$$\tau(\sigma(t)) = (Bb)\sigma^{-\kappa} \tag{7.125}$$

Or,

$$\tau(\sigma(t)) = C\sigma^{-\kappa} \tag{7.126}$$

where $C = Bb$, a constant at a constant temperature. In addition, the quantity γ in eq. (7.119) is also constant since there is no significant volume change for the segments of the polymer at a constant temperature.

Introducing eq. (7.126) in Baily criteria described by eq. (7.118), we get,

$$\int_0^{t^*} \frac{dt}{C\sigma^{-\kappa}} = 1 \tag{7.127}$$

Let t_r be the time to failure at a stress σ_r. By denoting $\sigma_r = w_r t_r$, where w_r is the stretching rate (MPa/min) for σ_r and substituting it in eq. (7.127), we obtain, after integration:

Therefore,

$$\int_0^{\tau} \frac{dt}{C\sigma^{-\kappa}} = \int_0^{t_r} \frac{dt}{C(wt)^{-\kappa}} = \frac{(w_r t_r)^{\kappa+1}}{Cw_r(\kappa+1)} = 1 \tag{7.128}$$

Rearranging and taking logarithm, we get the workable expression as

$$\log \sigma_r = \frac{1}{\kappa+1}(\log C + \log(\kappa+1)) + \frac{1}{\kappa+1}\log w_r \tag{7.129}$$

Equation (7.129) represents a straight line when $\log(\sigma_r)$ is plotted against $\log(w_r)$. These values can be obtained from a series of experiments of tensile/compression of the composite.

7.4.14 Lifetime estimation

In order to determine the lifetime, the polymer composite sample should be tested for tensile/compression property at various (at least four) stretching rates (w_r) and the failure stresses (σ_r) should be recorded. In case of thermoplastic polymer composites, sometimes yield stress is taken. Time to break t_r (failure) is also recorded for each experiment. The same sets of experiments are to be done with isothermal ageing at various temperatures to use in eq. (7.120).

Experimental data obtained from the above procedure are plotted according to eq. (7.129), and from the slope and intercept, C and κ are calculated. However, to find out lifetime at some other stress, eq. (7.129) as such is used with the slope and the intercept. From the relation $\sigma_r = w_r . t_r$, the time to fail (t_r) is calculated.

7.4.15 Example

Consider a composite consisting of epoxy resin, modified with 10% by weight of a chemically bonded liquid rubber and cured by a diamine to make a strong and tough cured thermoset filled with 10% chopped glass fibre. The uncured resin-fibre mixture is a dough moulding compound (DMC). It is to be found out if the material as such is suitable for making a 6″ industrial FRP ball valve. The valve has to continuously work under 10 bar hydrostatic pressure and the corresponding Hoop stress on the ball is 8 MPa. The service life of the ball valve has to be determined using the data from tensile measurement. Further, the total strain after 10 years should be found out.

The dimensions of the item are fixed as per design. Hence, the durability has to be determined only on the basis of continuous constant stress.

7.4.16 Solution

The tensile test of the cured samples was done with a universal testing machine (UTM) using 45 mm gauge length of the samples, using four cross-head speeds, e.g. 1, 2, 5 and 10 mm/min. It was ensured that there was no slip at the grips during any experiment, which means that the cross-head travel is the elongation of the sample due to tension. In a DMA experiment with a fixed frequency of 1 Hz, the temperature at which the storage modulus dropped from a constant value, was recorded as a maximum useable temperature.

Table 7.7 shows the tensile test data. The yield stress, corresponding strain and the time to reach yield point is taken here since the toughened epoxy shows a slight plastic deformation after stress maxima. The rate of stress (w_r) as MPa/min is calculated from the time to reach yield point.

Figure 7.20 shows the plot of $\log(\sigma_r)$ against $\log(w_r)$ according to eq. (7.129). The data has some scatters, but with fairly good linear fit ($r^2 = 0.926$) and the equation is

$$\log(\sigma_r) = 0.0929 \log(w_r) + 1.4984 \qquad (7.130)$$

The above equation can be used to determine w_r for various stress (σ_r) levels.

According to the above relationship, the stretching rate w_r at 8MPa stress is calculated as $w_r = 3.91\text{E-}07$ MPa/min, and correspondingly, the time to reach yield stress of 8MPa is:

$$t_r = \sigma_r/w_r = 77.9 \text{ years.}$$

At 8 MPa, the initial strain of the material in tension is found out from stress–strain plot as $\varepsilon = 0.45\%$.

Neglecting the physical ageing effect, additional strain after 10 years of use (due to creep) will be

$$\delta\varepsilon = (4.8 - 0.45) \times 10/78 = 0.5584\%, \text{ assuming a linear variation with time.}$$

Hence, total strain will be $= \varepsilon + \delta\varepsilon = 1.0084\%$.

However, since the composite is used at ambient temperature of about 20–30 °C, far below the DMA T_g (70 °C), the strain after 10 years will be somewhat less than the

Tab. 7.7: Tensile test data for toughened epoxy-glass composite.

Gauge length	Yield stress	Cross–head speed	Yield strain and time		Stress rate	Plot parameters	
L_0, mm	σ_r, MPa	mm/min	%dL	t_r, min	w_r, MPa/min	log(w_r)	log(σ_r)
45	53.35	10	4.85	0.218	244.55	2.388	1.727
45	54.03	10	4.62	0.208	259.81	2.415	1.733
45	55.28	10	4.85	0.218	253.41	2.404	1.743
45	52.20	10	4.81	0.217	241.00	2.382	1.718
45	51.01	10	4.87	0.219	232.76	2.367	1.708
45	48.20	5	4.78	0.430	112.10	2.050	1.683
45	47.55	5	4.50	0.405	117.43	2.070	1.677
45	48.44	5	4.58	0.412	117.51	2.070	1.685
45	46.18	5	4.96	0.446	103.51	2.015	1.664
45	48.75	5	5.19	0.467	104.46	2.019	1.688
45	45.38	2	5.23	1.178	38.54	1.586	1.657
45	43.35	2	5.63	1.266	34.23	1.534	1.637
45	43.16	2	4.43	0.996	43.32	1.637	1.635
45	45.33	2	4.18	0.940	48.24	1.683	1.656
45	46.66	2	4.53	1.020	45.74	1.660	1.669
45	45.38	1	5.23	1.178	38.54	1.586	1.657
45	42.16	1	5.03	2.263	18.63	1.270	1.625
45	40.58	1	4.46	2.005	20.24	1.306	1.608
45	40.28	1	4.94	2.225	18.10	1.258	1.605
45	42.12	1	4.54	2.044	20.61	1.314	1.624
45	42.82	1	5.03	2.263	18.92	1.277	1.632

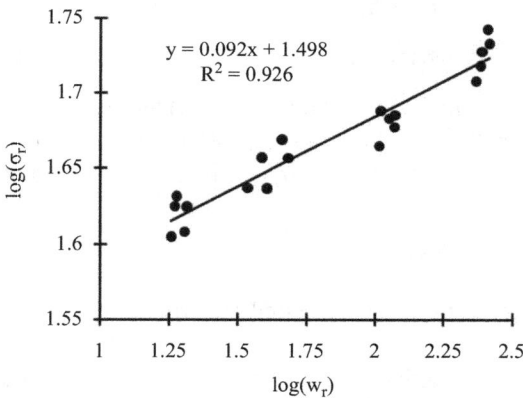

Fig. 7.20: Plot of UTS vs stretching rate according to eq. (7.129) and data of Tab. 7.7.

above estimation due to physical ageing, but not significantly since the relaxation time as such is very high and will continuously be delayed. Further, since the DMA T_g of this material is 70 °C, it can be said that the material can be used as load bearing item at ambient condition. Therefore, the material as such with all dimensions as per design can be used safely for a long service life based on yield stress only.

Using eq. (7.130) and the above procedure, durability of the toughened epoxy-chopped glass fibre composite is determined for various Hoop stress levels and plotted in Fig. 7.21. The plot can be used as a calibration curve and also as a first-hand information to decide on suitability for any application of this composite under tensile stress.

Similarly, under the constant Hoop stress, the strain will increase with time. Hence the approximate strain (% elongation) at initial stress and at various lapse of time is shown in Fig. 7.22. It is seen that the strain is 1% at 10 years and hence it is desirable to define the service life of maximum 10 years so that the dimensional tolerances are mostly satisfied.

7.4.17 Conclusion

All the methods of life estimation discussed above have one aspect in common. The calculated life is never accurate due to inaccuracy in interpolation to applicable environmental conditions and effect of various other factors which are not considered in the studies. The prediction is more accurate, when the temperature of laboratory experiment is nearer to the actual application temperature. This condition cannot be easily satisfied, because in that case the study would take very long time, especially for rigid thermosets. However, the superimposition of more than one stimulant for ageing is possible, but the life estimation would then be completely empirical, and applicable to that particular case only. This type of study with combined effects are done in many critical items such as rocket propellants, load bearing structures made of FRP composites, high temperature insulation by thermoset coatings, electrical insulators, etc. where serious accidents can take place due to degradation.

Apart from the factors considered in this chapter, there are many functional factors due to which the composite may become unsuitable for use after a certain period. Simple examples are structural damping composites, adhesive joints under fluctuating/dynamic force, low cycle fatigue of an FRP beam and dielectric failure of a composite insulator.

Additives play a very important role in ageing process. In composites, interaction of the reinforcing material with its own molecules (Filler-Filler) and with the host polymer (Filler-Matrix) is of prime importance. The second most effective parameter is aspect ratio of the filler. The third important parameter is the size of the filler, since the relative surface area decides the extent of secondary valence bond of the filler with the polymer, such as hydrogen bonding, polar interaction and van der Waals and also

the extent of agglomeration (clustering). In addition, the additives which are suscepti-
ble to strong interaction with the prepolymer or curing agent or any other ingredient
may cause the deviation in rate of change of properties when compared to the ageing
process of the polymer alone. Nanoparticles provide more complicated scenario due to
extremely small size and corresponding high surface energy. The complications are
the difficulty in modelling the ageing behaviour taking into account the micromechanics of stress-strain and other physical characteristics such as diffusional constraints, threshold limit for thermal and electrical conduction, frictional contribution
to dynamic property and changes in α-relaxation rate due to change in thermo-
fluctuation below the glass transition. However, the normal bulk behaviour in thermal
analysis of degradation kinetics, physical ageing and dynamic viscoelastic properties
can be fairly mapped with time to decide service and storage life.

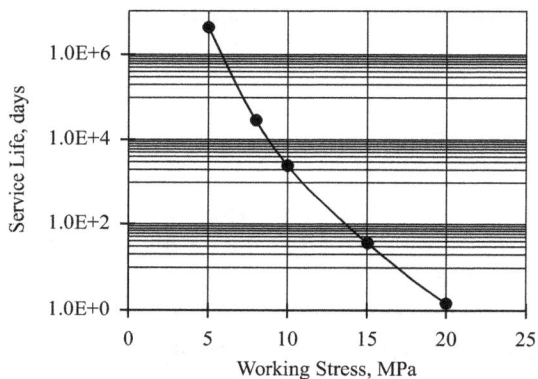

Fig. 7.21: Plot of service life at various stresses for the toughened epoxy-glass composite for use in the
ball valve in example 7.4.15.

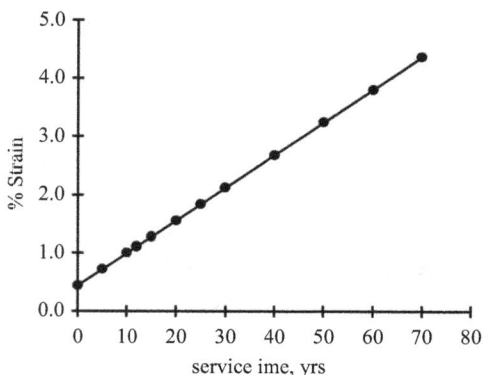

Fig. 7.22: Increase in % strain with service time under a reference constant stress of 8 MPa.

References

[1]   Eftekhari, M., Fatemi, A. Creep behavior and modeling of neat, talc-filled, and short glass fiber reinforced thermoplastics. Compos. B: Eng. 2016, 97, 68–83.

[2]   Struik, L. C. E. Physical Aging in Amorphous Polymersand other Materials. Amsterdam, Elsevier, 1978.

[3]   Struik, L. C. E. Physical aging in plastics and other glassy materials. Polym. Eng. Sci. 1977, 17(3),165–173.

[4]   Zheng, S. F., Weng, G. J. A new constitutive equation for the long-term creep of polymersbased on physical aging. Eur. J. Mech. A: Solids. 2002, 21, 411–421.

[5]   Lai, C. H. Physical ageing and dimensional changes in acrylic polymers. Ph.D. Thesis. Dept. of Chem Engg. And Physical and inorganic chemistry, The university of Adelaide, USA, 1992.

[6]   Minguez, R., Barrenetxea, L., Solaberrieta, E., Lizundia, E. A simple approach to understand the physical aging in polymers. Eur. J. Phys. 2018. https://doi.org/10.1088/1361-6404/aaf244

[7]   Patil, P. N., Rath, S. K., Sharma, S. K., Sudarshan, K., Maheshwari, P., Patri, M., Praveen, S., Khandelwal, P., Pujari, P. K. Free volumes and structural relaxations in diglycidyl ether of bisphenol-A based epoxy–polyether amine networks. Soft. Matter., 2013, 9, 3589–3599.

[8]   Schwarzl, F. R. The numerical calculation of storage and loss compliance from creep data for linear viscoelastic materials. Rheol Acta. 1969, 8(I), 6–17.

[9]   Eftekhari, M., Fatemi, A. Creep behavior and modeling of high-density polyethylene (HDPE). Polym. Test. 2021, 94, Article No. 107031. https://doi.org/10.1016/j.polymertesting.2020.107031

[10] Chang, T. D., Brittain, J. O. Studies of epoxy resin systems: Part C: Effect of sub-Tg aging on the physical properties of a fully cured epoxy resin. Polym. Eng. Sci. 1982, 22, 1221–1227.

[11] Kong, E. S. W. Physical ageing in epoxy matrices and composites. Adv. Polym. Sci. 1986, 80, 125–171.

[12] Le´ve^que, D., Schieffer, A., Mavel, A., Maire, J.-F. Analysis of how thermal aging affects the long-term mechanical behavior and strength of polymer–matrix composites. Compos. Sci. Technol. 2005, 65, 395–401.

[13] Frigione, M., Naddeo, C., Acierno, D. Cold-curing epoxy resins: aging and Environmental effects. I – thermal properties. J. Polym. Eng. 2001, 21(1), 23–51.

[14] Frigione, M., Naddeo, C., Acierno, D. Cold-curing epoxy resins: aging and Environmental effects. II – mechanical properties. J. Polym. Eng. 2001, 21(4), 349–367.

[15] Lizundia, E., Sarasua, J. R. Physical Aging in Poly(L-lactide) and its Multi-Wall Carbon Nanotube Nanocomposites. Macromol. Symp. 2012, 118(23),321–322.

[16] McKenna, G. B. On the Physics Required for the Prediction of Long Term Performance of Polymers and Their Composites. J. Res. NIST, 1994, 99, 169–189.

[17] McKenna, G. B. Physical aging in glasses and composites. In: Pochiraju, K. V., Tandon, G. P., Schoeppner, G. A. eds. Long-term durability of polymeric matrix composites. Springer, Boston, MA, 2012, pp. 237–309.

[18] Lu, H., Nutt, S. Enthalpy relaxation of layered silicate-epoxy nanocomposites. Macromol. Chem. Phys., 2003, 204, 1832–1841.

[19] Hu, H., Sun, C. T. The characterization of physical aging in polymeric composites. Compos. Sci. Technol. 2000, 60, 2693–2698.

[20] Odegard, G. M., Bandyopadhyay, A. Physical Aging of Epoxy Polymers and Their Composites. J. Polym. Sci. Part B: Polym. Phys. 2011, 49, 1695–1716.

[21] Bradshaw, R. D., Brinson, L. C. "Physical Aging in Polymers and Polymer Composites: AnAnalysis and Method for Time-Aging Time Superposition," Polym. Eng. Sci., 37, 31–44 (1997).

[22] Lee, A., Lichtenhan, J. D. Viscoelastic Responses of Polyhedral Oligosilsesquioxane Reinforced Epoxy Systems. Macromolecules, 1998, 31, 4970–4974.

[23] Lu, H. B., Nutt, S. Enthalpy Relaxation in Layered Silicate-Epoxy nanocomposites, Macromol. Chem. Phys. 2003, 204, 1832–1841.

[24] Rittigstein, P., Torkelson, J. M. Polymer–nanoparticle interfacial interactions in polymer nanocomposites: confinement effects on glass transition temperature and suppression of physical aging. J. Polym. Sci., Part B: Polym. Phys. 2006, 44, 2935–2943.

[25] Cangialosi, D., Boucher, V. M., Alegría, A., Colmenero, J. Enhanced physical aging of polymer nano composites: The key role of the area to volume ratio. Polymer. 2012, 53 (6), 1362–1372.

[26] Boucher, V. M., Cangialosi, D., Alegría, A., Colmenero, J., Rev, E.: Stat., Nonlinear. Soft. Matter. Phys., 2012, 86, 041501.

[27] Gates, T. S., Veazie, D. R., Brinson, L. C. Creep and physical aging in a polymericcomposite: comparison of tension and compression. J. Comp. Matls., 1997, 31, 2478–2505.

[28] Hu, H. W. Physical aging in long term creep of polymeric composite laminates. J. Mechanics, 2007, 23, 245–252.

[29] Kissinger, H. E. Reaction kinetics in differential thermal analysis. Anal. Chem. 1957, 29, 11, 1702–1706.

[30] Doyle, C. D. Kinetic analysis of thermogravimetric data, J. Appl. Polym. Sci. 1961, V (15), 285–292.

[31] Coats, A. W., Redfern, J. P. Kinetics parameters from thermogravimetric data. Nature, 1964, 201, 68–69.

[32] Nikolaidis, A. K., Achilias, D. S. Thermal degradation kinetics and viscoelastic behavior of poly(methyl methacrylate)/organomodified montmorillonite nanocomposites prepared via in situ bulk radical polymerization. Polymers 2018, 10, 491; doi:10.3390/polym10050491

[33] Vyazovkin, S., Dranca, I., Fan, X., Advincula, R. Kinetics of the thermal and thermo-oxidative degradation of a polystyrene–clay nanocomposite. Macromol. Rapid Commun. 2004, 25, 498–503.

[34] Menezes, J., Loureiro, D. S., Calado, F. A. M., V.m., D. A., Maria, A. Thermogravimetric study and kinetic modeling of semi-interpenetrating polymer network protonic conductive membranes to PEMFC. J. Therm. Anal. Calorim. 2022, 147, 9469–9486.

[35] Qui, T., Ge, F., Li, C., Łu, S. Study of the thermal degradation of flame retardant polyester GRPF using TGA and TG-FTIR-GC/MS. J. Therm. Anal. Calom. 2021. https://doi.org/10.1007/s10973-021-10895-z.

[36] Raveshtian, A., Fasihi, M., Norouzbeigi, R., Rasouli, S. Curing and thermal degradation reactions of Nano-Alumina filled natural rubber latex foams. Thermochim. Acta. 2022, 707, 179108.

[37] Zhang, X., Huang, R. Thermal decomposition kinetics of basalt fiber-reinforced wood polymer composites. Polymers, 2020, 12(10), Article No. 2283. doi: 10.3390/polym12102283.

[38] Yousef, S., Eimontas, J., Striūgas, N., Subadra, S. P., Abdelnaby, M. A. Thermal degradation and pyrolysis kinetic behaviour of glass fibre-reinforced thermoplastic resin by TG-FTIR, Py-GC/MS, linear and nonlinear isoconversional models. J. Mater. Res. Technol. 2021, 15, 5360–5374.

[39] Tranchard, P., Duquesne, S., Samyn, F., Estèbe, B., Bourbigo, S. Kinetic analysis of the thermal decomposition of a carbon fibre-reinforced epoxy resin laminate. J. Anal. Appl. Pyrolysis. 2017, 126, 14–21.

[40] Narayanankutty, S. K. Thermal degradation of short kevlarfibre thermoplastic polyurethane composite. Polym. Degrad. Stab. 1992, 38, 187–192.

[41] Friedman, H. L. New methods for evaluating kinetic parameters from thermal analysis data. J. Polym. Sci. Part B Polym. Lett. 1969, 7, 41–46.

[42] Freeman, E. S., Carroll, B. The application of thermoanalytical techniques to reaction kinetics- the thermogravimetric evaluation of the kinetics of the decomposition of calcium oxalate monohydrate. J. Phys. Chem. 1958, 62, 394–397.

[43] Kim, S., Park, J. K. Characterization of thermal reaction by peak temperature and height of DTG curves. Thermochim. Acta 1995, 264, 137–156.

[44] Madhu, G., Mandal, D. K., Bhunia, H., Bajpai, P. K. Thermal degradation kinetics and lifetime of HDPE/PLLA/pro-oxidant blends. J. Polym. Eng. 2016. doi: 10.1515/polyeng-2015-2019.

[45] Doyle, C. D. Estimating Isothermal Life from Thermogravimetric Data. J. Appl. Polym. Sci. 1962, 6(24),639–642.

[46]   Murray, P., White, J. Kinetics of the thermal dehydration of clays. Part IV. Interpretation of the differential thermal analysis of the clay minerals. Trans. Brit. Ceram. Soc. 54 (1955d), 204–238.

[47]   Senum, G. I., Yang, R. T. Rational approximations of integral of Aarrhenius function. J. Therm. Anal. 1977, 11, 445–449.

[48]   Perez-Maqueda, L. A., Criado, J. M. The accuracy of senum and yang's approximations to the Arrhenius integral. J. Therm. Anal. Calorim. 2000, 60, 909–915.

[49]   Toop, D. J. Theory of Life Testing and Use of Thermogravimetric Analysis to Predict the Thermal Life of Wire Enamels. IEEE Trans. Electr. Insul, 1971, EI-6(1), 2–14.

[50]   Flynn, J. H., Wall, L. A. A quick, direct method for the determination of activation energy from thermogravimetric data, Polym. Lett., 1966, 4, 323–328.

[51]   Ozawa, T. A new method of analyzing thermogravimetric data. Bull. Chem. Soc. Japn. 1965, 38, 1881.

[52]   Kissinger, H. E. Reaction kinetics in differential thermal analysis. Anal. Chem. 1957, 29, 1702–1706.

[53]   Starink, M. J. The determination of activation energy from linear heating rate experiments: A comparison of the accuracy of isoconversion methods. Thermochim. Acta. 2003, 404, 163–176.

[54]   Coats, A., Redfern, J. Kinetic Parameters from Thermogravimetric Data. Nature, 1964, 201, 68–69.

[55]   Phadnis, A. B., Deshpande, V. V. Determination of the kinetics and mechanism of a solid-state reaction. A simple approach. Thermochim. Acta. 1983, 62, 361–367.

[56]   Vyazovkin, S., Dollimore, D. Linear and nonlinear procedures in isoconversional computations of the activation energy on non-isothermal reactions in solids. J. Chem. Inf. Comput. Sci. 1996, 36, 42–45.

[57]   Vyazovkin, S. Evaluation of activation energy of thermally stimulated solid state reactions under arbitrary variation of temperature, J. Comput. Chem. 1997, 18, 393–402.

[58]   Vyazovkin, S. Model-free kinetics: staying free of multiplying entities without necessity. J. Therm. Anal. Calorim. 2006, 83, 45–51.

[59]   Vafayan, M., Beheshty, M. H., Ghoreishy, M. H. R., Abedini, H. Advanced integral isoconversional analysis for evaluating and predicting the kinetic parameters of the curing reaction of epoxy prepreg. Thermochim. Acta. 2013, 557, 37–43.

[60]   Cai, J., Chen, S. A New iterative linear integral isoconversional method for the determination of the activation energy varying with the conversion degree. J. Comput. Chem. 2009, 30, 1986–1991.

[61]   Budrugeac, P. An iterative model-free method to determine the activation energy of non-isothermal heterogeneous processes. Thermochim. Acta. 2010, 511, 8–16.

[62]   Flynn, J. H. The "temperature integral" – its use and abuse. Thermochim. Acta. 1997, 300, 83–92.

[63]   Dubaj, T., Cibulkova, Z., Šimon, P. An incremental isoconversional method for kinetic analysis based on the orthogonal distance regression. J. Comput. Chem. 2015, 36, 392–398.

[64]   Vyazovkin, S., Burnham, A. K., Criado, J. M., Pérez-Maqueda, L. A., Popescu, C., Sbirrazzuoli, N., Kinetics, I. C. T. A. C. Committee recommendations for performing kinetic computations on thermal analysis data. Thermochim. Acta. 2011, 520, 1–19.

[65]   Šimon, P., Dubaj, T., Cibulková, Z. Frequent flaws encountered in the manuscripts of kinetic papers. J. Therm. Anal. Calorim. 2022, 147, 10083–10088.

[66]   Pe'rez-maqueda, L. A., Criado, J. M., Gotor, F. J., Ma'lek, J. Advantages of combined kinetic analysis of experimental data obtained under any heating profile. J. Phys. Chem. A, 2002, 106, 2862–2868.

[67]   Budrugeac, P. On the use of the model-free way method for kinetic analysis of thermoanalytical data – advantages and limitations. Article 179063 Thermochim. Acta, Volume 706, 2022. https://doi.org/10.1016/j.tca.2021.179063

[68]   Maxwell, A. S., Broughton, W. R., Dean, G., Sims, G. D. Review of accelerated ageing methods and lifetime prediction techniques for polymeric materials, NPL Report DEPC MPR 016, 2005.

[69]   Ahmad, M. B., Gharayebi, Y., Salit, M. S., Hussein, M. Z., Ebrahimiasl, S., Dehzangi, A. Preparation, characterization and thermal degradation of polyimide (4-APS/BTDA)/SiO_2 composite films. Int. J. Mol. Sci. 2012, 13, 4860–4872.

[70] Saha, T., Bhowmick, A. K. Influence of nanofiller on thermal degradation resistance of hydrogenated nitrile butadiene rubber, Rubber Chem. Technol. 2019, 92 (2), 263–285.

[71] Malkin, A. Y., Petrie, C. J. S. Some conditions for rupture of polymer liquids in extension. J. Rheol. 1997, 41(1),1–26.

[72] Fukushima, K., Cai, H., Nakada, M., Miyano, Y. Determination of time-temperature shift factor for long-term life prediction of polymer composites, ICCM International Conferences on Composite Materials, 2009.

[73] Miyano, Y., Nakada, M., Cai, H. Long-Term Life Prediction of CRPF Structures Based on MMF/ATM Method. In: Conf. Proceed. Society for Experimental Mechanics Series, 2011, 3, 257–265.

[74] Cai, H., Miyano, Y., Nakada, M. Prediction of long-term flexural fatigue strength of honeycomb sandwich composites. J. Reinf. Plast. Compos. 2010, 29(2),266–277.

[75] Pazur, R. J., Cormier, J. G., Taymaz, K. K. Service life determination of nitrile O-rings in hydraulic fluid. Rubber Chem. Technol. 2014, 87(2),239–249.

[76] Batra, S. Creep rupture and life prediction of polymer composites, Graduate Theses, Dissertations, and Problem Reports. West Virginia University, College of Engineering and Mineral Resources, 2009. https://researchrepository.wvu.edu/etd/2035.

[77] Spathis, G., Kontou, E. Creep failure time prediction of polymers and polymer composites. Compos. Sci. Technol. 2012, 72, 959–964.

[78] Trachenko, K. The Vogel–Fulcher–Tammann law in the elastic theory of glass transition, J. Non-Cryst. Solids 2008, 354, 3903–3906.

[79] Ferry, J. D. Viscoelastic Properties of Polymers, 3rd ed. New York, John Wiley, 1980, 276.

[80] Luo, W. B., Wang, C., Hu, X. Long-term creep assessment of viscoelastic polymer by time-temperature-stress superposition, Acta Mech. Sol. Sinica 25 (2012) 571–578.

[81] Pap, J. S., Kästner, M., Muller, S., Jansen, I. Experimental characterization and simulation of the mechanical behavior of an epoxy adhesive. Proc. Mater. Sci. 2013, 78, 234–242.

[82] Emara, M., Torres, L., Baena, M., Barris, C., Moawad, M. Effect of sustained loading and environmental conditions on the creep behavior of an epoxy adhesive for concrete structures strengthened with CFRP laminates, Compos. B. 2017, 129, 88–96.

[83] Li, H., Luo, Y., Hu, D. Long Term Creep Assessment of Room-temperature Cured Epoxy Adhesive by Time-stress Superposition and Fractional Rheological Model. Appl. Rheology, 2018, 28(6), 201864796. https://doi.org/10.3933/applrheol-28-64796

[84] Jiang, C., Jiang, H., Zhu, Z., Zhang, J. Application of time–temperature–stress superposition principle on the accelerated physical aging test of polycarbonate, Polym. Eng. Sci. 2015, 55, 2215–2221.

[85] Suhir, E. Boltzmann-Arrhenius-zhurkov equation and its application in aerospace electronics and photonics reliability physics problems: review. Int. J. Aeronaut. Aerosp. Res. 2020, 7(1),210–223.

[86] Zhurkov, S. N. The Problem of the Strength of Solids. Bull. USSR Acad. Sci. 1957, 11, 78–82.

[87] Zhurkov, S. N. Kinetic concept of the strength of solids. Int. J. Fract. Mech. 1965, 1, 311–323.

[88] Sirota, D., Ivanov, V. The Determination and Practical Application the Kinetic Constants of Destruction of Rocks with Modified Zhurkov's Formula. In: E3S Web of Conferences. IIIrd International Innovative Mining Symposium. 2018, 41, 01030.

[89] Wineman, A., Shaw, J. Combined deformation- and temperature-induced scission in a rubber cylinder in torsion. Int. J. Non-Linear Mech. 2007, 42, 330–335.

[90] Vettegren, V. I., Kulik, V. B., Bronnikov, S. V. Temperature dependence of the tensile strength of polymers and metals at elevated temperatures. Tech. phys. lett., 2005, 31(11),969–972.

[91] Bailey, J. An attempt to correlate some tensile strength measurements on glass: III. Glass. Ind. 1939, 20, 95–99.

[92] Leonov, A. I., Basov, N. I., Kazanko, V., Yu, V. Fundamentals of Injection Molding of Thermosets and Rubbers. (In Russian), Moscow, Chimiya, 1977, pp. 56–59.

Index

2-hydroxyethyl methacrylate 6
3D fabric 177
4, 4'- dithiodibutyric acid 22
4,4'-bismaleimidodiphenylmethane 142

accuracy 308–309, 330, 337, 351
acrylic acid 16, 30
acrylonitrile butadiene styrene (ABS) 47
acrylonitrile-butadiene-styrene (ABS) 275
activation energy 357, 364, 374–375, 377, 379,
    381–387, 389–390, 392–393, 405–406
active constrained layer 269
active-passive constrained layer damping 262, 267
ageing 353, 355, 360, 371, 405
agglomerated 346–347
agglomeration 75, 78–79, 87, 98–99, 200, 203, 206,
    223, 227, 237, 240, 242
aminopropyltriethoxysilane 41
amorphous 110, 355–356, 359
angular 113
application 193, 246, 371, 386, 394–395, 403,
    405, 411
aramid fibre 95
aspect ratio 75, 78, 86, 89–90, 93, 98–99, 102, 112,
    115–116, 203, 206, 209–212, 214, 237
atmospheric 192, 254
attraction 98, 103, 115
autoclave moulding 173
automated fibre placement process (AFP), 41

benzoxazine resin 138
Bingham Model 106
bio-based phloretic acid 138
biocompatible 96–97, 128
bis(2-benzothiazole)disulfide 15
bis(4-maleimidophenyl) methane (BDM) 142
black filler 85
blending 104–105, 127

carbon black 75, 77–78, 83, 85, 87, 92–93, 120,
    123, 192
carbon fabric 177, 179, 182
carbon fiber reinforced epoxy composite
    (CFRP) 298
carbon fiber-epoxy composites (CFRP) 293
carbon nanotube composites 226

carboxyl terminated polyethylene glycol
    adipate 57
carboxylated nitrile butadiene ionic elastomer 16
carboxylated nitrile rubber (XNBR) 9
carboxyl-terminated poly(2-ethyl hexyl acrylate) 57
Casson Model 107
Cellulose 96
cellulosein 73
CFRP 262, 353, 367
chopped 86
chopped fibre 86
chord reinforced rubber composite 216
clay 223–225, 247
CNT 102–103, 120
Coats and Redfern 380
coefficient 73, 78, 83, 85, 87, 95, 108–109, 113
combination 84, 97
complex 74, 86, 105, 113–115
compliance 359–363, 365, 368–370, 394–395, 402
compliance calibration 53
composite 2–3, 4, 5, 12, 17, 23–24, 26, 37–47,
    49–50, 53–55, 58–59, 63–67, 75–88, 90,
    92–99, 101–104, 112–113, 115–120, 123–124,
    190, 194, 197–202, 207–210, 215, 220, 238, 241,
    244, 246–248, 252–255
composites 196, 229, 242–243, 308, 310–313,
    315–316, 318–319, 323, 330, 332, 335–336,
    339–341, 343–344, 346, 351, 354, 356, 366,
    371–372, 387, 394, 402, 408, 411
compression test 310, 316, 326
computational 385
concentration 73, 75, 83–84, 87, 109–110, 113, 197,
    203–204, 225, 237, 244, 255
conducting polymer bio-nanocomposites
    (CBNCs) 297
conducting polymers 82
conductivity 76–77, 80–85, 124
conventional 95
core shell rubber (CSR) 178
covalent adaptable networks 18
covalently crosslinked elastomer 14
crack opening displacement (COD) 53
creep 354, 356–363, 365–366, 368–371, 394–398,
    402–405, 409
critical length 86–87, 99
critical pigment volume concentration (CPVC) 197
critical strain energy factor 176–177

https://doi.org/10.1515/9783110781571-008

cross-link density 365–366
crosslinking 6, 8, 11–12, 14, 16, 18
CRTM 167–168
CTPEHA 347
curves 362–363, 369, 386, 389, 398
cyanate ester 11
cyanate ester resin 144

damage tolerance 176
damping 86, 108, 365–366, 411
decomposition 333
deformation 105, 116, 121, 362, 394, 409
degradation 354–356, 366–367, 371–377, 385,
        387–395, 406, 411–412
deterioration 354, 371, 394
dicyanate ester 11
dielectric analysis 306, 340
differential scanning calorimetry 306, 338
diffractometry 306–307
diglycidyl ether of bisphenol A (DGEBA) 22
diglycidylether of bisphenol-A (DGEBA) 134
diisocyanate 85
dipentamethylene thiuram tetrasulfide 15
discontinuous 83, 89
displacementof 53
distribution 74, 80, 84, 87–88, 98, 114
DMA 353, 357, 359, 368, 371, 409
DMTA 364
double cantilever beam 53
drop weight impact test 324
dual shear test 322
dynamic 86, 102–103, 105, 109, 113–114
dynamic mechanical analyser 306, 329
dynamic mechanical analysis 329
dynamic viscoelasticity 234

E-glass 368
Einstein, Mooney, Graham and Frankel &
        Activos 112
elastic modulus 199, 366, 369, 394, 399
elastomer /nanoclay composites 223
elastomer blend 366
elastomer matrices 189
elastomer nanocomposite 222
elastomeric 192, 199, 201, 205, 207, 229, 237,
        242–243, 245–247, 249
elastomer-nanocomposites 247
electrical properties 81
electrically conducting composites 272

electromagnetic compatibility (EMC) 300, 307
electromagnetic (EM) 276
electromagnetic induction 307
EMI shielding 82, 85, 248–249, 253–254, 354
energy-dispersive spectrometer 346
engineering thermoplastic 104
enthalpy 355–356, 364, 372
environmental exposure 367
environmental stress cracking (ESC) 63
environmental stress cracking resistance (ESCR). 66
epichlorohydrin 134
epoxy 357–358, 365–368, 372, 398, 408–411
epoxy resins 134
equation 357–358, 361–362, 364–366, 370–371,
        374, 376–377, 380, 382–383, 395, 397–401,
        403, 405–407, 409
equilibrium 355–358, 361, 364–365, 370, 394
ESR 306–307
ethylene vinyl acetate copolymer (EVA), 270
ethylene-vinyl acetate 83
exothermic 332–333, 338–339
expansion 332, 336–337
experiments 200, 236
extruder 74, 95, 98, 101, 105, 109–110, 116–118,
        120, 122
extrusion 86, 95, 97, 101, 105, 109, 116–117,
        121–123, 128

ferromagnetic resonance (FMR) 307
ferulic acid 139
fiber 366, 368, 372, 409
fibre 387
fibre-reinforced plastic 164
fibre-reinforced plastic (FRP) 307
fibres 5, 23–24, 26, 31, 86–87, 91, 94–97, 101,
        122–124, 128
filament winding 170
filaments 216, 218
filler 77, 80, 202
flexibility 85, 94, 104, 127
flexure test 312
fluid 105–109, 111, 118
fluids 105, 108
fluorinated ethynyl-terminated imide (FETI) 11
Flynn-Wall-Ozawa 379
fracture mechanics 49
fracture toughness 50, 54
fragmentation 332
free volume 355–356, 364–365, 397

frequency selective surface (FSS) 289
Friedman's method 150
FTIR 306–307, 334
functional 307, 351
functionalized interleaf technology (FIT) 276

glass fibre 93–95, 115, 120
glass fibre, carbon fibre 95
glass transition temperature, T_g. 239
glycidoxy- propyltriethoxysilane 41
GNP 73, 102–103, 111–113
gradient boosting regression 164
graphene 215, 223, 226–228, 238, 240–242,
    244–245, 249, 253–254
graphene nanoplates (GNP) 276
graphene oxide (GO) 179
graphite 73, 76, 80–81, 83–85, 90, 92–93, 120, 190,
    192–194, 198, 200–203, 226–228, 240, 242,
    244, 247–248, 253–254

halloysite nanotubes (HNT) 62
Halpin Tsai 97
hardness 193, 197–201, 208, 366, 403
HDPE 353, 366, 372
hexamethylene tetramine 137, 161
high density polyethylene (HDPE) 76, 123
high-resolution (HR) 348
homopolymers 74
hoop stress 42
hydrodynamic 111
hydrogen bonding 74, 91, 115
hydrothermal 94

impact strength 143, 176–178
important 76, 81, 84, 86, 98, 108, 110, 116, 122, 127
improvement 76–77, 85–86, 94, 96, 113, 192,
    196, 203, 207, 214, 217, 221, 223
ingredients 96
injection molding 74, 86, 94, 96–97, 105, 116
Injection Stretch Blow Moulding 120
in-plane shear 318, 321
interaction 75, 78–79, 84, 86, 88, 91, 93, 98,
    105, 116
interface 86, 89, 91, 96, 119
interfacial 77, 88
interlaminar shear failure 272
interlaminar stresses 40
interleaving 176–177, 180
intermolecular 93

interparticle 87, 99
interpenetrating polymer network 6
interpolation 393, 411
investigated 328, 343, 346
ionic elastomer 16
isoconversional 372, 381, 385
isothermal 364–368, 370–372, 374, 380, 386–387,
    392–394, 403, 406, 408
Izod and Charpy tests 322

Kelvin-Voigt 360, 369, 395
kernel ridge regression 164
kinetics 355, 372–374, 376, 378, 381, 385–387, 389,
    392, 394, 412
Kissinger-Akahira-Sunose (KAS) method 152
KV model 230, 233

laminar shear strength 24
lead zirconate titanate 269
life estimation 354, 372, 387, 389, 411
light microscopy 342
lightning strike protection (LSP) 298
linear elastic fracture mechanics 39
liquid composite moulding 165, 172
liquid composites moulding 177
load 79, 87–90, 95, 98–99, 101
loss factor 366, 368

manufacturing 307
mapping 363, 367
master curve 357, 361–363, 397, 399, 403, 405
mathematical 209, 218
Maxwell model 230
measurement 365
mechanical 190, 193–194, 196–198, 208, 219,
    226–227, 235, 238, 240, 242, 246
mechanical properties 75, 85, 94, 96, 103, 114, 116
mechanical property 77
mechanical testing 309
mechanism 354–355, 373–375, 377, 387, 389,
    395, 400
melt flow index (MFI) 80
membranes 372
methacryloxypropyltrimethoxysilane 41
methyl ethyl ketone peroxide 133
methyl methacrylate 6
methyl methacrylate (MMA) 11
microcapsules 264–265, 297–298
micrograph 76, 115

microhetergeneous 9
micromechanics of failure 46
microspheres 99
mineral particulate fillers 194
model 78–82, 87, 90, 93–94, 97, 106–107, 113–114
Modulus 81, 90, 357, 367–368, 394–395, 398–400, 402–403
molecular weight 74, 105, 108
montmorillonite clay 372
montmorillonite nanoclay (OMMT)) 241
morphology 307, 327–328, 343–344, 346–347, 349
multilayer graphene 63
multiwall 226
multiwall carbon nanotube (MWCNT) 275
MWCNT 346

N,N-dimethyl-*p*-toluidine (DMPT) 297
Nanocomposite 98, 103, 265, 274, 281, 291, 297, 301, 368, 372–373, 390, 398
nanocomposites 4, 44, 46, 61–63, 98, 104–105, 113, 123, 127, 307, 319, 335, 343, 346
nanofibers 291
nanofiller 254
nanoparticles 367
nanoplates 73, 85, 102, 111, 113, 123
natural fibre 96–97
natural rubber 1, 14, 57, 66–67
network 365–366, 368, 372
Newtonian 106–107, 109–110, 113, 118
nitrile butadiene rubber (NBR) 16
nitrile rubber (NBR), 270

of polylactic acid (PLA) 276
OMMT 73, 113–115
optical microscopy 40
optimization 349
orientation 87, 93–94, 100–101, 108, 123

PA6 115, 117
packaging 74, 97, 127
packing factor 84
paint 82, 108
parameters 76, 105–106, 109–111, 116–117, 122
particulate composites 197
particulate rubber composite 192
passive constrained layer damping 266
PEEK) 74–75, 95
PEEK, PEKK, PC, ABS 127
percolation 76–77, 80, 83–85

performance 216–217, 222–223, 254
periodic pattern surface (PPS) 290
permeability 83
permittivity 83
phenol formaldehyde 137
phenolic resin 137
phenomena 108
phenyl-ethynyl-terminated imide (PETI) 11
phthalonitrile resins 145
phthalonitriles 145
PMR resin 142
Poisson's 78–79, 81, 89–90
poly (vinyl chloride) 356
polyamide 76, 85, 93, 115, 366
polycaprolactone 97
poly(ethyl methacrylate) (PEMA) 9
polyethylene, 98
poly(ethylene oxide) 6
polyhydroxyalkanoates (PHA), 128
polyimides 140
poly(lactic acid) 31, 97
polylactic acid (PLA) 128
polylactide 365
polymer composite *See*
polymer matrices 6
poly(methyl methacrylate) 353, 364
polyphenylene sulfide (PPS) 47, 275
polypropylene 73–75, 78, 117
polypropylene (PP) 31, 275
polystyrene 76, 83, 356, 366, 372
polytetrafluoroethylene (PTFE). 299
poly(thiophene) 274
polyurethane 11, 57
polyurethanes 82
positron annihilation life time spectroscopy 356, 365
positron annihilation lifetime spectroscopy (PALS) 12
POSS 368
post-impact testing 326
Power Law 105, 107
prediction 354, 371, 386–387, 394–395, 401, 406, 411
predictive 97, 106
predominantly 319
prepreg 17, 38, 40, 165, 174–175
pressure 74, 108, 110, 117–118, 120
process technology 74
processing 189, 193, 243

pseudoplastic, 105, 107
pultrusion 168–169

radar absorbing sheets (RASHs) 291
radar reflecting composite structure (RRCS) 278
radar-absorbing composite structure (RACS) 278
radar-absorbing composite structures (RACSs) 291
random orientation 215
reduced graphene oxide (rGO) 281
reinforced 366, 368, 372
reinforcement 3–4, 5, 24, 44, 75–77, 81, 83–84, 86–87, 91, 93, 96, 98–99, 101, 103, 115, 128
reinforcements 309, 316, 332
reinforcing 194
relation 118
relatively 190, 230, 245–246
relaxation time 355–359, 361–362, 364–365, 368–371, 394, 396, 407, 411
resin film infusion bolding 172
resin film infusion (RFI), 172
resin transfer moulding 131, 165–167, 178
resistivity 76, 82–83, 85
resonance 306–307
reversible 91, 108
review 97, 104
rheological 105, 110
Rheometer 108
Rheopectic 108
ring-opening metathesis polymerization (ROMP) 18
rubber 189–190, 192–194, 196, 198–200, 202–203, 206–212, 215–220, 222, 225, 227–228, 230, 232–233, 237, 240–241, 243–247, 253–255, 353–355, 372, 390, 395, 403, 408

scanning electron microscope 344
scanning electron microscopy 344, 355
scanning electron microscopy (SEM) 7, 31
screw 95, 98, 105, 116–118
segmental 331, 336
segmental mobility 358, 364–365, 368
self-healing 16, 20
self-healing composites 263
self-healing polyurethane elastomers (SHEs) 296
SEM 76, 90, 115
semiconductors 82
semicrystalline 339, 364
semi-IPN 6
separation 308, 317, 343, 347

sequential method 8–9, 12
sewage sludge ash (SSA), 59
shear strain 315, 318, 322
shear test 318–319
shift factor 357, 361, 363, 369, 396–400, 403
shift factors 363, 369, 371, 399, 401, 403–405
short beam test 321
short carbon fiber (SCF) 282
short fiber 324
short fiber rubber composites 206
short fibre 96
silicon carbide (SiC) 59
single end notch bend test (SENB) 59
single wall carbon nanotubes (SWCNT) 275
single-walled carbon nanotube (SWCNT), 179
sisal fibre 96
spherulites 343
spinneret 123
Starink method 153, 157
static 105
stearic acid (SA) 16
stimuli 354, 395
stitching 176–177
strain 356, 358–360, 366–368, 394–396, 398–399, 402–403, 405, 409–412
strength 307, 309, 311–312, 315–319, 321–323, 327, 340, 354, 366–367, 387, 394, 402–403
Stress Intensity Factor 1, 50–51
stress relaxation 354, 394–395, 398–399, 401, 403
structural health monitoring(SHM) 271
styrene-butadiene rubber 13, 15
subsequently 191, 197–198, 203
superposition 355–356, 366, 371, 394, 398–399, 401, 407
surface 75, 85, 87–88, 96, 98–99, 103, 115, 119, 124
suspension 110–111
synthesized 190, 228

temperature 73–74, 85, 94, 105, 109–110, 113, 116–117, 119–120, 122
tensile behavior 79
tensile cracking 37
tetramethyl thiuram monosulfide 15
theoretical model 369
thermal analysis 306, 331–332
thermal conductivity 80
thermoforming 74, 105, 116, 128
thermomechanical analysis 306, 333, 336

thermoplastic 73–78, 80–82, 85–87, 90, 93, 96–98, 104–106, 109, 113, 115–116, 122–124, 127
thermoplastic composite 75–76, 87, 105, 115–116
Thermoplastic elastomers (TPE) 103
Thermoplastic polyurethane 73, 85, 372
thermoplastics 74–75, 78, 82, 85–86, 98, 103, 105, 119, 127
thermo-reversible Diels-Alder reaction 265
thermoset 357–358, 365–368, 387–389, 398, 400, 407, 409, 411
thermoset composites 132, 164, 176
thermoset matrix 132
thermoset nanocomposites 178
thermosets 356, 371, 411
thioetherdiphthalic anhydride 140
Thixotropic 108
three-point bending test 313
Time 353, 366, 387, 394–397, 401, 408
toughened thermoset composites 176
toughening methods 56
toughness 5, 10–11, 30, 37–38, 43, 50, 52, 54, 56–59, 61–63, 66–67
TPU 73–74, 85–86, 104, 127
transformation 372, 405
transition temperature 355, 364, 368, 395–396, 406–407
transmission electron microscope 306, 347
transmission electron microscopy 344, 355
transmission electron microscopy (TEM), 307
transverse 88, 90, 92, 108
treatment 94, 96
triazabicyclodecene (TBD) 21
triazobicyclodecene (TBD) 22
tribological 95
triethylene glycol dimethacrylate (TEGDM) 6
triethyleneglycol dimethacrylate (TEGDMA) 297

underwater noise 4
universal testing machine (UTM) 309–310

unmanned aerial vehicle (UAVs) 307
unsaturated polyester 132–133, 161, 168
unsaturated polyester resin 133

vacuum bagging 165, 173, 176
van der Waals 17, 63
variation 309, 321, 330, 332, 350
vector network analyser (VNA) 285
vector network analyzer (VNA) 252
VFT 353, 371, 396
vibration damping 193, 246
vinyl ester 10, 58
vinyl ester resin 136
viscoelastic 6, 43–44, 50, 55, 62, 85–87, 266–269, 271, 295, 300, 316, 329, 351
viscoelasticity 359
viscosity 87, 98, 104–111, 113–120, 122
vitrimer 21–22
Vogel-Fulcher-Tammann (VFT) 236
Volume 112
volume fraction 76–78, 81, 83–85, 88, 94–95, 99–103, 110, 113, 115
vulcanization 15, 189–192, 207, 215–217, 229, 244
vulcanized 189, 192, 211, 219, 227–228, 233, 243, 248
vulcanized rubber strip 266
Vyazovkin 372, 381–382, 384–385

wet lay-up moulding 165
William, Landel and Ferry 236
WLF 2, 49, 353, 355, 371, 394, 397–399

X-ray computed tomography 39
X-ray diffraction analysis 349

α-relaxation 412
α-relaxation 357, 359
α-relaxation temperature 235, 239
(β-transition) 355

www.ingramcontent.com/pod-product-compliance
Lightning Source LLC
Chambersburg PA
CBHW080137220326
41598CB00032B/5095